CONFORMAL MAPPING

p 214+ distortion theorems.

CONFORMAL MAPPING

Zeev Nehari
PROFESSOR OF MATHEMATICS
CARNEGIE-MELLON UNIVERSITY
PITTSBURGH, PENNSYLVANIA

DOVER PUBLICATIONS, INC.
NEW YORK

Published in Canada by General Publishing Company, Ltd., 30 Lesmill Road, Don Mills, Toronto, Ontario.
Published in the United Kingdom by Constable and Company, Ltd., 10 Orange Street, London WC 2.

This Dover edition, first published in 1975, is an unabridged and unaltered republication of the work originally published by the McGraw-Hill Book Company, Inc. in 1952.

International Standard Book Number: 0-486-61137-X
Library of Congress Catalog Card Number: 74-27513

Manufactured in the United States of America
Dover Publications, Inc.
180 Varick Street
New York, N. Y. 10014

PREFACE

In the preface to the first edition of Courant-Hilbert's "Methoden der mathematischen Physik," R. Courant warned against a trend discernible in modern mathematics in which he saw a menace to the future development of mathematical analysis. He was referring to the tendency of many workers in this field to lose sight of the roots of mathematical analysis in physical and geometric intuition and to concentrate their efforts on the refinement and the extreme generalization of existing concepts. In the intervening years, this trend has become much more pronounced, and it has led to an increasing division of the workers in mathematical analysis into "pure" and "applied" mathematicians. One may deplore this development or welcome it—and many, notably among the "pure" variety, do welcome it—but it seems inevitable that fields like conformal mapping, whose methods are mathematical while their subject matter is derived from physical and geometric intuition, are bound to suffer in such a division.

The present book tries to bridge the gulf between the theoretical approach of the pure mathematician and the more practical interest of the engineer, physicist, and applied mathematician, who are concerned with the actual construction of conformal maps. Both the theoretical and the practical aspects of the subject are covered, the discussion ranging from the fundamental existence theorems to the various techniques available for the conformal mapping of given geometric figures. It has not been the author's aim to prove every statement in its utmost generality and under the weakest possible conditions. Wherever such a procedure would have interfered with the clarity of presentation and would have encumbered the proof with a mass of detail likely to obscure the essential ideas, a slightly less general formulation was preferred.

The book is designed for the reader who has a good working knowledge of advanced calculus. No other previous knowledge is required. The potential theory and complex function theory necessary for a full treatment of conformal mapping are developed in the first four chapters, thus making the reader independent of any other texts on complex variables. These four chapters are also suitable for use as a text for a one-term first course on functions of a complex variable. The remainder of the book may be covered in a one-year graduate course. There is a large number of problems and exercises, which should make the book suitable for both classroom use and self-study. A special chapter is devoted to a detailed discussion of the conformal-mapping properties of a large number of analytic functions which are of importance in the applications.

The book covers many recent advances in the theory which so far have not yet been incorporated into textbooks, notably in Chap. VII. In keeping with the character of this book as a text, no references to the literature have been given. The interested reader will find detailed bibliographies in S. Bergman's "The Kernel Function and Conformal Mapping" (American Mathematical Society, New York, 1950) and in the Appendix by M. Schiffer in R. Courant's "Dirichlet's Principle" (Interscience, New York, 1950).

The author has tried to present the material as simply and clearly as possible. With this end in view, the proofs of many known results have been simplified either in form or in substance. No claim of completeness is made, and a number of interesting subjects which do not lend themselves to a concise presentation have been omitted.

ZEEV NEHARI

CONTENTS

CHAPTER I

HARMONIC FUNCTIONS

1. Definitions and Preliminary Remarks. In this section we list a number of elementary definitions and results concerning point sets in the plane which will be required in what follows. Any reader who has been exposed to a course in advanced calculus—and it is to such readers that this book is addressed—will be familiar with the concepts and facts in question; they are listed here only for convenience of reference.

A *point* in the xy-plane is given by a pair of real numbers (a,b) which are to be interpreted as rectangular coordinates in this plane. A *neighborhood*—also called ϵ-neighborhood—of a point (a,b) is the set of points (x,y) such that $(x - a)^2 + (y - b)^2 < \epsilon^2$, where ϵ is a positive number. A point (a,b) is said to be a *limit point* or *accumulation point* of a set of points S if every neighborhood of (a,b) contains a point of S distinct from (a,b). This definition clearly implies that every neighborhood of a limit point contains an infinite number of points of S.

A limit point of S may, or may not, be a point of S. If every limit point of a set belongs to the set, we say that the set is *closed*. There are two distinct types of limit points of a set, *interior points* and *boundary points*. A limit point (a,b) is said to be an interior point of a set S if there exists a neighborhood of (a,b) which consists entirely of points of S. A limit point which is not an interior point of S, that is, a point (a,b) such that any neighborhood of (a,b) contains both points of S and points which do not belong to S, is called a boundary point of S. A set all of whose points are interior points is called an *open set*. A simple example of an open set is the interior of the unit circle, that is, the set of points (x,y) for which $x^2 + y^2 < 1$. By adding to a set S all its limit points, we obtain the *closure* of S; obviously, the closure of any set is a closed set. The *complement* of a set S is the set of all points (x,y) which do not belong to S.

The set of points (x,y), where $x = x(t)$, $y = y(t)$, and $x(t)$ and $y(t)$ are continuous functions of t in an interval $t_1 \leq t \leq t_2$, is called a continuous arc. If a continuous arc has no multiple points, that is, if it does not happen that for two different points t' and t'' in the above interval we have $x(t') = x(t'')$, $y(t') = y(t'')$, it is said to be a *Jordan arc;* intuitively speaking, a Jordan arc is a continuous arc which does not cut itself. A simple example of a Jordan arc is a nonintersecting polygonal line consisting of a finite chain of linear segments. A continuous arc which has exactly one multiple point, namely, a double point corresponding to the end points t_1 and t_2 of the interval $t_1 \leq t \leq t_2$, is called a *simple closed Jordan curve.* For example, the circle $x = \cos t$, $y = \sin t$, $0 \leq t \leq 2\pi$, is a simple closed Jordan curve, the double point in question being the point (0,1) which corresponds to both $t = 0$ and $t = 2\pi$.

A set S of points is said to be *connected* if any two of its points can be joined by a Jordan arc all of whose points belong to S; here, the Jordan arc may also be replaced by a polygonal arc. An open connected set is called a *domain.* The *Jordan curve theorem* states that a simple closed Jordan curve divides the plane into two domains which have the curve as their common boundary. Although the truth of this theorem seems to be evident, its rigorous demonstration is a lengthy and difficult task.* One of the two domains into which the plane is divided by a closed Jordan curve is *bounded, i.e.,* there exists a positive constant C such that all the points (x,y) of this domain satisfy $x^2 + y^2 < C^2$; this domain is called the *interior* of the curve. Correspondingly, the *exterior* of the curve is characterized by the fact that it contains points (x,y) for which $x^2 + y^2$ is arbitrarily large.

A *differentiable arc* is an arc which possesses tangents at all its points; analytically speaking, such an arc is given by the parametric representation $x = x(t)$, $y = y(t)$, $t_1 \leq t \leq t_2$, where the functions $x(t)$, $y(t)$ have derivatives $x'(t)$, $y'(t)$ which do not both vanish at the same time. If, moreover, these derivatives are continuous—or, to use geometrical language, if the arc possesses a continuously turning tangent—we speak about a *smooth arc.* A continuous chain of a finite number of smooth arcs will be called a *piecewise smooth arc;* if this chain of arcs forms a closed nonintersecting curve, we obtain a *piecewise smooth curve.*

A *crosscut* of a domain is a simple Jordan arc which, apart from its end points, lies entirely in the domain; obviously, the end points of the crosscut coincide with boundary points of the domain. A *simply-connected* domain D is defined by the property that all points in the interior of any simple closed Jordan curve which consists of points of D are also points of D; in particular, the interior of a simple closed Jordan curve is simply-connected. A crosscut connecting two different boundary points of a simply-connected domain D divides D into two simply-connected domains without common points (the points of the crosscut having been removed). If there exist simple closed Jordan curves which consist entirely of points of a domain D and in whose interior there are points not belonging to D, then D is *multiply-connected.* D is said to be

* The interested reader will find a proof of this theorem, and references to other proofs, in Dienes, "The Taylor Series," Oxford, 1931.

of connectivity n if there exist no more than $n - 1$ such Jordan curves whose interiors have no points in common and which contain in their interiors points not belonging to D. The circular ring $1 < x^2 + y^2 < 4$ is clearly of connectivity 2 or, as we shall also say, the circular ring is *doubly-connected.* To use intuitive language, a bounded domain of connectivity n is a bounded domain with $n - 1$ "holes." By $n - 1$ appropriately chosen crosscuts, a domain of connectivity n can be transformed into a simply-connected domain. For example, the crosscut $1 < x < 2$, $y = 0$ transforms the circular ring $1 < x^2 + y^2 < 4$ into a simply-connected domain.

The restriction to bounded domains made so far can be easily lifted by introducing the *point at infinity* whose neighborhood is defined as the exterior of a circle of arbitrarily large radius. To the beginner, the assigning of the character of a point to the infinitely distant may seem strange; however, the situation is easily visualized by the well-known stereographic projection of the sphere onto the plane. A sphere of radius unity is placed on the xy-plane in such a fashion that its "south pole" rests on the point (0,0). To each point P in the plane, a point P' on the surface of the sphere is made to correspond in the following manner: the "north pole" is connected with P by a straight line, and P' is taken to be the point at which this line pierces the surface of the sphere. This clearly establishes a one-to-one correspondence between the points of the xy-plane and the points on the surface of the sphere. The only apparent exception is the north pole of the sphere. Obviously, no finitely distant point in the plane corresponds to this point; however, the outside of a circle $x^2 + y^2 > M^2$ of sufficiently large radius M corresponds to the inside of an arbitrarily small "arctic zone" in the neighborhood of the north pole. The correspondence between the plane and the surface of the sphere is therefore made complete if we define the "point at infinity," or "infinitely distant point," of the plane as the "point" corresponding to the north pole of the sphere. Statements of the character of those made in this section which involve the point at infinity are to be understood in the sense that they are true with respect to the "map" of the plane onto the surface of the sphere by means of the stereographic projection. For example, the Jordan curve theorem may now be formulated as follows: A simple closed Jordan curve divides the plane into two simply-connected domains, one of them bounded and the other unbounded. The unbounded domain, *i.e.*, the exterior of the curve, can be characterized by the fact that it contains the point at infinity. An example of a doubly-connected domain containing the point at infinity is given by the set of points of the xy-plane which remain after removal of the points of the circles $(x + 2)^2 + y^2 \leq 1$ and $(x - 2)^2 + y^2 \leq 1$. To use intuitive language, a domain of connectivity n containing the point at infinity is obtained by punching n "holes" out of the plane.

2. Elementary Properties of Harmonic Functions. A function $u(x,y)$ is said to be harmonic in a domain D if the partial derivatives

$$\frac{\partial u}{\partial x}, \quad \frac{\partial u}{\partial y}, \quad \frac{\partial^2 u}{\partial x^2}, \quad \frac{\partial^2 u}{\partial y^2}$$

exist and are continuous, and if

$$\Delta u \equiv \frac{\partial^2 u}{\partial x^2} + \frac{\partial^2 u}{\partial y^2} = 0 \tag{1}$$

at all points of D. U is said to be harmonic at a point P if it is harmonic in a neighborhood of P.

The partial differential equation (1), which is known as the *Laplace equation*, and its three-dimensional equivalent

$$\frac{\partial^2 u}{\partial x^2} + \frac{\partial^2 u}{\partial y^2} + \frac{\partial^2 u}{\partial z^2} = 0 \tag{2}$$

for functions $u(x,y,z)$ of the three variables x, y, z are of fundamental importance in many branches of applied mathematics and mathematical physics. The gravitational, electrostatic, and magnetostatic potentials and many other functions of physical significance are solutions of the equation (2), where x, y, z denote rectangular coordinates in space. If the physical situation described by the equation (2) is such that the function $u(x,y,z)$ does not change in the z-direction, then u is in reality only a function of the two variables x and y, and (2) reduces to (1).

From the linear character of the equation (1) it is immediately apparent that a linear combination $au(x,y) + bv(x,y)$, where a and b are constants, is a harmonic function if the same is true of the functions $u(x,y)$ and $v(x,y)$ separately. Another elementary consequence of (1) which is easily verified is the fact that the function $u(x + a,\ y + b)$ is harmonic in a domain obtained by translating D in the x- and y-directions by the amounts $-a$ and $-b$, respectively, if $u(x,y)$ is harmonic in D.

An important harmonic function is obtained by asking for those solutions of (1) which depend on the distance from a given point (a,b) only and are independent of the direction in which we proceed from this point. In view of what was said at the end of the last paragraph, we may assume that the point in question is the origin, *i.e.*, the point (0,0). If we introduce polar coordinates r, θ, we are thus asking for solutions of (1) which depend on r only. Transforming (1) to polar coordinates, we obtain

$$\frac{\partial^2 u}{\partial r^2} + \frac{1}{r}\frac{\partial u}{\partial r} + \frac{1}{r^2}\frac{\partial^2 u}{\partial \theta^2} = 0. \tag{1'}$$

Since the desired solutions are to depend on r only and not on θ, this reduces to

$$\frac{d^2u}{dr^2} + \frac{1}{r}\frac{du}{dv} = 0.$$

The general solution of this differential equation is easily found to be $u = A \log r + B$, where A and B are arbitrary constants. Since a constant is trivially a harmonic function and, as pointed out before, a linear combination of harmonic functions is likewise harmonic, we obtain the result that there is essentially only one harmonic function with the desired properties, namely, the function

$$u = \log r = \log \sqrt{(x-a)^2 + (y-b)^2}. \tag{3}$$

This function is harmonic at every finite point of the plane, with the obvious exception of the point (a,b), where the harmonicity of the function breaks down owing to the fact that the required partial derivatives cease to exist. A point of this type is called a *singular point*, or a *singularity* of the harmonic function in question.

Since the harmonic function (3) depends on the arbitrary parameters a and b, we can construct from it other harmonic functions by differentiating it with respect to these parameters. Indeed since

$$\log \sqrt{(x-a)^2 + (y-b)^2} \qquad \text{and} \qquad \log \sqrt{(x-a_1)^2 + (y-b)^2}$$

are harmonic functions, the same is true of the linear combination

$$\frac{1}{a_1 - a}[\log \sqrt{(x-a)^2 + (y-b)^2} - \log \sqrt{(x-a_1)^2 + (y-b)^2}].$$

Letting a_1 tend to a, we obtain the function

$$-\frac{\partial \log r}{\partial a} = \frac{x-a}{(x-a)^2 + (y-b)^2}. \tag{4}$$

Instead of justifying the passage to the limit we may also confirm directly that (4) is harmonic if $(x-a)^2 + (y-b)^2 \neq 0$. A similar result is obtained by differentiating $\log r$ with respect to b. The point (a,b) again is a singular point of (4), since the required partial derivatives are not defined there. It is worth noting the different character of the singularity of the function (4) at the point (a,b), as compared with that of the function (3) at the same point. The function (3) tends to $-\infty$ if (x,y) approaches the point (a,b) along any path. The function (4) shows quite a different behavior. If (x,y) approaches (a,b) along the line $x = a$, the function obviously vanishes for any value of y, and the same is therefore true in the limit $y \to b$. If, on the other hand, (x,y) approaches (a,b) along $y = b$, the function reduces to $(x - a)^{-1}$ and the limit for $x \to a$ is ∞ or $-\infty$, depending on the direction of approach.

EXERCISES

1. Show that for any rectilinear nonhorizontal approach to the point (a,b) the function (4) tends to either ∞ or $-\infty$.

2. Show that $e^x \sin y$, $e^x \cos y$, $\tan^{-1} [(y - b)/(x - a)]$ are harmonic functions.

3. Show that $u(xr^{-2},yr^{-2})$, where $r = \sqrt{x^2 + y^2}$, is a harmonic function, if the same is true of $u(x,y)$. *Hint:* Introduce polar coordinates and use the form (1′) of the equation (1); observe that the transformation in question consists merely in replacing r by r^{-1} in the polar form.

4. If r, θ are polar coordinates and n is an arbitrary real number, show that $r^n \cos n\theta$ and $r^n \sin n\theta$ are harmonic functions.

5. Show that

$$\frac{(R^2) - r^2}{R^2 - 2rR \cos (\theta - \varphi) + r^2}$$

(R, φ const.) is a harmonic function, and find its singularity.

3. Green's Formula. One of the fundamental formulas of the ordinary calculus is

$$\int_a^b f'(x)\, dx = f(b) - f(a).$$

The analogous result in two dimensions—known as *Gauss' theorem*—is as follows:

If the functions $p(x,y)$ and $q(x,y)$ are continuous and have continuous first partial derivatives in the closure of a domain D bounded by a piecewise smooth curve Γ, then

$$\iint_D [p_x(x,y) + q_y(x,y)]\, dx\, dy = \int_\Gamma [p(x,y)\, dy - q(x,y)\, dx], \tag{5}$$

where the integral on the right-hand side is a line integral extended over the boundary Γ of D in the positive sense, that is, in such a way that the interior of the domain remains at the left if the boundary is traversed.

The line integral on the right-hand side of (5) is defined by the ordinary integral

$$\int_{t_1}^{t_2} \left[p(x,y) \frac{dy}{dt} - q(x,y) \frac{dx}{dt} \right] dt,$$

where

$$x = x(t), \qquad y = y(t), \qquad t_1 \le t \le t_2,$$

is a parametric representation of the curve Γ. We omit here the proof of Gauss' theorem since it can be found in any text on advanced calculus.

Gauss' theorem remains true if D is a multiply-connected domain. If D is of connectivity n and bounded by n piecewise smooth simple closed curves, it is possible—as pointed out in Sec. 1—to transform D by $n - 1$ suitably chosen crosscuts into a simply-connected domain D'; these crosscuts may be taken to be piecewise smooth arcs. Since D' is simply-connected, we may apply to it Gauss' theorem, where the integrations on both sides of (5) are now to be extended over D' and its boundary Γ', respectively. The area integral on the left-hand side of (5) is obviously not affected if D' is replaced by D. As to the line integral, we observe that while those parts of Γ' which also belong to Γ are described only once, the crosscuts are described twice, in accordance with the fact that on both "edges" at the crosscuts there are points of D'. Since Γ' is to be described in such a way that the interior of D' remains at the left, it is clear that the two edges are traversed in different directions. Figure 1 illustrates the situation in the case of a triply-connected domain. For greater clarity, the two "edges" of the crosscuts have been separated; the sense in which the boundary of D' is described is indicated by arrows. Since the sign of a line integral is inverted if the direction of integration is changed, it follows that the line integrals over the crosscuts cancel each other. The only surviving line integrals are therefore those extended over the boundary Γ of D. We have thus proved that the identity (5) holds for an arbitrary multiply-connected domain bounded by piecewise smooth curves.

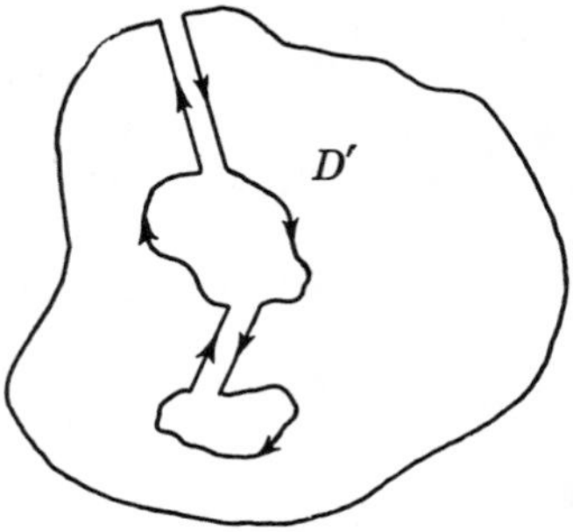

FIG. 1.

Consider now a function $u(x,y)$ which has continuous first partial derivatives in $D + \Gamma$ and a function $v(x,y)$ which has continuous partial derivatives of the first and second order there. The functions

$$p(x,y) = u\frac{\partial v}{\partial x}, \qquad q(x,y) = u\frac{\partial v}{\partial y}$$

will then satisfy the hypotheses of Gauss' theorem. In view of

$$p_x = \frac{\partial}{\partial x}\left(u\frac{\partial v}{\partial x}\right) = uv_{xx} + u_xv_x, \qquad q_y = \frac{\partial}{\partial y}\left(u\frac{\partial v}{\partial y}\right) = uv_{yy} + u_yv_y,$$

it follows from (5) that

$$\iint_D [uv_{xx} + u_xv_x + uv_{yy} + u_yv_y]\,dx\,dy = \int_\Gamma [uv_x\,dy - uv_y\,dx],$$

or

$$(6) \quad \iint_D u(v_{xx} + v_{yy})\,dx\,dy + \iint_D (u_x v_x + u_y v_y)\,dx\,dy = \int_\Gamma u[v_x\,dy - v_y\,dx].$$

The line integral on the right-hand side of (6) can be written in a simpler form if we introduce the differentiation in the direction of the outward pointing normal, denoted by $\partial/\partial n$. By the rules of partial differentiation

$$\frac{\partial v}{\partial n} = v_x \cos(x,n) + v_y \cos(y,n),$$

where (x,n) and (y,n) denote the angles between the outward pointing normal and the positive x- and y-axis, respectively. Since the direction cosines of the tangent are dx/ds and dy/ds, where s is the arc-length parameter, and the outer normal is obtained by turning the tangent clockwise by $\frac{1}{2}\pi$, it follows that

$$\cos(x,n) = \frac{dy}{ds}, \qquad \cos(y,n) = -\frac{dx}{ds}.$$

Hence

$$\frac{\partial v}{\partial n}\,ds = v_x\,dy - v_y\,dx.$$

Using this identity and the abbreviation $v_{xx} + v_{yy} = \Delta v$, we can finally bring (6) into the form

$$(7) \quad \iint_D u\,\Delta v\,dx\,dy + \iint_D (u_x v_x + u_y v_y)\,dx\,dy = \int_\Gamma u\,\frac{\partial v}{\partial n}\,ds.$$

This identity, which is of fundamental importance in the theory of harmonic functions, is known as *Green's formula;* the names Green's identity and Green's theorem are also used. Sometimes, (7) is referred to as Green's first formula, to distinguish it from another formula—Green's second formula—which is an immediate consequence of (7) If both $u(x,y)$ and $v(x,y)$ are supposed to possess continuous first and second derivatives, the roles of these two functions in (7) may be interchanged. Doing so, we obtain the companion formula

$$\iint_D v\,\Delta u\,dx\,dy + \iint_D (u_x v_x + u_y v_y)\,dx\,dy = \int_\Gamma v\,\frac{\partial u}{\partial n}\,ds.$$

If we subtract this formula from (7), we arrive at Green's second formula

$$\iint_D (u\,\Delta v - v\,\Delta u)\,dx\,dy = \int_\Gamma \left(u\frac{\partial v}{\partial n} - v\frac{\partial u}{\partial n}\right) ds. \tag{8}$$

Both (7) and (8) are generally referred to as "Green's formula."

4. Applications of Green's Formula. As a first application, we identify the function $v(x,y)$ in (7) with a function harmonic in $D + \Gamma$, and set $u(x,y) \equiv 1$. Since, in view of (1), $\Delta v = 0$ in $D + \Gamma$, it follows from (7) that

$$\int_\Gamma \frac{\partial v}{\partial n}\,ds = 0. \tag{9}$$

An easy consequence of this identity is the *mean value theorem: If $v(x,y)$ is harmonic in a closed circle, the value of v at the center of the circle is the arithmetic mean of its values on the circumference of the circle. Analytically expressed,*

$$v(a,b) = \frac{1}{2\pi}\int_0^{2\pi} v(a + r\cos\theta,\, b + r\sin\theta)\,d\theta, \tag{10}$$

where $v(x,y)$ is harmonic in the circle $(x-a)^2 + (y-b)^2 \leq r^2$.

To prove this theorem, we apply (9) to the case in which Γ is the circumference $(x-a)^2 + (y-b)^2 = \rho^2$, $0 \leq \rho \leq r$. Since, clearly, $\frac{\partial}{\partial n} = \frac{\partial}{\partial \rho}$ and $ds = \rho\,d\theta$, we have

$$\int_0^{2\pi} \frac{\partial}{\partial \rho} v(a + \rho\cos\theta,\, b + \rho\sin\theta)\,d\theta = 0,$$

or

$$\frac{\partial}{\partial \rho}\int_0^{2\pi} v(a + \rho\cos\theta,\, b + \rho\sin\theta)\,d\theta = 0,$$

which shows that the integral is independent of ρ. Its value for $\rho = 0$ is $2\pi v(a,b)$; if we set this equal to its value for $\rho = r$, the identity (10) follows.

Another result which can be easily deduced from (9) is the *maximum principle: If the function $v(x,y)$ is harmonic in a domain D, it cannot attain its absolute maximum or minimum at an interior point of D unless $v(x,y)$ reduces to a constant.*

Proof: Suppose that $v(x,y)$ does attain its maximum M at some interior point or points of D, and denote by S the set of all points of D at which $v(x,y) = M$. If $v(x,y)$ is not constant, S cannot contain all points of D. Accordingly, there must exist a boundary point of S, say P, which is an interior point of D. If r, θ denote polar coordinates with reference to the point P, we have $\frac{\partial v}{\partial r} \leq 0$ for arbitrary values of θ and sufficiently small r, since v is supposed to attain its absolute maximum at P. Integrating over this small circle, we obtain

$$\int_0^{2\pi} \frac{\partial v}{\partial r} \, d\theta \leq 0.$$

According to (9), however, this integral is equal to zero. Clearly, this is only possible if $\partial v/\partial r = 0$ on the whole circumference of the small circle in question and hence also on any smaller circle. It follows that $v(x,y)$ is constant in a small circle surrounding P, which contradicts our assumption that P is a boundary point of S. This proves the theorem for the case of the maximum. The reader will have no difficulty in modifying the argument slightly in order to prove the corresponding result for the minimum.

The above statement of the maximum principle is of a negative character; it denies the possibility of a harmonic function attaining its maximum at an interior point. It is in the nature of things that nothing more can be said in the case in which the harmonic function is defined only in an open set. If, however, it is known in addition that v is continuous in the closure of D, we can obtain more precise information. From the well-known result that a function which is continuous in a closed set attains its maximum at a point of the set it follows then that v must attain its maximum either in D or on its boundary. Since we have proved that the first alternative is excluded, it follows that the maximum must be attained at a point of the boundary. We have thus proved the following corollary of the maximum principle:

If $v(x,y)$ is harmonic in a domain D and continuous in the closure of D, then both the maximum and minimum values of v in the closure of D are attained on the boundary.

The maximum principle leads to an important uniqueness property of harmonic functions. Suppose that $u(x,y)$ and $v(x,y)$ are harmonic in a domain D and continuous in the closure of D and that $u = v$ on the boundary of D; the latter hypothesis is also expressed by saying that u and v have the same boundary values. Consider now the harmonic function $w = u - v$ which, by hypothesis, vanishes on the boundary of D. By the maximum principle, w attains both its maximum and its minimum on the boundary. Since it vanishes there identically, both the maximum and the minimum of w in D are zero; hence, w is identically zero throughout D and the harmonic functions u and v are identical. In other words, a harmonic function is completely determined by its boundary values.

EXERCISES

1. Multiplying (10) by r and integrating, with respect to r, from 0 to R, show that the mean-value theorem remains true if the arithmetic mean of the values of v on the circumference of the circle $(x - a)^2 + (y - b)^2 = R^2$ is replaced by the arithmetic mean of the values of v in the circle $(x - a)^2 + (y - b)^2 \leq R^2$.

2. If the function v is harmonic in a domain D and on its boundary Γ, show by means of Green's formula that

$$\int_\Gamma v \frac{\partial v}{ds} \geq 0,$$

and that equality will hold only if v reduces to a constant.

3. Use the result of the preceding exercise in order to show that (*a*) a harmonic function with vanishing boundary values is identically zero and that (*b*) a harmonic function with vanishing normal derivatives at all points of the boundary reduces to a constant.

4. Let $u(x,y)$ be a nonnegative solution of the partial differential equation

$$\Delta u = p(x,y)u,$$

where $p(x,y)$ is continuous and $p(x,y) > 0$ in a domain D. Using the ordinary necessary conditions for the existence of a local maximum, show that $u(x,y)$ cannot attain its maximum in D in the interior of this domain.

5. Show that the function u of the preceding exercise satisfies the inequality $\Delta(u^2) \geq 0$ in D.

5. The Green's Function and the Boundary Value Problem of the First Kind. Let D be a smoothly bounded domain and Γ its boundary, and let the function $w = w(x,y)$ be harmonic in D and have continuous first partial derivatives in $D + \Gamma$. If $r = \sqrt{(x - \xi)^2 + (y - \eta)^2}$ denotes the distance of (x,y) from a point (ξ,η) in D, then $\log r$ is harmonic in $D + \Gamma$ except at the point (ξ,η), and the same is therefore true of the function

$$h = h(x,y) = -\log r + w(x,y). \tag{11}$$

The result we are about to derive depends on the application of Green's formula (8) to a function of the type (11). Since $h(x,y)$ has a singularity at the point (ξ,η), it is clearly not permissible to take the entire domain D as the domain of integration. We circumvent this difficulty by deleting from D a small circle of center (ξ,η) and radius ϵ. In the remaining domain, which we shall denote by D_ϵ, $h(x,y)$ is harmonic.

We shall now apply the special form of (8) which is obtained if both u and v are harmonic functions in D; since, in this case, $\Delta u = 0$, $\Delta v = 0$, the left-hand side of (8) vanishes. The resulting formula is of such frequent occurrence as to warrant formulation as a separate result:

If u and v are harmonic in a smoothly bounded domain D and have continuous first derivatives in the closure $D + \Gamma$ of D, then

$$\int_\Gamma \left(u \frac{\partial v}{\partial n} - v \frac{\partial u}{\partial n} \right) ds = 0. \tag{12}$$

We now identify v with the function h defined in (11) and take u to be an arbitrary function which is harmonic in D and has continuous first derivatives in $D + \Gamma$. The domain D_ϵ to which we apply (12) is bounded by Γ and by the circumference C_ϵ of the circle of radius ϵ and center (ξ,η). Since this circle is outside D_ϵ and the boundary of D_ϵ is to be traversed in the positive sense—that is, leaving D_ϵ at the left—C_ϵ is clearly described in the clockwise direction. It thus follows from (12) that

$$\int_\Gamma \left(u \frac{\partial h}{\partial n} - h \frac{\partial u}{\partial n} \right) ds = \int_{C_\epsilon} \left(u \frac{\partial h}{\partial n} - h \frac{\partial u}{\partial n} \right) ds, \tag{13}$$

where the circle C_ϵ is now to be described counterclockwise. Introducing polar coordinates in the integral over C_ϵ and using (11), we obtain

$$\int_{C_\epsilon} \left(u \frac{\partial h}{\partial n} - h \frac{\partial u}{\partial n} \right) ds = \epsilon \int_0^{2\pi} \left[u \left(-\frac{1}{\epsilon} + \frac{\partial w}{\partial r} \right) + \frac{\partial u}{\partial r} (\log \epsilon - w) \right] d\theta$$
$$= -\int_0^{2\pi} u \, d\theta + \epsilon \int_0^{2\pi} \left(u \frac{\partial w}{\partial r} - w \frac{\partial u}{\partial r} \right) d\theta + \epsilon \log \epsilon \int_0^{2\pi} \frac{\partial u}{\partial r} \, d\theta, \qquad r = \epsilon.$$

The first integral is, by the mean-value theorem, equal to $2\pi u(\xi,\eta)$. The other two integrals are bounded in the neighborhood of (ξ,η); if we let ϵ tend to zero, we therefore obtain

$$\int_{C_\epsilon} \left(u \frac{\partial h}{\partial n} - h \frac{\partial u}{\partial n} \right) ds = -2\pi u(\xi,\eta).$$

We did not have to write "lim for $\epsilon \to 0$" in front of the integral since, in view of (13), its value does not really depend on ϵ. Comparing the last formula with (13), we thus have arrived at the following result:

If h is of the form (11), *where w is harmonic in D, and u is likewise harmonic in D and both u and w have continuous first derivatives in the closure $D + \Gamma$ of D, then*

$$u(\xi,\eta) = -\frac{1}{2\pi} \int_\Gamma \left(u \frac{\partial h}{\partial n} - h \frac{\partial u}{\partial n} \right) ds, \tag{14}$$

where (ξ,η) is any point of D. This result is sometimes referred to as *Green's third formula.*

A striking application of the identity (14) is obtained if $h(x,y)$ is identified with the *Green's function* $g(x,y;\xi,\eta)$ of D. This function, which is of fundamental importance both in the theory of harmonic functions and in the theory of conformal mapping, is defined as follows:

The Green's function $g(x,y;\xi,\eta)$ *of a domain D with respect to a point* (ξ,η) *in D is of the form*

$$(15)\qquad g(x,y;\xi,\eta) = -\log r + g_1(x,y;\xi,\eta), \qquad r = \sqrt{(x-\xi)^2 + (y-\eta)^2},$$

where g_1 *is harmonic in D; if* (x,y) *tends to any point of the boundary of D, g tends to zero.*

The proof for the existence of the Green's function of an arbitrary domain will be postponed to a later chapter (Sec. 4, Chap. V); it will then be obtained as a by-product of an existence theorem in the theory of conformal mapping. We shall meanwhile proceed on the assumption that our domain D does have a Green's function and that—except at the point (ξ,η)—this function has continuous first derivatives in the closure of the smoothly bounded domain D.

The Green's function g is zero on the boundary Γ of D. Identifying g with the function h in (11), we thus conclude from (14) that

$$u(\xi,\eta) = -\frac{1}{2\pi}\int_\Gamma u\,\frac{\partial g}{\partial n}\,ds.$$

Before giving a formal statement of this remarkable result, we introduce a notation which will permit us to formulate much of the following work in a more compact fashion. A point (x,y) will be denoted by the symbol z which stands for the complex number $z = x + iy$; similarly, the point (ξ,η) will be denoted by $\zeta = \xi + i\eta$. At this stage, this so-called *complex notation* is purely formal and its use does not require familiarity with the properties of complex numbers or variables. The symbol $g(x,y;\xi,\eta)$ will thus be replaced by $g(z,\zeta)$, but it should be perfectly clear that this does not imply that g is a function of the variables z and ζ. We shall also occasionally employ z, ζ, etc., as subscripts—such as in ds_z, $\partial/\partial n_z$—in order to emphasize the variables to which these differentials or differentiations refer. With these notations the above result reads as follows:

Let $u(z)$ *be harmonic in the closure* $D + \Gamma$ *of a smoothly bounded domain D, and let* $g(z,\zeta)$ *denote the Green's function of D with respect to a point* ζ *of D.* If $u(z)$ *is the boundary value of u at a point* $z \in \Gamma$, *then*

$$(16)\qquad u(\zeta) = -\frac{1}{2\pi}\int_\Gamma u(z)\,\frac{\partial g(z,\zeta)}{\partial n_z}\,ds_z.$$

Formula (16) makes it possible to compute a harmonic function if its boundary values—and, of course, the Green's function of the domain in question—are known. A question which presents itself naturally in this connection is the following: If, subject to certain regularity conditions, an arbitrary function $U(z)$ is given on the boundary Γ, does there always exist a harmonic function $u(\zeta)$ in D whose boundary values coincide with $U(z)$? We shall show at a later stage that in the case in which the boundary function $U(z)$ is piecewise continuous on Γ the answer is in the affirmative. The problem of constructing a harmonic function with a given set of boundary values is known as the *boundary value problem of the first kind*, or the *Dirichlet problem.* If the Green's function of a domain is known, the corresponding Dirichlet problem is solved by (16). Everything depends therefore on the explicit knowledge of the Green's function for the domain under discussion. In the case of a general domain D, the determination of the Green's function can be a matter of considerable difficulty.

It is worth noting that the construction of the Green's function is equivalent to the solution of a particular boundary value problem. Indeed, since $g(z,\zeta)$ vanishes for $z \in \Gamma$, the boundary values of the function $g_1(z,\zeta)$ in(15) coincide with those of $\log r$, where r is the distance between ζ and z. Hence, if we construct a function $u(z,\zeta)$ with the boundary values $\log r$, then the function $-\log r + u(z,\zeta)$ vanishes on the boundary. Since it further has the prescribed singularity at the point ζ, it necessarily is identical with the Green's function.

We now give the proofs of two important properties of the Green's function. The first of these is:

The Green's function $g(z,\zeta)$ of a domain D is positive throughout D.

By (15), $g(z,\zeta)$ is of the form $-\log r + g_1(z,\zeta)$, where $g_1(z,\zeta)$ is harmonic throughout D and r denotes the distance between the points z and ζ. Near the point ζ, $g_1(z,\zeta)$ is bounded while $-\log r$ takes arbitrarily large positive values. If ϵ ($\epsilon > 0$) is taken small enough, $g(z,\zeta)$ will therefore be positive on the circumference C_ϵ of radius ϵ and center ζ. Consider now the values of $g(z,\zeta)$ in the domain D_ϵ obtained from D by deleting from it the circle of radius ϵ and center ζ. $g(z,\zeta)$ is harmonic in D. The total boundary of D_ϵ consists of the boundary Γ of D and the circumference C_ϵ. On Γ, $g(z,\zeta) = 0$; on C_ϵ, $g(z,\zeta) > 0$. By the maximum principle, $g(z,\zeta)$ cannot attain its minimum inside D_ϵ. Hence, $g(z,\zeta)$ is positive throughout D_ϵ; since ϵ can be made arbitrarily small, the above statement is proved.

Next, we derive the so-called *symmetry property* of the Green's function: *The Green's function $g(z,\zeta)$ of a domain D is symmetric with respect to the two points z and ζ; that is,*

$$g(z,\zeta) = g(\zeta,z). \tag{17}$$

Let ζ and t be two given points of D. The Green's functions $g(z,\zeta)$ and $g(z,t)$ are harmonic in D with the exception of the points ζ and t, respectively. If we delete from D two small circles $C_\epsilon(\zeta)$ and $C_\epsilon(t)$ of radius ϵ

and centers ζ and t, respectively, both functions will therefore be harmonic in the domain D_ϵ thus obtained. If Γ_ϵ denotes the boundary of D_ϵ—consisting of the boundary Γ of D and the two circles $C_\epsilon(\zeta)$ and $C_\epsilon(t)$—it follows from the identity (12) that

$$\int_{\Gamma_\epsilon}\left[g(z,\zeta)\,\frac{\partial g(z,t)}{\partial n} - g(z,t)\,\frac{\partial g(z,\zeta)}{\partial n}\right] ds = 0.$$

In this integration, the boundary Γ is described in the positive sense, while the circumferences $C_\epsilon(\zeta)$ and $C_\epsilon(t)$ are described in the negative sense. The integral over Γ vanishes, since both $g(z,\zeta)$ and $g(z,t)$ vanish there. With the abbreviations $g(z,\zeta) = g_1$, $g(z,t) = g_2$, we have therefore

$$\int_{C_\epsilon(\zeta)}\left(g_1\,\frac{\partial g_2}{\partial n} - g_2\,\frac{\partial g_1}{\partial n}\right) ds + \int_{C_\epsilon(t)}\left(g_1\,\frac{\partial g_2}{\partial n} - g_2\,\frac{\partial g_1}{\partial n}\right) ds = 0.$$

g_1 has a logarithmic singularity of the type (11) at the point ζ, while g_2 is harmonic in the interior of $C_\epsilon(\zeta)$. By (14), the value of the integral over $C_\epsilon(\zeta)$ is therefore $2\pi g_2(\zeta)$. Similarly, the value of the integral over $C_\epsilon(t)$, as obtained from (14), is $-2\pi g_1(t)$, the negative sign arising from the fact that g_1 and g_2 appear in the same order in both integrals although the roles of ζ and t are interchanged. Hence, $g_2(\zeta) = g_1(t)$ or, i . view of the definition of the functions g_1 and g_2, $g(\zeta,t) = g(t,\zeta)$. This proves (17).

We end this section with the explicit determination of the Green's function in the case in which D is the circle C_R of radius R with center at the origin. If r, θ are polar coordinates in the (x,y) plane and ρ,φ the polar coordinates of a point ζ inside $C_R(\rho < R)$, we shall show that the Green's function of C_R is of the form

$$g(z,\zeta) = \frac{1}{2}\log\frac{R^2 - 2\rho r\cos(\theta - \varphi) + \rho^2 r^2 R^{-2}}{r^2 - 2\rho r\cos(\theta - \varphi) + \rho^2}. \tag{18}$$

Let r_1 denote the distance between the points z and ζ, and denote by r_2 the distance between z and the point inverse to ζ with respect to the circumference C_R, that is, the point with the polar coordinates R^2/ρ, φ. By elementary trigonometry, we have

$$r_1^2 = r^2 - 2\rho r\cos(\theta - \varphi) + \rho^2,$$
$$r_2^2 = \frac{R^4}{\rho^2} - 2\,\frac{R^2 r}{\rho}\cos(\theta - \varphi) + r^2.$$

Comparison with (18) shows that the function $g(z,\zeta)$ defined by (18) can be written in the form

$$g(z,\zeta) = -\log r_1 + \log r_2 + \log\frac{\rho}{R}. \tag{18'}$$

It was shown in Sec. 2 that the function $\log r$, where r denotes the distance of the variable point z from a fixed point a, is a harmonic function for $z \neq a$. Hence, $\log r_1$ and $\log r_2$, and therefore also the function $g(z,\zeta)$ of (18′), are harmonic functions. The possible singularities of $g(z,\zeta)$ are the points from which the distances r_1 and r_2 are measured, that is, the points of polar coordinates ρ, φ and R^2/ρ, φ, respectively. Since $\rho < R$, only the first of these points is situated inside C_R. Comparison of (18′) and (15) shows that the function (18′) has there precisely the singularity prescribed for the Green's function. In order to identify fully (18′), or its equivalent (18), with the Green's function, it therefore only remains to be shown that this function vanishes on the circumference C_R. That this is indeed the case is easily confirmed by setting $r = R$ in (18).

EXERCISES

1. By setting $\rho = 0$—thus obtaining the Green's function of C_R with the center as the point of reference—and using (16), obtain a new proof of the mean value theorem (10).

2. Show that

$$g(z,\zeta) = \frac{1}{2}\log\frac{(x+\xi)^2 + (y-\eta)^2}{(x-\xi)^2 + (y-\eta)^2}$$

is the Green's function of the half-plane $x > 0$, with (ξ,η) as the point of reference $(\xi > 0)$.

3. From the fact that the Green's function $g = g(z,\zeta)$ of a piecewise smoothly bounded domain D is positive in D while it vanishes on the boundary Γ of D, deduce that

$$\frac{\partial g}{\partial n} \leq 0, \qquad z \in \Gamma,$$

with the exception of the possible "corners" of Γ, where $\partial g/\partial n$ is not defined.

4. Use (16) and the result of the preceding exercise to obtain a new proof of the maximum principle. *Hint:* Write $u(z) \leq M$, and deduce from (16) that $u(\zeta)$ $(\zeta \in D)$ cannot be larger than M.

5. If $g = g(z,\zeta)$ is the Green's function of D and $u = u(z)$ is harmonic in D, show that

$$\iint_D (g_x u_x + g_y u_y)\, dx\, dy$$

exists and that the value of this integral is zero.

6. Let D^* be a domain contained in another domain D and let $g^*(z,\zeta)$ and $g(z,\zeta)$ denote the Green's functions of D^* and D, respectively. If z and ζ are points of D^*, show that

$$g^*(z,\zeta) \leq g(z,\zeta).$$

6. The Poisson Formula. We can use the explicit knowledge of the Green's function of a circle obtained at the end of the preceding section in order to derive, by means of (16), a representation of a harmonic function inside a circle in terms of its values on the circumference. If the center of the circle is the origin and its radius is R, its Green's function is given by (18). What is required in (16) is the normal derivative of $g(z,\zeta)$ for points on the circumference. Since, in the case of the circle, the derivative with respect to the outward pointing normal coincides with the partial derivative with respect to the radius, we obtain, from (18),

$$\frac{\partial g}{\partial r} = \frac{\rho^2 r R^{-2} - \rho \cos(\theta - \varphi)}{R^2 - 2\rho r \cos(\theta - \varphi) + \rho^2 r^2 R^{-2}} - \frac{r - \rho\cos(\theta - \varphi)}{r^2 - 2\rho r \cos(\theta - \varphi) + \rho^2},$$

whence

$$\left(\frac{\partial g}{\partial r}\right)_{r=R} = \frac{1}{R}\,\frac{\rho^2 - R^2}{R^2 - 2\rho R\cos(\theta - \varphi) + \rho^2}. \tag{19}$$

If $u(\rho,\varphi)$ denotes the value of the harmonic function u at the point whose polar coordinates are ρ, φ, it follows therefore from (16) that

$$u(\rho,\varphi) = \frac{R^2 - \rho^2}{2\pi}\int_0^{2\pi}\frac{u(R,\theta)\,d\theta}{R^2 - 2\rho R\cos(\theta - \varphi) + \rho^2}. \tag{20}$$

(20) *expresses the values of a harmonic function u in the interior of the circle $\rho < R$ in terms of its values on the circumference $\rho = R$.* This relation is known as the *Poisson formula.*

The question naturally arises whether (20) can also be interpreted in the following wider fashion: Is it true that for an arbitrarily given "boundary value function" $u(R,\theta)$—which, of course, has to satisfy certain regularity requirements—there always exists in $\rho < R$ a harmonic function, given by (20), which has the desired boundary values $u(R,\theta)$ on the circumference? We shall answer this question in the affirmative in the case in which the boundary value function $u(R,\theta)$ is piecewise continuous.

Before we do so, however, we shall first derive some important consequences of the formula (20). For the sake of convenience, we rewrite (20) in the form

$$u(\rho,\varphi) = \frac{1}{2\pi}\int_0^{2\pi} P(\rho,\varphi;R,\theta)u(R,\theta)\,d\theta, \tag{21}$$

where

$$P(\rho,\varphi;R,\theta) = \frac{R^2 - \rho^2}{R^2 - 2\rho R\cos(\theta - \varphi) + \rho^2} \tag{22}$$

Poisson Kernel

is the so-called *Poisson kernel.* By Exercise 5, Sec. 2, P is a harmonic function of the variables x, y for which $x = \rho\cos\varphi$, $y = \rho\sin\varphi$; it is easily verified that its only singularity is the point with the polar coordinates R, θ. It is further easily confirmed that P has derivatives of all orders with respect to x and y except at the point R, θ and that these derivatives likewise satisfy the equation (1), that is, they are also harmonic functions. Since all derivatives of P are continuous in the parameter θ as long as $r < R$, it is permissible to differentiate (21) under the sign of integration. All these derivatives of u are also harmonic functions of x and y; indeed, the derivatives of P are harmonic, and the integration of a harmonic function with respect to a parameter clearly leads again to a harmonic function. We have thus proved the following result:

A function which is harmonic in a domain possesses there derivatives of all orders; these derivatives are again harmonic functions.

The generality of this statement, due to our replacing a circle by a general domain, is only apparent. By definition, a function is harmonic at a point if it is harmonic in a small circle surrounding this point; but in this circle we may apply the Poisson integral, and the existence of derivatives of all orders follows.

We next show that the Poisson formula (20) can also be cast into the form of a series expansion. Our point of departure is the identity

$$(23) \quad \left(1 + 2\sum_{\nu=1}^{n-1} r^\nu \cos \nu\psi\right)(1 - 2r\cos\psi + r^2) = 1 - r^2 - 2r^n[\cos n\psi - r\cos(n-1)\psi]$$

which is easily confirmed by term-by-term multiplication and use of the addition theorem of the cosine function. (23) is equivalent to

$$(23') \quad \frac{1 - r^2}{1 - 2r\cos\psi + r^2} = 1 + 2\sum_{\nu=1}^{n-1} r^\nu \cos\nu\psi + \frac{2r^n[\cos n\psi - r\cos(n-1)\psi]}{1 - 2r\cos\psi + r^2}.$$

If $r < 1$, the second term on the right-hand side of (23′) is smaller than $2r^n(1 + r)(1 - r)^{-2}$. As a result, we have the identity

$$(23'') \quad \frac{1 - r^2}{1 - 2r\cos\psi + r^2} = 1 + 2\sum_{\nu=1}^{\infty} r^\nu \cos\nu\psi, \qquad r < 1,$$

where the expansion converges absolutely and uniformly in ψ $(0 \leq \psi \leq 2\pi)$ If we replace r by ρR^{-1} and ψ by $\theta - \varphi$, we obtain the expansion

$$\frac{R^2 - \rho^2}{R^2 - 2\rho R\cos(\theta - \varphi) + \rho^2} = 1 + 2\sum_{\nu=1}^{\infty}\left(\frac{\rho}{R}\right)^\nu (\cos\nu\theta\cos\nu\varphi + \sin\nu\theta\sin\nu\varphi)$$

for the Poisson kernel (22) which, for $\rho < R$, converges absolutely and uniformly in θ and φ. Inserting this expression in (20) and integrating term by term—which is permissible in view of the uniform convergence—we obtain the following alternative form of the Poisson formula:

$$u(\rho,\varphi) = \frac{a_0}{2} + \sum_{\nu=1}^{\infty} \rho^\nu (a_\nu \cos \nu\varphi + b_\nu \sin \nu\varphi), \qquad \rho < R, \tag{24}$$

where

$$a_\nu = \frac{1}{\pi R^\nu} \int_0^{2\pi} u(R,\theta) \cos \nu\theta \, d\theta \tag{25}$$

and

$$b_\nu = \frac{1}{\pi R^\nu} \int_0^{2\pi} u(R,\theta) \sin \nu\theta \, d\theta. \tag{25'}$$

Since, by Exercise 4, Sec. 2, each term of (24) is a harmonic function, (24) constitutes an expansion of an arbitrary harmonic function into a series of particularly simple harmonic functions.

We shall now show that, in addition to representing known harmonic functions in terms of their boundary values, the Poisson formula also solves the boundary value problem of the first kind in the case of a circle. The following theorem holds.

If $U(\theta)$ is a piecewise continuous function for $0 \leq \theta \leq 2\pi$, then the function

$$u(\rho,\varphi) = \frac{R^2 - \rho^2}{2\pi} \int_0^{2\pi} \frac{U(\theta)\, d\theta}{R^2 - 2R\rho \cos(\theta - \varphi) + \rho^2}, \tag{26}$$

which can also be expanded into the series

$$u(\rho,\varphi) = \frac{a_0}{2} + \sum_{\nu=1}^{\infty} \rho^\nu (a_\nu \cos \nu\varphi + b_\nu \sin \nu\varphi)$$

with

$$a_\nu = \frac{1}{\pi R^\nu} \int_0^{2\pi} U(\theta) \cos \nu\theta \, d\theta,$$

$$b_\nu = \frac{1}{\pi R^\nu} \int_0^{2\pi} U(\theta) \sin \nu\theta \, d\theta, \qquad \nu = 0, 1, 2, \ldots,$$

is harmonic in the circle $\rho < R$; on the circumference $\rho = R$, u takes the boundary values $U(\theta)$, except at the points at which $U(\theta)$ is discontinuous.

We have shown before that the function $u(\rho,\varphi)$ defined by the Poisson formula (26) is harmonic in $\rho < R$. All that remains to be shown, therefore, is the fact that this function indeed takes the correct boundary values on $\rho = R$. If θ_0 is a point of continuity of $U(\theta)$, we thus have to prove that

$$\lim_{\substack{\rho\to R\\ \varphi\to\theta}} u(\rho,\varphi) = U(\theta). \tag{27}$$

Applying (20) to the harmonic function 1, we obtain

$$1 = \frac{R^2 - \rho^2}{2\pi}\int_0^{2\pi} \frac{d\theta}{R^2 - 2\rho R\cos(\theta - \varphi) + \rho^2}.$$

We now multiply this identity with $U(\theta_0)$ and subtract the result from (26). This yields

$$u(\rho,\varphi) - U(\theta_0) = \frac{R^2 - \rho^2}{2\pi}\int_0^{2\pi} \frac{[U(\theta) - U(\theta_0)]\,d\theta}{R^2 - 2\rho R\cos(\theta - \varphi) + \rho^2}. \tag{28}$$

Since $U(\theta)$ is continuous at $\theta = \theta_0$, there exists an α such that for a given arbitrarily small positive ϵ we have $|U(\theta) - U(\theta_0)| < \epsilon$, provided $|\theta - \theta_0| < \alpha$. It follows that

$$\begin{aligned} \left|\frac{R^2 - \rho^2}{2\pi}\int_{\theta_0-\alpha}^{\theta_0+\alpha} \frac{[U(\theta) - U(\theta_0)]\,d\theta}{R^2 - 2\rho R\cos(\theta - \varphi) + \rho^2}\right| &\le \epsilon\,\frac{R^2 - \rho^2}{2\pi}\int_{\theta_0-\alpha}^{\theta_0+\alpha} \frac{d\theta}{R^2 - 2\rho R\cos(\theta - \varphi) + \rho^2} \\ &\le \epsilon\,\frac{R^2 - \rho^2}{2\pi}\int_0^{2\pi} \frac{d\theta}{R^2 - 2R\cos(\theta - \varphi) + \rho^2} = \epsilon. \end{aligned} \tag{29}$$

Since the point (ρ,φ) approaches the point (R,θ_0), we shall ultimately have $|\varphi - \theta_0| < \frac{1}{2}\alpha$. For such values of φ, and for values of θ for which $|\theta - \theta_0| > \alpha$, we have

$$|\theta - \varphi| \ge |\theta - \theta_0| - |\theta_0 - \varphi| \ge \alpha - \tfrac{1}{2}\alpha = \tfrac{1}{2}\alpha,$$

whence $\cos(\theta - \varphi) \le \cos\frac{1}{2}\alpha$ and

$$R^2 - 2\rho R\cos(\theta - \varphi) + \rho^2 \ge R^2 - 2\rho R\cos\tfrac{1}{2}\alpha + \rho^2 > 4\rho R\sin^2\tfrac{1}{4}\alpha.$$

Denoting the last expression by A and using the notation

$$\int_0^{2\pi} |U(\theta) - U(\theta_0)|\,d\theta = M,$$

we therefore obtain

$$\left|\frac{R^2 - \rho^2}{2\pi}\int_{\theta_0+\alpha}^{\theta_0+2\pi-\alpha} \frac{[U(\theta) - U(\theta_0)]\,d\theta}{R^2 - 2\rho R\cos(\theta - \varphi) + \rho^2}\right| \le \frac{M}{2\pi A}(R^2 - \rho^2).$$

Comparison of this inequality with (28) and (29) yields

$$|u(\rho,\varphi) - U(\theta_0)| \leq \epsilon + \frac{M}{2\pi A}(R^2 - \rho^2).$$

Since ϵ tends to zero if φ tends to θ_0 and $R^2 - \rho^2 \to 0$ for $\rho \to R$, (27) follows and our theorem is proved.

At a point of the circumference $\rho = R$ which corresponds to a discontinuity of the boundary value function $U(\theta)$, the boundary value problem loses its meaning. However, the following information regarding the behavior of $u(\rho,\varphi)$ in the neighborhood of such a point is easily obtained:

Let θ_0 be a discontinuity of $U(\theta)$ and let m and M $(m < M)$ be the two limits of $U(\theta)$ at this point. By suitable approach to the point (R,θ_0) from the interior of the circle $\rho = R$, the function $u(\rho,\varphi)$ can be made to tend to any limit L for which $m < L < M$; in particular, $u(\rho,\varphi)$ tends to $\frac{1}{2}(m + M)$, if the point R, θ_0 is approached radially.

This result is an immediate consequence of the properties of the harmonic function

$$p(x,y) = \tan^{-1}\frac{y - b}{x - a}$$

of Exercise 2, Sec. 2. Since $(y - b)(x - a)^{-1}$ is the slope of the straight line connecting the points (a,b) and (x,y), the value of $p(x,y)$ coincides with that of the angle which this line forms with the positive axis. If (a,b) is a point of the circumference $\rho = R$, this angle clearly jumps by the amount π if the point (x,y) passes through (a,b) while describing R. Furthermore, the limit of $p(x,y)$, if (x,y) tends to (a,b) along a straight line from within the circle, will be any number between these two values, depending on the angle of approach.

Let now $U(\theta)$ be the above-mentioned boundary value function, and let (a,b) be the rectangular coordinates of the point of angle θ_0 on the circle $\rho = R$. If $M - m$ is the jump of $U(\theta)$ at this point, then $U(\theta) - \frac{M - m}{\pi}p(x,y)$, or $U(\theta) + \frac{M - m}{\pi}p(x,y)$, will be continuous there, since the discontinuities of both functions cancel out. From the continuity of this auxiliary function and the properties of $p(x,y)$ the above result then follows easily; the details of the proof are left to the reader as an exercise.

EXERCISES

1. If the function $u(\rho,\varphi)$ is harmonic for $\rho \leq R$ and if $|u(\rho,\varphi)| \leq 1$ in this circle, deduce from the Poisson integral that

$$|u(\rho,\varphi) - u(0)| \leq \frac{\rho}{\pi}\int_0^{2\pi}\frac{|R\cos(\theta - \varphi) - \rho|}{R^2 - 2\rho R\cos(\theta - \varphi) + \rho^2}\,d\theta;$$

by evaluating the integral, show that

$$|u(\rho,\varphi) - u(0)| \leq \frac{4}{\pi} \sin^{-1} \frac{\rho}{R},$$

where the sign of equality can occur only if u has the boundary value 1 and -1 for $\cos(\theta - \varphi) > \frac{\rho}{R}$ and $\cos(\theta - \varphi) < \frac{\rho}{R}$, respectively.

2. If $u(\rho,\varphi)$ satisfies the same hypotheses as in the preceding exercise, show that the coefficients a_ν, b_ν of its expansion (24) are subject to the inequalities

$$|a_\nu| \leq \frac{4}{\pi R^\nu}, \qquad |b_\nu| \leq \frac{4}{\pi R^\nu}.$$

3. Show that the function $u(\rho,\varphi)$ which is harmonic in R and has the boundary values $u(R,\theta) = 1$ for $-\alpha < \theta < \alpha (0 < \alpha < \pi)$ and $u(R,\theta) = 0$ for $\alpha < \theta < 2\pi - \alpha$ has the expansion

$$u(\rho,\varphi) = \frac{\alpha}{\pi} + \frac{2}{\pi} \sum_{\nu=1}^{\infty} \frac{1}{\nu} \left(\frac{\rho}{R}\right)^\nu \sin \nu\alpha \cos \nu\varphi, \qquad \rho < R.$$

4. If $u(\rho,\varphi)$ is harmonic for $\rho < R$, show that

$$\frac{1}{\pi} \int_0^{2\pi} u^2(\rho,\varphi)\, d\varphi = \frac{a_0^2}{2} + \sum_{\nu=1}^{\infty} (a_\nu^2 + b_\nu^2)\rho^{2\nu}, \qquad \rho < R,$$

where a_ν, b_ν are the coefficients of the expansion (24).

5. Show that, for $\rho > R$, the Poisson formula (20) represents a function $u(\rho,\varphi)$ which is harmonic in the exterior of the circle $\rho = R$ and takes the boundary values $-U(R,\theta)$ on $\rho = R$.

7. The Neumann Function and the Boundary Value Problem of the Second Kind. It was shown before (Exercise 3, Sec. 4) that a harmonic function which has vanishing normal derivatives on the smooth boundary of a domain in which the function is harmonic reduces to a constant. This result can also be formulated as a uniqueness theorem:

If u and v are harmonic in a smoothly bounded domain D and their normal derivatives on the boundary of D coincide, then $u = v + c$, where c is a constant.

Indeed, $u - v$ is harmonic in D and has vanishing normal derivatives on the boundary of D; by what was said before, it therefore reduces to a constant. We remark here that the above statement would also remain true if "smooth" is replaced by "piecewise smooth." The restriction to smoothly bounded domains is only introduced in the interest of conciseness of formulation; the occurrence of corners in the boundary of D—at which the normal derivatives of the functions in question would cease to be defined—would require a special discussion in each case.

The uniqueness theorem may also be expressed by saying that, up to an additive constant, a harmonic function is completely determined by the values of its normal derivative on the boundary of a domain D. This uniqueness property leads in a natural way to the question whether it is always possible to construct a harmonic function if the values of its normal derivative on the boundary of D are arbitrarily prescribed. We shall show in Sec. 8 that this question, the *boundary value problem of the second kind,* can be reduced to a boundary value problem of the first kind. In the present section we leave aside the question of existence and we shall only be concerned with the representation of a harmonic function in terms of its normal derivatives on the boundary of D. This representation is achieved by means of the *Neumann function* of D, a function which plays here a part analogous to that of the Green's function in the boundary value problem of the first kind. This function is defined as follows:

The Neumann function $N(z,\zeta)$ of a smoothly bounded domain D with respect to a point ζ of D is of the form

$$N(z,\zeta) = -\log r + N_1(z,\zeta), \tag{30}$$

where r is the distance between z and ζ and $N_1(z,\zeta)$ is harmonic in D; on the boundary Γ of D, $N(z,\zeta)$ has a constant normal derivative, i.e.,

$$\frac{\partial N(z,\zeta)}{\partial n_z} = \text{const.}, \qquad z \in \Gamma. \tag{31}$$

$N(z,\zeta)$ is further normalized by the condition

$$\int_\Gamma N(z,\zeta)\, ds_z = 0. \tag{32}$$

A condition of the type (32) is necessary if $N(z,\zeta)$ is to be defined in a unique manner; without this requirement, $N(z,\zeta)$ is clearly only determined up to an arbitrary additive constant. We might, of course, dispose of this constant by a normalization condition different from (32); for instance, we might require that $N(z,\zeta)$ vanish at a specified point of D. The condition (32) was chosen because it leads to the most simple formulas. The existence of the Neumann function is equivalent to the possibility of solving a particular boundary value problem of the second kind. If $N_2(z,\zeta)$ is the harmonic function with the boundary derivatives

$$\frac{\partial \log r}{\partial n} + \text{const.},$$

then, clearly, $-\log r + N_2(z,\zeta)$ is the Neumann function.

The constant in (31) cannot be chosen arbitrarily; in fact, its value is completely determined by the other conditions imposed on $N(z,\zeta)$. By

(30) and (31), we have

$$\text{const.} \int_\Gamma ds = -\int_\Gamma \frac{\partial \log r}{\partial n}\, ds + \int_\Gamma \frac{\partial N_1(z,\zeta)}{\partial n}\, ds.$$

Since $N_1(z,\zeta)$ is harmonic in D, the last integral vanishes by virtue of (9). Setting $u = 1$ in (14), we find that the first integral on the right-hand side has the value 2π. Since, further, $\int_\Gamma ds = L$ where L is the length of Γ, we finally obtain

$$\frac{\partial N(z,\zeta)}{\partial n} = -\frac{2\pi}{L}, \qquad z \in \Gamma. \tag{31'}$$

Turning now to the question of representing a harmonic function in terms of its normal derivatives on the boundary, we observe that since such a representation can determine a function only up to an arbitrary additive constant, it becomes necessary to adopt a normalization convention if we want to make this representation unique. We choose the normalization corresponding to (32), that is, we assume that the harmonic functions we consider are normalized by the requirement

$$\int_\Gamma u(z)\, ds = 0; \tag{33}$$

for a given harmonic function, (33) can clearly be achieved by adding a suitable constant. We now identify the function h in (14) with the Neumann function $N(z,\zeta)$ of D—which is permissible, since $N(z,\zeta)$ is of the form (11)—and apply (14) to a harmonic function u which is normalized by (33). We obtain

$$u(\zeta) = -\frac{1}{2\pi} \int_\Gamma \left[u(z) \frac{\partial N(z,\zeta)}{\partial n} - N(z,\zeta) \frac{\partial u(z)}{\partial n} \right] ds.$$

Since, by (31′) and (33),

$$\int_\Gamma u(z) \frac{\partial N(z,\zeta)}{\partial n}\, ds = -\frac{2\pi}{L} \int_\Gamma u(z)\, ds = 0,$$

this reduces to

$$u(\zeta) = \frac{1}{2\pi} \int_\Gamma N(z,\zeta) \frac{\partial u(z)}{\partial n}\, ds.$$

This is the desired representation of a harmonic function in terms of its normal derivatives on the boundary of a domain. Because of its importance, we state this result in the form of a theorem.

Let $u(z)$ be harmonic in the closure $D + \Gamma$ of a smoothly bounded domain D and let $u(z)$ be normalized by the condition

$$\int_\Gamma u(z)\, ds = 0.$$

If $N(z,\zeta)$ denotes the Neumann function of D with respect to a point ζ of D, then

$$u(\zeta) = \frac{1}{2\pi} \int_\Gamma N(z,\zeta) \frac{\partial u(z)}{\partial n}\, ds. \tag{34}$$

As in a similar case in Sec. 5, the question arises whether the following wider interpretation of the formula (34) is permissible: If $p(z)$ $(z \in \Gamma)$ is an arbitrary function satisfying certain regularity conditions, is it true that the function

$$u(\zeta) = \frac{1}{2\pi} \int_\Gamma N(z,\zeta) p(z)\, ds \tag{35}$$

is harmonic in D and that, on Γ, $\partial u/\partial n = p(z)$? The answer to the first question is not difficult. We shall show presently that $N(z,\zeta)$ is also a harmonic function of the variable ζ; hence, if $p(z)$ is integrable, it follows easily that the function $u(\zeta)$ defined by (35) is harmonic in D. The second question is identical with the *boundary value problem of the second kind*—or, as it is also occasionally called, the *Neumann problem*—mentioned above; as already said, it will be shown in Sec. 8 that it can be reduced to a boundary value problem of the first kind.

The fact that $N(z,\zeta)$ is also a harmonic function of the variable ζ—the reader is reminded that "the variable ζ" is short for "the variables ξ, η"—is a consequence of the following *symmetry property* of the Neumann function:

If $N(z,\zeta)$ denotes the Neumann function of a smoothly bounded domain D with respect to a point ζ of D, then

$$N(z,\zeta) = N(\zeta,z). \tag{36}$$

The proof of (36) is modeled on that of the corresponding symmetry property (17) of the Green's function. Let ζ and t be two given points of D. The Neumann functions $N(z,\zeta)$ and $N(z,t)$ are harmonic in D with the exception of the points ζ and t, respectively. If we delete from D two small circles $C_\epsilon(\zeta)$ and $C_\epsilon(t)$ of radius ϵ and centers ζ and t, respectively, both functions will therefore be harmonic in the domain D_ϵ thus obtained. If Γ_ϵ denotes the boundary of D_ϵ, consisting of the boundary Γ of D and the two circles $C_\epsilon(\zeta)$ and $C_\epsilon(t)$, it follows from Green's formula (12) that

$$\int_{\Gamma_\epsilon} \left[N(z,\zeta) \frac{\partial N(z,t)}{\partial n} - N(z,t) \frac{\partial N(z,\zeta)}{\partial n} \right] ds = 0.$$

In this integration, the boundary Γ is described in the positive sense, while the circumferences $C_\epsilon(\zeta)$ and $C_\epsilon(t)$ are described in the negative sense. The integral over Γ vanishes; indeed, by (31′) and (32), we have

$$\int_\Gamma N(z,\zeta)\frac{\partial N(z,t)}{\partial n}\,ds = -\frac{2\pi}{L}\int_\Gamma N(z,\zeta)\,ds = 0,$$

and the same is true if the roles of ζ and t are interchanged. With the abbreviations $N(z,\zeta) = N_1$, $N(z,t) = N_2$, we have therefore

$$\int_{C_\epsilon(\zeta)}\left(N_1\frac{\partial N_2}{\partial n} - N_2\frac{\partial N_1}{\partial n}\right)ds + \int_{C_\epsilon(t)}\left(N_1\frac{\partial N_2}{\partial n} - N_2\frac{\partial N_1}{\partial n}\right)ds = 0.$$

N_1 has a logarithmic singularity of the type (11) at the point ζ, while N_2 is harmonic in the interior of $C_\epsilon(\zeta)$. By (14), the value of the integral over $C_\epsilon(\zeta)$ is therefore $2\pi N_2(\zeta)$. Similarly, the value of the integral over $C_\epsilon(t)$ is found from (14) to be $-2\pi N_1(t)$. Hence, $N_2(\zeta) = N_1(t)$ or, in view of the definition of N_1 and N_2, $N(\zeta,t) = N(t,\zeta)$. This proves (36).

We end this section with the explicit determination of the Neumann function in the case in which D is the circle C_R of radius R with center at the origin. If r, θ are polar coordinates in the z-plane and ρ, φ the polar coordinates of a point ζ inside $C_R(\rho < R)$, we shall show that the Neumann function of C_R with respect to the point ζ is of the form

$$(37)\quad N(z,\zeta) = -\log\,[r^2 - 2\rho r\cos(\theta-\varphi) + \rho^2] - \log\left[\frac{R^4}{\rho^2} - 2\frac{R^2r}{\rho}\cos(\theta-\varphi) + r^2\right] + 2\log\frac{R}{\rho}.$$

If r_1 denotes the distance between the points z and ζ, and r_2 denotes the distance between z and the point inverse to ζ with respect to the circumference C_R, that is, the point with the polar coordinates R^2/ρ, φ, it follows by elementary trigonometry that (37) is equivalent to

$$N(z,\zeta) = -\log r_1 - \log r_2 + 2\log\frac{R}{\rho}.$$

This shows that $N(z,\zeta)$ is a harmonic function of z in C_R, with the exception of the point ζ at which $\log r_1$ has the prescribed singularity (30) of the Neumann function. For the complete identification of (37) with the Neumann function of C_R, we thus have only to show that it has a constant normal derivative on the circumference $r = R$. This, however, is easily confirmed by differentiating (37) with respect to r and setting $r = R$ in the result. We obtain

$$\left[\frac{\partial N(z,\zeta)}{\partial r}\right]_{r=R} = -\frac{1}{R},$$

in agreement with (31′). The determination of the additive constant 2 log (R/ρ) by means of the normalization condition (32) is left as an exercise to the reader.

EXERCISES

1. Verify the symmetry property (36) in the case of the circle.

2. Derive an explicit formula representing a harmonic function in a circle by means of its radial derivatives on the circumference.

3. Let $u(\rho,\varphi)$ be harmonic for $\rho \leq R$ and let $p(\theta)$ be the normal derivative of u at the point (R,θ) of the circle $\rho = R$; if u has the normalization (33), show that for $\rho \leq R$, u can be expanded into the series

$$u(\rho,\varphi) = \sum_{\nu=1}^{\infty} \rho^{\nu}(a_\nu \cos \nu\theta + b_\nu \sin \nu\theta),$$

where

$$a_\nu = \frac{1}{\pi\nu R^{\nu-1}} \int_0^{2\pi} p(\theta) \cos \nu\theta \, d\theta,$$

$$b_\nu = \frac{1}{\pi\nu R^{\nu-1}} \int_0^{2\pi} p(\theta) \sin \nu\theta \, d\theta.$$

4. Let $K(z,\zeta)$ be defined by

$$K(z,\zeta) = \frac{1}{2\pi} [N(z,\zeta) - g(z,\zeta)],$$

where $N(z,\zeta)$ and $g(z,\zeta)$ are, respectively, the Neumann function and the Green's function of a smoothly bounded domain D. If $u(z)$ is harmonic in the closure $D + \Gamma$ of D and normalized by (33), show that

$$u(\zeta) = \int_\Gamma u(z) \frac{\partial K(z,\zeta)}{\partial n} \, ds,$$

$$u(\zeta) = \int_\Gamma \frac{\partial u(z)}{\partial n} K(z,\zeta) \, ds.$$

5. Show that the function $K(z,\zeta)$ of the preceding exercise is harmonic at all points of D and that it has the "reproducing property"

$$u(\zeta) = \iint_D (K_x u_x + K_y u_y) \, dx \, dy,$$

where $K = K(z,\zeta)$, and $u = u(z)$ is harmonic in $D + \Gamma$ and normalized by (33).

8. The Conjugate Harmonic Function.

If the function $u = u(x,y)$ is harmonic in a domain D, we can associate with it another function $v = v(x,y)$ by means of the following system of partial differential equations:

$$\frac{\partial u}{\partial x} = \frac{\partial v}{\partial y}, \tag{38′}$$

$$\frac{\partial u}{\partial y} = -\frac{\partial v}{\partial x}. \tag{38″}$$

This system of equations is known as the *Cauchy-Riemann differential equations* and plays a fundamental part in the theory of conformal mapping. The function v defined by the equations (38) is harmonic in D; indeed, differentiating (38′) with respect to y and (38″) with respect to x, and subtracting the results from each other, we obtain

$$\frac{\partial^2 v}{\partial x^2} + \frac{\partial^2 v}{\partial y^2} = 0.$$

v is called the *harmonic conjugate* of u. It is worth noting that in deriving the harmonicity of v we did not use the fact that u is harmonic; we only made use of the existence of two continuous derivatives. Indeed, it easily follows from (38) that two functions u and v which have two continuous derivatives and are connected by the system of equations (38) are necessarily both harmonic.

The definition (38) for the partial derivatives of the harmonic conjugate v can be replaced by a definition for v itself. If dv is the total differential of v, we have by (38)

$$dv = \frac{\partial v}{\partial x}\,dx + \frac{\partial v}{\partial y}\,dy = -\frac{\partial u}{\partial y}\,dx + \frac{\partial u}{\partial x}\,dy.$$

Hence, integration along a curve C connecting two points (x,y) and (x_0,y_0) yields

$$v(x,y) - v(x_0,y_0) = \int_C \left(-\frac{\partial u}{\partial y}\,dx + \frac{\partial u}{\partial x}\,dy\right). \tag{39}$$

The fact that the line integral on the right-hand side of (39) is independent of the curve C and that its value depends only on the terminals (x,y) and (x_0,y_0) can also be verified directly. From (1), it follows that

$$\frac{\partial}{\partial y}\left(-\frac{\partial u}{\partial y}\right) = \frac{\partial}{\partial x}\left(\frac{\partial u}{\partial x}\right),$$

and this is the well-known condition for the independence of a line integral of the integration path, applied to the integral (39). It is important to note that the conjugate harmonic function is only determined up to an arbitrary additive constant which plays the part of an integration constant in (39) and was expressed there as the value of the function v at the arbitrary point (x_0,y_0).

The Cauchy-Riemann equations can also be formulated with respect to two arbitrary perpendicular directions which do not necessarily have to be parallel to the axes of the coordinate system. With a view to later applications in which these directions will be those of the normal and the tangent of a smooth curve, we denote these perpendicular directions by n and s, respectively. Here, a word has to be said regarding the relative orientation of two mutually perpendicular directions. In a rectangular coordinate system, the positive x-direction and the positive y-direction play two slightly different parts. While a rotation of $\frac{1}{2}\pi$ transforms the positive x-axis into the positive y-axis, the inverse transformation requires a rotation of $-\frac{1}{2}\pi$. In our (n,s) system, we shall assume that the relative orientations of the n- and s-directions are like those of the x- and y-axes, respectively. This is in agreement with our conventions regarding the positive direction of the normal and the positive sense in which the boundary of a domain is described (see Fig. 2). If $\partial/\partial n$ and $\partial/\partial s$ denote

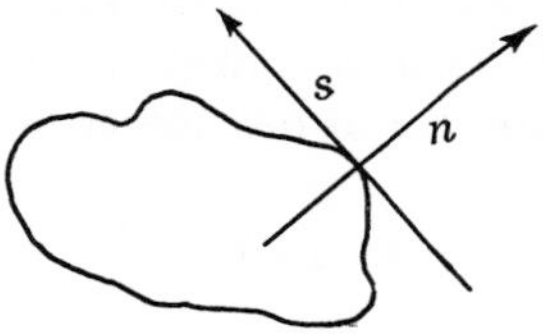

FIG. 2.

differentiations with respect to the positive n-direction and the positive s-direction, respectively, we have

$$\frac{\partial u}{\partial n} = \frac{\partial u}{\partial x} \cos (x,n) + \frac{\partial u}{\partial y} \cos (y,n), \tag{40}$$

where (x,n) and (y,n) are the angles between the positive directions indicated. Similarly,

$$\frac{\partial v}{\partial s} = \frac{\partial v}{\partial x} \cos (x,s) + \frac{\partial v}{\partial y} \cos (y,s).$$

Since $(x,n) = (y,s)$ and $(x,s) + (y,n) = \pi$, this can also be written

$$\frac{\partial v}{\partial s} = -\frac{\partial v}{\partial x} \cos (y,n) + \frac{\partial v}{\partial y} \cos (x,n)$$

or, in view of the Cauchy-Riemann equations (38),

$$\frac{\partial v}{\partial s} = \frac{\partial u}{\partial y} \cos (y,n) + \frac{\partial u}{\partial x} \cos (x,n).$$

It therefore follows from (40) that

$$\frac{\partial u}{\partial n} = \frac{\partial v}{\partial s}, \tag{41}$$

which is the desired general form of the Cauchy-Riemann equations; (41) can also be brought into the form

$$\frac{\partial u}{\partial s} = -\frac{\partial v}{\partial n}, \tag{42}$$

which follows from the observation that the relative orientation of the n- and s-directions is reversed if the positive and negative directions of n are interchanged. A comparison of (41) and (42) incidentally brings out the fact that the relation between the functions u and v is not quite symmetric. If v is the harmonic conjugate of u, then $-u$, and not u, is the harmonic conjugate of v.

(41) can be used in order to compute the harmonic conjugate of a given harmonic function u. If C is a smooth curve connecting two points (x,y) and (x_0,y_0) and s is the arc-length parameter of this curve, then

$$v(x,y) - v(x_0,y_0) = \int_{s_0}^{s} \frac{\partial v}{\partial s}\, ds,$$

where s and s_0 are the values of the parameters corresponding to the points in question. It therefore follows from (41) that

$$v(x,y) = v(x_0,y_0) + \int_{s_0}^{s} \frac{\partial u}{\partial n}\, ds. \tag{43}$$

It is left to the reader as an exercise to show that (43) is essentially equivalent to (39). In a practical application of (43), the curve C will, of course, be so chosen as to make the computation of the normal derivative of u as simple as possible. As an example, consider the harmonic function $u = \log r$, where r, θ are polar coordinates. Choosing the curves C to be circular arcs $r = \text{const.}$, we have $\dfrac{\partial}{\partial n} = \dfrac{\partial}{\partial r}$ and $ds = r\, d\theta$. Hence

$$v = \int_{\theta_0}^{\theta} \frac{1}{r}\, r\, d\theta + \text{const.} = \theta,$$

where a suitable value of the constant has been taken. We thus find that the angle θ is the harmonic conjugate of $\log r$. (As for θ being a harmonic function, compare Exercise 2, Sec. 2.)

Using the same integration curves, the reader will verify without difficulty that the conjugate of the harmonic function $r^n \cos n\theta$ is $r^n \sin n\theta$, and that the harmonic conjugate of $r^n \sin n\theta$ is $-r^n \cos n\theta$. This result is especially important since, by (24), any function u which is harmonic in a circle $r < R$ can be expanded into a series whose terms are constant multiples of the functions $r^n \cos n\theta$ and $r^n \sin n\theta$. Since, clearly, the conjugate of the sum of two functions is the sum of their conjugates, we can therefore immediately write down the conjugate of a harmonic function if this function is given in the form of its expansion (24).

The conjugate of the harmonic function

$$u(r,\theta) = \frac{a_0}{2} + \sum_{\nu=1}^{\infty} r^\nu(a_\nu \cos \nu\theta + b_\nu \sin \nu\theta), \qquad r < R, \tag{44}$$

is

$$v(r,\theta) = \sum_{\nu=1}^{\infty} r^\nu(a_\nu \sin \nu\theta - b_\nu \cos \nu\theta). \tag{44'}$$

It hardly needs to be pointed out that the region of convergence of (44′) is identical with that of (44).

We next use (41) in order to show that the Neumann problem of the preceding section is equivalent to a suitably posed Dirichlet problem. If it is desired to construct a harmonic function v which, on the boundary Γ of a smoothly bounded domain D, satisfies

$$\frac{\partial v}{\partial n} = p(s), \tag{45}$$

where $p(s)$ is a piecewise continuous function of the arc length parameter s, we proceed as follows: We define a function $q(s)$ by

$$q(s) = \int_{s_0}^{s} p(s)\, ds, \tag{46}$$

where s_0 corresponds to an arbitrary fixed point of Γ. Next, we construct a harmonic function u which has the boundary values $-q(s)$ on Γ, that is,

$$u(z) = -q(s), \qquad z(s) \in \Gamma.$$

By (46), we have

$$\frac{\partial u}{\partial s} = -p(s)$$

on Γ. If v denotes the harmonic conjugate of u, it therefore follows from (42) that

$$\frac{\partial v}{\partial n} = p(s),$$

which shows that the function v solves the Neumann problem in question. It should be noted that, in accordance with (9), the function $p(s)$ has to satisfy the condition

$$\int_\Gamma p(s)\, ds = 0.$$

If this were not the case, the function $q(s)$ of (46) would not be single-valued on Γ.

EXERCISES

1. Show that the harmonic conjugate of the harmonic function

$$u = \frac{x}{x^2 + y^2} \ (x^2 + y^2 \neq 0) \text{ is } -\frac{y}{x^2 + y^2}.$$

2. Let $u(x,y)$ be a harmonic function of x, y, and let $x = x(\xi,\eta)$ and $y = y(\xi,\eta)$ be harmonic conjugates with respect to the variables ξ, η, that is,

$$x_\xi = y_\eta, \qquad x_\eta = -y_\xi.$$

Show that $u_1(\xi,\eta) = u[x(\xi,\eta), y(\xi,\eta)]$ is a harmonic function of ξ,η.

3. If $u = u(x,y)$ and $v = v(x,y)$ are harmonic conjugates, show that the Jacobian $\dfrac{\partial(u,v)}{\partial(x,y)}$ cannot be negative.

4. If polar coordinates are introduced by $x = \rho \cos \theta$, $y = \rho \sin \theta$, show that the Cauchy-Riemann equations are transformed into

$$\rho u_\rho = v_\theta, \qquad u_\theta = -\rho v_\rho.$$

5. If $u = u(x,y)$ is harmonic in a domain D and $v = v(x,y)$ is the harmonic conjugate of u, show that the expression $w = u^2 + v^2$ cannot attain its maximum in D at an interior point of D. *Hint:* Use the fact, following from Exercise 2, that log w is a harmonic function of x, y if $w \neq 0$.

9. Multiply-connected Domains. In the results concerning harmonic functions which we have derived so far, no reference was made to the connectivity of the domains D involved; these results are true regardless of whether the domains in question are simply-connected or multiply-connected. There are certain properties of harmonic conjugates, however, which are decisively influenced by the connectivity of the domain in which these functions are defined. Consider first the case of a simply-connected domain D in which a harmonic function u is defined. It is easy to see that the harmonic conjugate v of u as defined by (39) is a single-valued function in D, that is, it has a uniquely defined value at each point of D [provided, of course, the lower limit of the integral in (39) is kept fixed]. This follows from the fact, pointed out above, that the integral in (39) does not depend on the particular curve C connecting the points (x,y) and (x_0,y_0), and it does not change its value if C is continuously deformed into another curve C' connecting these points in such a manner that all intermediate curves are entirely within C. Since in a simply-connected domain any two curves connecting the same two points clearly can be deformed into each other in the manner described, it follows that the function v defined in (39) has in this case a uniquely determined value at each point of D. A slightly different way of proving the single-valuedness of the harmonic conjugate in a simply-connected domain is the following. Let C_1 and C_2 be two nonintersecting curves (except for their terminals) connecting (x,y) and (x_0,y_0), and compute v by means of (39) along these two curves. If we denote the results by v_1 and v_2, we have by (39)

$$v_1 - v_2 = \int_{C_1}\left(-\frac{\partial u}{\partial y}\,dx + \frac{\partial u}{\partial x}\,dy\right) - \int_{C_2}\left(-\frac{\partial u}{\partial y}\,dx + \frac{\partial u}{\partial x}\,dy\right).$$

The curves C_1 and C_2 enclose a domain B which, in view of the fact that D is simply-connected, consists entirely of points of D. Since one of the two curves, say C_1, is traversed in the positive sense with respect to B and the other in the negative sense, the last formula can also be written in the form

$$v_1 - v_2 = \int_b \left(-\frac{\partial u}{\partial y}\,dx + \frac{\partial u}{\partial x}\,dy\right), \tag{47}$$

where b denotes the positively oriented boundary of B. By Gauss' theorem (5), this is equivalent to

$$v_1 - v_2 = \iint_B \left(\frac{\partial^2 u}{\partial x^2} + \frac{\partial^2 u}{\partial y^2}\right) dx\,dy,$$

and this vanishes by (1). It thus follows that $v_1 = v_2$, that is, the value of v does not depend on the integration curve. If the curves C_1 and C_2 intersect at a number of points, the same argument is applied to each of the domains surrounded by parts of C_1 and C_2, and we again obtain $v_1 = v_2$.

The preceding argument stands and falls with the assumption that the domain D is simply-connected. We used the fact that the interior of a simple closed curve all of whose points are in D consists entirely of points of D; as mentioned in Sec. 1, this is a characteristic property of a simply-connected domain. To illustrate the failure of the preceding argument in a multiply-connected case, consider the circular ring $1 < x^2 + y^2 < R^2$. In this doubly-connected domain there exist different types of curves connecting two given points which cannot be continuously deformed into each other within the domain; two such types are illustrated by the curves C_1 and C_2 in Fig. 3. Obviously, it is also not true any more that all points in the interior of the closed curve formed by C_1 and C_2 are points of the circular ring, and the above argument fails. We are therefore unable to show that the integral on the right-hand side of (47), taken along the simple closed curve consisting of the arcs C_1 and C_2, vanishes. Indeed, there is no reason to suppose that it will vanish in the general case. Denoting the value of this integral by p, that is,

$$p = \int_b \left(- \frac{\partial u}{\partial y} dx + \frac{\partial u}{\partial x} dy \right)$$

or, using (43),

$$p = \int_b \frac{\partial u}{\partial n} ds, \tag{48}$$

we now ask the following question: If the function v defined by (43) is not single-valued in the circular ring, what are the possible different values this function can take at one and the same point? By the argument used in the simply-connected case, it is clear that two different integration paths in (43) will lead to the same value of the function v if they can be continuously deformed into each other within the domain. We can

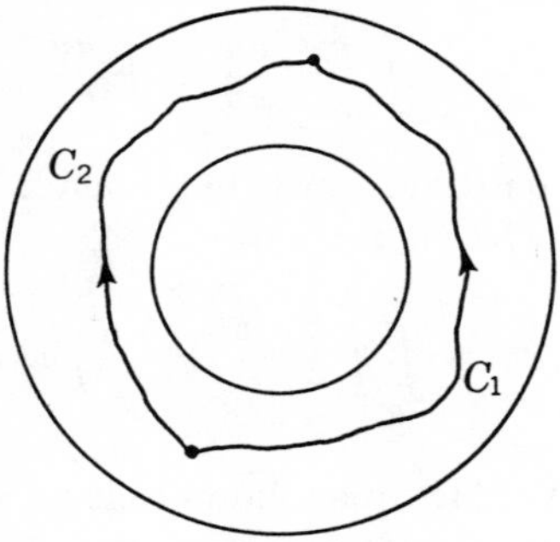

FIG. 3.

therefore expect different values of v only in the case of two paths for which such a continuous deformation is impossible. In order to give a complete classification of all essentially different paths connecting two points, it is only necessary to observe that the only obstacle in the way of such a continuous deformation is the circular "hole" in our domain. Two such paths will therefore be essentially different from each other if they surround the hole a different number of times; here, of course, it is essential to note the sense—positive or negative—in which the hole is surrounded.

To make our statement more precise, consider two different paths C_1 and C_2 which connect the same two points P and P'. As mentioned before, C_1 and C_2 form together a closed curve b if C_1 is described starting from P and ending at P', and C_2 in the opposite direction. The radius vector connecting a point of b with the center of the ring will sweep out an angle of magnitude $2\pi m$, where m is an integer, if the point describes the entire curve b. If m is zero, C_1 and C_2 are equivalent in the above sense, since the hole has not been surrounded at all by the curve b, and C_1 and C_2

are therefore deformable into each other. If, however, m is not zero, C_1 and C_2 are essentially different. In order to find the difference in the value of the function v corresponding to these two paths by means of (43), we observe that

$$\int_{C_1} \frac{\partial u}{\partial n}\,ds - \int_{C_2} \frac{\partial u}{\partial n}\,ds = \int_b \frac{\partial u}{\partial n}\,ds.$$

Denoting the two different values of v by v_1 and v_2, respectively, we thus have, in view of (43),

$$v_1(x,y) = v_2(x,y) + \int_b \frac{\partial u}{\partial n}\,ds.$$

The integral over the closed curve b is clearly equal to m times its value for a simple circuit, that is, one for which the angular variation just mentioned is 2π. With the notation (48), where the curve b was precisely of this type, we have

$$v_1(x,y) = v_2(x,y) + mp, \qquad m = 0, \pm 1, \pm 2, \ldots; \tag{49}$$

(49) shows that the various values which the harmonic conjugate v of a single-valued harmonic function u can take at one and the same point differ by integral multiples of a number p, the *period* of the function v. It is important to keep in mind that the function v is continuous at all points of the domain; its many-valuedness is expressed in the fact that a complete circuit around the "hole" leads us to a different value of the function. As an example, consider the conjugate of $u = \log r$, $r = \sqrt{x^2 + y^2}$. As shown before, $v = \theta = \tan^{-1}(y/x)$. Both functions are harmonic in the circular ring mentioned before, since their only singularity is situated at the origin. θ is not single-valued in the ring; indeed, the angle is only determined up to an integral multiple of 2π, and a closed positive circuit about the origin adds 2π to the value of θ. The function $v = \theta$ thus has the period 2π; if θ is meant to denote an angle between 0 and 2π, the conjugate of $\log r$ will therefore be $v = \theta + 2\pi m$, where m is an arbitrary integer.

Before we pass to the case of a domain of connectivity n, where n is an arbitrary positive integer, we point out that our use of the word "period" is not identical with that in the phrase "the function $\sin x$ has the period 2π," meaning that $\sin(x + 2\pi) = \sin x$. However, no confusion can arise as long as both meanings of the word "period" are not used in the same context. If there is a possibility of confusion, we shall replace the word "period"—as in "the period of the harmonic conjugate"—by the expression "modulus of periodicity." Turning now to a domain D of connectivity n, we see that the different values of the function v defined by

(43) depend on the number of times each of the $n-1$ "holes" is surrounded by the integration path. There will therefore be $n-1$ independent periods of v, corresponding to a complete circuit about each of these holes. Attaching to these holes the subscripts $1, 2, \ldots, n-1$, we thus have the independent periods $p_1, p_2, \ldots, p_{n-1}$, where the word "independent" means that none of these periods is, in general, expressible as the sum of integral multiples of the other periods. If b_ν is a simple smooth curve which surrounds the hole of subscript ν but does not contain in its interior points of any other hole, we have

$$(50) \qquad p_\nu = \int_{b_\nu} \frac{\partial u}{\partial n}\, ds, \qquad \nu = 1, 2, \ldots, n-1,$$

corresponding to (48) in the doubly-connected case. Since, apart from continuous deformations inside D, the various integration paths connecting two points of D differ only by the number of times they surround the various holes, we thus obtain the result, corresponding to (49), that the various values of $v(x,y)$ are of the form

$$(51) \qquad v(x,y) = v_0(x,y) + \sum_{\nu=1}^{n-1} m_\nu p_\nu,$$

where the m_ν are arbitrary integers and $v_0(x,y)$ is any one of the values of $v(x,y)$. It is, of course, possible that in some cases one or more of the periods p_ν are zero; if all periods p_ν, $\nu = 1, \ldots, n-1$, vanish, v will be single-valued in D.

It is finally pointed out that there also exists a period p_n of the function v associated with the outer boundary Γ_n of D. The exceptional position of the outer boundary of D as compared with the inner boundaries $\Gamma_1, \ldots, \Gamma_{n-1}$ is of a rather superficial nature; if we map the plane onto the surface of the sphere by means of the stereographic projection of Sec. 1, D is transformed into a domain D' which is bounded by n closed curves $\Gamma_1', \ldots, \Gamma_n'$ none of which deserves to be called "outer boundary" any more. In addition to the closed circuits about the curves $\Gamma_1', \ldots, \Gamma_{n-1}'$ which lead to the periods $p_1, \ldots, p_{n-1}$, we therefore also have to consider the period related to a closed circuit about Γ_n'. If we return to the (x,y) plane, the period p_n of v will thus be equal to the increment of v if the point (x,y) describes a closed curve b_n in D which contains the $n-1$ inner boundaries in its interior, that is,

$$(52) \qquad p_n = \int_{b_n} \frac{\partial u}{\partial n}\, ds.$$

Now it is clear that the effect of surrounding all interior boundaries along b_n is identical with that of surrounding each $\Gamma_\nu (\nu = 1, \ldots, n-1)$ separately along the curves $b_1, \ldots, b_{n-1}$ of (50). Since the sense in which b_n is described in (52) is necessarily opposite to that in which the b_ν in (50) are described—all the b_ν, $\nu = 1, \ldots, n$ have to be described in the positive sense with respect to the interior of D—it follows from (50) and (52) that $-p_n = p_1 + p_2 + \cdots + p_{n-1}$, or

$$p_1 + p_2 + \cdots + p_n = 0. \tag{53}$$

The sum of the periods of v associated with all the boundary components of D is zero. Thus, the period associated with Γ_n contributes nothing new; we still have only $n - 1$ independent periods. If we are dealing with a finite domain, it is natural to regard the periods $p_1, \ldots, p_{n-1}$ associated with the interior boundary as a fundamental set of independent periods. However, in the case of a domain of connectivity n which contains the point at infinity, that is, the full plane furnished with n finite holes, there is no distinguished boundary; in such a case, we shall arbitrarily disregard one of the periods and take the remaining $n - 1$ periods as the fundamental set. The relation (53) connecting the periods associated with all the boundary curves is, of course, also true in this case.

EXERCISES

1. Let D be a domain bounded by n smooth curves $\Gamma_1, \Gamma_2, \ldots, \Gamma_n$; if $M(z,\zeta)$ is the harmonic conjugate of the Neumann function $N(z,\zeta)$ of D, show that the period of $M(z,\zeta)$ corresponding to a complete circuit about Γ_ν is $-2\pi \frac{L_\nu}{L}$ $(\nu = 1, \ldots, n)$, where L_ν is the length of Γ_ν and $L = \sum_{\mu=1}^{n} L_\mu$.

2. Show that the harmonic function

$$u = \log \frac{\sqrt{(x+1)^2 + y^2}}{\sqrt{(x-1)^2 + y^2}}$$

has a single-valued harmonic conjugate in the ring $2 < x^2 + y^2 < 3$.

10. The Harmonic Measure. Let D be a smoothly bounded domain and Γ its boundary; if α denotes an open arc which forms a part of Γ and if β denotes the remainder of Γ with the exception of the end points of α, then the harmonic measure of α with respect to D is defined as follows:

The harmonic measure $\omega(z;\alpha)$ is the function which is harmonic in D and takes the boundary values 1 *and* 0 *if z approaches α or β, respectively.*

At the points at which α and β meet, the boundary values remain undefined; as in a similar case in Sec. 6, it is easy to show that by suitable approach of these points from within D, any limiting value between 0 and

1 can be obtained. As regards the existence of the harmonic measure in the case of a general domain, we observe that $\omega(z;\alpha)$ has been defined as the solution of a particular boundary value problem of the first kind; hence, its existence is a consequence of the fact, to be proved later, that the Dirichlet problem can be solved for an arbitrary piecewise continuous boundary value function.

Since $\omega(z;\alpha)$ solves a particularly simple boundary value problem, it can easily be written down explicitly with the help of the general formula (16). $\omega(z;\alpha)$ takes the boundary values 1 and 0 on α and β, respectively; hence, it follows from (16) that

$$\omega(\zeta;\alpha) = -\frac{1}{2\pi}\int_{\alpha}\frac{\partial g(z,\zeta)}{\partial n_z}\,ds_z, \tag{54}$$

where $g(z,\zeta)$ is the Green's function of D.

The concept of the harmonic measure of an arc is of importance in many problems in the theory of functions of a complex variable which do not come within the province of this book. The only case which will be important for our purposes is that in which D is a domain of connectivity n, bounded by n smooth simple closed curves $\Gamma_1, \ldots, \Gamma_n$, and α coincides with one of the boundary components Γ_ν. With the abbreviated notation $\omega_\nu(z) = \omega(z;\Gamma_\nu)$, we have the following definition:

The harmonic measure $\omega_\nu(z)$ of the domain D, bounded by the simple smooth closed curves $\Gamma_1, \ldots, \Gamma_n$, is the function which is harmonic in D and has the boundary values 1 and 0 on Γ_ν and Γ_μ ($\mu \neq \nu$), respectively.

By (54), we have

$$\omega_\nu(\zeta) = -\frac{1}{2\pi}\int_{\Gamma_\nu}\frac{\partial g(z,\zeta)}{\partial n_z}\,ds_z. \tag{55}$$

If we compare this formula with (50) and (52), we arrive at the following important connection between the harmonic measures and the Green's function of a multiply-connected domain:

The period of the harmonic conjugate of the Green's function $g(z,\zeta)$ of a multiply-connected domain D with respect to a circuit about the boundary component Γ_ν is $-2\pi\omega_\nu(\zeta)$, where $\omega_\nu(\zeta)$ is the harmonic measure of Γ_ν.

Since the Green's function $g(z,\zeta)$ has a singularity at the point ζ, we cannot expect that our result of the preceding section, according to which the sum of all n periods of a harmonic function is zero, will also hold in this case. However, something similar is still true. We have

$$\omega_1(z) + \omega_2(z) + \cdots + \omega_n(z) \equiv 1. \tag{56}$$

The proof is very simple. Since $\omega_\nu(z)$ is equal to 1 on Γ_ν and equal to zero on Γ_μ, $\mu \neq \nu$, it follows that $\omega_1(z) + \cdots + \omega_n(z)$ takes the value 1 at all points of the boundary of D. By the maximum principle, both the maximum and the minimum of this expression in D are therefore equal to 1. Hence, this function is identically equal to 1 throughout D.

We next turn our attention to the harmonic conjugates of the functions $\omega_\nu(z)$ and, especially, to their periods. If $p_{\mu\nu}$ denotes the period of the harmonic measure $\omega_\nu(z)$ with respect to a circuit about the boundary component Γ_μ, we have by (50)

$$p_{\mu\nu} = \int_{\Gamma_\mu} \frac{\partial \omega_\nu}{\partial n}\, ds. \tag{57}$$

These periods have the important symmetry property

$$p_{\mu\nu} = p_{\nu\mu}, \tag{58}$$

that is, the period of $\omega_\nu(z)$ over Γ_μ is the same as the period of $\omega_\mu(z)$ over Γ_ν. The proof of (58) is as follows: Since $\omega_\mu = 1$ on Γ_μ and $\omega_\mu = 0$ on the rest of the boundary of D, (57) may also be written

$$p_{\mu\nu} = \int_{\Gamma} \omega_\mu \frac{\partial \omega_\nu}{\partial n}\, ds, \tag{59}$$

where $\Gamma = \Gamma_1 + \Gamma_2 + \cdots + \Gamma_n$ is the complete boundary of D. By Green's formula (12), this is identical with

$$p_{\mu\nu} = \int_{\Gamma} \omega_\nu \frac{\partial \omega_\mu}{\partial n}\, ds.$$

Since $\omega_\nu = 1$ on Γ_ν and $\omega_\nu = 0$ on $\Gamma - \Gamma_\nu$, this is the same as

$$p_{\mu\nu} = \int_{\Gamma_\nu} \frac{\partial \omega_\mu}{\partial n}\, ds.$$

By (57), the right-hand side is equal to $p_{\nu\mu}$, and this proves (58).

Another property of the periods $p_{\nu\mu}$ which we shall require later is the following:

The symmetric quadratic form

$$\sum_{\nu=1}^{n-1} \sum_{\mu=1}^{n-1} p_{\nu\mu}\lambda_\nu\lambda_\mu$$

is positive-definite, that is,

$$\sum_{\nu=1}^{n-1} \sum_{\mu=1}^{n-1} p_{\nu\mu}\lambda_\nu\lambda_\mu > 0 \tag{60}$$

unless all the numbers $\lambda_1, \ldots, \lambda_{n-1}$ *are zero.*

Before we prove (60), we remark that the exclusion of the subscript n from the summation in (60) is only due to the resulting convenience of notation; we might as well have excluded any other subscript. One subscript has to be omitted; we shall see that otherwise the result will not be true. Consider now the expression

$$\omega(z) = \lambda_1\omega_1(z) + \lambda_2\omega_2(z) + \cdots + \lambda_{n-1}\omega_{n-1}(z),$$

where $\lambda_1, \lambda_2, \ldots, \lambda_{n-1}$ are arbitrary real numbers, not all zero. On Γ_ν, $\nu = 1, 2, \ldots, n-1$, the harmonic function $\omega(z)$ takes the boundary values λ_ν, and on Γ_n we have $\omega = 0$. Since not all λ_ν are zero, it follows that $\omega(z)$ cannot reduce to a constant. By Green's formula (7), we have

$$\iint_D (\omega_x^2 + \omega_y^2)\, dx\, dy = \int_\Gamma \omega \frac{\partial \omega}{\partial n}\, ds.$$

The left-hand side of this expression is nonnegative. Since ω is not constant, it cannot be zero, and we have therefore

$$\int_\Gamma \omega \frac{\partial \omega}{\partial n}\, ds > 0.$$

Written in full, this reads

$$\int_\Gamma \left(\sum_{\nu=1}^{n-1} \lambda_\nu \omega_\nu\right)\left(\sum_{\mu=1}^{n-1} \lambda_\mu \frac{\partial \omega_\mu}{\partial n}\right) ds > 0$$

or

$$\sum_{\nu=1}^{n-1}\sum_{\mu=1}^{n-1} \lambda_\nu\lambda_\mu \int_\Gamma \omega_\nu \frac{\partial \omega_\mu}{\partial n}\, ds > 0.$$

In view of (59), this is equivalent to (60). We add the remark that (60) would not be true if the summation were extended from 1 to n. Since the sum of all n periods of ω_ν is zero, we have

$$\sum_{\nu=1}^{n}\sum_{\mu=1}^{n} p_{\nu\mu} = \sum_{\nu=1}^{n}\left(\sum_{\mu=1}^{n} p_{\nu\mu}\right) = 0,$$

which shows that (60) would be wrong for $\lambda_1 = \lambda_2 = \cdots = \lambda_n = 1$ if $n - 1$ were replaced by n.

From (60) we conclude that

$$(61) \qquad \begin{vmatrix} p_{11} & p_{12} & \cdots & p_{1,n-1} \\ p_{21} & p_{22} & \cdots & p_{2,n-1} \\ \cdots & \cdots & \cdots & \cdots \\ p_{n-1,1} & p_{n-1,2} & \cdots & p_{n-1,n-1} \end{vmatrix} \neq 0.$$

To prove (61), we note that, by the theory of linear equations, the vanishing of this determinant would imply the existence of a nontrivial solution $\lambda_1, \ldots, \lambda_{n-1}$ of the homogeneous system of linear equations

$$\sum_{\mu=1}^{n-1} p_{\nu\mu}\lambda_\mu = 0, \qquad \nu = 1, 2, \ldots, n-1.$$

Multiplying by λ_ν and summing from 1 to $n - 1$, we obtain

$$\sum_{\nu=1}^{n-1}\sum_{\mu=1}^{n-1} p_{\nu\mu}\lambda_\nu\lambda_\mu = 0,$$

which contradicts (60).

EXERCISES

1. Let D be a domain of connectivity n and let $u(z)$ be harmonic in D and on its boundary components $\Gamma_1, \ldots, \Gamma_n$. If, on $\Gamma_\nu (\nu = 1, 2, \ldots, n)$, $u(z) \leq M_\nu$, show that the inequality

$$u(z) \leq M_n + \sum_{\nu=1}^{n-1} (M_\nu - M_n)\omega_\nu(z)$$

holds at all points of D.

2. If D is a domain of connectivity n and if $u(z)$ is harmonic in D, show that it is always possible to find $n - 1$ constants $A_1, \ldots, A_{n-1}$ so that the harmonic conjugate of $u_1(z) = u(z) + \sum_{\nu=1}^{n-1} A_\nu\omega_\nu(z)$ is free of periods in D.

3. A domain D is enlarged by adding to it a piece α of its boundary component Γ_1 and a domain—outside D—whose common boundary with D is α. If the harmonic measures of D and the enlarged domain D^* are denoted by $\omega_\nu(z)$ and $\omega_\nu^*(z)$ $(\nu = 1, \ldots, n)$, respectively, show that

$$\omega_1^*(z) < \omega_1(z), \qquad \omega_\nu^*(z) > \omega_\nu(z), \qquad \nu = 2, \ldots, n, \ z \in D.$$

4. If the function $u(z)$ is harmonic in D and on its boundary components $\Gamma_1, \ldots, \Gamma_n$ $(\Gamma_1 + \Gamma_2 + \cdots + \Gamma_n = \Gamma)$, show that the period P_ν of its harmonic conjugate $v(z)$ about Γ_ν is

$$P_\nu = \int_\Gamma u(z) \frac{\partial\omega_\nu(z)}{\partial n}\, ds.$$

5. Show that the harmonic measures associated, respectively, with the outer and inner boundary of the ring $a^2 < x^2 + y^2 < b^2$ are

$$\omega_1(z) = \frac{\log r - \log a}{\log b - \log a},$$
$$\omega_2(z) = \frac{\log r - \log b}{\log a - \log b}, \qquad r = \sqrt{x^2 + y^2}.$$

Find their periods.

11. Dependence of the Green's Function on the Domain. If $p(x)$ is a differentiable function of x in an interval $a \leq x \leq b$ and ϵ a small positive number, then, by the rules of the ordinary calculus,

$$p(x + \epsilon) = p(x) + \epsilon p'(x) + o(\epsilon), \tag{62}$$

where $o(\epsilon)$ denotes a quantity for which

$$\lim_{\epsilon \to 0} \frac{o(\epsilon)}{\epsilon} = 0. \tag{63}$$

(63) can also be expressed by saying that $o(\epsilon)$ is a term of higher than first order in ϵ. If we call the expression $p(x + \epsilon) - p(x)$ the variation of $p(x)$ with respect to the increment ϵ, it follows from (62) that, except for terms of higher order in ϵ, the variation of $p(x)$ is $\epsilon p'(x)$.

A similar situation arises if it is desired to compute, up to first-order terms, the variation of the Green's function $g(z,\zeta)$ of a domain D if D is made subject to a slight variation whose magnitude is measured by a small positive parameter ϵ. There is, however, a significant difference. While $p(x)$ is a point function, that is, it only depends on the point x, $g(z,\zeta)$ is a *domain function* depending upon the domain D. If, for instance, D is a simply-connected domain and $x = x(t)$, $y = y(t)$, $t_1 \leq t \leq t_2$, is a parametric representation of the boundary Γ of D, then $g(z,\zeta)$ will depend on the values of the functions $x(t)$ and $y(t)$ throughout the interval $t_1 \leq t \leq t_2$. Our problem is therefore of the same general character as those treated in the classical calculus of variations.

In what follows we shall assume that the domain D is bounded by closed *analytic curves* $x = x(t)$, $y = y(t)$, $t_1 \leq t \leq t_2$, that is, curves such that for every $t_1 < t_0 < t_2$ the functions $x(t)$ and $y(t)$ can be expanded into power series $x(t) = \sum_{\nu=0}^{\infty} a_\nu(t - t_0)^2$, $y(t) = \sum_{\nu=0}^{\infty} b_\nu(t - t_0)^\nu$ which converge in a certain neighborhood of t_0. The reason for this restriction is the following result which will be proved at a later stage (Sec. 5, Chap. V): *The Green's function of a domain D which is bounded by analytic curves Γ is harmonic at the points of* Γ. As shown in Sec. 6 of this chapter, this implies that the Green's function has derivatives of all orders at the points of Γ.

Let now s be the parameter which measures the arc length of Γ in such a way that the whole boundary Γ is described if s grows from 0 to L; if the domain D is multiply-connected, the various components will be traversed in a certain order. Let D^* be a domain whose boundary Γ^* is obtained as the result of a slight deformation of Γ. In order to obtain a complete description of this deformation, we construct the normal to Γ at every point $z(s)$ of Γ and measure along the normal the distance $\delta n(s)$ between

$z(s)$ and the first point of intersection $z^*(s)$ of the normal with Γ^*. Here, the "normal displacement" $\delta n(s)$ is to be taken positive if the vector starting from $z(s)$ and terminating at $z^*(s)$ coincides with the outward pointing normal, and negative if it has the opposite direction. Intuitively speaking, $\delta n(s)$ is positive at points at which Γ is "pushed outward," and negative if it is pushed inward. If D^* contains D, $\delta n(s)$ will obviously be nonnegative at all points of Γ. The fact that we are concerned with "small" deformations is expressed analytically by the condition that, for $0 \leq s \leq L$, $|\delta n(s)| < \epsilon M$, where M is a constant and ϵ a small positive parameter. If we write $\delta n(s) = \epsilon p(s)$, then the function $p(s)$ has to satisfy $|p(s)| < M$; in addition, we shall assume that $p(s)$ is a piecewise continuous function of s. It is understood that the variation of Γ is to be such that, apart from the points of Γ which were not moved at all, Γ and Γ^* have no common points; for a given function $p(s)$, this can always be achieved by choosing ϵ small enough.

We temporarily assume that $\delta n(s)$ is negative at all points of Γ; this restriction is easily removed once the result for the special case has been obtained. If $\delta n(s)$ is negative throughout Γ, it follows that D^* is a subdomain of D, and, consequently, that $g(z,\zeta)$ is, apart from its singularity at $z = \zeta$, harmonic in D^*; we remark that ϵ has to be taken sufficiently small so as to ensure that ζ is in the interior of D^*. If $g(z,\zeta)$ and $g^*(z,\zeta)$ denote the Green's functions of D and D^*, respectively, the function $g^*(z,\zeta) - g(z,\zeta)$ will be harmonic in D^* since the singularities of these two functions at $z = \zeta$ cancel out. Hence, by (16),

$$(64) \qquad g^*(z,\zeta) - g(z,\zeta) = -\frac{1}{2\pi}\int_{\Gamma^*}[g^*(\eta,\zeta) - g(\eta,\zeta)]\,\frac{\partial g^*(\eta,z)}{\partial n_\eta}\,ds_\eta.$$

On the other hand, using the abbreviation

$$(65) \qquad D_B[u,v] = \iint_B (u_x v_x + u_y v_y)\,dx\,dy,$$

we have by Green's formula (7)

$$\frac{1}{2\pi}D_{D-D^*}[g(\eta,\zeta),g(\eta,z)] = \frac{1}{2\pi}\int_\Gamma g(\eta,\zeta)\,\frac{\partial g(\eta,z)}{\partial n_\eta}\,ds_\eta - \frac{1}{2\pi}\int_{\Gamma^*} g(\eta,\zeta)\,\frac{\partial g(\eta,z)}{\partial n_\eta}\,ds_\eta.$$

Since $g(\eta,\zeta)$ and $g^*(\eta,\zeta)$ vanish for $\eta \in \Gamma$ and $\eta \in \Gamma^*$, respectively, this may also be written

$$\frac{1}{2\pi}D_{D-D^*}[g(\eta,\zeta),g(\eta,z)] = \frac{1}{2\pi}\int_{\Gamma^*}[g^*(\eta,\zeta) - g(\eta,\zeta)]\,\frac{\partial g(\eta,z)}{\partial n_\eta}\,ds_\eta.$$

Adding this identity to (64), we obtain

$$g^*(z,\zeta) - g(z,\zeta) + \frac{1}{2\pi} D_{D-D^*}[g(\eta,\zeta),g(\eta,z)]$$
$$= -\frac{1}{2\pi}\int_{\Gamma^*} [g^*(\eta,\zeta) - g(\eta,\zeta)] \frac{\partial}{\partial n_\eta} [g^*(\eta,z) - g(\eta,z)]\, ds,$$

or, in view of (7),

$$(66)\quad g^*(z,\zeta) - g(z,\zeta) = -\frac{1}{2\pi} D_{D-D^*}[g(\eta,\zeta),g(\eta,z)]$$
$$-\frac{1}{2\pi} D_{D^*}[g^*(\eta,\zeta) - g(\eta,\zeta), g^*(\eta,z) - g(\eta,z)].$$

We now proceed to show that the second term on the right-hand side of (66) is of the order of magnitude of ϵ^2. To this end we observe that, by the mean-value theorem of the differential calculus,

$$g(z^*,\zeta) - g(z,\zeta) = \delta n(s)\left(\frac{\partial g}{\partial n}\right)_{z'},$$

where z^* is that point of Γ^* which is obtained from the point $z(s)$ of Γ by means of the variation $\delta n(s)$ and z' a certain point on the linear segment connecting z and z^*. $g(z,\zeta)$ is harmonic in the closure of the domain enclosed between Γ and Γ^*; hence its derivatives in this region exist and are uniformly bounded. Since, moreover, $\delta n(s) = \epsilon p(s)$, where $p(s)$ is bounded, it follows that

$$|g(z^*,\zeta) - g(z,\zeta)| < M\epsilon,$$

where M is a suitable constant. But $g(z,\zeta)$ vanishes for $z \in \Gamma$, and we have therefore

$$|g(z^*,\zeta)| < M\epsilon, \qquad z^* \in \Gamma^*.$$

On Γ^*, $g^*(z,\zeta)$ vanishes. The last inequality may therefore also be written in the form

$$|g^*(z,\zeta) - g(z,\zeta)| < \epsilon M, \qquad z \in \Gamma^*.$$

But $g^*(z,\zeta) - g(z,\zeta)$ is harmonic at all points of D^*. By the maximum principle, it therefore follows that this inequality holds at all points of D^*, that is,

$$|g^*(z,\zeta) - g(z,\zeta)| < \epsilon M, \qquad z \in D^*.$$

If $u(z) = g^*(z,\zeta) - g(z,\zeta)$, the function $u(z)$ thus has the order of magnitude of ϵ. In view of (16), the same is true of the derivatives u_x and u_y.

Indeed, differentiating (16) under the integral sign, we obtain

$$u_x(z) = -\frac{1}{2\pi}\int_{\Gamma^*} u(\eta)\,\frac{\partial^2 g^*(\eta,z)}{\partial x\,\partial n_\eta}\,ds_\eta,$$

whence

$$|u_x(z)| \leq \frac{\epsilon M}{2\pi}\int_{\Gamma^*}\left|\frac{\partial^2 g^*(\eta,z)}{\partial x\,\partial n_\eta}\right|ds_\eta,$$

and similarly for u_y. In view of (65), it therefore follows that the last term on the right-hand side of (66) is of the order of magnitude of ϵ^2, and (66) can be replaced by

$$g^*(z,\zeta) - g(z,\zeta) = -\frac{1}{2\pi}D_{D-D^*}[g(\eta,\zeta),g(\eta,z)] + o(\epsilon). \tag{67}$$

To evaluate the first term on the right-hand side of (67) up to first-order terms in ϵ, we introduce a new coordinate system (σ,ν), where ν is measured along the normals to Γ and σ is measured along the orthogonal trajectories to these normals. In particular, the curve Γ belongs to the family of curves ν = const. In a sufficiently small neighborhood of Γ, the normals to Γ do not intersect one another and the transformation from the (x,y) coordinates to the (σ,ν) coordinates is uniquely determined. At the points of Γ, the Jacobian of the transformation is

$$\frac{\partial(x,y)}{\partial(\sigma,\nu)} = \begin{vmatrix} x_\sigma & x_\nu \\ y_\sigma & y_\nu \end{vmatrix} = \begin{vmatrix} x_\sigma & y_\sigma \\ y_\sigma & -x_\sigma \end{vmatrix} = x_\sigma^2 + y_\sigma^2 = 1.$$

In an ϵ-neighborhood of Γ, we therefore have, for reasons of continuity,

$$\frac{\partial(x,y)}{\partial(\sigma,\nu)} = 1 + o(1),$$

where $o(1)$ is small if ϵ is small. Hence, in view of (65),

$$\begin{aligned} A &= D_{D-D^*}[g(\eta,\zeta),g(\eta,z)] \\ &= \iint_{D-D^*} [g_x^{(1)}g_x^{(2)} + g_y^{(1)}g_y^{(2)}]\,d\nu\,d\sigma + o(\epsilon), \end{aligned} \tag{68}$$

where the abbreviations $g^{(1)} = g(\eta,\zeta)$ and $g^{(2)} = g(\eta,z)$ have been used. The term $o(\epsilon)$ is due to the correction term of the Jacobian and the fact that the width of the domain of integration is of the order of magnitude of ϵ. It follows that

$$A = \int_0^L d\sigma \int_{\delta n}^0 [g_x^{(1)}g_x^{(2)} + g_y^{(1)}g_y^{(2)}]\,d\nu + o(\epsilon).$$

The value of the inner integral is, by the mean-value theorem of the differential calculus,

$$\int_{\delta n}^{0} [g_x^{(1)}g_x^{(2)} + g_y^{(1)}g_y^{(2)}]\, d\nu = -[g_x^{(1)}g_x^{(2)} + g_y^{(1)}g_y^{(2)}]\, \delta n + o(\epsilon),$$

where the derivatives of the Green's functions may be taken at the points of Γ. Hence,

$$A = -\int_0^L [g_x^{(1)}g_x^{(2)} + g_y^{(1)}g_y^{(2)}]\delta n(s)\, ds + o(\epsilon).$$

On Γ, we have

$$g_s = g_x x_s + g_y y_s, \qquad g_n = g_x x_n + g_y y_n = g_x y_s - g_y x_s,$$

whence, by a formal computation,

$$\begin{aligned} g_n^{(1)}g_n^{(2)} + g_s^{(1)}g_s^{(2)} &= [g_x^{(1)}g_y^{(2)} + g_y^{(1)}g_y^{(2)}](x_s^2 + y_s^2) \\ &= g_x^{(1)}g_x^{(2)} + g_y^{(1)}g_y^{(2)}. \end{aligned}$$

Since $g^{(1)}$ and $g^{(2)}$ are zero on Γ, both $g_s^{(1)}$ and $g_s^{(2)}$ are zero. We thus obtain

$$A = -\int_0^L \frac{\partial g(\eta,\zeta)}{\partial n} \frac{\partial g(\eta,z)}{\partial n} \delta n(s)\, ds_\eta + o(\epsilon)$$

and, in view of (67) and (68),

$$\delta g(z,\zeta) = \frac{1}{2\pi} \int_\Gamma \frac{\partial g(\eta,z)}{\partial n_\eta} \frac{\partial g(\eta,\zeta)}{\partial n_\eta} \delta n(s)\, ds_\eta + o(\epsilon), \tag{69}$$

where

$$\delta g(z,\zeta) = g^*(z,\zeta) - g(z,\zeta) \tag{70}$$

is the variation of the Green's function if the boundary Γ of D is made subject to the normal variation $\delta n(s) = \epsilon p(s)$.

Hadamard's formula (69) has so far only been proved under the restriction that δn is negative or zero. However, this restriction is easily lifted. Suppose first that δn is positive or zero throughout Γ. Then the roles of Γ and Γ^* are interchanged, and we have, by (69) and (70),

$$g(z,\zeta) - g^*(z,\zeta) = \frac{1}{2\pi} \int_{\Gamma^*} \frac{\partial g^*(\eta,z)}{\partial n} \frac{\partial g^*(\eta,\zeta)}{\partial n} \delta n^*(s)\, ds + o(\epsilon).$$

Since δn^* is the variation leading from Γ^* to Γ, we have $\delta n^* = -\delta n + o(\epsilon)$. Moreover, replacement in this formula of g^* by g and Γ^* by Γ results only in corrections of the order of magnitude of $o(\epsilon)$. It therefore follows that Hadamard's formula (69) also remains true in this case. Finally, if δn may take both positive and negative values, we decompose Γ into two sets Γ_1 and Γ_2 such that δn is positive on Γ_1 and negative or zero on Γ_2.

The total variation of Γ may then be regarded as the result of two successive variations. The first of these coincides with δn on Γ_1 and is zero on Γ_2; the second coincides with δn on Γ_2 and vanishes on Γ_1. Since (69) has been proved for variations which are either nonnegative or nonpositive, it follows therefore that (69) holds in the general case.

As an important special case of (69) we obtain the variation formula for the domain constant $d(\zeta)$ defined by

$$d(\zeta) = \lim_{z \to \zeta} [g(z,\zeta) + \log r]. \tag{71}$$

By (69) and (70), we have

$$\delta d(\zeta) = \frac{1}{2\pi} \int_\Gamma \left[\frac{g(\eta,\zeta)}{\partial n} \right]^2 \delta n(s)\, ds + o(\epsilon). \tag{72}$$

$d(\zeta)$ grows monotonically with the domain D. Indeed, if D grows, δn is nonnegative, and it follows from (72) that $\delta d(\zeta) \geq 0$. This monotonicity property of $d(\zeta)$ can also be easily established with the help of the maximum principle. If $D' \supset D$, and we denote the Green's function of D' by $g'(z,\zeta)$, we have $g'(z,\zeta) - g(z,\zeta) \geq 0$ on Γ, since $g(z,\zeta)$ vanishes there and $g'(z,\zeta)$ is positive in the interior of D'. But $g'(z,\zeta) - g(z,\zeta)$ is harmonic in D and it therefore follows from the maximum principle that $g'(z,\zeta) - g(z,\zeta) > 0$ throughout D. Letting $z = \zeta$ and observing (71), we thus obtain $d'(\zeta) - d(\zeta) > 0$, which is the above result.

If D is a circle, (69) enables us to compute, up to first-order terms, the Green's function of a nearly circular domain. The Green's function of a circle of radius R about the origin was earlier found to be of the form (18), and the value of its normal derivative on the circumference was found to be (19). If we denote the polar coordinates of z and ζ by r, ϕ and ρ, φ, respectively, and insert the expression (19) in (69), we obtain

$$\delta g(z,\zeta) = \frac{(R^2 - r^2)(R^2 - \rho^2)}{2\pi R} \int_0^{2\pi} \frac{\delta n(\theta)\, d\theta}{[R^2 - 2rR\cos(\theta - \phi) + r^2][R^2 - 2\rho R\cos(\theta - \varphi) + \rho^2]} + o(\epsilon). \tag{73}$$

For $\zeta = 0$, we have, in particular,

$$\delta g(z,0) = \frac{R^2 - r^2}{2\pi R} \int_0^{2\pi} \frac{\delta n(\theta)\, d\theta}{R^2 - 2rR\cos(\theta - \phi) + r^2} + o(\epsilon).$$

Since the Green's function $g(z,0)$ of the circle is $\log (R/r)$, this result can also be formulated as follows:

If D is a nearly circular domain bounded by the curve whose polar equa-

tion is $\rho = R + \epsilon p(\theta)$, *then the Green's function of D with respect to the center of the circle is of the form*

$$g(z,0) = \log\frac{R}{r} + \frac{\epsilon(R^2 - r^2)}{2\pi R}\int_0^{2\pi}\frac{p(\theta)\,d\theta}{R^2 - 2rR\cos(\theta - \phi) + r^2} + o(\epsilon).$$

In Sec. 11, Chap. V, this result will be applied to a problem in the theory of conformal mapping.

EXERCISES

1. Show that the nearly circular ellipse of half axes $1 + \epsilon$ and 1, where ϵ is a small positive number, has the polar equation $\rho = 1 + \epsilon\cos^2\theta + o(\epsilon)$ and derive the expression

$$g(z,0) = -\log r + \tfrac{1}{2}\epsilon(1 + r^2\cos 2\varphi) + o(\epsilon)$$

for the Green's function of this ellipse.

2. Let D be a domain which is bounded by n closed analytic curves $\Gamma_1, \ldots, \Gamma_n$ and let $\omega_\nu(z)$ denote the harmonic measure of D associated with Γ_ν. Using (69) and the fact that $-2\pi\omega_\nu(\zeta)$ is the period of $g(z,\zeta)$ if z surrounds Γ_ν, derive the variation formula

$$\delta\omega_\nu(z) = \frac{1}{2\pi}\int_\Gamma \frac{\partial\omega_\nu(\eta)}{\partial n_\eta}\frac{\partial g(\eta,z)}{\partial n_\eta}\,\delta n(s)\,ds_\eta.$$

3. Use the result of the preceding exercise to show that the periods $p_{\nu\mu}$ of $\omega_\nu(z)$ have the variation formula

$$\delta p_{\nu\mu} = -\int_\Gamma \frac{\partial\omega_\nu(\eta)}{\partial n_\eta}\frac{\partial\omega_\mu(\eta)}{\partial n_\eta}\,\delta n(s)\,ds_\eta.$$

4. Use the preceding result to show that the quadratic form $A = \sum_{\nu,\mu=1}^{n-1} p_{\nu\mu}\lambda_\nu\lambda_\mu$ has the variation formula

$$\delta A = -\int_\Gamma\left[\sum_{\nu=1}^{n-1}\lambda_\nu\frac{\partial\omega_\nu(\eta)}{\partial n}\right]^2\delta n(s)\,ds$$

and deduce the fact that A decreases if the domain D grows.

5. If $g(z,\zeta) = -\log r + h(z,\zeta)$, where $h(z,\zeta)$ is harmonic in D, show that $h(z,\zeta)$ has the variation formula

$$\delta h(z,\zeta) = \frac{1}{2\pi}\int_\Gamma \frac{\partial g(\eta,z)}{\partial n_\eta}\frac{\partial g(\eta,\zeta)}{\partial n_\eta}\,\delta n(s)\,ds_\eta,$$

and conclude that

$$\delta[h(z,z) + h(\zeta,\zeta) - 2h(z,\zeta)] = \frac{1}{2\pi}\int_\Gamma\left[\frac{\partial g(\eta,z)}{\partial n_\eta} - \frac{\partial g(\eta,\zeta)}{\partial n_\eta}\right]^2\delta n(s)\,ds_\eta.$$

Deduce that the expression $h(z,z) + h(\zeta,\zeta) - 2h(z,\zeta)$ grows monotonically with the domain.

6. Enclosing a given finite domain D in a large circle of radius R and using the form (18′) for the Green's function of this circle, show that the expression $h(z,z) + h(\zeta,\zeta) - 2h(z,\zeta)$ tends to zero for $R \to \infty$. Using this and the result of the preceding exercise, show that for any finite domain D we have the inequality

$$h(z,z) + h(\zeta,\zeta) \leq 2h(z,\zeta).$$

CHAPTER II

ANALYTIC FUNCTIONS

1. Complex Numbers. The language of the theory of conformal mapping is that of complex-valued functions of a complex variable. The use of complex numbers, however, is of a much earlier date. It is due to the observation that certain algebraic equations do not admit of solutions if the algebraic operations are confined to ordinary numbers, or, as they came to be called later, *real numbers.** For instance, the equation $x^2 = -1$ has no real solution, since the square of a real number is either positive or zero, but never negative. The rule that every algebraic equation of second order has two solutions can, however, be saved by a simple expedient. Since there is no real number whose square is -1, we conclude that this is a new type of number which is defined by just this property; if this number is denoted by i, it is thus defined by the equation

$$i^2 = -1. \tag{1}$$

Although the number i was introduced in order to give a solution to the particular second-order equation $x^2 + 1 = 0$, it is immediately seen that the adjoining of the number i to the real-number system guarantees two solutions (which may coincide) to any second-order equation with real coefficients. Indeed, the equation $x^2 + px + q = 0$ is formally solved by the elementary formula

$$x = -\frac{p}{2} \pm \sqrt{\left(\frac{p}{2}\right)^2 - q},$$

which shows that its solutions are either real or else they are numbers of the form $a + bi$, where a and b are real numbers. A number of the type $a + bi$ is called a *complex number*. If b is zero, the complex number reduces to a real number; conversely, a real number may be regarded as a complex number in which i has the coefficient zero. If a is zero, that is, if the complex number reduces to the form bi, it is called an *imaginary number*. This unfortunate name, which seems to imply that there is something unreal about these numbers and that they only lead a precarious existence in some people's imagination, has contributed much

* It is assumed that the reader is familiar with the basic properties of the real-number system.

toward making the whole subject of complex numbers suspect in the eyes of generations of high school students.

The fact that the introduction of complex numbers simplifies certain theorems in the theory of algebraic equations would not be of much consequence, if it were not permissible to apply to complex numbers the same algebraic manipulations as are applied to real numbers. That this is indeed the case can be easily confirmed by regarding the number $a + bi$ as the sum of the terms a and bi and by making it subject to the usual algebraic operations while observing (1). The sum of the two complex numbers $a + bi$ and $c + di$ will thus be the complex number $(a + b) + (c + d)i$, the product $(a + bi)(c + di)$ will be $(ac - bd) + (ad + bc)i$, etc. Both the addition and multiplication of complex numbers are obviously commutative and associative, and it is easily confirmed that the multiplication is distributive with respect to addition.

While originally the algebra of complex numbers was introduced along the lines indicated above, it has now become customary to introduce it in the following, somewhat more abstract, manner. A complex number α is defined as an ordered pair (a,b) of real numbers; if $b = 0$, α reduces to the real number a, that is, $(a,0) = a$. The sum of two complex numbers (a,b) and (c,d) is defined by

$$(a,b) + (c,d) = (a + c,\ b + d).$$

If b and d are zero, it follows that $(a,0) + (c,0) = (a + c,\ 0) = a + c$, which shows that our definition of a sum does not violate the rules of addition of real numbers. Multiplication of two complex numbers is defined by

$$(a,b)(c,d) = (ac - bd,\ bc + ad). \tag{2}$$

If $b = d = 0$, we have $(a,0)(c,0) = (ac,0) = ac$; hence, this operation reduces to ordinary multiplication if both complex numbers are real. The reader will verify that these rules of addition and multiplication of complex numbers have the same basic properties as the corresponding operations in the theory of real numbers, that is, we have $\alpha + \beta = \beta + \alpha$, $\alpha \cdot \beta = \beta \cdot \alpha$, $(\alpha + \beta) + \gamma = \alpha + (\beta + \gamma)$, $(\alpha \cdot \beta) \cdot \gamma = \alpha \cdot (\beta \cdot \gamma)$, $\alpha \cdot (\beta + \gamma) = \alpha \cdot \beta + \alpha \cdot \gamma$. If the complex number $(0,1)$ is denoted by the symbol i we have, in view of the multiplication rule,

$$i^2 = (0,1)(0,1) = (-1,0) = -1,$$

in accordance with (1). Moreover, it follows from the addition rule that

$$(a,b) = (a,0) + (0,b) = a + b \cdot (0,1) = a + bi,$$

which shows that our ordered pairs of real numbers are indeed equivalent to the complex numbers which are obtained by naïvely operating with the square root of -1 and observing the usual rules of algebra.

The definition of complex numbers as ordered pairs of real numbers has the advantage of showing clearly that the essential feature of this algebra is not the emergence of mysterious "imaginary" numbers; the crucial step is to realize that what we write as one complex number α is in reality a pair of two real numbers and that it is possible to define for these pairs algebraic operations which obey the same laws as the customary algebraic operations defined for real numbers. We shall, however, continue to use the notation $\alpha = a + bi$ for the complex number (a,b). As shown above, formal algebraic manipulation of the expression $a + bi$ and observance of the rule (1) will always lead to results identical with those obtained by means of the addition and multiplication rules for ordered pairs; the notation $a + bi$, seeming to imply that we are dealing with the sum of a real and an imaginary number, is therefore retained because of its extreme manipulative convenience.

a is called the *real part* and b the *imaginary part* of the complex number $\alpha = a + bi$; in symbols, $a = \text{Re}\,\{\alpha\}$, $b = \text{Im}\,\{\alpha\}$. Two complex numbers are said to be equal if, and only if, they are identical, that is, if they have both the same real parts and the same imaginary parts. The number $\bar{\alpha} = a - ib$ is called the *complex conjugate* of $\alpha = a + ib$; obviously, the complex conjugate of $\bar{\alpha}$ is α itself. A complex number is real, if its imaginary part is zero; if the real part of a complex number is zero, this number is said to be *pure imaginary*. It is clearly characteristic of real numbers that $\alpha = \bar{\alpha}$; pure imaginary numbers α, on the other hand, are characterized by the relation $\bar{\alpha} = -\alpha$.

The complex number zero, defined as the number β for which $\alpha + \beta = \alpha$ is, of course, identical with the real number zero, as the reader will have no difficulty in verifying. This can also be expressed by saying that a complex number is zero if, and only if, both its real and imaginary parts are zero. The difference $\gamma = \alpha - \beta$ of two complex numbers is defined by the equation $\alpha = \gamma + \beta$; evidently γ is a uniquely determined complex number.

The product of two complex numbers can only be zero if one of the factors is zero. Indeed, from $\alpha\beta = 0$ we obtain $\alpha\bar{\alpha}\beta\bar{\beta} = \bar{\alpha}\bar{\beta}\alpha\beta = 0$; if $\alpha = a + bi$, $\beta = c + di$ we thus have, in view of

$$(a + ib)(a - ib) = a^2 + b^2,$$

$(a^2 + b^2)(c^2 + d^2) = 0$, which is only possible if either $a = b = 0$ or $c = d = 0$. There is only one complex number γ such that $\alpha\gamma = \alpha$ (if $\alpha \neq 0$), namely, the real number 1; indeed, subtracting the equation $\alpha \cdot 1 = \alpha$ from $\alpha \cdot \gamma = \alpha$ we obtain $\alpha(\gamma - 1) = 0$, whence $\gamma = 1$.

Division of complex numbers is a uniquely determined operation resulting again in a complex number, provided the divisor is not zero. Defining $\gamma = \alpha/\beta$ by the equation $\beta\gamma = \alpha$, we have, multiplying both sides with $\bar{\beta}$, $(c^2 + d^2)\gamma = \beta\bar{\beta}\gamma = \alpha\bar{\beta}$, whence

$$\gamma = \frac{\alpha\bar{\beta}}{c^2 + d^2},$$

except when $c = d = 0$.

We close this section with a simple but useful remark. It is easily verified that equations between complex numbers of the type $\alpha + \beta = \gamma$, $\alpha - \beta = \gamma$, $\alpha\beta = \gamma$, $(\alpha/\beta) = \gamma$ remain true if all complex numbers appearing in these equations are replaced by their complex conjugates; for instance, it follows from $\alpha + \beta = \gamma$ that $\bar{\alpha} + \bar{\beta} = \bar{\gamma}$, and similarly for the other equations. As an immediate consequence, we obtain the following result: *Any equation between complex numbers which only involves the operations of addition, subtraction, multiplication, and division remains true if all complex numbers appearing in the equation are replaced by their complex conjugates.*

EXERCISES

1. If $\alpha = \cos t + i \sin t$, where $0 < t < 2\pi$, show that

$$\frac{1 + \alpha}{1 - \alpha} = i \cot \frac{t}{2}.$$

2. If $\alpha = a + bi$ and $b \neq 0$, show that the expression

$$\frac{\alpha}{1 + \alpha^2}$$

will be real only if $a^2 + b^2 = 1$.

3. Show that $\sqrt{\alpha} = \sqrt{a + bi}$ is a complex number $c + di$, and find c and d.

4. If $\alpha = a + bi$, show that α^4 can be positive only if either a or b, or both, are zero; show further that α^4 can be imaginary only if one of the four relations

$$a = \pm(1 \pm \sqrt{2})b$$

holds.

5. Show that $(\cos \theta + i \sin \theta)(\cos \varphi + i \sin \varphi) = \cos (\theta + \varphi) + i \sin (\theta + \varphi)$, and deduce the formula $(\cos \theta + i \sin \theta)^n = \cos n\theta + i \sin n\theta$, where n is a positive integer.

6. Show that the reciprocal of $\cos \theta + i \sin \theta$ is $\cos \theta - i \sin \theta$, and deduce that the formula of the preceding exercise remains true in the case in which n is a negative integer.

2. The Complex Plane. Since a complex number $\alpha = a + bi$ is uniquely determined by a pair of real numbers, and the same is true for a point in a plane, it is possible to establish a one-to-one correspondence between all complex numbers and the points of a given plane. Indeed, all we have to do is to associate with the complex number $\alpha = a + bi$ the point in the plane whose rectangular coordinates are a, b; it is obvious that by this rule each complex number determines precisely one point in the plane, and vice versa. For the sake of simplicity, this point will also be denoted by α. The plane whose points represent the complex numbers is called the *complex plane*.

The beginner is apt to think that the introduction of the complex plane is a mere technical device, designed to facilitate the operations with complex numbers by enlisting the powers of geometrical visualization. This is not so. In the following chapters, it will become increasingly clear that both the theory of functions of a complex variable and the theory of conformal mapping are pervaded by geometrical ideas, the key to which is the representation of complex numbers by the points of the complex plane. This is particularly true of the theory of conformal mapping; in fact, it is one of the characteristics of this theory that it emphasizes the geometrical aspects of complex numbers and that it tends to regard their arithmetic properties as mere tools.

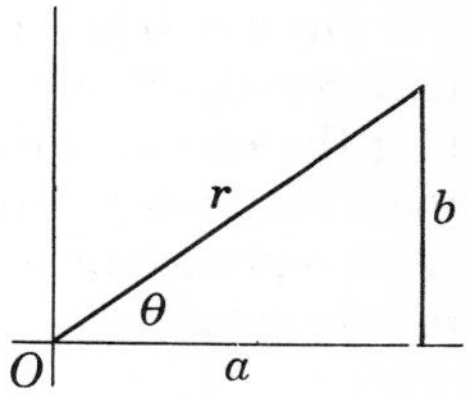

FIG. 4.

If we introduce polar coordinates r, θ in the complex plane (Fig. 4), we obtain a different representation of the complex numbers.

In view of

$$a = r\cos\theta, \qquad b = r\sin\theta,$$

we have

$$\alpha = a + bi = r(\cos\theta + i\sin\theta). \tag{3}$$

$r = \sqrt{a^2 + b^2}$ is the distance of α from the origin. This quantity is called the *absolute value*, or the *modulus*, of α and is denoted by the symbol $|\alpha|$; the angle θ is called the *argument* of α and is denoted by $\arg\{\alpha\}$. We clearly have

$$|\alpha| = \sqrt{\alpha\bar{\alpha}}.$$

If α and β are two complex numbers, it follows that

$$|\alpha\beta| = \sqrt{\alpha\bar{\alpha}\beta\bar{\beta}} = \sqrt{\alpha\bar{\alpha}}\sqrt{\beta\bar{\beta}},$$

whence

$$|\alpha\beta| = |\alpha||\beta|. \tag{4}$$

If, in (4), we replace α by $\frac{\alpha}{\beta}$ $(\beta \neq 0)$, we obtain $\left|\frac{\alpha}{\beta}\beta\right| = \left|\frac{\alpha}{\beta}\right| |\beta|$, whence

$$\left|\frac{\alpha}{\beta}\right| = \frac{|\alpha|}{|\beta|}. \tag{5}$$

The identity (4) is easily extended to a product of an arbitrary number of complex numbers $\alpha_1, \alpha_2, \ldots, \alpha_n$; for such a product we obtain

$$|\alpha_1\alpha_2 \cdots \alpha_n| = |\alpha_1||\alpha_2| \cdots |\alpha_n|.$$

Taking, in particular, $\alpha_1 = \alpha_2 = \cdots = \alpha_n = \alpha$, we have

$$|\alpha^n| = |\alpha|^n, \qquad n = 1, 2, \ldots .$$

The distance between the points $\alpha = a + bi$ and $\beta = c + di$ is, by elementary geometry, equal to $\sqrt{(a - c)^2 + (b - d)^2}$; since this is also the absolute value of the complex number $\alpha - \beta$, we thus have the result: *The distance between the points α and β is $|\alpha - \beta|$.* Since the sides of the triangle $(0,\alpha,\beta)$ are $|\alpha|$, $|\beta|$, $|\alpha - \beta|$, it follows from the elementary fact that the sum of two sides of a triangle is larger than the third side (unless the triangle degenerates into a linear segment) that $|\alpha - \beta| \leq |\alpha| + |\beta|$. If we replace β by $-\beta$ and observe that $|-\beta| = |\beta|$, we obtain the important inequality

$$|\alpha + \beta| \leq |\alpha| + |\beta|. \tag{6}$$

The reader will have no difficulty in verifying that the sign of equality in (6) can only occur if $\arg \{\alpha\} = \arg \{\beta\}$. A lower bound for $|\alpha + \beta|$ is provided by the inequality

$$|\alpha + \beta| \geq |\alpha| - |\beta|, \tag{7}$$

which follows easily by replacing α in (6) by $\alpha - \beta$. We obtain

$$|\alpha| \leq |\alpha - \beta| + |\beta|;$$

substituting $-\beta$ for β, we arrive at (7).

The addition of two complex numbers can be carried out graphically with the help of the fact that the four points 0, α, $\alpha + \beta$, β form a parallelogram such that 0 and $\alpha + \beta$ occupy opposite vertices. The proof is left to the reader. There is also a construction for the product of two complex numbers which follows from the representation (3). If α and β have the representations

$$\alpha = r(\cos \theta + i \sin \theta), \qquad \beta = \rho(\cos \varphi + i \sin \varphi),$$

we have

$$\alpha\beta = r\rho[(\cos \theta \cos \varphi - \sin \theta \sin \varphi) + i(\sin \theta \cos \varphi + \cos \theta \sin \varphi)].$$

In view of the addition theorems of the trigonometric functions, this is equivalent to

$$\alpha\beta = r\rho[\cos (\theta + \varphi) + i \sin (\theta + \varphi)].$$

It follows, in agreement with (4), that $|\alpha\beta| = |\alpha||\beta|$, and that

$$\arg\{\alpha\beta\} = \arg\{\alpha\} + \arg\{\beta\}.$$

Hence, the point $\alpha\beta$ is found by adding the arguments of α and β and, in the direction thus obtained, taking the point whose distance from the origin is the product of $|\alpha|$ and $|\beta|$.

We finally point out that the complex plane is completed by the infinitely distant point, denoted by the symbol ∞, discussed at the end of Sec. 1, Chap. I; as shown there, this concept is given a definite meaning with the help of the stereographic projection of the plane onto a sphere.

EXERCISES

1. Show that the area of the triangle whose vertices are α, β, γ is given by the absolute value of

$$\frac{1}{2}|\gamma - \beta|^2 \operatorname{Im}\left\{\frac{\gamma - \alpha}{\gamma - \beta}\right\}.$$

2. Show that the three points α, β, γ are situated on the same straight line if, and only if, the expression

$$\frac{\alpha - \beta}{\alpha - \gamma}$$

is real.

3. Find the loci of the points α for which

(a) $\left|\dfrac{1 + \alpha}{1 - \alpha}\right| = 1.$

(b) $\left|\dfrac{1 + \alpha}{1 - \alpha}\right| = 2.$

(c) $|1 - \alpha^2| = \mu \ (\mu > 0).$

(d) $|\alpha + 1| + |\alpha - 1| = \mu > 2.$

4. Show that if the real parameter λ grows from $-\infty$ to ∞, the point $\alpha + \lambda(\beta - \alpha)$ describes the straight line determined by the points α and β.

5. If α and β are two fixed points, show that the locus of the points γ for which $\arg\{(\gamma - \alpha)/(\gamma - \beta)\}$ is constant is a circle passing through α and β.

3. Sequences and Series of Complex Numbers. The sequence of complex numbers $\alpha_1 = a_1 + ib_1, \alpha_2 = a_2 + ib_2, \ldots, \alpha_n = a_n + ib_n, \ldots$ will be said to converge if the two real sequences

$$a_1, a_2, \ldots, a_n, \ldots$$
$$b_1, b_2, \ldots, b_n, \ldots$$

converge. If

$$\lim_{n\to\infty} a_n = a, \qquad \lim_{n\to\infty} b_n = b,$$

we say that $\alpha = a + ib$ is the limit of the sequence $\{\alpha_n\}$, that is,

$$\lim_{n\to\infty} \alpha_n = \alpha.$$

The well-known necessary and sufficient convergence criteria for real sequences are easily translated into criteria applying to complex sequences. The real sequences $\{a_n\}$ and $\{b_n\}$ converge to the limits a and b, respectively, if, and only if, for every arbitrarily small positive ϵ there exists an integer $n = n(\epsilon)$ such that

$$(8) \qquad |a - a_m| < \epsilon, \qquad |b - b_m| < \epsilon,$$

provided $m > n$. The corresponding criterion for complex sequences is the following: *The complex sequence $\{\alpha_n\}$ converges to a limit α if, and only if, for every arbitrarily small positive ϵ there exists an integer $n = n(\epsilon)$ such that*

$$(9) \qquad |\alpha - \alpha_m| < \epsilon$$

for $m > n$. The proof is immediate. From (8) it follows that

$$|\alpha - \alpha_m| = \sqrt{(a - a_m)^2 + (b - b_m)^2} < \epsilon\sqrt{2},$$

which shows that the criterion is necessary. It is, however, also sufficient, since the inequality

$$|\alpha - \alpha_m| = \sqrt{(a - a_m)^2 + (b - b_m)^2} < \epsilon$$

entails $|a - a_m| < \epsilon$ and $|b - b_m| < \epsilon$, that is, the inequalities (8).

As in the case of real sequences, this criterion may be replaced by a different type of necessary and sufficient condition in which no explicit reference to the limit α is made. *The complex sequence $\{\alpha_n\}$ converges if, and only if, for every arbitrarily small positive ϵ there exists an integer $n = n(\epsilon)$ such that*

$$|\alpha_m - \alpha_k| < \epsilon$$

if both k and m are larger than n. The proof of this criterion from the corresponding result for real sequences is left as an exercise to the reader.

(9) has a simple geometric interpretation. Since, as shown in Sec. 2, $|\alpha - \alpha_m|$ is the distance between the points α and α_m, the convergence criterion is equivalent to the following geometric condition: If we describe about α a circle of arbitrarily small radius ϵ, then ultimately, that is, starting from a certain subscript n, all points α_m must lie in the interior of this circle. α is therefore a limit point of the set of points $\alpha_1, \alpha_2, \ldots$ in the sense of the definition of Sec. 1, Chap. I. Furthermore, α is clearly the only limit point of the set.

We now turn to the consideration of the infinite series

$$\alpha_1 + \alpha_2 + \cdots + \alpha_n + \cdots$$

whose terms are complex numbers $\alpha_n = a_n + ib_n$. The series is said to converge if its partial sums

$$s_1 = \alpha_1, \qquad s_2 = \alpha_1 + \alpha_2, \qquad s_3 = \alpha_1 + \alpha_2 + \alpha_3, \qquad \ldots$$

form a converging sequence. If $\lim_{n \to \infty} s_n = s$, then s is called the sum of the series and we write

$$\alpha_1 + \alpha_2 + \cdots + \alpha_n + \cdots = s.$$

Obviously, this series converges if, and only if, the series of real terms

$$\begin{aligned} &a_1 + a_2 + \cdots + a_n + \cdots \\ &b_1 + b_2 + \cdots + b_n + \cdots \end{aligned} \tag{10}$$

converge; if the sums of these series are a and b, respectively, then $s = a + ib$. If we apply the convergence criterion for sequences to the sequence $\{s_n\}$, we obtain the following result: *For the convergence of the series*

$$\alpha_1 + \alpha_2 + \cdots + \alpha_n + \cdots$$

it is necessary and sufficient that for every arbitrarily small positive ϵ *there exist an* $n = n(\epsilon)$ *such that*

$$|\alpha_{n+1} + \alpha_{n+2} + \cdots + \alpha_{n+m}| < \epsilon$$

for every positive integer m.

It is well known that the real series (10) converge "unconditionally," that is, they remain converging and their sums are not affected if the order of their terms is changed, if, and only if, they also converge absolutely, that is, the series

$$\begin{aligned} &|a_1| + |a_2| + \cdots + |a_n| + \cdots \\ &|b_1| + |b_2| + \cdots + |b_n| + \cdots \end{aligned} \tag{11}$$

converge. Since, by (6),

$$|\alpha_n| = |a_n + ib_n| \leq |a_n| + |b_n|,$$

the convergence of the series (11) entails that of the series

$$|\alpha_1| + |\alpha_2| + \cdots + |\alpha_n| + \cdots$$

On the other hand, both $|a_n|$ and $|b_n|$ are not larger than $|\alpha_n| = \sqrt{a_n^2 + b_n^2}$. We thus have the following result: *A series* $\alpha_1 + \alpha_2 + \cdots + \alpha_n + \cdots$ *converges unconditionally if, and only if, it converges absolutely, i.e., if the series of the absolute values*

$$|\alpha_1| + |\alpha_2| + \cdots + |\alpha_n| + \cdots$$

converges. In the applications it is generally not the unconditionality of the convergence, guaranteed by this result, that matters but the convergence itself. Since absolute convergence is a sufficient condition for convergence, we are thus able to test the convergence of a complex series by means of the known criteria for real series.

EXERCISES

1. Let $\alpha_0, \alpha_1, \alpha_2, \ldots$ and $\beta_0, \beta_1, \beta_2, \ldots$ be two sequences, and set

$$\alpha_0 + \alpha_1 + \cdots + \alpha_n = s_n.$$

Show that, for $m \geq 0$ and $p \geq 1$,

$$\sum_{n=m+1}^{m+p} \alpha_n\beta_n = \sum_{n=m+1}^{m+p} s_n(\beta_n - \beta_{n+1}) - s_m\beta_{m+1} + s_{m+p}\beta_{m+p+1}.$$

2. Use the result of the preceding exercise to show that the series $\sum_{n=0}^{\infty} \alpha_n\beta_n$ converges if $\beta_0 > \beta_1 > \cdots > 0$, $\lim_{n\to\infty} \beta_n = 0$, and if there exists a constant M such that $|s_n| < M$ for all n.

3. Show that the convergence of $\sum_{n=0}^{\infty} \alpha_n\beta_n$ is also assured by the convergence of $\sum_{n=0}^{\infty} \alpha_n$ and the conditions $\beta_0 > \beta_1 > \cdots > A$, where A is a real constant.

4. Analytic Functions. If S is a set of points in the complex plane and the point represented by the complex number $z = x + iy$ may be identified with every point of S, we say that z is a *complex variable* defined in S. If we are given a certain rule according to which a complex value $w = u + iv$ is associated with every point z of S, we say that $w = f(z)$ is a complex *function* of the complex variable z. We are mainly interested in the case in which the set S is a *domain* D, as defined in Sec. 1, Chap. I; in this case, D will be called the *domain of definition* of the complex-valued function $w = f(z)$.

In this generality, the concept of a complex-valued function of a complex variable is not very useful. Clearly, the statement that $w = u + iv$ is a function of $z = x + iy$ does not amount to more than to the fact that $u = u(x,y)$ and $v = v(x,y)$ are two real functions of the two real variables x and y. This situation is also not altered if we restrict ourselves to *continuous functions*, that is, to functions $f(z)$ for which $|f(z) - f(a)| < \epsilon$, where ϵ is a given, arbitrarily small positive number, provided $|z - a| < \delta(\epsilon)$; the reader will have no difficulty in verifying that this continuity is equivalent to the continuity of the two real functions $u = u(x,y)$ and $v = v(x,y)$ of the real variables x and y. The crucial step is the requirement that our complex functions be *differentiable* at the points of D. The definition of differentiability of functions of a complex variable is formally the same as in the case of functions of a real variable. We say that the function $f(z)$ is differentiable at the point $z_0 \in D$ if

$$f'(z_0) = \lim_{h\to 0} \frac{f(z_0 + h) - f(z_0)}{h} \tag{12}$$

exists; the value $f'(z_0)$ of this limit is called the derivative of the function $f(z)$ at the point $z = z_0$. There is, however, a fundamental difference between this derivative and the derivative of a function of a real variable. While in the latter case the value of the parameter h is either positive or negative, h may now be an arbitrary complex parameter which ultimately tends to zero. Geometrically speaking, the point $z_0 + h$ may approach the point z_0 along an arbitrary curve ending at z_0. By the statement that the limit (12) exists, we mean that the limit exists, and is the same, regardless of the path along which z_0 is approached.

Since, formally, the definition (12) is identical with that of a function of a real variable, it follows that the formal rules of differentiation, such as the formulas for the derivatives of sums, products, quotients, etc., are exactly the same as in the real case. Moreover, those differentiable functions known from real analysis whose domain of definition can be extended to complex values and which possess a derivative in the sense of (12) for real values of z will clearly retain their old differentiation formulas.

We now give a number of definitions. A complex function which in some domain D of the complex plane is single-valued, that is, has a uniquely determined value at each point of D, and which has a derivative in the sense of (12) at each point of D is called an *analytic function*. The analytic function $f(z)$ is called *regular in the domain* D if it is single-valued and has a derivative at each point of D; $f(z)$ is called *regular at the point* z_0 if it is regular in a small neighborhood $|z - z_0| < \epsilon$ of z_0. A point z_0 at which $f(z)$ is not regular is called a *singular point* of the function $f(z)$. In the literature on the subject, one meets occasionally a slightly different terminology. Some writers use "analytic" also in the sense of "regular," as in the statement "$f(z)$ is analytic in D"; others, influenced by French usage, replace "regular" by "holomorphic."

As an example of an analytic function, consider the function

$$f(z) = z^n = (x + iy)^n,$$

where n is a positive integer. By (12), we have

$$f'(z) = \lim_{h\to 0} \frac{f(z + h) - f(z)}{h} = \lim_{h\to 0} \frac{(z + h)^n - z^n}{h},$$

provided the limit exists and is independent of the manner in which h approaches 0. In view of the binominal theorem, which is a purely formal algebraic identity and therefore holds for complex as well as for real numbers, we have

$$f'(z) = \lim_{h \to 0} \left[nz^{n-1} + \frac{h}{2} n(n-1)z^{n-2} + \cdots + h^{n-1} \right]$$

where, with the exception of the first term, all terms contain powers of h. If z is a fixed finite complex number, all these terms will therefore tend to zero if h tends to zero. It follows that $f'(z) = nz^{n-1}$, in accordance with the formula known from the elementary calculus. We have thus proved that the function $f(z) = z^n$, where n is a positive integer, is an analytic function of the complex variable z; this function is regular at all finite points of the complex plane. Its differentiation formula agrees with that of the real function $f(x) = x^n$; as mentioned before, this is a consequence of the fact that $f(z) = z^n$ is regular on the real axis, that is, for real values of z, and that the rule (12) for calculating the derivative of an analytic function is formally identical with that used in the ordinary calculus.

We have likewise pointed out above that, because of this formal identity of the real and complex differentiation rules, the differentiation rules for sums, products, quotients, etc., remain valid for complex differentiation. For example, if the functions $f(z)$ and $g(z)$ have derivatives in the sense of (12) at the point $z = z_0$, it follows that the same is true for the function $af(z) + bg(z)$, where a and b are constants, and that the value of this derivative is $af'(z_0) + bg'(z_0)$. Hence, if $f(z)$ and $g(z)$ are analytic functions, regular in a domain D, the same is true of the function $af(z) + bg(z)$. Similarly, the product of two analytic functions, regular in D, is again an analytic function, regular in the same domain. The same statement holds for the quotient of two regular functions; however, in this case we have to except those points of the complex plane at which the denominator is zero, since there the division ceases to be defined. If we apply these principles to the analytic functions $f(z) = z^n$, which were shown before to be regular at all finite points of the complex plane, we obtain the following results: *A polynomial*

$$f(z) = a_0 + a_1 z + a_2 z^2 + \cdots + a_n z^n$$

is an analytic function of z, regular at all finite points of the complex z-plane. A rational function

$$f(z) = \frac{a_0 + a_1 z + a_2 z^2 + \cdots + a_n z^n}{b_0 + b_1 z + b_2 z^2 + \cdots + b_m z^m},$$

that is, the ratio of two polynomials, is an analytic function which is regular at all finite points of the complex z-plane, except at those points at which the denominator is zero.

EXERCISES

1. Show that if the function $f(z)$ is differentiable at $z = z_0$, it is also continuous there.

2. Show that the functions

$$f(z) = \text{Re}\,\{z\}, \qquad f(z) = \text{Im}\,\{z\}, \qquad f(z) = |z|$$

are not differentiable in the sense of (12) and that, therefore, they are not analytic functions of z.

3. Let $\zeta = f(z)$ be an analytic function of z, regular in a domain D in the z-plane, such that, for $z \in D$, its values are situated in the interior of a domain D' in the ζ-plane; if the analytic function $w = g(\zeta)$ is regular in D', show that the derivative of w with respect to z can be computed by means of the formula

$$w'(z) = \frac{dw}{dz} = \frac{dw}{d\zeta}\frac{d\zeta}{dz}.$$

5. The Cauchy-Riemann Equations. It is to be expected that the condition of differentiability (12) will establish certain relations between the real part $u = u(x,y)$ and the imaginary part $v = v(x,y)$ of an analytic function $f(z) = f(x + iy) = u(x,y) + iv(x,y)$. In order to find these relations, we compute the value of $f'(z)$ in two different ways. In the first case we let the parameter h in (12) tend to zero through positive values; in the second case h will be taken to be of the form $h = ik$, where k is positive. Since the derivative is independent of the manner in which h approaches zero, both procedures must lead to the same result. In the first case, $f'(z)$ is obtained as the limit of the expression

$$\frac{f(z + h) - f(z)}{h} = \frac{u(x + h, y) - u(x)}{h} + i\,\frac{v(x + h, y) - v(x,y)}{h}.$$

Since h tends to zero through positive values, we thus have

$$f'(z) = \frac{\partial u}{\partial x} + i\,\frac{\partial v}{\partial x}. \tag{13}$$

In the second case, $f'(z)$ is the limit of

$$\frac{f(z + ik) - f(z)}{ik} = \frac{u(x, y + k) - u(x,y)}{ik} + \frac{v(x, y + k) - v(x,y)}{k},$$

if k approaches zero through positive values. Hence,

$$f'(z) = \frac{1}{i}\frac{\partial u}{\partial y} + \frac{\partial v}{\partial y} = -i\,\frac{\partial u}{\partial y} + \frac{\partial v}{\partial y}.$$

Comparing this with (13), we obtain

$$\frac{\partial u}{\partial x} + i\frac{\partial v}{\partial x} = -i\frac{\partial u}{\partial y} + \frac{\partial v}{\partial y}.$$

Since the equality of two complex numbers implies that they agree in both their real and their imaginary parts, we have thus arrived at the following result.

At a point at which an analytic function $f(z) = u + iv$ is regular, its real part $u = u(x,y)$ and its imaginary part $v = v(x,y)$ are connected by the Cauchy-Riemann differential equations

$$\frac{\partial u}{\partial x} = \frac{\partial v}{\partial y}, \qquad \frac{\partial u}{\partial y} = -\frac{\partial v}{\partial x}. \tag{14}$$

If $f(z)$ is regular in a domain D, the equations (14) *hold throughout D.*

The existence of $f'(z)$ implies the existence of the partial derivatives u_x, u_y, v_x, v_y. We shall see later (Sec. 3, Chap. III) that this assumption is also sufficient in order to guarantee the existence of derivatives of all orders. Anticipating this result to the extent that we assume the existence of continuous second derivatives, we can draw an interesting conclusion from the equations (14). Differentiating the first equation (14) with respect to x and the second equation with respect to y, and adding the results, we obtain

$$u_{xx} + u_{yy} = 0.$$

Interchanging the roles of x and y in these differentiations, we have similarly

$$v_{xx} + v_{yy} = 0.$$

Since the second derivatives of u and v are continuous and u and v satisfy the Laplace equation, it therefore follows that both u and v are harmonic functions (see Sec. 2, Chap. I). Moreover, as a comparison of (14) with (38) (Sec. 8, Chap. I) shows, v is the harmonic conjugate of u. We thus have proved the following result:

Both the real part $u = u(x,y)$ and the imaginary part $v = v(x,y)$ of an analytic function $f(z) = u + iv$ are harmonic functions in a domain in which $f(z)$ is regular; moreover, $v(x,y)$ is the harmonic conjugate of $u(x,y)$.

However, the relations between the theory of analytic functions and the theory of harmonic functions are even closer than indicated by this statement. We shall show that the following converse statement holds:

If $u = u(x,y)$ is a harmonic function and $v = v(x,y)$ its harmonic conjugate defined by (14), *then $u(x,y) + iv(xy)$ is an analytic function of the complex variable $z = x + iy$.*

We shall prove a slightly stronger result, namely, that $u + iv$ is an analytic function of $x + iy$ if u and v are continuous, have continuous first derivatives, and satisfy the Cauchy-Riemann equations (14); this result is stronger than the above statement because the definition of a harmonic function requires, in addition, the existence and continuity of the derivatives of the second order.

We define the function $f(z)$ by

$$f(z) = u(x,y) + iv(x,y) = u + iv,$$

where u and v have continuous first derivatives and satisfy (14). In order to prove that $f(z)$ is an analytic function of z, we have to show that it has a derivative in the sense of (12). Since u and v have continuous first derivatives, it follows from the mean-value theorem of the ordinary calculus that, with the notation $h = h_1 + ih_2$,

$$\begin{aligned}\frac{f(z+h) - f(z)}{h} &= \frac{u(x+h_1,\, y+h_2) - u(x,y)}{h} \\ &\qquad + i\,\frac{v(x+h_1,\, y+h_2) - v(x,y)}{h} \\ &= \frac{u_x(x+\theta h_1,\, y+\theta h_2)h_1 + u_y(x+\theta h_1,\, y+\theta h_2)h_2}{h} \\ &\qquad + i\,\frac{v_x(x+\theta' h_1,\, y+\theta' h_2)h_1 + v_y(x+\theta' h_1,\, y+\theta' h_2)h_2}{h}. \\ &= \frac{u_x(x,y)h_1 + u_y(x,y)h_2}{h} \\ &\qquad + i\,\frac{v_x(x,y)h_1 + v_y(x,y)h_2}{h} + \epsilon_1(h)\,\frac{h_1}{h} + \epsilon_2(h)\,\frac{h_2}{h},\end{aligned}$$

where $0 \leq \theta,\ \theta' \leq 1$, and $\epsilon_1(h)$, $\epsilon_2(h)$ are quantities that tend to zero if $h \to 0$. We now use the Cauchy-Riemann equations (14) and the fact that $|h_1| \leq |h|$, $|h_2| \leq |h|$. We obtain

$$\begin{aligned}\frac{f(z+h) - f(z)}{h} &= u_x\,\frac{h_1 + ih_2}{h} + iv_x\,\frac{h_1 + ih_2}{h} + \epsilon(h) \\ &= u_x + iv_x + \epsilon(h),\end{aligned}$$

where $\epsilon(h) \to 0$ if $h \to 0$. Letting $h \to 0$, we thus find that $f(z)$ has a derivative as defined by (12) and that its value is $f'(z) = u_x(x,y) + iv_x(x,y)$, irrespective of the manner in which h approaches zero. The function $f(z)$ is therefore analytic and our above statement is proved.

A comparison of the last two italicized statements shows that the theory of analytic functions of a complex variable and the theory of harmonic functions of two real variables are essentially one and the same thing. It would therefore be possible to avoid the use of complex quantities altogether and to obtain all the results of complex function theory without leaving the field of real analysis. That this is generally not done is due to two reasons. The first of these is the extraordinary elegance and convenience of manipulation inherent in the use of complex algebra. The second reason—already pointed out in Sec. 2—is that operations with complex quantities are the analytic equivalents of certain geometric facts. "Complex thinking" is, essentially, geometric thinking; the translation of complex function theory into the language of real analysis would obscure many of the geometric ideas which play a vital part in the theory of analytic functions and in the theory of conformal mapping. There are, however, situations which are "real" in character and in which the appropriate tools are furnished by the theory of harmonic functions rather than by the theory of analytic functions. In these cases, we shall decompose our complex expressions into their real and imaginary parts and apply the techniques developed in Chap. I. This procedure will also be adopted in cases in which the methods of Chap. I yield shorter and more direct proofs than corresponding complex manipulations.

EXERCISES

1. Show by means of the Cauchy-Riemann equations that an analytic function which is regular in a domain D and whose values at all points of D are real reduces to a constant.

2. Show that an analytic function which is regular and has a constant absolute value in a domain D reduces to a constant.

3. If $f(z) = u + iv$ is regular in a domain D, show that the Jacobian $u_xv_y - u_yv_x$ of the transformation $(x,y) \to (u,v)$ is equal to $|f'(z)|^2$ $(z = x + iy \in D)$.

4. Find the real and imaginary parts of the analytic functions

$$(a)\ \frac{1+z}{1-z}, \qquad (b)\ z + \frac{1}{z},$$

and verify that, with the exception of certain points [$z = 1$ in (a), $z = 0$ in (b)], the Cauchy-Riemann equations are satisfied.

5. Show that the function

$$f(z) = e^x \cos y + ie^x \sin y$$

is an analytic function of $z = x + iy$ which is regular at every finite point of the complex z-plane.

6. Show that

$$f(z) = \log \sqrt{x^2 + y^2} + i \tan^{-1} \frac{y}{x} = \log r + i\theta, \qquad 0 \leq \theta < 2\pi,$$

is an analytic function of z which has its only finite singularity at $z = 0$.

7. Show that

$$\left(\frac{\partial^2}{\partial x^2} + \frac{\partial^2}{\partial y^2}\right) |f(z)|^2 = 4|f'(z)|^2.$$

8. Show that the function $\log |f(z)|$ is harmonic in a domain in which $f(z)$ is regular and $f(z) \neq 0$.

9. If $f(z)$ is regular in a domain and $f'(z) \neq 0$, $|f(z)| \neq 1$, show that the function

$$w = \log \frac{|f'(z)|}{1 - |f(z)|^2}$$

= log ρ_D for f : D → Δ conf.

satisfies the partial differential equation

$$\frac{\partial^2 w}{\partial x^2} + \frac{\partial^2 w}{\partial y^2} = 4e^{2w}.$$

6. Power Series. A power series is an infinite series of the form

$$a_0 + a_1(z - a) + a_2(z - a)^2 + \cdots + a_n(z - a)^n + \cdots, \tag{15}$$

where z is a complex variable and a, a_0, a_1, . . . are fixed complex numbers. If a is zero, the series (15) reduces to the particularly simple form

$$a_0 + a_1 z + a_2 z^2 + \cdots + a_n z^n + \cdots. \tag{16}$$

We shall show in this section that, in its region of convergence, a power series represents a regular analytic function of z. At a later stage (Sec. 4, Chap. III) we shall see that, conversely, every analytic function can be represented by means of power series. This explains the great importance of power series as computational tools for the practical evaluation of analytic functions.

Turning now to the convergence properties of power series, we first derive the following result.

If a power series (15) *converges at a point* $z = z_0$, *then it converges absolutely at all points situated in the interior of the circle whose center is at* a *and which passes through* z_0.

Proof: Since

$$a_0 + a_1(z_0 - a) + \cdots + a_n(z_0 - a)^n + \cdots$$

converges, the terms of this series must tend to zero. There exists therefore an integer N such that for $n > N$ the terms of this series are, in absolute value, smaller than 1. The terms for which $n \leq N$ are finite in number, and it follows therefore that there exists a positive number M such that

$$|a_n(z_0 - a)^n| < M \tag{17}$$

for $n = 0, 1, 2, \ldots$. Let now z be a point inside the circle about a which passes through z_0. We clearly have $|z - a| < |z_0 - a|$. Hence

$$\sum_{n=0}^{\infty} |a_n(z - a)^n| = \sum_{n=0}^{\infty} |a_n(z_0 - a)^n| \left|\frac{z - a}{z_0 - a}\right|^n$$

$$< M \sum_{n=0}^{\infty} \left|\frac{z - a}{z_0 - a}\right|^n = M \sum_{n=0}^{\infty} \rho^n.$$

The latter series converges since, in view of $|z - a| < |z_0 - a|$, $\rho < 1$. It follows therefore that the series (15) indeed converges absolutely for this value of z, and the above statement is proved.

A moment's reflection shows that our result implies the following alternative: *If a power series* (15) *converges at a point* $z_0 \neq a$, *then either it converges for all finite values of* z, *or else there exists a positive number* r *such that* (15) *converges for* $|z - a| < r$ *and diverges for* $|z - a| > r$. r is called the *radius of convergence* of the power series, and the circle of center a and radius r is its *circle of convergence.* Nothing general can be said regarding convergence of a power series at points of the circumference of its circle of convergence; there it may, or may not, converge.

The concept of *uniform convergence* of series whose terms are functions of the same variable is defined in the case of functions of a complex variable in exactly the same way as with respect to functions of a real variable. If the functions $u_1(z)$, $u_2(z)$, . . . are defined in a domain D and the series

$$u_1(z) + u_2(z) + \cdots + u_n(z) + \cdots$$

converges at all points of D, then, for a given arbitrarily small ϵ, there exists an integer $N = N(\epsilon;z)$ such that

$$|u_{n+1}(z) + u_{n+2}(z) + \cdots| < \epsilon, \qquad z \in D,\ n \geq N.$$

In general, N will depend on z and it will not always be possible to find an integer N such that the latter inequality is satisfied for all points of D at the same time. However, if this is the case, that is, if for every $\epsilon > 0$ there exists an $N = N(\epsilon)$ that depends on ϵ but not on z such that the absolute value of the nth remainder ($n \geq N$) of the series is smaller than ϵ, then we say that the series converges uniformly in D. As in the case of functions of a real variable, the usefulness of the concept of uniform convergence is due to the fact that *the sum of a series of continuous functions of* z, *which converges uniformly in* D, *is also a continuous function of* z *in* D. The proof is almost identical with that employed in the case of a real variable and is left as an exercise to the reader.

It was pointed out before that an analytic function $f(z)$ which is regular in a domain D is also continuous in D. It therefore follows that a series whose terms are analytic functions regular in a domain D, and which converges uniformly in D, has a sum which is a continuous function in D. The terms of the power series (15) are regular for all finite values of z; hence, the sum of this series will be a continuous function of z in any domain D in which it converges uniformly. It is easy to see that D may be taken to be the interior of any circle concentric with the circle of convergence and contained in it. Indeed, if R is the radius of convergence and $\rho = |z - a| < |z_0 - a| = r < R$, then, by (17),

$$\left|\sum_{n=N}^{\infty} a_n(z-a)^n\right| \leq \sum_{n=N}^{\infty} |a_n(z-a)^n| = \sum_{n=N}^{\infty} |a_n(z_0-a)^n| \left|\frac{z-a}{z_0-a}\right|^n$$

$$\leq M \sum_{n=N}^{\infty} \left(\frac{\rho}{r}\right)^n = \frac{\left(\frac{\rho}{r}\right)^N}{1-\frac{\rho}{r}}.$$

If N is taken large enough, the last expression can obviously be made smaller than any given positive ϵ. Since this N does not depend on z, it follows that *a power series converges uniformly in the interior of any circle which is concentric with the circle of convergence and contained in it.* As pointed out before, this implies the fact that the sum of a power series is a continuous function of z in the interior of the circle of convergence. However, we want to prove more than this; we want to show that the sum of a power series is a regular analytic function throughout the interior of its circle of convergence. This is an immediate consequence of a general theorem according to which the sum of a series of regular analytic functions which converges uniformly in a domain D is again regular in D. However, since this theorem will be proved only at a later stage (Sec. 3, Chap. III), we shall give here a simple direct proof of the desired result for power series. For greater simplicity of writing, we shall carry through the proof for power series of the form (16); the reader will have no difficulty in verifying that a power series of the general form (15) yields to the same treatment.

Our aim is to show that the power series

$$f(z) = \sum_{n=0}^{\infty} a_n z^n, \tag{18}$$

which is supposed to converge in the circle $|z| < R$, is a regular analytic function for $|z| < R$ and that its derivative can be obtained by means of term-by-term differentiation, that is,

$$f'(z) = \sum_{n=1}^{\infty} n a_n z^{n-1}. \tag{19}$$

We first observe that the series

$$g(z) = \sum_{n=1}^{\infty} n a_n z^{n-1}$$

likewise converges for $|z| < R$. Indeed, let $|z| = r < R$ and $r < r_0 < R$. Since (18) converges for $|z_0| = r_0$, it follows from (17) that

$$\sum_{n=1}^{\infty} n|a_n z^{n-1}| = \sum_{n=1}^{\infty} n|a_n z_0{}^{n-1}| \left|\frac{z}{z_0}\right|^{n-1} \leq \frac{M}{r_0} \sum_{n=1}^{\infty} n \left(\frac{r}{r_0}\right)^{n-1},$$

which shows that the series in question converges for all $|z| < R$.

Let now h be a complex number such that $r + |h| < \rho < R$, and denote the absolute value of h by ϵ. We then have

$$\left|\frac{(z+h)^n - z^n}{h} - nz^{n-1}\right| = \left|\frac{n(n-1)}{1 \cdot 2} z^{n-2}h + \cdots + h^{n-1}\right|$$
$$\leq \frac{n(n-1)}{1 \cdot 2} r^{n-2}\epsilon + \cdots + \epsilon^{n-1} = \frac{(r+\epsilon)^n - r^n}{\epsilon} - nr^{n-1}.$$

Hence,

$$\left|\frac{f(z+h) - f(z)}{h} - g(z)\right| = \left|\sum_{n=0}^{\infty} a_n \left[\frac{(z+h)^n - z^n}{h} - nz^{n-1}\right]\right|$$
$$\leq \sum_{n=0}^{\infty} |a_n| \left[\frac{(r+\epsilon)^n - r^n}{\epsilon} - nr^{n-1}\right]$$
$$= \sum_{n=0}^{\infty} |a_n|\rho^n \left[\frac{\left(\frac{r+\epsilon}{\rho}\right)^n - \left(\frac{r}{\rho}\right)^n}{\epsilon} - \frac{n}{r}\left(\frac{r}{\rho}\right)^n\right]$$
$$\leq M \sum_{n=0}^{\infty} \left\{\frac{1}{\epsilon}\left[\left(\frac{r+\epsilon}{\rho}\right)^n - \left(\frac{r}{\rho}\right)^n\right] - \frac{n}{r}\left(\frac{r}{\rho}\right)^n\right\}$$
$$= M \left\{\frac{1}{\epsilon}\left[\frac{\rho}{\rho - r - \epsilon} - \frac{\rho}{\rho - r}\right] - \frac{\rho}{(\rho - r)^2}\right\}$$
$$= \frac{M\epsilon}{(\rho - r)^2(\rho - r - \epsilon)}.$$

Since this tends to 0 if $\epsilon = |h| \to 0$, it follows that $f(z)$ has a derivative and that this derivative is of the form (19). We have thus arrived at the following result: *In the interior of its circle of convergence, a power series represents a regular analytic function; its derivative can be found by term-by-term differentiation.* Since a derivative of a power series is again a power series which converges in the same circle, we can, in turn, differentiate this power series in order to obtain the second derivative of the original series. By continuing in this fashion, it is seen that a power series has an arbitrary number of derivatives, all of which are again power series which converge in the same circle as the original series. It is therefore permissible to determine the coefficient a_n of the series

$$f(z) = \sum_{n=0}^{\infty} a_n(z-a)^n, \qquad |z-a| < R,$$

by differentiating $f(z)$ n times and then setting $z = a$. It follows that $f^{(n)}(a) = n!a_n$ and, therefore,

$$f(z) = \sum_{n=0}^{\infty} \frac{f^{(n)}(a)}{n!}(z-a)^n.$$

Hence, *a power series is the Taylor series of the function which is represented by it.* On the other hand, the Taylor expansions known from the calculus are obviously power series and the definitions of the functions represented by them may be extended to the complex domain by the simple expedient of regarding the independent variable as a complex variable. As an example, consider the Taylor series

$$\log(1+x) = x - \frac{x^2}{2} + \frac{x^3}{3} - \cdots$$

known from the elementary calculus. This series converges for $-1 < x \leq 1$. We can now "extend" this function to the complex domain by writing

$$\log(1+z) = z - \frac{z^2}{2} + \frac{z^3}{3} - \cdots,$$

where z is a complex variable. Obviously, this new series converges in the circle $|z| < 1$ and represents there a regular analytic function. Because, for real values of z, this function coincides with the real function $\log(1+x)$, this function is also denoted by $\log(1+z)$. In the next section, we shall see more examples of analytic functions which, for real values of z, coincide with well-known elementary real functions.

We finally derive Cauchy's rule for the determination of the radius of convergence of a power series. *If* $a_0, a_1, \ldots$ *are the coefficients of the power series* (16) *and* $A = \limsup_{n\to\infty} \sqrt[n]{|a_n|}$, *then the radius of convergence of* (16) *is given by*

$$R = \frac{1}{A}.$$

If z is a fixed value, then

$$\limsup_{n\to\infty} \sqrt[n]{|a_n z^n|} = A|z|.$$

Hence, if $|z| > (1/A)$, then for an infinity of subscripts n we have $\sqrt[n]{|a_nz^n|} > 1$, that is, $|a_nz^n| > 1$. In this case, the series diverges because we cannot have $\lim_{n\to\infty} a_nz^n = 0$. If, on the other hand, $|z| < (1/A)$, then there exists an index n_0 such that, for $n > n_0$, $\sqrt[n]{|a_nz^n|} < t$, where $A|z| < t < 1$. For $n > n_0$, the terms of (16) are therefore smaller in absolute value than the corresponding terms of the converging geometric series $\sum_{n=0}^{\infty} t^n$ $(0 < t < 1)$. Hence, the series (16) converges.

EXERCISES

1. Find the radii of convergence of the power series $\sum_{n=0}^{\infty} a_nz^n$, where

(*a*) $a_n = \dfrac{1}{n^n}$.

(*b*) $a_n = \left(1 - \dfrac{1}{n}\right)^{n^2}$.

(*c*) $\dfrac{1}{n^m}$.

2. If the power series $f(z) = \sum_{n=0}^{\infty} a_nz^n$ and $g(z) = \sum_{n=0}^{\infty} b_nz^n$ both converge for $|z| < R$, show that the series $h(z) = \sum_{n=0}^{\infty} c_nz^n$, where $c_n = \sum_{\nu=0}^{n} a_\nu b_{n-\nu}$, also converges for $|z| < R$ and that $h(z) = f(z)g(z)$.

3. If $f(z) = \sum_{n=0}^{\infty} \dfrac{z^n}{n!}$, show with the help of the result of the preceding exercise that $[f(z)]^2 = f(2z)$.

4. Find a power series which satisfies the differential equation $f'(z) = zf(z)$ and show that it converges for all finite values of z.

5. Find the most general power series (involving two arbitrary constants) which satisfies the differential equation $f''(z) + f(z) = 0$.

6. Show that the value of a power series cannot be constant in the interior of its circle of convergence unless it reduces to its first term.

7. Using the result of the preceding exercise, obtain the power series expansion of the function $f(z) = (1 - z)^{-2}$ by means of a recursion formula derived from the identity $(1 - z)^2f(z) = 1$. Find the radius of convergence of the series in question.

7. Preliminary Study of Some Elementary Functions. As already pointed out in the preceding section, the Taylor series expansions known from the elementary calculus can be made the starting point for the definition of many analytic functions. In the present section we shall discuss the properties of some of these functions. We begin with the study of the *exponential function* which is defined by the power series

$$e(z) = 1 + \frac{z}{1!} + \frac{z^2}{2!} + \cdots = \sum_{n=0}^{\infty} \frac{z^n}{n!}.$$

Since, by Cauchy's test, this series converges for all finite values of z, it follows that the analytic function $e(z)$ is regular at all finite points of the complex z-plane. By term-by-term differentiation it is seen that the function $e(z)$ satisfies the differential equation $e'(z) = e(z)$. If a is a constant complex number, it is immediately verified that the derivative of the analytic function $e(z)e(a - z)$ is zero. Hence, this function is necessarily a constant, whose value is found to be $e(a)$ by setting $z = 0$. Thus, $e(z)e(a - z) = e(a)$ or, setting $a - z = w$,

$$e(z)e(w) = e(z + w).$$

The *addition theorem* of the real function e^x—to which $e(z)$ reduces if z is real—is therefore also verified for the analytic function $e(z)$. With the notation $e(1) = \sum_{n=0}^{\infty} \frac{1}{n!} = e$, it follows from the addition theorem that $e(2) = e^2$, $e(3) = e^3$, . . . , $e(n) = e^n$. Because of this property of the function $e(z)$, we shall henceforth write e^z instead of $e(z)$. The reader will, however, do well to remember that this is a purely symbolical notation; by no stretch of the imagination can the number $e^{1+i} = \sum_{n=0}^{\infty} \frac{(1 + i)^n}{n!}$ be regarded as the "$(1 + i)$th power" of e.

An immediate consequence of the addition theorem of e^z is the fact that the exponential function does not take the value zero at any finite point of the complex plane. Indeed, we have

$$e^z e^{-z} = e^0 = 1;$$

since the value of e^{-z} is a well-determined finite number, it follows that e^z cannot be zero as otherwise its product with e^{-z} would also be zero.

Consider now the function

$$e^{iz} = 1 + \frac{(iz)}{1!} + \frac{(iz)^2}{2!} + \frac{(iz)^3}{3!} + \cdots.$$

In view of $i^2 = -1$, this may also be written

$$e^{iz} = 1 - \frac{z^2}{2!} + \frac{z^4}{4!} - \cdots + i\left(z - \frac{z^3}{3!} + \frac{z^5}{5!} - \cdots\right). \tag{20}$$

The two power series appearing here are formally identical with the well-known Taylor expansions of the real trigonometric functions cosine and sine. The analytic functions which are obtained if z is regarded as a complex variable will also be denoted by the symbols $\cos z$ and $\sin z$, although it goes without saying that, for complex values of z, the trigonometric associations of these functions lose their meaning. The analytic functions $\cos z$ and $\sin z$ are thus defined by the power series

$$\cos z = 1 - \frac{z^2}{2!} + \frac{z^4}{4!} - \cdots,$$

$$\sin z = z - \frac{z^3}{3!} + \frac{z^5}{5!} - \cdots.$$

Since both these series clearly converge for all finite values of z, $\cos z$ and $\sin z$ are regular at all finitely distant points of the z-plane. It is immediately confirmed that

(21) $$\cos(-z) = \cos z, \qquad \sin(-z) = -\sin z,$$

that is, $\cos z$ is an *even function of* z and $\sin z$ is an *odd function of* z. It is further apparent from the expansions for $\cos z$ and $\sin z$ that these functions have the differentiation formulas

$$\frac{d}{dz}(\cos z) = -\sin z, \qquad \frac{d}{dz}(\sin z) = \cos z.$$

(20) shows that $\cos z$ and $\sin z$ are related to the exponential function by the identity

(22) $$e^{iz} = \cos z + i \sin z.$$

Substituting $-z$ instead of z in (22) and using (21), we obtain

(23) $$e^{-iz} = \cos z - i \sin z.$$

Multiplication of (22) and (23) yields

(24) $$\cos^2 z + \sin^2 z = 1$$

which shows that the relation which, in the trigonometric interpretation of the real functions sine and cosine, expresses the theorem of Pythagoras, remains true if the definition of these functions is extended to complex values of the variable. (22) and (23) can also be used to give alternative definitions of the functions $\cos z$ and $\sin z$. We evidently have

(25) $$\cos z = \frac{1}{2}(e^{iz} + e^{-iz}), \qquad \sin z = \frac{1}{2i}(e^{iz} - e^{-iz}).$$

With the help of (22) and the addition theorem, the function e^z can be decomposed into its real and imaginary parts. We obtain

$$e^z = e^{x+iy} = e^x e^{iy} = e^x(\cos y + i \sin y),$$

that is,

$$e^z = e^x \cos y + ie^x \sin y. \tag{26}$$

The absolute value of e^z is therefore

$$|e^z| = e^x(\cos^2 y + \sin^2 y)^{\frac{1}{2}} = e^x.$$

Since the real function e^x assumes the value 1 only for $x = 0$ (e^x increases monotonically with x), the absolute value of e^z will be 1 if, and only if, $x = \text{Re}\{z\} = 0$, that is, if z is pure imaginary. We further observe that, with the help of (22), the polar representation (3), Sec. 2, of a complex number can be cast into a more compact form. If the absolute value r and the argument θ ($r \geq 0, 0 \leq \theta < 2\pi$) of the complex number $z = x + iy$ are defined by

$$x = r \cos \theta, \qquad y = r \sin \theta,$$

then, by virtue of (22),

$$z = re^{i\theta}. \tag{27}$$

The real trigonometric functions $\cos y$ and $\sin y$ are known to have the period 2π, that is, they satisfy the relations $\cos (y + 2\pi) = \cos y$ and $\sin (y + 2\pi) = \sin y$. As shown by (26), the function e^z will therefore also remain unchanged if 2π is added to the imaginary part y of the variable z. Hence, *the function e^z has the period $2\pi i$*, that is,

$$e^{z+2\pi i} = e^z. \tag{28}$$

This periodicity of e^z can also be derived in a direct fashion which, at the same time, shows that e^z has no periods which are not integral multiples of $2\pi i$. By the addition theorem, we have

$$e^{z+p} = e^p e^z.$$

p will therefore be a period of e^z if, and only if,

$$e^p = 1. \tag{29}$$

As shown above, the absolute value of e^z is equal to 1 only if z is pure imaginary. Hence, $p = it$, where t is real. It thus follows from (22) that we must have $\cos t + i \sin t = 1$ which, of course, is only possible if $\cos t = 1$, $\sin t = 0$. It is known from elementary trigonometry that this occurs for $t = 2\pi$ and for no other value between 0 and 2π. $2\pi i$ and its integral multiples are therefore periods of e^z. If there were a period

$p = it$ which is not of the form $2\pi im$, then there would exist an integer m such that $2\pi m < t < 2\pi(m + 1)$, and we would have

$$e^{z+i(t-2\pi m)} = e^z e^{it} e^{-2\pi im} = e^z.$$

The number $i(t - 2\pi m)$, where $0 < t - 2\pi m < 2\pi$, would therefore also be a period, but this was shown to be impossible. Hence, the equation (29) has only solutions of the form $2\pi im$, and these are all the periods of e^z.

The addition theorems of the real trigonometric functions also hold for the analytic functions $\cos z$ and $\sin z$. We thus have

$$\cos (z + w) = \cos z \cos w - \sin z \sin w, \tag{30}$$
$$\sin (z + w) = \sin z \cos w + \cos z \sin w. \tag{31}$$

The proof is an immediate consequence of (25) and (22) and is left as an exercise for the reader.

We now turn to the function $\log z$ which, exactly as in the case of the real logarithm, is defined as the function inverse to the exponential function. The fact that the exponential function and the logarithm are inverse to each other is expressed by saying that the two equations

$$w = \log z, \qquad z = e^w \tag{32}$$

are equivalent to each other. We first prove that $\log z$ is indeed an analytic function of z by showing that, for $z \neq 0$, $f(z) = \log z$ has a derivative. We have, by (32),

$$f'(z) = \lim_{z_1 \to z} \frac{f(z_1) - f(z)}{z_1 - z} = \lim_{w_1 \to w} \frac{w_1 - w}{e^{w_1} - e^w}$$
$$= \frac{1}{\lim\limits_{w_1 \to w} \dfrac{e^{w_1} - e^w}{w_1 - w}} = \frac{1}{\dfrac{d}{dw}(e^w)} = \frac{1}{e^w}.$$

In view of $z = e^w$, we finally obtain

$$\frac{d}{dz}(\log z) = \frac{1}{z},$$

which shows that $\log z$ has a derivative, provided $z \neq 0$. The function $\log z$ is therefore regular at all points of the z-plane except at the origin.

The function $\log z$ has a peculiar property which was not displayed by the analytic functions considered so far. It is *many-valued*. In order to understand this property, we recall the periodicity of the exponential function, that is, the identity

$$e^{w+2\pi im} = e^w, \qquad m = 0, \pm 1, \pm 2 \dots.$$

The inverse of the exponential function defined by (32) may therefore be taken to be any of the functions $\log z + 2\pi i m$ ($m = 0, \pm 1, \pm 2, \ldots$); in other words, the function $\log z$ is only defined up to an integral multiple of $2\pi i$. This may appear to the beginner as a source of confusion; however, order is readily restored by means of a more penetrating study of the properties of the function $\log z$. Writing $w = u + iv$, we find from (32) and (26) that

$$z = x + iy = e^u(\cos v + i \sin v),$$

whence $|z| = e^u$ and $(y/x) = \tan v$. If "log" now denotes the real logarithm of a positive number, as known from real analysis (it is somewhat inconsistent to use this notation for both this real function and the many-valued analytic function $\log z$; however, confusion can hardly arise because of this practice), we have $\log (e^u) = u$, whence

$$u = \log |z| = \log r, \qquad z = re^{i\theta}.$$

In view of $(y/x) = \tan v$, we have $v = \tan^{-1} (y/x) = \theta$, where the angle θ is only determined up to an integral multiple of 2π. With the expressions for $u = \text{Re} \{\log z\}$ and $v = \text{Im} \{\log z\}$ thus obtained, we have

$$\log z = \log r + i\theta, \qquad z = re^{i\theta}. \tag{33}$$

The indeterminacy of the logarithm is therefore simply an expression of the fact that the polar coordinates of a point in the plane are not uniquely determined; we can always add an integral multiple of 2π to the angle θ without changing the position of the point which is described by the polar coordinates. In analytic geometry, this many-valuedness of the polar coordinate system is decreed out of existence by the simple expedient of prescribing that the angle θ is always to be taken between 0 and 2π. We may employ the same device in order to turn the function $\log z$ into a single-valued analytic function; however, it is customary to restrict the value of θ to the interval $-\pi < \theta \leq \pi$, rather than to the interval $0 \leq \theta < 2\pi$. The value of the function $\log z$ obtained by this restriction is called the *principal value* of $\log z$. It is defined by (33), where $\log r$ denotes the real logarithm of the positive number r, with the additional condition that $-\pi < \theta \leq \pi$. In order to understand the connection between the principal value of $\log z$ and its other values, namely, those obtained by adding $2\pi m$ to θ, we consider a point $z = re^{i\theta}$ and let this point traverse the circle $|z| = r$ until it returns to its original position. In this operation, the angle θ will have increased or decreased by 2π, according as the point traveled in the positive or negative direction. (33) shows that, at the same time, the value of $\log z$ was changed by an amount of $\pm 2\pi i$, the sign depending on whether the origin was surrounded in the positive or negative direction. If the point z surrounds the origin m times (where m may be positive, negative, or zero), we shall therefore arrive at the value

$$\log z = \log r + i\theta + 2\pi i m, \qquad z = re^{i\theta},\ -\pi < \theta \leq \pi. \tag{34}$$

This situation may also be described by saying that the function $\log z$ has an infinity of *branches* all of which are single-valued; in order to get from one branch to another in a continuous manner, we have to surround the origin a suitable number of times.

The condition $-\pi < \theta \leq \pi$ which determined the single-valued principal value of $\log z$ can be restated in a more graphic fashion. Since both $\theta = \pi$ and $\theta = -\pi$ can be regarded as the equation of the negative axis in the z-plane, the condition $-\pi < \theta \leq \pi$ "forbids" the passing of the point z across the negative axis. The function $\log z$ will therefore be a single-valued function of z if a "cut" along the whole negative axis is applied to the z-plane: In accordance with the restriction $-\pi < \theta \leq \pi$, the "upper edge" of the cut belongs to the cut plane and the "lower edge" does not. The placing of the cut along the negative axis is, of course, a pure matter of convention; the reader will easily verify that any curve that does not intersect itself and connects the origin with the point at infinity will serve just as well. In the cut plane, not only the principal value but also all the other values of the function $\log z$ are single-valued functions. The desire to define a region in which the function $\log z$ in its entirety is single-valued may therefore suggest the following device: To each branch of the function $\log z$ we assign one replica of the cut plane in which this branch is single-valued; these cut planes can be ordered according to the integer m ($m = \ldots, -1, 0, 1, \ldots$) in (34) by which they are characterized. We now stack all these cut planes upon each other in such a fashion that the $(m + 1)$st plane lies immediately on top of the mth plane and that the corresponding points in all planes lie exactly on top of each other. In order to take account of the fact that we pass from the mth branch of $\log z$ to the $(m + 1)$st branch if the point z surrounds the origin in the positive direction, we now connect the upper edge of the cut in the mth plane with the lower edge of the cut in the $(m + 1)$st plane. If these connections are effected throughout the whole infinite set of planes (m runs from $-\infty$ to $+\infty$), we arrive at what is known as *the Riemann surface of the function log z*. On this Riemann surface, $\log z$ is a single-valued function. It is interesting to note that this surface has only two boundary points, namely, the points $z = 0$ and $z = \infty$. All the other possible boundary points satisfy $0 > z > -\infty$; but, since each sheet of the Riemann surface is connected with its two adjacent sheets along the two edges of the cut $0 > z > -\infty$, these points are—in all sheets—situated in the interior of the surface.

We close this chapter with a discussion of the function $f(z) = z^\alpha$, where α is an arbitrary complex number. If α is a positive integer, say n, then z^n is defined as a product of n factors all of which are equal to z. The easiest way of generalizing this definition to the case in which α is an arbitrary complex number is by means of the function log z. We define

$$z^\alpha = e^{\alpha \log z}. \tag{35}$$

The reader will verify without difficulty that if α is an integer or a rational number, this definition coincides with the usual definition of a power of z. Clearly, the many-valuedness of the function log z will generally result in the many-valuedness of the function z^α. If, in (35), log z denotes a definite branch, say, the principal value, of the logarithm, the various possible values of z^α will be of the form

$$z^\alpha = e^{\alpha(\log z + 2\pi i m)} = e^{\alpha \log z} e^{2\pi i \alpha m},$$

where m is an arbitrary integer. The various different values of z^α will therefore be obtained from any one of them by multiplication with a factor $e^{2\pi i \alpha m}$ ($m = \ldots, -1, 0, 1, \ldots$). As a result, z^α will in general have an infinity of different values. However, there are cases in which the number of the different values of z^α is finite; obviously, these correspond to the cases in which only a finite number of the values

$$e^{2\pi i \alpha m}, \qquad m = \ldots, -1, 0, 1, \ldots$$

are different from each other. If this occurs, then there must be two integers m and m' ($m' \neq m$) such that $e^{2\pi i \alpha m} = e^{2\pi i \alpha m'}$ or, in view of the addition theorem of the exponential function,

$$e^{2\pi i \alpha (m - m')} = 1.$$

Since e^z takes the value 1 only if $z = 2\pi i n$, where n is an integer, this is equivalent to $\alpha(m - m') = n$. It follows that α is a rational number. On the other hand, if α is a rational number of the form m/n (m and n having no common factors and $n \geq 1$), then the set

$$e^{2\pi i \frac{m}{n} \mu}, \qquad \mu = 0, \pm 1, \pm 2, \ldots,$$

clearly contains only n different numbers; as such we may take those corresponding to $\mu = 0, 1, \ldots, n - 1$. Hence, the function z^α is n-valued if, and only if, α is a rational number m/n, where m and n have no common factors; if α is not rational, z^α has an infinity of values.

In view of (35), the function z^α is single-valued on the Riemann surface of the logarithm; each branch can be made single-valued in the z-plane by the introduction of a cut along the negative axis. However, if α is rational, it is possible to construct a Riemann surface which has only a finite number of sheets and on which z^α is single-valued. As an example, consider the function $f(z) = z^{\frac{1}{2}} = \sqrt{z}$. In view of the foregoing, this function is two-valued; it has two branches which, for $z = 1$, reduce to 1 and -1, respectively. We have

$$\sqrt{z} = \sqrt{r}\, e^{\frac{1}{2}i\theta}, \qquad z = re^{i\theta},$$

where $\sqrt{r}$ is the positive square root of r. If we surround the origin in the positive direction, that is, if θ grows by the amount 2π, $\sqrt{z}$ will assume the value

$$\sqrt{r}\, e^{\frac{1}{2}i(\theta+2\pi)} = \sqrt{r}\, e^{\frac{1}{2}i\theta}e^{\pi i} = -\sqrt{r}\, e^{\frac{1}{2}i\theta} \equiv -\sqrt{z}.$$

Another closed circuit will again result in multiplication with the factor -1. Hence, two closed circuits about the origin will bring us back to the original value of the function. In order to construct a Riemann surface that takes account of these facts, we proceed as follows: We take two replicas of the z-plane both of which have been furnished with a cut from $z = \infty$ to $z = 0$ along the negative axis and place one on top of the other; to make things definite we suppose that the upper plane corresponds to the branch of $\sqrt{z}$ for which $\sqrt{1} = 1$ and the lower plane to that for which $\sqrt{1} = -1$. We connect the upper edge of the cut in the upper plane with the lower edge of the cut in the lower plane and, at the same time, we connect the lower edge of the cut in the upper plane with the upper edge with the cut in the lower plane. It should be noted that the two connections of the two sheets penetrate each other and that it is consequently impossible to build a correct paper model of this Riemann surface. The points $z = 0$ and $z = \infty$ deserve special attention. Obviously, the function $f(z) = \sqrt{z}$ is two-valued in any domain which contains either of these points and therefore, in view of our definition of regularity, $f(z)$ cannot be regular throughout these domains. Taking, in particular, the domains $|z| < \epsilon$ and $|z| > \dfrac{1}{\epsilon}$ (ϵ arbitrarily small) which contain the points $z = 0$ and $z = \infty$, respectively, we conclude that $z = 0$ and $z = \infty$ are singular points of the function $f(z) = \sqrt{z}$. A singular point of this type, that is, a point a such that there exists no neighborhood $|z - a| < \epsilon$ in which the function $f(z)$ is single-valued, is called a *branch point of* $f(z)$. If, for arbitrary small ϵ, the function is n-valued in $|z - a| < \epsilon$, then we say that a is a *branch point of order* $n - 1$ *of* $f(z)$. $z = 0$ and $z = \infty$ are thus branch points of first order of $f(z) = \sqrt{z}$. The singularities of the function $f(z) = \log z$ at $z = 0$ and $z = \infty$ may be regarded as branch points of infinite order.

The construction of a Riemann surface for the n-valued function $z^{\frac{1}{n}} = \sqrt[n]{z}$ is effected along the same lines as in the case $n = 2$. We take n replicas of the z-plane and cut them along the negative axis. We then place them one upon another and connect the edges of the cuts in the following manner: We connect the upper edge of the uppermost sheet with the lower edge of the second sheet, the upper edge of the second sheet with the lower edge of the third sheet, etc. If this has been done for each pair of adjacent sheets, there remain only two unconnected edges, namely, the lower edge of the first sheet and the upper edge of the last sheet. We then connect these two edges and thus obtain a closed n-sheeted Riemann surface. The proof that the function $f(z) = \sqrt[n]{z}$ is indeed single-valued on this Riemann surface is left as an exercise to the reader.

EXERCISES

1. The hyperbolic functions are defined by $\cosh z = \cos iz$ and $i \sinh z = \sin iz$. Show that the functions are real for real values of z, and use your result and the addition theorems of the trigonometric functions in order to decompose $\sin z$ and $\cos z$ into their real and imaginary parts.

2. Obtain the power series expansion of the function $f(z) = \dfrac{1+z}{1-z}$ and, by taking real parts and using (27), derive the identity (23″) of Sec. 6, Chap. I.

3. By applying a similar procedure to the series obtained in Exercise 7 of the preceding section, obtain closed expressions for $\sum_{n=1}^{\infty} nr^n \cos n\theta$ and $\sum_{n=1}^{\infty} nr^n \sin n\theta$ $(0 < r < 1)$.

4. Prove the periodicity of $\sin z$ and $\cos z$ from their addition theorems, the identity (24), and the fact that $\sin \frac{1}{2}\pi = 1$.

5. Use (22) and (25) to prove the trigonometric identities

(a) $\dfrac{1}{2} + \cos \theta + \cos 2\theta + \cdots + \cos n\theta = \dfrac{\sin \frac{1}{2}(2n+1)\theta}{2 \sin \frac{1}{2}\theta}$,

(b) $\cos \theta + \cos 3\theta + \cdots + \cos (2n-1)\theta = \dfrac{\sin 2n\theta}{2 \sin \theta}$,

(c) $\sin \theta + \sin 2\theta + \cdots + \sin n\theta = \dfrac{\sin \frac{1}{2}n\theta \sin \frac{1}{2}(n+1)\theta}{\sin \frac{1}{2}\theta}$.

6. If t is a real parameter, growing from $-\infty$ to ∞, and a, b are constants, show that the points $z = \cos (a + it)$ and $z = \cos (t + ib)$ describe, respectively, hyperbolas and ellipses whose foci are at $z = \pm 1$.

7. Prove the addition theorem of the function e^z by direct multiplication of the series for e^z and e^w, and suitable rearrangement of terms.

8. Solving the differential equation $f''(z) + f(z) = 0$ by means of a power series with indeterminate coefficients, show that the most general analytic solution of this equation can be obtained in either of the forms $A \cos z + B \sin z$ or $Ce^{iz} + De^{-iz}$, where A, B, C, D are arbitrary constants. Use your result in order to deduce the identity (22).

9. Use the addition theorem of the exponential function in order to prove the identity

$$\log (zw) = \log z + \log w.$$

[This formula is to be understood in the following manner: If $\log z$ is one of the values (34) and $\log w$ is one of the infinitely many possible values denoted by this symbol, then their sum is one of the infinitely many possible values of $\log (zw)$.]

10. Use Taylor's formula in order to find a power series expansion of the form $\sum_{n=0}^{\infty} a_n(z - a)^n$ $(a \neq 0)$ for one particular branch of the function $\log z$; show that the circumference of the circle of convergence of this series passes through the origin.

11. Show that the usual formula for the derivative of a power also holds for the general power of z defined in (35), and find a power series expansion of the form $\sum_{n=0}^{\infty} a_n(z - a)^n$ $(a \neq 0)$ for one branch of the function z^α; show that the circle of convergence of this series coincides with that of the series in Exercise 10.

12. Find all roots of the equation $z^8 = 1$.

13. Show that the function $f(z) = \sqrt{1 - z^2}$ is single-valued on a Riemann surface consisting of two sheets which are crosswise connected along the linear segment $-1 \leq z \leq 1$; show that $z = \pm 1$ are simple branch points of $f(z)$ and that $z = \infty$ is *not* a branch point.

14. Show that the Riemann surface of the function $f(z) = \sqrt{(1 - z^2)(1 - k^2z^2)}$ $(0 < k < 1)$ can be obtained in the following manner: Two replicas of the z-plane are both cut along the linear segments $-(1/k) \leq z \leq -1$ and $1 \leq z \leq (1/k)$ and placed upon each other. Then, the two sheets are crosswise [in the manner described in the text for $f(z) = \sqrt{z}$] connected with each other along the two linear segments. Show that the only branch points of this surface are simple branch points situated at $z = \pm 1, \pm 1/k$.

15. Show that the analytic function $w = \sin^{-1} z$, defined by $z = \sin w$, is regular at all finite points of the z-plane with the exception of the points $z = \pm 1$, and that its derivative is $(dw/dz) = (1 - z^2)^{-\frac{1}{2}}$; show further that $\sin^{-1} z$ is an infinitely many-valued function of z which is made single-valued by cutting the z-plane along the rays $-\infty \leq z \leq -1$ and $1 \leq z \leq \infty$.

16. Show that the function $w = \tan^{-1} z$, defined by $z = \tan w = [(\sin w)/(\cos w)]$, is regular for $z \neq \pm i$ and has the derivative $(dw/dz) = (1 + z^2)^{-1}$; prove that this infinitely many-valued function is made single-valued by cutting the z-plane along the parts of the imaginary axis extending from i to $i\infty$ and $-i$ to $-i\infty$, respectively (the meaning of the symbols $i\infty$ and $-i\infty$ is obvious).

17. Prove the identities

$$\sin^{-1} z = \frac{1}{i} \log [\sqrt{1 - z^2} + iz]$$

and

$$\tan^{-1} z = \frac{1}{2i} \log \frac{1 + iz}{1 - iz}.$$

CHAPTER III

THE COMPLEX INTEGRAL CALCULUS

1. Complex Integration. In the ordinary calculus, the concept of integration is introduced in two a priori entirely unrelated ways. Indefinite integration is defined as the reverse of differentiation, and definite integration is defined by the limit of a process of summation in which the individual terms tend to zero and their number tends to infinity. The fact that both definitions lead to one and the same thing is one of the most fundamental results of the calculus and also one of the main reasons for its usefulness. If we wish to extend the concept of integration to analytic functions of a complex variable, we encounter no difficulties as far as indefinite integration is concerned. As in real analysis, the indefinite integral $F(z)$ of a regular analytic function $f(z)$ is defined by the equation $F'(z) = f(z)$. By means of this rule, we can immediately write down the integral of a converging power series; we integrate the series term by term and observe that the integral of z^n is $\frac{1}{n+1} z^{n+1}$. The indefinite integral $F(z)$ of $f(z)$ will be denoted by $F(z) = \int f(z)\, dz$; the true reason for this notation will become apparent in the next section.

The situation is, however, not so simple if we wish to extend the concept of definite integration to the complex domain and, at the same time, preserve the connection between definite and indefinite integration known from the calculus. It is in the very nature of definite integration that it involves an integration path along which the integration is carried out. In the case of a function $g(x)$ of a real variable x, the integration path along which the integral $\int_a^b g(x)\, dx$ $(a < b)$ is computed is the linear segment $a \leq x \leq b$ connecting the lower and upper limits of the integral. In the case of a function of the complex variable z, the integration limits a and b may be any two points of the complex plane and, moreover, the integration path may be any curve C connecting a and b which satisfies certain smoothness requirements. It might therefore be expected that the definite integral between the limits a and b depends on the particular path C along which it is computed. This is, however, not the case, as we shall see in the next section. It is this independence of the integral of the particular path along which it is computed that makes complex integration one of the most powerful tools of mathematical analysis.

We now proceed to give a formal definition of the complex integration process. Let C be a smooth arc connecting the two distinct points a and b of the z-plane, that is, let C be given by a parametric representation

$$(1) \qquad x = x(t), \qquad y = y(t), \qquad 0 \leq t \leq 1, \qquad x(0) + iy(0) = a, \\ x(1) + iy(1) = b,$$

where $x(t)$ and $y(t)$ have continuous derivatives $x'(t)$ and $y'(t)$, respectively. Let further $z_1, z_2, \ldots, z_n$ be n distinct points on the curve C, where $z_1 = a$, $z_n = b$, and z_ν is situated on the section of C which connects $z_{\nu-1}$ and $z_{\nu+1}$; finally, let $f(z)$ be an analytic function which is regular at all points of C, including its end points a and b. Consider now the expression

$$S_n = \sum_{\nu=1}^{n-1} f(\zeta_\nu)(z_{\nu+1} - z_\nu),$$

where ζ_ν is an arbitrary point on the section of C which connects z_ν and $z_{\nu+1}$, and denote by Δ_n the largest among the numbers $|z_2 - z_1|$, $|z_3 - z_2|$, $\ldots$, $|z_n - z_{n-1}|$. If the number of points z_ν on C grows and finally tends to infinity in such a way that the largest distance between two consecutive points on C tends to zero, S_n tends to a limit which is called the integral of $f(z)$ along the curve C and denoted by $\int_C f(z)\,dz$. Thus,

$$(2) \qquad \int_C f(z)\,dz = \lim_{\substack{n \to \infty \\ \Delta n \to 0}} \sum_{\nu=1}^{n-1} f(\zeta_\nu)(z_{\nu+1} - z_\nu).$$

The existence of the limit and its independence of the particular sets of partition points z_ν chosen is shown in exactly the same way as in the case of the integral of a continuous function of a real variable, whose definition is formally identical with (2); the adaptation of the "real" proof to the case of a regular, and therefore continuous, analytic function is recommended to the reader as a useful exercise. It also follows from the formal identity of (2) with the definition of a real integral that in those cases in which $f(z)$ is real for real values of z, and C is a segment of the real axis, real and complex integration will give identical results.

The integral (2) has some elementary properties which follow immediately from its definition. If both $f(z)$ and $g(z)$ are regular on C, then

$$\int_C [f(z) + g(z)]\,dz = \int_C f(z)\,dz + \int_C g(z)\,dz.$$

If α is a point of C other than a and b, then

$$\int_a^\alpha f(z)\,dz + \int_\alpha^b f(z)\,dz = \int_a^b f(z)\,dz.$$

Finally, if the direction of the integration along C is reversed, then

$$\int_b^a f(z)\,dz = -\int_a^b f(z)\,dz.$$

The integral (2) is subject to an important inequality. If L is the length of the curve C and $|f(z)| < M$ at all points of C, then

$$\left|\int_C f(z)\,dz\right| < ML. \tag{3}$$

This is an immediate consequence of (2). Indeed, we have

$$\left|\sum_{\nu=1}^{n-1} f(\zeta_\nu)(z_{\nu+1} - z_\nu)\right| < M \sum_{\nu=1}^{n-1} |z_{\nu+1} - z_\nu|,$$

and the latter sum cannot be larger than L since the chord connecting the points $z_{\nu+1}$ and z_ν cannot be larger than the section of C between these points; hence, (3) follows.

For many purposes it is important to decompose the integral (2) into its real and imaginary parts. If we write $f(z) = u(z) + iv(z)$ and $z = x + iy$, we may bring (2) into the form

$$\int_C f(z)\,dz = \lim_{n\to\infty} \sum_{\nu=1}^{n-1} [u(\zeta_\nu)(x_{\nu+1} - x_\nu) - v(\zeta_\nu)(y_{\nu+1} - y_\nu)]$$
$$+ i \lim_{n\to\infty} \sum_{\nu=1}^{n-1} [u(\zeta_\nu)(y_{\nu+1} - y_\nu) + v(\zeta_\nu)(x_{\nu+1} - x_\nu)],$$

where, in view of $\Delta_n \to 0$, the maxima of both $|x_{\nu+1} - x_\nu|$ and $|y_{\nu+1} - y_\nu|$ tend to zero. If we denote by t_ν the value of the curve parameter t in (1) which corresponds to the point x_ν, y_ν, and observe that because of the differentiability of the functions $x(t)$ and $y(t)$ we have

$$x_{\nu+1} - x_\nu = x'(\tau_\nu)(t_{\nu+1} - t_\nu), \qquad y_{\nu+1} - y_\nu = y'(\tau_\nu^*)(t_{\nu+1} - t_\nu),$$

where $t_\nu \leq \tau_\nu$, $\tau_\nu^* \leq t_{\nu+1}$, we obtain

$$\int_C f(z)\,dz = \lim_{n\to\infty} \sum_{\nu=1}^{n-1} [u(\zeta_\nu)x'(\tau_\nu) - v(\zeta_\nu)y'(\tau_\nu^*)](t_{\nu+1} - t_\nu)$$
$$+ i \lim_{n-\infty} \sum_{\nu=1}^{n-1} [u(\zeta)y'(\tau_\nu^*) + v(\zeta_\nu)x'(\tau_\nu)](t_{\nu+1} - t_\nu),$$

where max $(t_{\nu+1} - t_\nu)$ tends to zero if $n \to \infty$. Since $x'(t)$ and $y'(t)$ are continuous functions, we therefore obtain

$$\int_C f(z)\,dz = \int_0^1 [u(x,y)x'(t) - v(x,y)y'(t)]\,dt$$
$$+ i\int_0^1 [u(x,y)y'(t) + v(x,y)x'(t)]\,dt,$$

or, in view of the definition of a real line integral,

$$\int_C f(z)\,dz = \int_C u\,dx - v\,dy + i\int_C u\,dy + v\,dx. \tag{4}$$

Occasionally, the identity (4)—which, in a purely formal way, seems to follow from the relation

$$f(z)\,dz = (u + iv)(dx + i\,dy) = u\,dx - v\,dy + i(u\,dy + v\,dx)$$

—is taken as the definition of the integral $\int_C f(z)\,dz$.

We finally mention the obvious fact that our definition of the integral can immediately be extended to curves C which are continuous and are the sum of a finite number of smooth, that is, continuously differentiable, arcs. Such piecewise smooth curves are also known as *contours*. A simple closed curve which consists of a finite number of smooth arcs is called a *closed contour*.

EXERCISES

1. Deduce from (2) that $\int_C dz = 0$, if C is a closed contour.

2. Show by means of (4) that $\int_C e^z\,dz = 0$ if C is the perimeter of the square whose corners are at the points $z = 0, 1, 1 + i, i$.

3. Show that $\int_C \cos z\,dz = 2$ if C is the polygonal line obtained by connecting, in that order, the points $z = \frac{1}{2}\pi$, $z = \frac{1}{2}\pi + i$, $z = -\frac{1}{2}\pi + i$, $z = -\frac{1}{2}\pi$.

4. By introducing polar coordinates, show that $\int_C z^n\,dz = 0$, where C is the circle $|z| = r$ $(r > 0)$ and n is an integer different from -1; show further that

$$\int_{|z|=\rho} \frac{dz}{z} = 2\pi i.$$

2. Cauchy's Theorem. The definition (2) of the complex integral involves the curve C along which the integral is to be computed. If it were true, as this fact might indicate, that the value of the integral depends on the particular curve C which connects the points a and b, complex integration would be a highly complicated affair and its usefulness would be very limited. We shall show that this is not the case and that the following theorem, due to Cauchy, holds.

Let $f(z)$ be regular in a simply-connected domain D and let a and b be two points of D. If C_1 and C_2 are two different contours, entirely situated in D, which connect a and b, then

$$\int_{C_1} f(z)\,dz = \int_{C_2} f(z)\,dz \tag{5}$$

Here, C_1 and C_2 have to be traversed in the same sense, that is, starting from a and ending at b. Cauchy's theorem is often stated in the following more compact form.

If $f(z)$ is regular within and on a closed contour C, then

$$\int_C f(z)\,dz = 0. \tag{6}$$

Since the two contours C_1 and C_2 in (5) form together a closed contour C as in (6), provided the direction of one of them is reversed, these two formulations of Cauchy's theorem are clearly equivalent. In (6) it makes no difference in which direction the integration over the closed contour C is carried out. Since, however, we shall also be concerned with integrals over closed contours which are not zero, it becomes necessary to adopt a convention regarding the direction in which such integrals are to be taken. In agreement with the convention of Sec. 3, Chap. I, we therefore stipulate that all integrals over closed contours are to be taken in the *positive direction*, that is, in such a way that the interior of the domain bounded by C remains at the left if C is traversed.

Cauchy's theorem would be easy to prove if it were known at this stage that the derivative $f'(z)$ of a regular analytic function $f(z)$ is continuous. In this case, it would follow from (4) and Gauss' theorem (5) (Sec. 3, Chap. I) that

$$\begin{aligned}\int_C f(z)\,dz &= \int_C u\,dx - v\,dy + i\int_C u\,dy + v\,dx \\ &= -\iint\limits_D \left(\frac{\partial u}{\partial y} + \frac{\partial v}{\partial x}\right) dx\,dy + i\iint\limits_D \left(\frac{\partial u}{\partial x} - \frac{\partial v}{\partial y}\right) dx\,dy,\end{aligned}$$

where D is the domain enclosed by C. Gauss' theorem requires that the partial derivatives of u and v be continuous, and this condition is satisfied if $f'(z)$ is continuous. Since, by the Cauchy-Riemann differential equations (14) (Sec. 5, Chap. II), the integrands of both area integrals vanish identically, it follows that $\int_C f(z)\,dz = 0$, which is Cauchy's theorem. However, we cannot use the fact that the derivative of a regular function is continuous, since so far we have not shown this to be true; in fact, the continuity of $f'(z)$—and, moreover, the fact that $f'(z)$ is also a regular analytic function of z—will be one of the results to be deduced from Cauchy's theorem. We therefore have to have recourse to another proof which, though it lacks the elegance and brevity of the above argument, avoids the use of the continuity of $f'(z)$.

We begin with the observation that if the domain D is divided by means of a number of contours into subdomains

$$D_1, \ldots, D_n \qquad D_1 + D_2 + \cdots + D_n = D,$$

then

$$\int_C f(z)\,dz = \sum_{\nu=1}^{n} \int_{C_\nu} f(z)\,dz, \tag{7}$$

where C_ν denotes the boundary of D_ν. Indeed, any part C' of a boundary C_ν which does not also belong to C is the common boundary of C_ν and another subdomain, say C_μ $(\nu \neq \mu)$. If C' is traversed in the positive direction with respect to D_ν, it clearly is traversed in the negative direction with respect to D_μ. Hence, since the integrals over C_ν and C_μ are taken in the positive sense with respect to the domains D_ν and D_μ, respectively, it follows that the contributions of the arc C' to these two integrals cancel out. This proves (7). For the proof of Cauchy's theorem it will therefore be sufficient to show that $\sum_\nu \int_{C_\nu} f(z)\,dz = 0$ if the C_ν are the boundaries corresponding to some subdivision of D. The particular subdivision which we shall use is obtained by intersecting D with two sets of equidistant straight lines, parallel to the x-axis and y-axis, respectively. If two adjacent straight lines have the distance d from each other, this procedure will divide D into a number of squares of side d and a number of irregularly shaped domains which are parts of such squares.

If S denotes the boundary of one of the complete squares and z_0 is a point in the interior of the square, then

$$\int_S |z - z_0|\,|dz| < 4\sqrt{2}\,d^2 = 4\sqrt{2} \text{ area of the square.} \tag{8}$$

This is an immediate consequence of the fact that $|z - z_0|$ cannot be larger than the diagonal $\sqrt{2}\,d$ of the square and that $\int_S |dz| = \int_S \sqrt{dx^2 + dy^2}$ is the length $4d$ of S.

Since $f(z)$ is regular, that is, differentiable, at all points of the closure $\bar{D}$ of D, there exists for each z_0 of $\bar{D}$ a certain neighborhood in which

$$\left| \frac{f(z) - f(z_0)}{z - z_0} - f'(z_0) \right| < \epsilon,$$

where ϵ is an arbitrary positive constant. This may also be written in the form

$$f(z) - f(z_0) = f'(z_0)(z - z_0) + \delta|z - z_0|, \tag{9}$$

where $|\delta| < \epsilon$. We shall now show that this "local" property can be extended throughout $\bar{D}$ in the following manner: *Given ϵ, then it is possible, in the manner indicated above, to divide $\bar{D}$ into a finite number of meshes—either complete or incomplete squares—such that within each mesh there is a point z_0 for which* (9) *holds, where z is any point in the same mesh and*

$|\delta| < \epsilon$. The truth of this statement, which is known as *Goursat's lemma*, is easily established. We first remark that with every point z_0 of D we can associate a positive number $d(z_0)$ such that (9) is true for any square parallel to the axes which contains z_0 and whose side is smaller than $d(z_0)$. This follows from the fact, mentioned before, that because of the differentiability of $f(z)$ at z_0 there exists a neighborhood of z_0 in which (9) is satisfied; we may therefore choose $d(z_0)$ such that $\sqrt{2}\, d(z_0)$ is smaller than the radius of this neighborhood. Consider now the set of the numbers $d(z)$, where z may denote any point of D. If the greatest lower bound of the $d(z)$ is a positive number d, that is, $0 < d \leq d(z)$, then, for a subdivision into squares of side d, (9) will be true, and Goursat's lemma is proved. If g.l.b. $d(z) = 0$, then there exists a sequence of points z_1, z_2, . . . of $\bar{D}$, converging to a point z_0, for which $\lim_{n\to\infty} d(z_n) = 0$; since $\bar{D}$ is a closed set, z_0 is also a point of $\bar{D}$. But this contradicts the fact that $f(z)$ is differentiable at z_0. Indeed, as shown above, the differentiability at z_0 entails the existence of a neighborhood of z_0 in which (9) is satisfied; this is clearly incompatible with the fact that there are points arbitrarily close to z_0 which cannot be encased in arbitrarily small squares in which (9) is true. This proves Goursat's lemma.

We now integrate (9) around the boundary μ of each mesh. We obtain

$$(10) \quad \int_\mu f(z)\, dz - f(z_0) \int_\mu dz = f'(z_0) \int_\mu z\, dz - z_0 f'(z_0) \int_\mu dz + \int_\mu \delta |z - z_0|\, dz.$$

The integrals $\int_\mu dz$ and $\int_\mu z\, dz$ both vanish. This is seen either by using the fact that the functions 1 and z have continuous derivatives and Cauchy's theorem has already been proved for such functions, or else as immediate consequences of the definition (2). In the case of $\int_\mu z\, dz$, we have

$$\int_\mu z\, dz = \lim \sum_\nu z_\nu(z_{\nu+1} - z_\nu) = \lim \sum_\nu z_{\nu+1}(z_{\nu+1} - z_\nu)$$
$$= \lim \tfrac{1}{2} \sum_\nu (z_\nu + z_{\nu+1})(z_{\nu+1} - z_\nu) = \lim \tfrac{1}{2} \sum_\nu (z_{\nu+1}^2 - z_\nu^2),$$

and this is zero, since μ is a closed contour and the first and the last point z_ν are identical. The latter remark also proves $\int_\mu dz = 0$. Hence (10) reduces to

$$\int_\mu f(z)\, dz = \int_\mu \delta |z - z_0|\, dz.$$

By adding the contributions of all meshes, we obtain, in view of (7),

$$\int_C f(z)\, dz = \sum \int_\mu \delta|z - z_0|\, dz,$$

whence

(11) $$\left| \int_C f(z)\, dz \right| < \epsilon \sum \int_\mu |z - z_0|\, dz.$$

Let now σ be a large square which contains D and whose sides also belong to the network of straight lines by means of which D was subdivided. The integrations in the right-hand side of (11) are extended over those parts of the network which are inside D and over the boundary C of D. We shall therefore enlarge the right-hand side of (11) if we add to it integrations over that part of the network which is outside D but contained in σ. Since, by (8), the value of the integral over each individual small square is smaller than $4\sqrt{2}$ times its area, it follows that

$$\left| \int_C f(z)\, dz \right| < 4\sqrt{2}\, A^2\epsilon + \epsilon \int_C |z - z_0|\, dz,$$

where A is the side of σ. On the other hand, z and z_0 are both in $\bar{D}$; $|z - z_0|$ is therefore not larger than the diagonal of σ. Since $\int_C |dz|$ is the length L of C, we obtain finally

$$\left| \int_C f(z)\, dz \right| < (4\sqrt{2}\, A^2 + \sqrt{2}\, AL)\epsilon.$$

Since ϵ is arbitrarily small, it follows that

$$\int_C f(z)\, dz = 0.$$

This completes the proof of Cauchy's theorem.

As a first application of Cauchy's theorem, we show that every analytic function $f(z)$ which is regular in a simply-connected domain D has an indefinite integral $F(z)$ which is again a regular analytic function in D, and that the connection between definite and indefinite integration is the same as in the ordinary calculus. *The function*

(12) $$F(z) = \int_a^z f(\zeta)\, d\zeta,$$

where z and a are both in D, is an analytic function regular in D, and

(13) $$F'(z) = f(z).$$

We note that it is sufficient to indicate the upper and lower limits of the integral (12) since, by Cauchy's theorem, the value of the integral is independent of the particular contour C in D along which the integral between these two points is taken. To prove (13), we observe that, for any $z_0 \in D$, it follows from (12) that

$$\frac{F(z) - F(z_0)}{z - z_0} - f(z_0) = \frac{\int_{z_0}^{z} [f(\zeta) - f(z_0)]\, d\zeta}{z - z_0}.$$

Since $f(z)$ is continuous at z_0, there exists a $\delta = \delta(\epsilon)$ such that $|f(\zeta) - f(z_0)| < \epsilon$ for $|\zeta - z_0| < \delta$. If $|z - z_0| < \delta$ and the circle $|z - z_0| < \delta$ is wholly contained in D, the integration may be carried out along the linear segment of length $|z - z_0|$ connecting z and z_0. Hence, by (3),

$$\left| \int_{z_0}^{z} [f(\zeta) - f(z_0)]\, d\zeta \right| < \epsilon\, |z - z_0|,$$

and it follows that

$$\left| \frac{F(z) - F(z_0)}{z - z_0} - f(z_0) \right| < \epsilon$$

for $|z - z_0| < \delta$. This shows that $F(z)$ is a regular analytic function in D and that its derivative is $f(z)$.

The validity of Cauchy's theorem is not confined to simply-connected domains. If D is a domain of connectivity n which is bounded by n closed contours, we can transform D into a simply-connected domain D by $n - 1$ appropriately chosen crosscuts (see Sec. 1, Chap. I), which may be taken to be contours. If $f(z)$ is regular and single-valued in D, and C^* denotes the boundary of D^*, then

$$\int_{C^*} f(z)\, dz = 0,$$

since Cauchy's theorem was shown to hold for simply-connected domains. Those parts of C^* which do not also belong to C, that is, the crosscuts, appear twice in the above integration, but in different directions; this situation is described in detail in Sec. 3, Chap. I, to which the reader may refer. Since $f(z)$ is single-valued in D, the values of $f(z)$ on both "edges" of the crosscut are the same and the two contributions of the crosscut to the integral cancel out. Hence, the integral over C has the same value as that over C and we have

$$\int_{C} f(z)\, dz = 0,$$

where C is the boundary of the multiply-connected domain D. The condition that $f(z)$ be single-valued in D deserves particular stress. If $f(z)$ is regular at all points of $D + C$ but not single-valued, that is, if we can continue $f(z)$ along a closed circuit about one of the "holes" of D and come back with a different value of $f(z)$, the above argument fails. The values of $f(z)$ at both edges of the crosscuts will not necessarily be the same, and Cauchy's theorem will, in general, cease to be true. In the case of a simply-connected domain D, we do not have to worry about a possible many-valuedness of the function $f(z)$; it will be shown in Sec. 5 that a function which is regular in a simply-connected domain D is necessarily also single-valued in D (monodromy theorem).

If the function $f(z)$ is single-valued in D but not regular at all points of D, then $\int_C f(z)\,dz$ will, in general, not be zero. As an example, consider

$$\int_C \frac{dz}{z},$$

where C is a simple closed contour surrounding the origin. In order to evaluate this integral, we observe that the function $1/z$ is regular in the doubly-connected domain D' which is bounded by C and a small circle $|z| = \rho$ situated in the interior of C. If the point z describes the boundary of D', this circle is obviously described in the negative direction. Applying Cauchy's theorem to the boundary of D', we obtain therefore

$$\int_C \frac{dz}{z} - \int_{|z|=\rho} \frac{dz}{z} = 0,$$

or

$$\int_C \frac{dz}{z} = \int_{|z|=\rho} \frac{dz}{z}. \tag{14}$$

The value of this integral thus remains unchanged if the contour C is "deformed" into the circle $|z| = \rho$. The procedure used in the derivation of this result is obviously of wider application. The reader is recommended to prove by similar considerations that the following general statement is true: *The value of a "contour integral," that is, an integral taken over a closed contour, remains unchanged if the contour over which it is taken is continuously deformed into another contour in such a way that at no time a singularity of the integrand is crossed.* To return now to our integral, we introduce polar coordinates in the integral on the right-hand side of (14). We have $z = \rho e^{i\theta}$ and, since ρ is constant, $dz = i\rho e^{i\theta}\,d\theta$. Hence

$$\int_{|z|=\rho} \frac{dz}{z} = i \int_0^{2\pi} d\theta = 2\pi i,$$

whence, by (14),

$$\int_C \frac{dz}{z} = 2\pi i. \tag{15}$$

We note that we also have

$$\int_{C_0} \frac{dz}{z - z_0} = 2\pi i, \tag{16}$$

if C_0 is a simple closed contour surrounding the point z_0. (16) follows from (15) by the substitution of $z - z_0$ instead of z and the observation that, under this substitution, a simple closed contour surrounding the origin is transformed into a similar contour surrounding z_0.

The considerable freedom we have in deforming a contour without altering the value of the corresponding contour integral can be utilized for the convenient computation of many real definite integrals. We shall confine ourselves here to the discussion of one famous example, namely, the integral

$$\int_0^\infty \frac{\sin x}{x}\, dx.$$

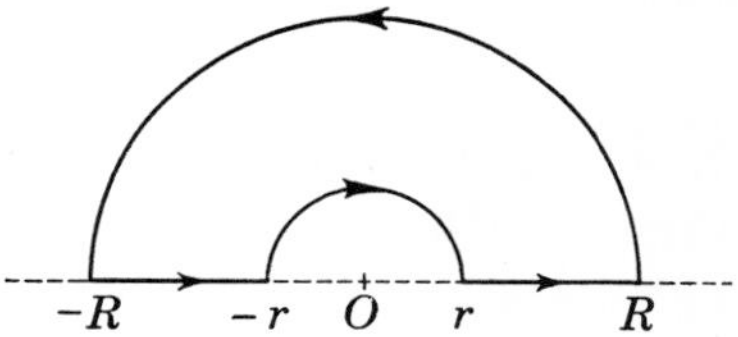

FIG. 5.

Consider the contour integral

$$\int_C \frac{e^{iz}}{z}\, dz,$$

taken along the contour C indicated in Fig. 5. Since the origin is in the exterior of C and e^{iz} is regular at all finite points of the z-plane, the integrand is regular in the interior of C and on C itself. By Cauchy's theorem, the value of the integral is therefore zero. Introducing polar coordinates on the two half circles, we thus have

$$\int_{-R}^{-r} \frac{e^{ix}}{x}\, dx + \int_r^R \frac{e^{ix}}{x}\, dx - i\int_0^\pi e^{-r\sin\theta + ir\cos\theta}\, d\theta + i\int_0^\pi e^{-R\sin\theta + iR\cos\theta}\, d\theta = 0. \tag{17}$$

In view of

$$\int_{-R}^{-r} \frac{e^{ix}}{x}\, dx = -\int_r^R \frac{e^{-ix}}{x}\, dx,$$

the sum of the first two integrals is equal to

$$\int_r^R [e^{ix} - e^{-ix}]\frac{dx}{x} = 2i\int_r^R \frac{\sin x}{x}\, dx.$$

We now let $r \to 0$ in (17). Since the third integral in (17) obviously has the limit $-i\int_0^\pi d\theta = -\pi i$, we obtain

$$2\int_0^R \frac{\sin x}{x}\, dx - \pi + \int_0^\pi e^{-R\sin\theta + iR\cos\theta}\, d\theta = 0. \tag{17'}$$

This is true for every $R > 0$ and therefore also in the limit for $R \to \infty$. We shall show that the last integral in (17′) tends to zero for $R \to \infty$. Since

$$\left|e^{-R\sin\theta + iR\cos\theta}\right| = e^{-R\sin\theta},$$

we have

$$|I| = \left|\int_0^\pi e^{-R\sin + iR\cos\theta}\,d\theta\right| \le \int_0^\pi e^{-R\sin\theta}\,d\theta.$$

In view of

$$\int_{\frac{1}{2}\pi}^{\pi} e^{-R\sin\theta}\,d\theta = \int_0^{\frac{1}{2}\pi} e^{-R\sin\theta}\,d\theta,$$

it follows further that

$$|I| \le 2\int_0^{\frac{1}{2}\pi} e^{-R\sin\theta}\,d\theta.$$

Now, for $0 \le \theta \le \frac{1}{2}\pi$, we have $\sin\theta \ge 2\theta/\pi$; this inequality simply expresses the fact that the linear segment connecting the origin and the point $(\frac{1}{2}\pi, 1)$ lies under the arc of the sine curve for $0 \le \theta \le \frac{1}{2}\pi$. Hence

$$|I| \le 2\int_0^{\frac{1}{2}\pi} e^{-\frac{2R}{\pi}\theta}\,d\theta = \frac{\pi}{R}(1 - e^{-R}),$$

and this obviously tends to zero if R tends to infinity. It follows therefore from (17′) that

$$2\lim_{R\to\infty}\int_0^R \frac{\sin x}{x}\,dx - \pi = 0,$$

or

$$\int_0^\infty \frac{\sin x}{x}\,dx = \frac{\pi}{2}.$$

EXERCISES

1. Show that

$$\int_C \frac{dz}{z^n} = 0, \qquad n = 2, 3, \ldots,$$

where C is a simple closed contour surrounding the origin. *Hint:* Consider the corresponding integral over the circle $|z| = R$, and introduce polar coordinates.

2. Using (15), Cauchy's theorem, and the result of the preceding exercise, show by evaluating

$$\int_{|z|=1}\left(z + \frac{1}{z}\right)^{2n}\frac{dz}{z},$$

both with the help of polar coordinates and as a contour integral, that

$$\int_0^{2\pi} \cos^{2n}\theta\,d\theta = 2\pi\,\frac{1\cdot 3\cdot 5\cdots(2n-1)}{2\cdot 4\cdot 6\cdots(2n)}.$$

3. Show, by means of (15), that the analytic function $\log z$ defined by

$$\log z = \int_1^z \frac{dz}{z}$$

is determined only up to an additive term $2\pi i m$ where m is an integer, and that m is the number of closed circuits about the origin made by the contour connecting 1 and z; show that a cut along the negative axis, as described in Sec. 7, Chap. II, makes this function single-valued.

4. Using the fact that $\int_z^{zw} \frac{d\zeta}{\zeta} = \int_1^w \frac{d\zeta}{\zeta}$, derive the addition theorem

$$\log (zw) = \log z + \log w$$

from the definition of $\log z$ given in the preceding exercise.

5. By deforming the contour connecting 1 and z into a piece of the positive axis and a circular arc, show that

$$\log z = \log r + i\theta$$

if $z = re^{i\theta}$.

6. The function $\tan^{-1} z$ is defined by the integral

$$\tan^{-1} z = \int_0^z \frac{dz}{1 + z^2}.$$

Show, by taking an integration contour similar to that used in the preceding exercise, that $\text{Re}\,\{\tan^{-1} z\} = \frac{1}{4}\pi$ if $|z| = 1$ and $|\arg z| < \frac{1}{2}\pi$.

3. The Cauchy Integral Formula. Let $f(z)$ be regular and single-valued in a domain D, which may be simply-connected or multiply-connected, and on the closed contour or contours C by which D is bounded. The integral

$$\int_C \frac{f(\zeta)}{\zeta - z}\, d\zeta$$

will in general not be zero if z is situated in the interior of D, since the integrand has a singularity at the point z. If we wish to find the value of this integral, we can proceed in the same fashion as in the evaluation of the integral (15) in the preceding section. We delete from D a small circle of radius r and center z; since, in the remainder D^* of the domain, the integrand is regular and single-valued, we conclude from Cauchy's theorem that

$$\int_C \frac{f(\zeta)}{\zeta - z}\, d\zeta - \int_{|\zeta - z| = r} \frac{f(\zeta)}{\zeta - z}\, d\zeta = 0. \tag{18}$$

Here, the negative sign is due to the fact that the positive sense with respect to the interior of the small circle is identical with the negative sense with respect to the domain D^*. In the second integral in (18) we introduce polar coordinates $\zeta = z + re^{i\theta}$; we obtain

$$\int_{|\zeta - z| = r} \frac{f(\zeta)}{\zeta - z} \, d\zeta = i \int_0^{2\pi} f(z + re^{i\theta}) \, d\theta.$$

If we let r tend to zero, this tends to

$$i \int_0^{2\pi} f(z) \, d\theta = if(z) \int_0^{2\pi} d\theta = 2\pi i f(z).$$

Indeed, because of the continuity of $f(\zeta)$, the expression $|f(z + re^{i\theta}) - f(z)|$ can be made smaller than an arbitrary positive ϵ, whence

$$\left| \int_0^{2\pi} f(z + re^{i\theta}) \, d\theta - \int_0^{2\pi} f(z) \, d\theta \right|$$
$$= \left| \int_0^{2\pi} [f(z + re^{i\theta}) - f(z)] \, d\theta \right| < \epsilon \int_0^{2\pi} d\theta = 2\pi\epsilon.$$

The second integral in (18) has thus the value $2\pi i f(z)$, and we obtain

$$f(z) = \frac{1}{2\pi i} \int_C \frac{f(\zeta)}{\zeta - z} \, d\zeta. \tag{19}$$

This is the *Cauchy integral formula*, which expresses an analytic function in a domain D in terms of its values on the boundary of D. This formula reveals the surprising fact that the values of a regular analytic function in a domain D are completely determined once its values on the boundary of D are known.

As a first application of (19), we shall show that the derivative of a regular analytic function is again a regular analytic function. If $f(z)$ is regular inside and on the contour C, we have by (19)

$$\frac{f(z + h) - f(z)}{h} = \frac{1}{2\pi i} \int_C \frac{1}{h} \left\{ \frac{1}{\zeta - z - h} - \frac{1}{\zeta - z} \right\} f(\zeta) \, d\zeta$$
$$= \frac{1}{2\pi i} \int_C \frac{f(\zeta)}{(\zeta - z)^2} \, d\zeta + I,$$

where

$$I = \frac{h}{2\pi i} \int_C \frac{f(\zeta) \, d\zeta}{(\zeta - z)^2(\zeta - z - h)}.$$

$f(\zeta)$ is regular on C and therefore it is bounded there; hence, there exists a positive M such that $|f(\zeta)| < M$ on C. If d is the smallest distance between z and C and $|h|$ is taken small enough so that $|h| < \frac{1}{2}d$, we have $|\zeta - z|^2 \geq d^2$ and $|\zeta - z - h| \geq \frac{1}{2}d$; hence, by (3),

$$|I| < |h| \frac{ML}{\pi d^3},$$

where L is the length of C. If $h \to 0$, we therefore obtain $I \to 0$, whence

$$f'(z) = \lim_{h \to 0} \frac{f(z + h) - f(z)}{h} = \frac{1}{2\pi i} \int_C \frac{f(\zeta)}{(\zeta - z)^2} \, d\zeta.$$

This shows that the derivative of $f(z)$ can be computed by differentiating the expression (19) under the integral sign. However, the most important conclusion from the formula

$$f'(z) = \frac{1}{2\pi i} \int_C \frac{f(\zeta)}{(\zeta - z)^2} \, d\zeta \tag{20}$$

is the following. If, in the expression

$$\frac{f'(z + h) - f'(z)}{h},$$

we express both $f'(z)$ and $f'(z + h)$ by means of (20) and repeat the procedure which was applied to (19), we find in the same way that $f'(z)$, in turn, has a derivative $f''(z)$ which is given by

$$f''(z) = \frac{2}{2\pi i} \int_C \frac{f(\zeta)}{(\zeta - z)^3} \, d\zeta.$$

But this means that the derivative $f'(z)$ of a regular analytic function $f(z)$ is again a regular analytic function. Applying the same reasoning to $f'(z)$, $f''(z)$, etc., we arrive at the following result.

A regular analytic function $f(z)$ possesses derivatives $f^{(n)}(z)$ of all orders, all of which are again regular analytic functions. If $f(z)$ is regular in the closure of a domain D which is bounded by a closed contour or closed contours C, then

$$f^{(n)}(z) = \frac{n!}{2\pi i} \int_C \frac{f(\zeta)}{(\zeta - z)^{n+1}} \, d\zeta. \tag{21}$$

We point out that this result fills a gap in the proof of one of the theorems of Sec. 5, Chap. II. We showed there that the real part u and the imaginary part v of a regular analytic function $f(z) = u + iv$ are harmonic functions, provided the second partial derivatives of both u and v exist. Since the existence of these derivatives is an immediate consequence of the differentiability of $f'(z)$, the proof of that statement is therefore now complete.

Another consequence of the fact that the derivative of a regular function is again a regular function is the following converse of Cauchy's theorem which is known by the name of *Morera's theorem.*

If the function $f(z)$ is coutinuous and single-valued in a domain D and if the value of

$$F(z) = \int_a^z f(z)\,dz, \tag{22}$$

where a, z, and the contour C connecting them are in the interior of D, is independent of the contour C, then $f(z)$ is a regular analytic function of z in D.

The reader will verify that in our proof of the fact that the function defined by (12) has the derivative (13) we made no use of the analyticity of $f(z)$. All we used was the existence of the integral and this is certainly guaranteed by the assumption that $f(z)$ is single-valued and continuous in D. The function $F(z)$ in (22) has therefore, the derivative $f(z)$, which shows that $F(z)$ is a regular analytic function in D. Consequently, the same is true of its derivative $f(z)$, and Morera's theorem is proved.

Morera's theorem leads to a short proof of the following very useful result which has already been mentioned in Sec. 6, Chap. II: *If the analytic functions $f_1(z), f_2(z), \ldots f_n(z), \ldots$ are all regular in the same domain D and if the series*

$$f(z) = \sum_{n=1}^{\infty} f_n(z) \tag{23}$$

converges uniformly in D, then its sum $f(z)$ is also a regular analytic function in D. To prove this statement, we first remark that, as the sum of a uniformly convergent series of continuous functions, $f(z)$ is certainly continuous in D; hence $\int_C f(z)\,dz$, where C is a closed contour situated entirely within D, exists. Moreover, because of the uniform convergence, the integral can be found by term-by-term integration of the series (23). Indeed, because of the uniform convergence there exists, for $\epsilon > 0$, an integer $N = N(\epsilon)$ such that

$$|r_N| = \left| \sum_{n=N+1}^{\infty} f_n(z) \right| < \epsilon, \qquad z \in D.$$

Hence,

$$\left| \int_C f(z)\,dz - \sum_{n=1}^{N} \int_C f_n(z)\,dz \right| = \left| \int_C r_N\,dz \right| < \epsilon L,$$

where L is the length of C. Since ϵ may be taken arbitrarily small, it follows that

$$\int_C f(z)\,dz = \sum_{n=1}^{\infty} \int_C f_n(z)\,dz.$$

The functions $f_n(z)$ are regular in D and, by Cauchy's theorem, their integrals over the closed contour C vanish. As a result, we have also

$$\int_C f(z)\,dz = 0.$$

As shown in connection with the proof of Cauchy's theorem, this means that the integral of $f(z)$ between two arbitrary points of D is independent of the contour over which it is taken. Hence, in view of Morera's theorem, $f(z)$ is a regular analytic function in D. This completes the proof of our theorem.

EXERCISES

1. If $f(z)$ is regular in the closed circle $|z| \leq R$, show by (19) and Cauchy's theorem that

$$\frac{R^2 - |z|^2}{2\pi i} \int_C \frac{f(\zeta)\,d\zeta}{(\zeta - z)(R^2 - \bar{z}\zeta)} = \frac{1}{2\pi i} \int_C \left[\frac{1}{\zeta - z} + \frac{\bar{z}}{R^2 - \bar{z}\zeta}\right] f(\zeta)\,d\zeta = f(z),$$

where $|z| < R$ and C is the circumference $|\zeta| = R$; introducing polar coordinates $z = re^{i\varphi}$, $\zeta = Re^{i\theta}$, show that this reduces to the relation

$$f(z) = \frac{R^2 - r^2}{2\pi} \int_0^{2\pi} \frac{f(Re^{i\theta})\,d\theta}{R^2 - 2rR\cos(\theta - \varphi) + r^2},$$

which is identical with the Poisson formula of Sec. 6, Chap. I.

2. Let $f(z)$ be regular in the unit circle and let $f(0) = 1$; by evaluating the integrals

$$\frac{1}{2\pi i} \int_{|z|=1} \left[2 \pm \left(z + \frac{1}{z}\right)\right] f(z) \frac{dz}{z}$$

both by means of (21) and by introducing polar coordinates, show that

$$\frac{2}{\pi} \int_0^{2\pi} f(e^{i\theta}) \cos^2 \frac{\theta}{2}\,d\theta = 2 + f'(0), \qquad \frac{2}{\pi} \int_0^{2\pi} f(e^{i\theta}) \sin^2 \frac{\theta}{2}\,d\theta = 2 - f'(0).$$

Use this result in order to prove the following statement: If $f(z)$ is regular in $|z| \leq 1$, $f(0) = 1$ and $\text{Re}\,\{f(z)\} \geq 0$ in $|z| \leq 1$, then $-2 \leq \text{Re}\,\{f'(0)\} \leq 2$.

3. With the help of (21), prove the following statement: If $f(z)$ is regular and $|f(z)| \leq 1$ in $|z| \leq 1$, then $|f'(0)| \leq 1$.

4. The Legendre polynomial $P_n(z)$ is defined by

$$P_n(z) = \frac{1}{2^n n!} \frac{d^n}{dz^n} [(z^2 - 1)^n];$$

show that $P_n(z)$ can be represented by the contour integral

$$P_n(z) = \frac{1}{2\pi i} \int_C \frac{(\zeta^2 - 1)^n}{2^n(\zeta - z)^{n+1}}\,d\zeta,$$

where C surrounds the point z. Taking, in particular, C to be the circle of center z and radius $\sqrt{|z^2 - 1|}$, deduce Laplace's formula

$$P_n(z) = \frac{1}{\pi} \int_0^{\pi} (z + \sqrt{z^2 - 1}\cos\theta)^n\,d\theta.$$

4. The Taylor Series. The function $(\zeta - z)^{-1}$ can be expanded into a converging power series in any circle in the z-plane which does not contain the point ζ. Indeed,

$$\frac{1}{\zeta - z} = \frac{1}{\zeta - a - (z - a)} = \frac{1}{\zeta - a} \frac{1}{1 - \left(\dfrac{z - a}{\zeta - a}\right)}$$

$$= \frac{1}{\zeta - a} \sum_{n=0}^{\infty} \left(\frac{z - a}{\zeta - a}\right)^n = \sum_{n=0}^{\infty} \frac{(z - a)^n}{(\zeta - a)^{n+1}};$$

since the geometric series converges if the absolute value of the variable is smaller than 1, this expansion is valid if $\left|\dfrac{z - a}{\zeta - a}\right| < 1$, that is, if $|z - a| < |\zeta - a|$. It will therefore converge in the interior of the circle C with center a which passes through the point ζ.

Let now $f(z)$ be an analytic function which is regular for $|z - a| \leq R$. If C denotes the circumference $|\zeta - a| = R$, we have by (19)

$$f(z) = \frac{1}{2\pi i} \int_C \frac{f(\zeta)}{\zeta - z}\, d\zeta = \frac{1}{2\pi i} \int_C \left[\sum_{n=0}^{\infty} \frac{(z - a)^n}{(\zeta - a)^{n+1}}\right] d\zeta,$$

if $|z - a| < R$. The convergence of the series under the integral sign is obviously uniform with respect to ζ if ζ is on C. Hence, we may integrate term by term; this yields

$$f(z) = \sum_{n=0}^{\infty} a_n(z - a)^n, \qquad |z - a| < R, \tag{24}$$

where

$$a_n = \frac{1}{2\pi i} \int_C \frac{f(\zeta)}{(\zeta - a)^{n+1}}\, d\zeta. \tag{25}$$

We thus have proved the following theorem.

An analytic function $f(z)$ which is regular in the interior of a circle $|z - a| < R$ can be expanded there into a power series (24) *whose coefficients are given by* (25), *where C may be taken to be any closed contour surrounding a within and on which $f(z)$ is regular.*

In this statement, we have replaced the regularity of $f(z)$ in $|z - a| \leq R$ by the weaker condition that $f(z)$ be regular in $|z - a| < R$; this is permissible because for any z such that $|z - a| = r < R$ we can find a number R_1 such that $r < R_1 < R$. For $|z - a| = R_1$, $f(z)$ is regular and the above argument applies. We also remark that the replacement of the circle by the general contour C is possible because of Cauchy's theorem.

It was shown in Sec. 6, Chap. II, that every power series is a Taylor series. We therefore necessarily have

$$a_n = \frac{f^n(a)}{n!}.$$

This can also be directly verified by comparing (25) and (21). However, the formula for the remainder of the Taylor series if it is broken off after a finite number of terms takes a form which is different from that known in the real case. If we write

$$f(z) = \sum_{n=0}^{m} a_n(z - a)^n + r_m(z),$$

it follows from (24) and (25) that

$$r_m(z) = \frac{1}{2\pi i} \int_C \left[\sum_{n=m+1}^{\infty} \left(\frac{z - a}{\zeta - a} \right)^n \right] \frac{f(\zeta)}{\zeta - a} \, d\zeta.$$

Summing the series, we therefore obtain

$$r_m(z) = \frac{(z - a)^{m+1}}{2\pi i} \int_C \frac{f(\zeta) \, d\zeta}{(\zeta - a)^{m+1}(\zeta - z)}. \tag{26}$$

From (26), we can deduce a convenient estimate for $r_m(z)$. If $f(z)$ is regular in $|z - a| \leq R$, then $|f(\zeta)|$ is bounded on $|\zeta - a| = R$, that is, $|f(\zeta)| < M$. If d denotes the shortest distance from z to the circumference $|\zeta - a| = R$, it follows therefore from (26) and (3) that

$$|r_m(z)| < \frac{|z - a|^{m+1}}{R^m d} M. \tag{27}$$

As shown above, any analytic function which is regular in the interior of a circle can be expanded there into a power series. On the other hand, we have seen in Sec. 6, Chap. II, that any power series represents in its circle of convergence a regular analytic function. As a result, regular analytic functions can also be defined as those complex functions which are represented by converging power series, and it is possible to develop the whole theory of analytic functions starting from this definition. Another consequence of this identity of the regular analytic functions with the functions represented by converging power series is the following result:

On the circumference of the circle of convergence of a power series there is at least one singular point of the analytic function represented by it.

The proof follows from the remark that the power series is identical with the expansion (24) of the function represented by it. It therefore must converge in the largest circle $|z - a| < R$ in which $f(z)$ is regular. If it does not converge in any circle $|z - a| < R + \epsilon$, this is only possible because $f(z)$ is not regular in these circles. Since $f(z)$ is regular in $|z - a| < R$, there must be at least one singularity in $R \leq |z - a| < R + \epsilon$. As this is true for any arbitrarily small positive ϵ, it follows that there is at least one singularity on $|z - a| = R$. This result can also be expressed by saying that *the circumference of the circle of convergence of the expansion* (24) *of an analytic function which is regular at a passes through the singularity of $f(z)$ which is closest to a.* For instance, the radius of convergence of the power series (24) for the function log z is a, since the only finite singularity of this function is situated at the origin.

From (25), we obtain a useful estimate for the absolute value of the coefficient a_n. Taking, for convenience, $a = 0$, we assume that $f(z)$ is regular for $|z| < R$ and that $|f(z)| \leq M(r)$ for $|z| = r < R$. In view of (3), it then follows from (25) that

$$|a_n| \leq \frac{M(r)}{r^n}; \tag{28}$$

(28) is known as *Cauchy's inequality.*

More precise information regarding the magnitude of the coefficients can be obtained by means of *Parseval's identity*

$$\frac{1}{2\pi} \int_0^{2\pi} |f(re^{i\theta})|^2 \, d\theta = \sum_{n=0}^{\infty} |a_n|^2 r^{2n}, \tag{29}$$

which has also many other useful applications in the theory of functions. To prove (29), we observe that

$$|f(re^{i\theta})|^2 = f(re^{i\theta})\overline{f(re^{i\theta})} = \left(\sum_{n=0}^{\infty} a_n r^n e^{in\theta}\right)\left(\sum_{n=0}^{\infty} \bar{a}_n r^n e^{-in\theta}\right)$$
$$= \sum_{n=0}^{\infty} \sum_{m=0}^{\infty} a_n \bar{a}_m r^{n+m} \, e^{i(n-m)\theta};$$

in view of the absolute convergence of both series, the term-by-term multiplication and the rearrangement of the terms are permissible. Hence,

$$\int_0^{2\pi} |f(re^{i\theta})|^2 \, d\theta = \sum_{n=0}^{\infty} \sum_{m=0}^{\infty} a_n \bar{a}_m r^{n+m} \int_0^{2\pi} e^{i(n-m)\theta} \, d\theta,$$

where the term-by-term integration is justified by the uniform convergence of the series. The integrals in which $n \neq m$ are obviously zero. If $n = m$, the integrals reduce to $\int_0^{2\pi} d\theta = 2\pi$. Hence, the only surviving terms are $2\pi \sum_{n=0}^{\infty} |a_n|^2 r^{2n}$. This proves (29).

If $|f(z)| \leq M(r)$ for $|z| = r$, then obviously

$$\frac{1}{2\pi} \int_0^{2\pi} |f(re^{i\theta})|^2 \, d\theta \leq M^2(r).$$

Hence, by (29),

$$\sum_{n=0}^{\infty} |a_n|^2 r^{2n} \leq M^2(r). \tag{30}$$

This inequality not only proves (28) but also shows under what conditions we can have equality in (28). A comparison of (28) and (30) shows that this is only possible if all coefficients other than a_n vanish. $f(z)$ thus reduces to the form $f(z) = a_n z^n$. The reader will easily verify that in this case there is indeed equality in (28).

EXERCISES

1. With the help of (21), show that

$$\left(\frac{z^n}{n!}\right)^2 = \frac{1}{2\pi i} \int_C \frac{z^n e^{z\zeta}}{n!\zeta^n} \frac{d\zeta}{\zeta},$$

where C surrounds the origin; using this identity, prove that

$$\sum_{n=0}^{\infty} \left(\frac{z^n}{n!}\right)^2 = \frac{1}{2\pi} \int_0^{2\pi} e^{2z \cos \theta} \, d\theta.$$

2. Show that the absolute value of the error committed by substituting for the function e^z the finite sum

$$1 + z + \frac{z^2}{2!} + \cdots + \frac{z^{n-1}}{(n-1)!}$$

is not larger than

$$\frac{e^R |z|^n}{R^n(R - |z|)},$$

where R is any number satisfying $R > |z|$. For a given z, how has R to be chosen in order to make this quantity as small as possible?

3. If $f(z)$ is regular for $|z| < 1$ and $|f(z)| < \dfrac{1}{1 - |z|}$ $(|z| < 1)$, show that the coefficients a_n in the expansion $f(z) = \sum_{n=0}^{\infty} a_n z^n$ are subject to the inequality

$$|a_n| \leq (n + 1)\left(1 + \frac{1}{n}\right)^n < e(n + 1).$$

4. Applying Parseval's identity to the function

$$\frac{1}{(1-z)^2} = \sum_{n=1}^{\infty} nz^{n-1}, \qquad |z| < 1,$$

show that

$$\frac{1}{2\pi}\int_0^{2\pi} \frac{d\theta}{(1 - 2\rho\cos\theta + \rho^2)^2} = \sum_{n=1}^{\infty} n^2\rho^{2n-2} = \frac{1+\rho^2}{(1-\rho^2)^3}, \qquad 0 \leq \rho < 1.$$

5. Applying Parseval's identity to the function

$$1 + z + \cdots + z^{n-1} = \frac{z^n - 1}{z - 1},$$

deduce the formula

$$\int_0^{2\pi} \left(\frac{\sin \frac{1}{2}n\theta}{\sin \frac{1}{2}\theta}\right)^2 d\theta = 2\pi n.$$

5. Analytic Continuation. As a means of representing an analytic function, a power series has one serious drawback. Beyond the circumference of the circle about the "center" of the series which passes through the nearest singularity of the function, it ceases to converge, and it therefore does not provide any direct information regarding the behavior of the function outside the circle of convergence. For example, the series $1 + z + z^2 + \cdots + z^n + \cdots$ represents the function $f(z) = 1/(1 - z)$ for $|z| < 1$. For $|z| \geq 1$, the series diverges (its general term does not tend to zero) and becomes useless; nevertheless, the function $f(z)$ is well defined for all values of z with the exception of the point $z = 1$.

The deeper issue involved here is the question of *what constitutes a definition of an analytic function*. The answer to this question is easy if we are given what we may call a *global definition* of a particular function. By this is meant a definition which enables us to do the following: Starting from a point z_0 at which the function, or a branch of it, is regular, we draw arbitrary continuous curves; along any such curve the function is regular and its values, which vary continuously along the curve, can be uniquely determined. This process of "continuation" will only come to an end if the curve we use meets a singularity of the function; it is part of the global definition of a function to provide complete information as to where, on any given curve, we first meet a singularity of the function. Examples of global definitions are the definitions

$$\log z = \log r + i\theta, \qquad r = |z|, \ \theta = \arg z,$$

and

$$\log z = \int_1^z \frac{dz}{z}$$

for the function $\log z$. As shown in detail in Sec. 7, Chap. II, these definitions enable us to calculate the value of $\log z$ if we continue this function in the above sense along any curve starting from, say, the point $z = 1$, at which we assume $\log 1 = 0$; continuation becomes impossible only if the curve meets the origin or the point at infinity. An even simpler example of a global definition is the function $f(z) = [1/(1 - z)]$ mentioned above. For every value of z, except $z = 1$, this function is uniquely defined; along every curve starting from, say, $z = 0$, it can be uniquely continued as long as the curve does not meet the point $z = 1$. Since this function is single-valued in the whole plane, it is, of course, immaterial along which curve we proceed in order to reach a given point; only in the case of a many-valued function, such as $\log z$, does the identity of the curve along which we continue the function become important.

A power series can also give a global definition of an analytic function if the function has no finite singularities and, therefore, the series converges for all finite values of z. The function is in this case single-valued and it is completely defined by the series. However, as already pointed out, a power series with a finite radius of convergence is an entirely different affair. The above described process of finding the values of the function represented by the series along an arbitrary curve breaks down the instant the curve crosses the circumference of the circle of convergence, and we are left in the dark as far as the behavior of the function outside the circle of convergence is concerned. It therefore would appear that a power series with a finite radius of convergence is not a suitable instrument for the global definition of an analytic function.

However, this is not the case. We shall show that a function which originally is given in the form of a power series with a finite radius of convergence can also be investigated beyond the circumference of the circle of convergence and that, by a procedure termed *analytic continuation*, the local definition afforded by the power series can be extended so as to yield a global definition of the function. Suppose, for the sake of simplicity, that a function $f(z)$ is originally defined by a power series of the form $f(z) = \sum_{n=0}^{\infty} a_n z^n$ which converges in the circle $|z| < R$. We choose an arbitrary point $a \neq 0$ such that $|a| < R$, and we compute $f(a)$ and the derivatives $f'(a), f''(a), \ldots$ by means of the given power series. With these values, we set up the new power series

$$f_1(z) = \sum_{n=0}^{\infty} \frac{f^n(a)}{n!} (z - a)^n.$$

In view of the results of the preceding section, this series converges in the largest circle about a in which the function $f(z)$ is regular and gives there a representation of $f(z)$. Since $f(z)$ is regular in $|z| < R$, this series will therefore certainly converge in the interior of the largest circle about a which is still contained in $|z| < R$; its radius of convergence R will thus be at least $R - |a|$, and for $|z - a| < R - |a|$ we have $f(z) = f_1(z)$. Now there are two possibilities. Either the value of R_1 is exactly equal to $R - |a|$, or else the circle $|z - a| < R$ has a crescent-shaped part A which is not contained in the circle $|z| < R$ (see Fig. 6). In the first case, both circles have only one point in common. This point is obviously a singularity of $f(z)$; indeed, there must be at least one singularity of $f(z)$ among the points of the circumference $|z - a| = R_1$, and all these points except the one mentioned lie in the interior of the circle $|z| < R$ and are thus regular points of $f(z)$. In the second case, we say that in the region A the series $f_1(z)$ provides an analytic continuation of the function $f(z)$ which latter was originally only defined in the interior of the circle $|z| < R$. If the function $f(z)$ is also defined by some other method, that is, not by power series, and this definition is valid in a domain which contains the circles of convergence of both series, it is clear from the expansion theorem of the preceding section that, in their respective regions of convergence, both series represent the function $f(z)$. If, however, the first power series is the only definition of $f(z)$ available to us, we *define* the function $f(z)$ in the region A by the process of analytic continuation described above. It is not difficult to see that this rule leads to a global definition of the function in the sense discussed above. Let $f(z)$ be given in $|z| < R$ by its power series expansion, and trace an arbitrary continuous curve C starting from the origin. With each point z_0 of this curve such that $|z_0| < R$ we associate a *function element* $f_{z_0}(z)$, namely, the power series expansion of $f(z)$ about the point z_0. Suppose now—and this is the only interesting case—that C crosses the circumference at a point α. If the radii of convergence of the series $f_{z_0}(z)$ tend to zero if z_0 approaches α, then α is a singular point of $f(z)$. Indeed, on the circumference of each of these circles there must be at least one singular point, and the entire circumferences, including these singular points, converge toward the point α. But, as a limit point of singularities, α is itself a singularity. If $f(z)$ were regular at α, it could be expanded in a certain neighborhood of α into a converging power series; this, however, is impossible since any neighborhood of α contains singularities of $f(z)$.

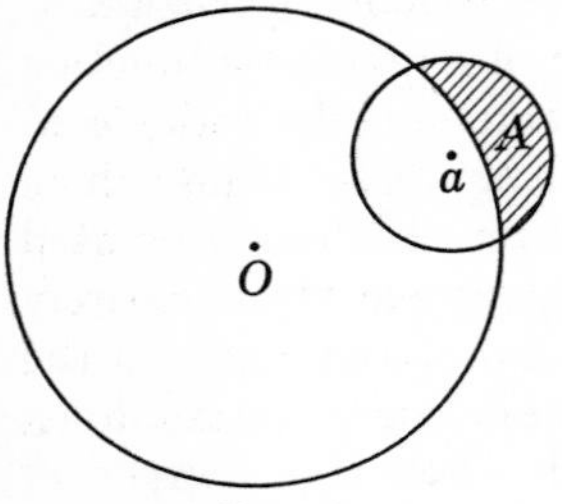

FIG. 6.

If α is not a singular point, the radii of convergence $R(z_0)$ of the function elements $f_{z_0}(z)$ will therefore not tend to zero if z_0 tends to α. If $|z_0| < R$

and z_0 is close enough to α, the circle about z_0 of radius $R(z_0)$ will therefore contain a finite portion of C which is not situated in $|z| < R$. In the domain in which this circle and the circle $|z| < R$ overlap, $f(z)$ and $f_{z_0}(z)$ coincide; we therefore say that along the part of C which protrudes from $|z| < R$ but is situated in $|z - z_0| < R(z_0)$, $f(z)$ is continued analytically by $f_{z_0}(z)$. In order to continue our function beyond the point β ($|\beta| > R$) at which C intersects $|z - z_0| = R(z_0)$, we now employ the same procedure which we used with respect to the point α. Again, there are two possibilities. Either β is a singular point of the function, or there exists an analytic continuation of $f(z)$ along a finite portion of C which is not situated in the circle $|z - z_0| < R(z_0)$. By repeated application of this process we obtain a chain of circles which, together, cover the curve C up to a certain point η. Each of these circles has a domain in common with the two adjacent circles, and wherever two circles overlap, the function elements belonging to these two circles take identical values. As a result, the analytic continuation of $f(z)$ along C is uniquely determined up to the point η. If the point η is a singularity of the function, analytic continuation beyond η becomes impossible; if η is a regular point, then $f(z)$ can be continued beyond η in the manner described before. However, before we conclude from this that $f(z)$ can be continued along C indefinitely unless C meets a singularity of $f(z)$, we have to rule out another possibility. It may happen that in spite of an infinite repetition of the above process we do not cover more than a finite portion of the curve C. Although each circle contributes a crescent-shaped domain which was not contained in the preceding circle, the radii of these circles may tend to zero and there may exist a point η on C such that, after a certain number of steps, all circles are contained in a given neighborhood of η. This, however, is only possible if η is a singularity of $f(z)$. Since on the circumference of each circle there is at least one singularity of the function, and since every neighborhood of η contains an infinity of circles, it follows that η is the limit point of an infinity of singularities of $f(z)$. As shown above, this means that η itself is a singularity of $f(z)$. We have thus proved that *along any continuous curve C we can continue $f(z)$ analytically; this process is interrupted only if we reach a singularity of $f(z)$ which is situated on C.* The same process of analytic continuation may, of course, also be applied in the case in which the function $f(z)$ is originally not given in the form of a power series but by some other method which does not provide a global definition of the function.

From now on, whenever we refer to "the analytic function $f(z)$" without further specification of its domain of definition, we shall mean *the totality of all function elements which may be obtained from a given function element of $f(z)$ by analytic continuation.* Since one function element is sufficient in order to determine all its analytic continuations in a unique manner, it follows that two analytic functions are completely identical if they agree in one function element. From this fact we can draw the following remarkable conclusion:

If $f(z_n) = g(z_n)$, $n = 1, 2, \ldots$, $z_n \neq z_0$, *and* $\lim_{n\to\infty} z_n = z_0$, *and if both* $f(z)$ *and* $g(z)$ *are regular at* z_0, *then* $f(z) \equiv g(z)$.

If $h(z) = f(z) - g(z)$, it is plainly sufficient to show that $h(z)$ must be identically zero if $h(z_n) = 0$, $n = 1, 2, \ldots$, and if $h(z)$ is regular at z_0. Because of the regularity of $h(z)$ at z_0 we have an expansion

$$h(z) = a_0 + a_1(z - z_0) + a_2(z - z_0)^2 + \cdots$$

which converges in a certain circle about z_0. $h(z)$ is continuous at z_0; hence, $h(z_0) = \lim_{n\to\infty} h(z_n) = 0$. It follows that the coefficient a_0 is zero. We may now write

$$h(z) = (z - z_0)[a_1 + a_2(z - z_0) + \cdots].$$

Since $h(z_n) = 0$ and $z_n \neq z_0$, the expression in the bracket vanishes for z_n, $n = 1, 2, \ldots$. Because of its continuity, we have $a_1 = 0$. Continuing in this fashion, we find successively that all coefficients vanish. In its circle of convergence, the function $h(z)$ is therefore identically zero. Since the constant zero is an analytic function defined in the whole z-plane, and since one function element of $h(z)$ is equal to zero, it follows therefore that the function $h(z)$ is identically zero.

An important consequence of the last result is the so-called principle of the *permanence of a functional equation*. Before we give a formal statement of this principle, we shall illustrate it by two examples. From elementary trigonometry it is known that the real function $\sin x$ has the addition theorem

$$\sin (x + u) = \sin x \cos u + \cos x \sin u,$$

where u is an arbitrary real value. Since $\sin z$, $\cos z$, and $\sin (z + u)$ are analytic functions which are regular for all finite values of z, and since the relation

$$\sin (z + u) = \sin z \cos u + \cos z \sin u$$

is satisfied if z is any point on the real axis, it follows by analytic continuation that the same relation must hold for all finite values of z. If we repeat the same argument with respect to the real variable u, we find that u may be replaced by a complex variable w without invalidating the relation in question. Hence, the addition theorem of the function $\sin z$ is true for arbitrary complex values of z and w. Another important example for the principle of permanence of a functional equation is afforded by functions satisfying differential equations. To take a particularly simple case, we consider the function $f(z) = \log(1 + z)$ which, for $|z| < 1$, is represented by the power series

$$f(z) = z - \frac{z^2}{2} + \frac{z^2}{3} - \cdots .$$

Its derivative is

$$f'(z) = 1 - z + z^2 - z^3 + \cdots = \frac{1}{1 + z}.$$

Although the identity

$$f'(z) = \frac{1}{1 + z}$$

has thus only been proved for $|z| < 1$, it follows that it must hold for all analytic continuations of the two functions involved. The "analytic function $\log(1 + z)$," in the sense defined above, has therefore everywhere the derivative $1/(1 + z)$.

We now formulate the general principle of the permanence of a functional equation.

Let $F(p,q,r)$ be an analytic function of the three variables p,q,r, which is regular for all finite values of p,q,r, and let $f(z)$ and $g(z)$ be analytic functions of z. If a relation of the form $F[f(z),g(z),z]$ holds between a function element of $f(z)$ and a function element of $g(z)$, this relation is also true for all analytic continuations of these function elements.

The proof of this principle is left to the reader. We remark that it may easily be generalized to functional equations involving more than two functions.

In connection with the problem of analytic continuation, the following remarks concerning the singularities of analytic functions are of interest. In all the examples of analytic functions considered so far, the singularities were isolated points. It is, however, easy to construct functions for which this is not the case. Consider, for example, the function

$$f(z) = \frac{1}{\sin(1/z)}.$$

For $(1/z) = \pi n$, where n is an integer, the denominator vanishes; hence, $f(z)$ is not continuous there and the points $z = (1/\pi n)$ are singular points of the function $f(z)$. These points converge to the point $z = 0$, which is therefore also a singular point of $f(z)$, but clearly not an isolated one. It is furthermore possible for the singular points of a function to fill an entire arc of a continuous curve; in this case, we speak of a *singular line* of the function. A particularly interesting situation presents itself if a function $f(z)$ has a closed singular line. If a function element of $f(z)$ is known in the interior of the domain bounded by the singular line C, then it is obviously impossible to continue $f(z)$ analytically across C. The entire domain of definition of $f(z)$ is therefore the interior of C, and we say that C is a *natural boundary* of $f(z)$. Such an occurrence is not so unusual as it may seem. Consider, for instance, the analytic function $f(z)$ defined by the power series

$$f(z) = z + z^2 + z^4 + z^8 + \cdots = \sum_{n=0}^{\infty} z^{2^n}.$$

By Cauchy's test, the circle of convergence of this series is $|z| < 1$. By the results of the preceding section, $f(z)$ must therefore have at least one singularity on $|z| = 1$. For the sake of simplicity, we shall assume that this singularity is situated at the point $z = 1$; a different location will cause a minor change of the argument which the reader will be able to carry out without difficulty. From the definition of $f(z)$ it follows that

$$f(z^2) = z^2 + z^4 + z^8 + \cdots = \sum_{n=1}^{\infty} z^{2^n} = f(z) - z.$$

By the principle of permanence, the functional equation $f(z) = z + f(z^2)$ is true for all analytic continuations of $f(z)$. Let now z tend to the point -1. In view of $f'(z) = 1 + 2zf'(z^2)$, it is clear that $f(z)$ cannot have a derivative at $z = -1$, because from $f'(-1) = 1 - 2f'(1)$ it would follow that $f'(1)$ exists; but this would imply that $z = 1$ is a regular point of $f(z)$, which is false. Thus, $z = -1$ is also a singular point of $f(z)$. In the same way, it follows from $f(z) = z + f(z^2) = z + z^2 + f(z^4)$ that the points z for which $z^4 = 1$ are singularities of $f(z)$; if this were not true, $z = 1$ would be a regular point. Continuing in this fashion, we conclude from $f(z) = z + z^2 + \cdots + z^{2^{n-1}} + f(z^{2^n})$ that all points z for which $z^{2^n} = 1$ are singularities of $f(z)$. But these are the points $e^{\frac{2\pi i}{2^n}}$ which divide the circumference $|z| = 1$ into 2^n equal parts. Since, for $n \to \infty$, all points on $|z| = 1$ are limits of these points, and since the limit point of singular points is also a singularity, it follows therefore that all points on $|z| = 1$ are singular points of $f(z)$. In other words, the unit circle is the natural boundary of the analytic function $f(z)$.

We close this section with the proof of an important result which is known as the *monodromy theorem*.

An analytic function $f(z)$ which is regular in a simply-connected domain D is also single-valued in D.

If z_1 and z_2 are any two points of D and if C' and C'' are two different curves connecting z_1 and z_2 which are entirely contained in D, then the monodromy theorem asserts that the analytic continuation of the function $f(z)$ along C' and C'' yields identical results. If the theorem were false, there would exist two points z_1, z_2 in D and two curves C' and C'' connecting them, such that the continuation of the same function elements $f_{z_1}(z)$ along these curves leads to two different values of $f(z_2)$. Since D is simply-connected, the points of the domain D_1 bounded by C' and C'' are also points of D, and $f(z)$ may be analytically continued along any curve in D_1 which connects z_1 and z_2. For any given positive ϵ, we can draw a finite family of curves R in D_1 which has the following properties: (1) R contains C' and C''; (2) all curves of R connect z_1 and z_2; (3) any curve of R has two neighboring curves whose distance from it is less than ϵ, with the exception of C' and C'' which have only one such neighboring curve. By saying that the distance between two curves is less than ϵ we mean that for any given point on one curve we can find a point on the other curve such that the distance of the two points is less than ϵ. We now consider the analytic continuation of $f(z)$ along all the curves of R. Obviously, there must exist two neighboring curves of R such that continuation of $f(z)$ along them yields different results. If this were not so, continuation along all curves of R would lead to the same value of $f(z_2)$, and this contradicts our assumption regarding the continuation along C' and C''.

From the assumption that the monodromy theorem is false it thus follows that there must exist two curves, say C_1 and C_2, whose distance is less than any given positive ϵ, such that continuation of $f(z)$ along C_1 and C_2 leads to different values of $f(z_2)$. But this involves a contradiction. Indeed, the analytic continuation of $f(z)$ along C_1 is carried out by means of a chain of circles S which, in view of the fact that $f(z)$ is regular on C_1 and its end points, can be taken to be finite. Hence, there exists a positive ϵ such that any circle of radius ϵ whose center is on C_1 is contained in S. It follows that a curve C_2 whose distance from C_1 is less than ϵ is also contained in S. But this means that the analytic continuation of $f(z)$ along C_2 can be carried out by means of the same chain of circles S that was used for continuation along C_1. Hence, continuation along these two curves must lead to the same result. This completes the proof of the monodromy theorem.

EXERCISES

1. The gamma function is defined, for values of z such that $\operatorname{Re}\{z\} > 0$, by the integral

$$\Gamma(z) = \int_0^\infty e^{-t}t^{z-1}\,dt,$$

where t is taken along the positive axis and $t^{z-1} = e^{(z-1)\log t}$ is formed with the real value of $\log t$. Show that $\Gamma(z)$ is an analytic function which is regular for $\text{Re}\,\{z\} > 0$. Integrating by parts, deduce the functional equation $z\Gamma(z) = \Gamma(1 + z)$. Using the principle of permanence, show that $\Gamma(z)$ can be continued analytically throughout the z-plane and that its only singularities are situated at the points $z = 0, -1, -2, -3, \ldots$.

2. Show that the sum of the series

$$\frac{1}{1-z} + \frac{z}{z^2-1} + \frac{z^2}{z^4-1} + \frac{z^4}{z^8-1} + \cdots$$

is 1 when $|z| < 1$, but is 0 when $|z| > 1$. This seems to contradict our results on analytic continuation. How is this possible?

3. By the example of the function $f(z) = \log z$ and the domain $1 < |z| < 2$, show that the monodromy theorem cannot be true for multiply-connected domains.

6. The Laurent Series. Consider an analytic function $f(z)$ which is regular and single-valued in the circular ring $0 < \alpha < |z - a| < \beta$. If z is an arbitrary given point in the interior of the ring, we can choose a positive ϵ such that $\alpha + \epsilon < |z - a| < \beta - \epsilon$. If we denote the circles $|z - a| = \beta - \epsilon$ and $|z - a| = \alpha + \epsilon$ by C_1 and C_2, respectively, then $f(z)$ is regular and single-valued in the closure of the circular ring bounded by C_1 and C_2. Since this ring contains the point z, Cauchy's integral formula yields

$$(31) \qquad f(z) = \frac{1}{2\pi i}\int_{C_1}\frac{f(\zeta)}{\zeta - z}\,d\zeta - \frac{1}{2\pi i}\int_{C_2}\frac{f(\zeta)}{\zeta - z}\,d\zeta;$$

indeed, C_1 and C_2 constitute the entire boundary of the ring, and the negative sign of the second integral takes account of the fact that the negative direction on the circle C_2 coincides with the positive direction with respect to the ring. In the first integral we have

$$|\zeta - a| = \beta - \epsilon > |z - a|.$$

Hence

$$\frac{1}{\zeta - z} = \frac{1}{\zeta - a}\,\frac{1}{1 - \dfrac{z-a}{\zeta-a}} = \sum_{n=0}^{\infty}\frac{(z-a)^n}{(\zeta-a)^{n+1}},$$

where the geometric series converges because of $|z - a| < |\zeta - a|$. Since, on C_1, this convergence is uniform with respect to ζ, we may integrate term by term. We obtain

$$(32) \qquad f_1(z) = \frac{1}{2\pi i}\int_{C_1}\frac{f(\zeta)}{\zeta - z}\,d\zeta = \sum_{n=0}^{\infty} a_n(z-a)^n,$$

where

$$a_n = \frac{1}{2\pi i} \int_{C_1} \frac{f(\zeta)}{(\zeta - a)^{n+1}} d\zeta.$$

On C_2, we have $|\zeta - a| = \alpha + \epsilon < |z - a|$. In this case, we write

$$\frac{1}{\zeta - z} = -\frac{1}{z - a} \frac{1}{1 - \dfrac{\zeta - a}{z - a}} = -\sum_{n=1}^{\infty} \frac{(\zeta - a)^{n-1}}{(z - a)^n},$$

and the geometric series converges because now $|\zeta - a| < |z - a|$. Hence, by integration,

$$f_2(z) = -\frac{1}{2\pi i} \int_{C_2} \frac{f(\zeta)}{\zeta - z} d\zeta = \sum_{n=1}^{\infty} \frac{a_{-n}}{(z - a)^n}, \tag{33}$$

where

$$a_{-n} = \frac{1}{2\pi i} \int_{C_2} (\zeta - a)^{n-1} f(\zeta)\, d\zeta.$$

We note that in the integrals defining a_n and a_{-n} we may replace the circles C_1 and C_2 by any closed curve which can be obtained from them by continuous deformation within the ring; this follows from the fact that both integrands are regular in the ring. In particular, we may replace both C_1 and C_2 by the same circle C which is concentric with them and situated in the interior of the ring. We have thus proved the following result.

If $f(z)$ is regular and single-valued in the circular ring $\alpha < |z - a| < \beta$, it can be expanded there into a series of the form

$$f(z) = \sum_{n=-\infty}^{\infty} a_n (z - a)^n, \tag{34}$$

where

$$a_n = \frac{1}{2\pi i} \int_C \frac{f(\zeta)}{(\zeta - a)^{n+1}} d\zeta, \qquad n = 0, \pm 1, \pm 2, \ldots, \tag{35}$$

and C is a circle $|\zeta - a| = r$, $\alpha < r < \beta$.

We may also say that the function $f(z)$ can be written as the sum of the two functions $f_1(z)$ and $f_2(z)$, defined by (32) and (33), which can be expanded into series proceeding by positive and negative powers of $z - a$, respectively. (32) is an ordinary power series; as such, it will converge in the interior of the circle $|z - a| < r$ if it converges at a point z_0 for which $|z_0 - a| = r$. It follows that the function $f_1(z)$ is regular for $|z - a| < \beta$. Similarly, (33) is a power series of the variable $(z - a)^{-1}$; if it converges at z_0, it converges for all values of z such that $|z - a|^{-1} < |z_0 - a|^{-1}$, that is, $|z - a| > |z_0 - a|$. (33) will therefore converge in the exterior of the circle $|z - a| > \alpha$ and represent there a regular function. Hence, *if $f(z)$ is regular for $\alpha < |z - a| < \beta$, then $f(z) = f_1(z) + f_2(z)$, where $f_1(z)$ and $f_2(z)$ are regular for $|z - a| < \beta$ and $|z - a| > \alpha$, respectively.* It should be noted that the domain of regularity of $f_2(z)$ includes the point at infinity. Generally speaking, a function $g(z)$ is said to be regular at $z = \infty$ if $g(1/z)$ is regular at $z = 0$.

If $|z - a| = \alpha$ and $|z - a| = \beta$ are the boundaries of the region of convergence of the Laurent series, it is clear that on both these circumferences there must be at least one singular point of $f(z)$. A particularly interesting case is obtained if $f(z)$ has only one singular point in $|z - a| < \beta$ and this point is situated at $z = a$. Clearly, the series (34) will then converge for $0 < |z - a| < \beta$. We now distinguish between two cases. In the first case, only a finite number of the coefficients $a_{-1}, a_{-2}, \ldots$ are different from zero, and the expansion (34) takes the form

$$(36) \quad f(z) = \frac{a_{-m}}{(z - a)^m} + \cdots + \frac{a_{-1}}{(z - a)} + a_0 + a_1(z - a) + \cdots, \qquad a_{-m} \neq 0;$$

in the second case there is an infinite number of nonvanishing coefficients with negative subscripts. If the expansion of $f(z)$ is of the form (36), we say that $f(z)$ has a *pole of order* m at $z = a$; in the second case, we say that $z = a$ is an *essential singularity of* $f(z)$. This terminology can also be applied to the case in which the singularity is situated at $z = \infty$. Since z^{-1} tends to zero if z tends to infinity, a pole of order m at $z = \infty$ will correspond to an expansion

$$f(z) = a_{-m}z^m + \cdots + a_{-1}z + \sum_{n=0}^{\infty} \frac{a_n}{z^n},$$

while an isolated essential singularity at $z = \infty$ will give rise to an infinity of different positive powers of z. Any power series of the form $\sum_{n=0}^{\infty} a_n z^n$ which converges for all finite values of z and which has an infinity of nonzero coefficients will therefore represent a function with an essential singularity at $z = \infty$.

The behavior of an analytic function in the neighborhood of a pole is easy to describe. It is clear from (36) that $f(z) \to \infty$ if $z \to a$, regardless of the path along which a is approached. It further follows from (36) that

$$g(z) = (z - a)^m f(z) = a_{-m} + a_{-(m-1)}(z - a) + \cdots$$

is regular in the neighborhood of $z = a$ and that $g(a) \neq 0$. Now, if $g(z)$ is regular in the vicinity of $z = a$ and if $g(a) \neq 0$, the function $h(z) = [1/g(z)]$ will also be regular in a certain neighborhood of a. To show this, we first remark that there is a circle $|z - a| < \gamma$ in which $g(z) \neq 0$. If this were not the case, the point $z = a$ would be a limit point of zeros of $g(z)$, that is, of points z_n for which $f(z_n) = 0$. Since $g(z)$ is regular at $z = a$, $g(z)$ would therefore, as shown in the preceding section, reduce to the constant 0; but this is impossible since $g(a) \neq 0$. For $|z - a| < \gamma$, the function will therefore be continuous and its derivative is found to be

$$h'(z) = -\frac{g'(z)}{g^2(z)}.$$

$h(z)$ is thus indeed a regular function in a certain neighborhood of a. It follows that the function $[(z - a)^m f(z)]^{-1}$ has there a converging Taylor expansion

$$\frac{1}{(z - a)^m f(z)} = b_0 + b_1(z - a) + b_2(z - a)^2 + \cdots;$$

hence,

$$\frac{1}{f(z)} = b_0(z - a)^m + b_1(z - a)^{m+1} + \cdots.$$

This shows that if $f(z)$ has a pole of order m at $z = a$, then $1/f(z)$ is regular at $z = a$ and has there a zero of order m. Here, a zero of order m is defined as a point at which both the function and its derivatives up to the $(m - 1)$st order vanish. Obviously, the converse of the italicized statement is also true: if $1/f(z)$ has a zero of order m, then $f(z)$ has a pole of order m. It is also worth noting that a pole is always an isolated singularity; this is an immediate consequence of the fact that the zeros of a regular function are isolated.

The behavior of a function in the neighborhood of an isolated essential singularity is vastly different from the behavior characteristic of a pole. An idea of the complicated character of a function near an essential singularity is given by the following theorem:

In any neighborhood of an isolated essential singularity, an analytic function approaches any given value arbitrarily closely.

To prove this theorem, we first remark that in the neighborhood of an isolated singular point a function cannot be bounded. Indeed, if $|f(z)| < M$ for $|z - a| \leq r$, then, by (35),

$$|a_{-n}| = \left| \frac{1}{2\pi i} \int_{|z-a|=r} (\zeta - z)^{n-1} f(\zeta)\, d\zeta \right| \leq Mr^n, \qquad n = 1, 2, \ldots .$$

Since r may be taken as small as we please, it follows that

$$a_{-1} = a_{-2} = \cdots = 0.$$

All the coefficients of the negative powers vanish, and the Laurent series reduces to a Taylor series, which shows that $f(z)$ is regular at $z = a$. Suppose now that, for $|z - a| < \epsilon$, $|f(z) - \gamma|$ does not become arbitrarily small. In this case, there exists a constant M such that

$$\left| \frac{1}{f(z) - \gamma} \right| < M.$$

The only possible singularity of the function $[f(z) - \gamma]^{-1}$ in $|z - a| < \epsilon$ is $z = a$. But this function is bounded there, and, as we have just seen, this means that $z = a$ is not a singularity. Hence, $[f(z) - \gamma]^{-1}$ is regular for $|z - a| < \epsilon$. Its reciprocal, that is, $f(z) - \gamma$, is therefore either regular in $|z - a| < \epsilon$ or else has a pole there. But this contradicts our assumption that $f(z)$ has an essential singularity at $z = a$. Since γ was an arbitrary value, this proves our theorem. We shall show at a later stage (Sec. 6, Chap. VI) that even more is true. In the neighborhood of an isolated essential singularity, an analytic function actually takes any arbitrary value, with one possible exception. That such an exception can occur is shown by the function $f(z) = e^z$ which has an essential singularity at $z = \infty$ but nevertheless never takes the value 0.

Returning to poles, we shall now derive the following result.

A rational function has no other singularities than poles; conversely, an analytic function which has no other singularities than poles is necessarily a rational function.

A rational function $f(z)$ is of the form

$$f(z) = \frac{P(z)}{Q(z)}, \tag{37}$$

where

$$P(z) = \alpha_0 + \alpha_1 z + \cdots + \alpha_n z^n$$

and

$$Q(z) = \beta_0 + \beta_1 z + \cdots + \beta_m z^m$$

are polynomicals of degree n and m, respectively. $P(z)$ and $Q(z)$ are regular at all finite points of the plane. We may further assume that $P(z)$ and $Q(z)$ have no common zeros. If z_0 is a zero of $P(z)$, then it follows from its Taylor expansion about z_0 that $P(z) = (z - z_0)^{k_1}P_1(z)$, where $P_1(z)$ is a polynomial, $P_1(z_0) \neq 0$ and k_1 is a positive integer. Similarly, it follows from $Q(z_0) = 0$ that $Q(z) = (z - z_0)^{k_2} Q_1(z)$, $Q_1(z_0) \neq 0$. In view of (37), it is therefore always possible, by canceling a suitable number of the $(z - z_0)$-factors, to write $f(z)$ as the quotient of two polynomials with no common zeros. Obviously, the only possible finite singularities are situated at the zeros of $Q(z)$. Since the zeros of $P(z)$ do not coincide with those of $Q(z)$, it follows that $1/f(z)$ is regular at these points. As shown before, such points can only be poles but not essential singularities. We have thus proved that all finite singularities of $f(z)$ are poles. At $z = \infty$, we have to distinguish between two cases, according as $n > m$ or $n \leq m$. If $n > m$, we have

$$f(z) = z^{n-m}\left(\frac{\alpha_n + \dfrac{\alpha_{n-1}}{z} + \cdots + \dfrac{\alpha_0}{z^n}}{\beta_m + \dfrac{\beta_{m-1}}{z} + \cdots + \dfrac{\beta_0}{z^m}}\right)$$

$$= z^{n-m}\left(\gamma_0 + \frac{\gamma_1}{z} + \frac{\gamma_2}{z^2} + \cdots\right),$$

where the power series in $1/z$ converges in some neighborhood $\left|\dfrac{1}{z}\right| < \epsilon$ of $z = \infty$. Hence,

$$f(z) = \gamma_0 z^{n-m} + \gamma_1 z^{n-m-1} + \cdots,$$

which shows that $f(z)$ has a pole of order $n - m$ at $z = \infty$. If $n \leq m$, the reader will confirm in a similar fashion that $f(z)$ is regular at $z = \infty$. This proves the first half of our theorem. In order to prove the converse, suppose that all the singularities of a function $f(z)$ are poles at the points $a_1, a_2, \ldots, a_n$; the orders of these poles may be denoted by $m_1, m_2, \ldots, m_n$, respectively. In the vicinity of the point a_ν, the function $f(z)$ has an expansion of the form

$$f(z) = \frac{A_{-m_\nu}{}^{(\nu)}}{(z - a_\nu)^{m_\nu}} + \cdots + \frac{A_{-1}{}^{(\nu)}}{(z - a_\nu)} + \sum_{\mu=0}^{\infty} b_\mu{}^{(\nu)}(z - z_0)^\mu.$$

This can be written in the form

$$f(z) = p_\nu(z) + r_\nu(z),$$

where

$$p_\nu(z) = \frac{A_{-m_\nu}{}^{(\nu)}}{(z - a_\nu)^{m_\nu}} + \cdots + \frac{A_{-1}{}^{(\nu)}}{(z - a_\nu)} \tag{38}$$

is the so-called *principal part* or *meromorphic part* of $f(z)$ at the pole $z = a$, and $r_\nu(z)$ is regular at $z = a$. If a_ν is the point at infinity, the principal part of $f(z)$ belonging to this point will be of the form

$$p_\nu(z) = A_{m_\nu}{}^{(\nu)} z^{m_\nu} + \cdots + A_1 z. \tag{39}$$

We now consider the expression

$$g(z) = f(z) - p_1(z) - p_2(z) - \cdots - p_n(z).$$

Since $f(z) - p_\nu(z)$ is regular at $z = a_\nu$ and $p_\nu(z)$ is regular everywhere except at $z = a_\nu$, it follows that $g(z)$ is regular at all points of the plane, including the point at infinity. Such a function is necessarily a constant. Indeed, since $g(z)$ is regular and single-valued everywhere, it must have a Taylor expansion $g(z) = \sum_{n=0}^{\infty} \gamma_n z^n$ which converges for all finite z. But, as shown before, such a power series has a singularity at $z = \infty$ unless all its coefficients except γ_0 vanish. Since $g(z)$ is regular at $z = \infty$, we have therefore identically $g(z) \equiv \gamma_0$, whence

$$f(z) = \gamma_0 + \sum_{\nu=1}^{n} p_\nu(z). \tag{40}$$

In view of (38), $f(z)$ can therefore be brought into the form (37). This completes the proof of our theorem. As a by-product we have also obtained the following result.

Any rational function (37) *possesses a decomposition* (40) *into partial fractions of the form* (38)*; if the degree of the numerator exceeds that of the denominator, then* (40) *includes a term of the form* (39).

EXERCISES

1. The Bessel function $J_n(z)$ is defined as the nth coefficient ($n \geq 0$) of the "generating function"

$$e^{\frac{z}{2}\left(\zeta - \frac{1}{\zeta}\right)} = \sum_{n=-\infty}^{\infty} J_n(z)\zeta^n.$$

Prove that $J_n(z)$ can be represented by the formula

$$J_n(z) = \frac{1}{\pi}\int_0^\pi \cos(n\theta - z\sin\theta)\,d\theta.$$

Use your result in order to find the power series expansion

$$J_n(z) = \sum_{\nu=0}^{\infty} \frac{(-1)^\nu \left(\frac{z}{2}\right)^{n+2\nu}}{\nu!(n+\nu)!}.$$

Deduce the same expansion by writing the generating function in the form

$$e^{\frac{z\zeta}{2}}e^{-\frac{z}{2\zeta}}$$

and computing the product of the two exponential series.

2. Show that, for $1 < |z| < 2$,

$$\frac{1}{(z-1)(2-z)} = \sum_{n=1}^{\infty} \frac{1}{z^n} + \sum_{n=0}^{\infty} \frac{z^n}{2^{n+1}}.$$

3. For $0 < \rho \leq |z| \leq 1$, the function $f(z)$ is regular and single-valued, and it satisfies the inequality $\text{Re}\,\{f(z)\} \geq 0$. If

$$f(z) = \sum_{n=1}^{\infty} \frac{a_{-n}}{z^n} + 1 + \sum_{n=1}^{\infty} a_n z^n, \qquad \rho < |z| < 1,$$

is its Laurent expansion, show that

$$2 - a_{-n} - a_n = \frac{1}{2\pi i}\int_{|z|=1}\left[2 - \zeta^n - \frac{1}{\zeta^n}\right] f(\zeta)\,\frac{d\zeta}{\zeta} = \frac{1}{\pi}\int_0^{2\pi} f(e^{i\theta}) \sin^2\frac{\theta}{2}\,d\theta,$$

$$\begin{aligned} 2\rho^n + a_{-n} + a_n\rho^{2n} &= \frac{\rho^n}{2\pi i}\int_{|z|=\rho}\left[2 + \left(\frac{\zeta}{\rho}\right)^n + \left(\frac{\rho}{\zeta}\right)^n\right] f(\zeta)\,\frac{d\zeta}{\zeta} \\ &= \frac{\rho^n}{\pi}\int_0^{2\pi} f(\rho e^{i\theta}) \cos^2\frac{\theta}{2}\,d\theta. \end{aligned}$$

By adding these two identities and observing that $\text{Re}\,\{f(z)\} \geq 0$, prove that $2(1 + \rho^n) - (1 - \rho^{2n})\,\text{Re}\,\{a_n\} \geq 0$, and hence

$$\text{Re}\,\{a_n\} \leq \frac{2}{1-\rho^n}, \qquad n = 1, 2, \ldots.$$

Verify that the function $f_1(z) = f(e^{i\alpha}z)$ $(0 \leq \alpha < 2\pi)$ satisfies the same conditions as $f(z)$, and use this fact in order to deduce the more general inequality

$$|a_n| \leq \frac{2}{1-\rho^n}, \qquad n = 1, 2, \ldots.$$

Show that $f_2(z) = f\left(\frac{\rho}{z}\right)$ satisfies the same hypotheses as $f(z)$ and that therefore

$$|a_{-n}| \leq \frac{2\rho^n}{1-\rho^n}, \qquad n = 1, 2, \ldots.$$

7. Liouville's Theorem. The following theorem, due to Liouville, is as important as it is easy to prove.

An analytic function cannot be single-valued, regular, and bounded at all finite points of the plane unless it reduces to a constant.

If $f(z)$ is single-valued and regular for all finite values of z and if everywhere $|f(z)| < M$, then, by (20),

$$|f'(z)| = \left|\frac{1}{2\pi i}\int_{|\zeta|=R} \frac{f(\zeta)}{(\zeta - z)^2}\,d\zeta\right| < \frac{MR}{(R-|z|)^2}, \qquad |z| < R,$$

where R is an arbitrary positive number. By choosing R large enough, we can obviously make the right-hand side of this inequality as small as we please. Hence, $f'(z) = 0$ for all finite values of z, and $f(z)$ must be a constant.

As an application of Liouville's theorem, we prove the *fundamental theorem of algebra.* This theorem states that an algebraic equation of the form $a_0 + a_1z + \cdots + a_nz^n = 0$ $(a_n \neq 0)$ has exactly n finite roots. We first remark that it is sufficient to demonstrate the existence of only one root of the equation. Indeed, if $P(z)$ is the polynomial $a_0 + \cdots + a_nz^n$ and if $P(z_1) = 0$, then, as shown in the preceding section,

$$P(z) = (z - z_1)P_1(z),$$

where $P_1(z)$ is a polynomial of degree $n - 1$. If $P_1(z)$, in turn, has one root z_2, we may conclude that $P_1(z) = (z - z_2)P_2(z)$, where $P_2(z)$ is now a polynomial of degree $n - 2$. Continuing in this fashion, we find that $P(z)$ will have n roots if it can be shown that every polynomial must have at least one root. In order to prove the latter, we consider the function

$$f(z) = \frac{1}{P(z)}.$$

$f(z)$ is a rational function whose only singularities are situated at the zeros of $P(z)$; $z = \infty$ is clearly a regular point of $f(z)$. If $P(z)$ had no zeros, $f(z)$ would therefore be regular in the whole plane, including at $z = \infty$. But this means that $f(z)$ must be bounded everywhere; indeed if there existed a sequence of point $a_1, a_2, \ldots,$ converging to a point a such that $\lim_{\nu = \infty} f(a_\nu) = \infty$, this would contradict the fact that $f(z)$ is regular at a and that therefore $f(a) = \lim_{\nu \to \infty} f(a_\nu)$. Hence, by Liouville's theorem, $f(z)$ must be a constant. But the conclusion

$$P(z) = a_0 + a_1z + \cdots + a_nz^n = \text{const.}$$

is absurd since $P^{(n)}(z) = n!a_n \neq 0$. Hence, $P(z)$ must have at least one zero.

An analytic function which is regular and single-valued at all finite points of the plane is called an *entire function.* Liouville's theorem is therefore equivalent to the statement that *a bounded entire function is necessarily a constant.* The simplest entire functions are polynomials. An entire function which is not a polynomial is called a *transcendental entire function;* examples of such functions are e^z, $\cos z$, $\sin z$. A transcendental entire function has necessarily an essential singularity at $z = \infty$. In the first place, it must have a singularity there, since otherwise it would have no singularities at all and would thus reduce to a constant. But this singularity cannot be a pole unless the function reduces to a polynomial.

By a suitable modification of the proof of Liouville's theorem we can derive the following more general result. *If $f(z)$ is an entire function and, for all z, $|f(z)| < A|z|^m$, where A is a positive constant and $m > 0$, then $f(z)$ is a polynomial of degree not exceeding m.* By (21), we have

$$|f^{(n)}(z)| = \left| \frac{n!}{2\pi i} \int_{|z|=R} \frac{f(\zeta)}{(\zeta - z)^{n+1}} \, d\zeta \right| < An! \frac{R^{m+1}}{(R - |z|)^{n+1}}, \qquad R > |z|.$$

If $n > m$, the last term evidently tends to zero if $R \to \infty$. Hence, the nth derivative of $f(z)$ is identically zero and $f(z)$ can only be a polynomical whose degree does not exceed $n - 1$.

EXERCISES

1. Prove both Liouville's theorem and its generalization by means of Cauchy's inequality.

2. If $n > 0$ and x is on the positive axis, show that $\lim_{x \to \infty} x^n e^{-x} = 0$.

3. Use the Poisson formula of Chap. I, Sec. 6 (see also Exercise **1**, Sec. 3 of this chapter) to derive the following result, analogous to Liouville's theorem: A function $u(x,y)$ which is harmonic and bounded at all points of the xy-plane is a constant.

8. The Maximum Principle. If the function $f(z)$ is regular and single-valued in a domain D, then its absolute value $|f(z)|$ is obviously a continuous function in D. We shall show that $|f(z)|$ is subject to the following *maximum principle.*

If $f(z)$ is regular in a domain D, then $|f(z)|$ cannot obtain its maximum in D at an interior point of D, unless $f(z)$ reduces to a constant.

This is also known as the *maximum modulus principle.* If $f(z)$ is regular in the closure $\bar{D}$ of D, then $|f(z)|$—as a continuous function—must obtain its maximum in $\bar{D}$ at some point of $\bar{D}$. Since, by the maximum principle, this point cannot be situated in D, it must lie on the boundary of D. Hence, *if $f(z)$ is regular in a domain D and on its boundary C, then $|f(z)|$ attains its maximum in $D + C$ at a point of C.*

Suppose the maximum principle is not true and that there are points $z = z_0$ in D such that $|f(z)| \leq |f(z_0)|$ throughout D. Let S denote the set of all points z_0 with this property. Unless $|f(z)|$, and therefore also $f(z)$, is constant, S cannot contain all points of D. Accordingly there must exist a boundary point of S which is an interior point of D. Let z_0 denote such a boundary point. If $r > 0$ is taken sufficiently small, then the circle $|z - z_0| = r$ is in the interior of D and, moreover, it contains points z for which $|f(z)| < |f(z_0)|$. By the Cauchy integral formula,

(41) $$f(z_0) = \frac{1}{2\pi i} \int_{|z - z_0| = r} \frac{f(\zeta)}{\zeta - z}\, d\zeta = \frac{1}{2\pi} \int_0^{2\pi} f(z_0 + re^{i\theta})\, d\theta.$$

Hence,

$$|f(z_0)| \leq \frac{1}{2\pi} \int_0^{2\pi} |f(z_0 + re^{i\theta})|\, d\theta,$$

or

$$\frac{1}{2\pi} \int_0^{2\pi} [|f(z_0)| - |f(z_0 + re^{i\theta})|]\, d\theta \leq 0.$$

Since $|f(z_0)| \geq |f(z_0 + re^{i\theta})|$ and the integrand is continuous, this is only possible if

$$|f(z_0 + re^{i\theta})| = |f(z_0)|, \qquad 0 \leq \theta < 2\pi.$$

It follows that $f(z_0 + re^{i\theta})$ is of the form $f(z_0)e^{i\varphi}$, $0 \leq \varphi < 2\pi$. Inserting this in (41), we obtain

$$1 = \frac{1}{2\pi} \int_0^{2\pi} e^{i\varphi}\, d\theta$$

and, by taking real parts,

$$1 = \frac{1}{2\pi} \int_0^{2\pi} \cos \varphi\, d\theta.$$

Since φ is a continuous function of θ and $\cos \varphi \leq 1$, this is only possible if $\cos \varphi = 1$ for all θ. Hence, $\varphi = 0$ and, in view of the definition of φ, $f(z_0 + re^{i\theta}) = f(z_0)$ for $0 \leq \theta < 2\pi$. But if $f(z)$ is constant on a circumference on which it is regular, it is constant everywhere. This completes the proof of the maximum principle.

A similar result will hold for the minimum of an analytic function which is regular in D, provided the function does not vanish in D. Indeed, if $f(z) \neq 0$ in D, then $[f(z)]^{-1}$ is regular there. Since the minimum of $|f(z)|$ is attained at the same points as the maximum of $|f(z)|^{-1}$, it follows from the maximum principle that $|f(z)|$ cannot attain its minimum in the interior of D.

Since the inability of a function to attain its maximum at an interior point is a "local" property, the maximum principle is also valid for analytic functions which are regular but not single-valued in a multiply-connected domain. As an application of this remark, we prove *Hadamard's three-circle theorem.*

If $f(z)$ is regular and single-valued in $\rho < |z| < R$ and $M(r)$ denotes the maximum of $|f(z)|$ on the circle $|z| = r$ $(\rho < r < R)$, then $\log M(r)$ *is a convex function of* $\log r$.

In other words, for $\rho < r_1 < r_2 < r_3 < R$ we have the inequality

$$\log M(r_2) \leq \log M(r_1) \frac{\log r_3 - \log r_2}{\log r_3 - \log r_1} + \log M(r_3) \frac{\log r_2 - \log r_1}{\log r_3 - \log r_1}. \tag{42}$$

To prove (42), consider the function $z^\alpha f(z)$, where

$$\alpha = \frac{\log M(r_3) - \log M(r_1)}{\log r_1 - \log r_3}. \tag{43}$$

Although $z^\alpha f(z)$ will in general not be single-valued in $r_1 \leq |z| \leq r_3$, it is regular there and, moreover, $|z^\alpha f(z)|$ is single-valued. $|z^\alpha f(z)|$ will therefore attain its maximum on one of the circles $|z| = r_1$, $|z| = r_3$. The reader will confirm without difficulty that α has been so chosen as to make the maxima of this function on these circles equal. If $r_1 < r_2 < r_3$, we have therefore

$$r_2{}^\alpha M(r_2) \leq r_1{}^\alpha M(r_1).$$

With the help of (43) and some elementary manipulation it is easily verified that this inequality is equivalent to (42).

EXERCISES

1. Let $f(z)$ be regular in a domain D and on its boundary C. If $|f(z)|$ is constant on C, show that unless $f(z)$ reduces to a constant there must be at least one zero of $f(z)$ in D.

2. Let $P_1, \ldots, P_n$ be n arbitrary points of the plane and let $\overline{PP_\nu}$ denote the distance between P_ν and a variable point P. If P is confined to a closed domain D, show that the product $\prod_{\nu=1}^{n} \overline{PP_\nu}$ attains its maximum if P is a point of the boundary of D.

3. Prove the fundamental theorem of algebra by applying the maximum principle to the function $f(z) = (a_0 + a_1 z + \cdots + a_n z^n)^{-1}$ in the closed domain $|z| \leq R$, and letting $R \to \infty$.

4. Show that the function $\log f(z)$ is a regular analytic function in any domain in which $f(z)$ is regular and different from zero; by considering the real part of $\log f(z)$, show that the maximum principle of this section and the maximum principle for harmonic functions (Sec. 4, Chap. I) are equivalent.

5. If the power series $p(z) = a_0 + a_m z^m + a_{m+1} z^{m+1} + \cdots$ $(a_m \neq 0)$ converges in a neighborhood of the origin, show that for sufficiently small positive ϵ there exists a point $|z_0| = \epsilon$ such that $|p(z_0)| > |a_0|$. Use this result to prove the maximum principle.

9. The Residue Theorem. If the analytic function $f(z)$ is single-valued in a domain D and is regular there except at a point a of D, then, as shown in Sec. 6, $f(z)$ may be expanded in the vicinity of a into a converging Laurent series

$$f(z) = \sum_{n=-\infty}^{\infty} a_n (z - a)^n$$

whose coefficients are given by (35). The coefficient a_{-1} of this expansion is of particular interest. By (35), we have

$$a_{-1} = \frac{1}{2\pi i} \int_C f(\zeta)\, d\zeta, \tag{44}$$

where C is a closed contour surrounding $z = a$ and, except at $z = a$, $f(z)$ is regular within and on C. a_{-1} is called the *residue* of $f(z)$ at the singular point $z = a$. (44) shows that, apart from a numerical factor, the residue is equal to the integral of $f(z)$ over a closed contour surrounding the singular point $z = a$. This result can easily be generalized to the case in which the contour C contains more than one singularity of $f(z)$. Suppose that $f(z)$ is single-valued and regular within and on the closed contour C, the only exception being the n points $\alpha_1, \alpha_2, \ldots, \alpha_n$ within C which are singular points of $f(z)$. We now describe about each point α_ν a circle C_ν which is contained in the domain D bounded by C and whose radius is taken sufficiently small in order to prevent the overlapping of two such circles. In the domain D^* which is obtained by deleting from D the circles $C_1, \ldots, C_n$, $f(z)$ is regular and single-valued. By Cauchy's theorem, the integral of $f(z)$, taken over the boundary of D^*, is therefore equal to zero. Hence

$$\int_C f(z)\, dz - \sum_{\nu=1}^{n} \int_{C_\nu} f(z)\, dz = 0,$$

where the negative sign is due to the fact that the positive direction with respect to D^* coincides with the negative direction with respect to C_ν. It follows that

$$\int_C f(z)\, dz = \sum_{\nu=1}^{n} \int_{C_\nu} f(z)\, dz.$$

If the residue of $f(z)$ at the point α_ν is denoted by R_{α_ν}, this can, as shown by (44), also be written in the form

$$\int_C f(z)\, dz = 2\pi i \sum_{\nu=1}^{n} R_{\alpha_\nu}. \tag{45}$$

The identity (45) is known as the *residue theorem.* The reader will have no difficulty in verifying that the residue theorem remains valid if D is a multiply-connected domain bounded by a finite number of closed contours.

Cauchy's residue theorem has numerous applications both in the theory of analytic functions and in the theory of conformal mapping. Besides, it renders useful services in the evaluation of a great many real definite integrals. We shall devote the remainder of this section to illustrations of the latter type of application.

Consider first the evaluation of integrals of the type

$$\int_0^{2\pi} F(\cos\theta, \sin\theta)\, d\theta,$$

where $F(\cos\theta, \sin\theta)$ is a rational function of $\cos\theta$ and $\sin\theta$. If we write $z = e^{i\theta}$, the integration path is transformed into the circle $|z| = 1$ in the complex z-plane. Since

$$\cos\theta = \frac{1}{2}(e^{i\theta} + e^{-i\theta}) = \frac{1}{2}\left(z + \frac{1}{z}\right),$$
$$\sin\theta = \frac{1}{2i}(e^{i\theta} - e^{-i\theta}) = \frac{1}{2i}\left(z - \frac{1}{z}\right),$$

and $dz = ie^{i\theta}\, d\theta = iz\, d\theta$, the integral in question is transformed into the contour integral

$$\frac{1}{i}\int_C F\left[\frac{1}{2}\left(z + \frac{1}{z}\right), \frac{1}{2i}\left(z - \frac{1}{z}\right)\right]\frac{dz}{z},$$

where C is the unit circle. The integrand is clearly a rational function of z, say $r(z)$. By the residue theorem, the value of the integral is therefore $2\pi i$ times the sum of the residues of $r(z)$ at the poles of $r(z)$ which are situated in the interior of the unit circle. As an example, consider the integral

$$A = \int_0^{2\pi} \frac{d\theta}{a + b\cos\theta}, \qquad a > b > 0.$$

Making the above substitutions, we obtain

$$A = \frac{2}{i}\int_C \frac{dz}{bz^2 + 2az + b} = \frac{2}{ib}\int_C \frac{dz}{(z-\alpha)(z-\beta)},$$
$$\alpha = \frac{-a + \sqrt{a^2 - b^2}}{b}, \qquad \beta = \frac{-a - \sqrt{a^2 - b^2}}{b}.$$

Since $\alpha\beta = 1$ and $|\beta| > |\alpha|$, only α is situated in the interior of the unit circle. Our task is therefore reduced to finding the residue of the integrand at $z = \alpha$, that is, the coefficient of $(z - \alpha)^{-1}$ in its Laurent expansion about $z = \alpha$. This is particularly easy because the singularity at $z = \alpha$ happens to be a pole of the first order. In general, if a function $g(z)$ has a pole of the first order—or, as we shall also say, a simple pole—

at $z = \alpha$, then its residue at $z = \alpha$ is $\lim_{z \to \alpha} (z - \alpha)\, g(z)$; this follows immediately from the expansion $g(z) = a_{-1}(z - \alpha)^{-1} + a_0 + a_1(z - a) + \cdots$, valid in the neighborhood of α. In our case, the residue of

$$\frac{2}{ib}\,\frac{1}{(z - \alpha)(z - \beta)}$$

at $z = \alpha$ will therefore be $2[ib(\alpha - \beta)]^{-1}$ or, using the values of α and β,

$$\frac{1}{i\sqrt{a^2 - b^2}}.$$

It thus follows from (45) that

$$\int_0^{2\pi} \frac{d\theta}{a + b\cos\theta} = \frac{2\pi}{\sqrt{a^2 - b^2}}, \qquad a > b > 0.$$

Another type of real integral which can be evaluated is

$$\int_{-\infty}^{\infty} R(x)\, dx,$$

where $R(z)$ is a rational function of z, and the integral $\int_{-\infty}^{\infty} |R(x)|\, dx$ exists. It is clear that the latter condition will be satisfied if, and only if, $R(z)$ has no poles on the real axis and the degree of the denominator of $R(z)$ exceeds that of the numerator of $R(z)$ by at least two. We consider the integral

$$\int_C R(z)\, dz,$$

where C consists of the part of the real axis between $-\rho$ and ρ $(\rho > 0)$ and the half-circle $\rho e^{i\theta}$, $0 \leq \theta \leq \pi$; ρ will be so chosen that there will be no poles of $R(z)$ on the half circle. By the residue theorem,

$$2\pi i \sum_\nu R_{\alpha_\nu} = \int_C R(z)\, dz = \int_{-\rho}^{\rho} R(x)\, dx + \int_{C'} R(z)\, dz,$$

where α_ν are the poles of $R(z)$ within C and C' is the above mentioned half circle. $R(z)$ is of the form

$$R(z) = \frac{a_0 + a_1 z + \cdots + a_n z^n}{b_0 + b_1 z + \cdots + b_m z^m} = \frac{1}{z^{m-n}} \left[\frac{a_n + \dfrac{a_{n-1}}{z} + \cdots + \dfrac{a_0}{z^n}}{b_m + \dfrac{b_{m-1}}{z} + \cdots + \dfrac{b_0}{z^m}} \right],$$

where $m - n \geq 2$. It follows that, for sufficiently large $|z|$, we shall have $|zR(z)| < \epsilon$, where ϵ is arbitrarily small. Hence, if ρ is large enough,

$$\left| \int_{C'} R(z)\, dz \right| = \left| i\rho \int_0^{\pi} R(\rho e^{i\theta})\, d\theta \right| < \pi\epsilon,$$

and therefore

$$\left| \int_{-\rho}^{\rho} R(x)\,dx - 2\pi i \sum_{\nu} R_{\alpha_\nu} \right| < \pi\epsilon.$$

For $\rho \to \infty$, we finally obtain

$$\int_{-\infty}^{\infty} R(x)\,dx = 2\pi i \sum_{\nu} R_{\alpha_\nu},$$

where the summation is extended over the residues belonging to all poles of $R(z)$ in the upper half-plane. As an example, we consider the rational function $R(z) = (z^2 + 1)^{-n}$, where n is a positive integer. Its poles are situated at $z = \pm i$. In the upper half-plane, $R(z)$ has therefore only the pole $z = i$. In order to find the residue of $R(z)$ at this point, we have to expand $R(z)$ by powers of $z - i$. Writing $\zeta = z - i$, we have for $|\zeta| < 2$

$$\frac{1}{(z^2+1)^n} = \frac{1}{[(i+\zeta)^2+1]^n} = \frac{1}{[\zeta(2i+\zeta)]^n} = \frac{1}{(2i)^n\zeta^n}\left(1 - \frac{i\zeta}{2}\right)^{-n}$$
$$= \frac{1}{(2i)^n\zeta^n}\left[1 + n\frac{i\zeta}{2} + \frac{n(n+1)}{2!}\left(\frac{i\zeta}{2}\right)^2 + \frac{n(n+1)(n+2)}{3!}\left(\frac{i\zeta}{2}\right)^3 + \cdots\right].$$

The coefficient of $1/\zeta$ in this expansion is

$$\frac{1}{(2i)^n}\frac{n(n+1)(n+2)\cdots(2n-2)}{(n-1)!}\left(\frac{i}{2}\right)^{n-1} = \frac{1}{i}\frac{1}{2^{2n-1}}\frac{(2n-2)!}{[(n-1)!]^2}.$$

Hence,

$$\int_{-\infty}^{\infty} \frac{dx}{(x^2+1)^n} = \frac{\pi}{2^{2n-2}}\frac{(2n-2)!}{[(n-1)!]^2}.$$

The last type of integral we shall consider here is

$$\int_0^{\infty} x^{a-1}R(x)\,dx,$$

where a is not an integer, and where $R(z)$ is a rational function which, of course, has to be free of poles on the real axis and, moreover, is such that $z^aR(z)$ tends to zero if z tends to 0 or ∞. We evaluate the integral

$$\int_C z^{a-1}R(z)\,dz$$

over the closed contour indicated in Fig. 7. A contour of this type is necessary since z^{a-1} is not single-valued in the z-plane (see Sec. 7, Chap.

II); in the domain bounded by C, z^a is single-valued since this domain is contained in the z-plane which has been cut along the entire positive axis. It is shown in the same way as in the preceding example that the integral over the circle of radius ρ tends to zero if $\rho \to \infty$. The integral over the circle of radius r tends to zero if $r \to 0$; indeed,

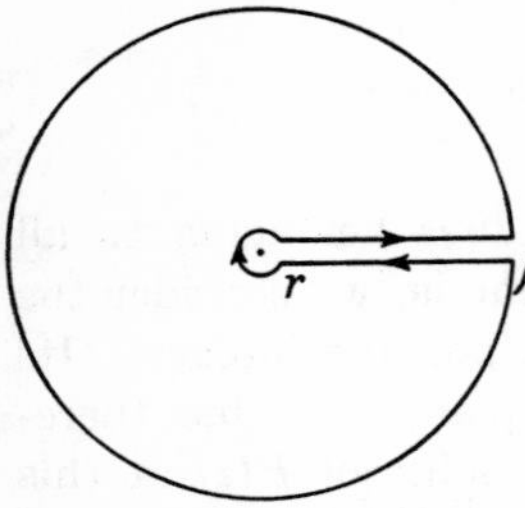

FIG. 7.

$$\int_{|z|=r} z^{a-1}R(z)\,dz = i\int_0^{2\pi} z^a R(z)\,d\theta,$$

and $z^a R(z) \to 0$ for $z \to 0$. If on the "upper edge" of the positive axis we take the positive value of z^{a-1}, then we have to take the value $z^{a-1}e^{2\pi i(a-1)}$ on the lower edge; this follows from the fact that, because of the cut, $\arg\{z\} = 0$ on the upper edge and $\arg\{z\} = 2\pi$ on the lower edge. For $\rho \to \infty$ and $r \to 0$, the contour integral therefore reduces to

$$\int_0^\infty x^{a-1}R(x)\,dx + \int_\infty^0 e^{2\pi i(a-1)}x^{a-1}R(x)\,dx.$$

Since $e^{2\pi i(a-1)} = e^{2\pi i a}$, we thus obtain

$$\int_0^\infty x^{a-1}R(x)\,dx = \frac{2\pi i \sum_\nu R_{\alpha_\nu}}{1 - e^{2\pi i a}},$$

where $\sum_\nu R_{\alpha_\nu}$ is the sum of the residues at the poles of $z^{a-1}R(z)$. To illustrate this formula, consider the integral

$$\int_0^\infty \frac{x^{a-1}}{1+x}\,dx, \qquad 0 < a < 1.$$

Since $\dfrac{x^a}{1+x} \to 0$, if $x \to 0$ or $x \to \infty$, the above procedure applies. The only pole is situated at $z = -1$, and the residue of $\dfrac{z^{a-1}}{1+z}$ at this point is $(-1)^{a-1} = e^{\pi i(a-1)} = -e^{\pi i a}$. Hence,

$$\int_0^\infty \frac{x^{a-1}}{1+x}\,dx = -2\pi i\frac{e^{\pi i a}}{1 - e^{2\pi i a}} = \frac{2\pi i}{e^{\pi i a} - e^{-\pi i a}},$$

and finally

$$\int_0^\infty \frac{x^{a-1}}{1+x}\,dx = \frac{\pi}{\sin \pi a}.$$

EXERCISES

Evaluate the integrals 1 to 6 by the residue method.

1. $\displaystyle\int_0^{2\pi} \frac{d\theta}{(a + b\cos\theta)^2} = \frac{2\pi a}{(a^2 - b^2)^{\frac{3}{2}}}, \qquad 0 < b < a.$

2. $\displaystyle\int_0^{2\pi} \frac{\cos^2 3\theta}{1 - 2a\cos 2\theta + a^2}\, d\theta = \pi\frac{1 - a + a^2}{1 - a}, \qquad 0 < a < 1.$

3. $\displaystyle\int_{-\infty}^{\infty} \frac{x^2 - x + 2}{x^4 + 10x^2 + 9}\, dx = \frac{5\pi}{12}.$

4. $\displaystyle\int_{-\infty}^{\infty} \frac{x^6}{(a^4 + x^4)^2}\, dx = \frac{3\sqrt{2}\,\pi}{16a}, \qquad a > 0.$

5. $\displaystyle\int_0^{\infty} \frac{x^a\, dx}{(1 + x^2)^2} = \frac{\pi(1 - a)}{4\cos\frac{1}{2}\pi a}, \qquad -1 < a < 2.$

6. $\displaystyle\int_0^{\infty} \frac{x^{-a}\, dx}{1 + 2x\cos\theta + x^2} = \frac{\pi}{\sin\pi a}\,\frac{\sin a\theta}{\sin\theta}, \qquad -1 < a < 1,\ -\pi < \theta < \pi.$

7. By evaluating the integral $\int e^{-z^2}\, dz$ around a rectangle whose corners are $-R$, R, $R + ai$, $-R + ai$ and letting $R \to \infty$, show that

$$\int_{-\infty}^{\infty} e^{-x^2}\cos 2ax\, dx = 2e^{-a^2}\int_0^{\infty} e^{-x^2}\, dx.$$

8. If $f(z)$ is regular in $|z| \leq 1$, show that

$$(1 - |z|^2)f(z) = \frac{1}{2\pi i}\int_{|\zeta|=1} f(\zeta)\,\frac{1 - \bar{z}\zeta}{\zeta - z}\, d\zeta;$$

verify that, for $|\zeta| = 1$, $|1 - \bar{z}\zeta| = |\zeta - z|$, and that therefore

$$(1 - |z|^2)|f(z)| \leq \frac{1}{2\pi}\int_0^{2\pi} |f(e^{i\theta})|\, d\theta.$$

9. Show that if

$$f(z) = \sum_{n=0}^{\infty} a_n z^n, \qquad |z| \leq r,$$

$$g(z) = \sum_{n=0}^{\infty} b_n z^n, \qquad |z| < \rho,$$

then

$$\sum_{n=0}^{\infty} a_n b_n z^n = \frac{1}{2\pi i}\int_{|\zeta|=r} f(\zeta)g\left(\frac{z}{\zeta}\right)\frac{d\zeta}{\zeta}, \qquad |z| < \rho r.$$

10. If $f(z) = \sum_{\nu=0}^{\infty} a_\nu z^\nu$ is regular in $|z| \leq R$ and $s_n(z) = \sum_{\nu=0}^{n} a_\nu z^\nu$ denotes the nth partial sum of the power series expansion of $f(z)$, show that

$$s_n(z) = \frac{1}{2\pi i}\int_{|z|=R} f(\zeta)\left(\frac{\zeta^{n+1} - z^{n+1}}{\zeta - z}\right)\frac{d\zeta}{\zeta^{n+1}}, \qquad |z| < R.$$

11. Let $z_1, z_2, \ldots, z_n$ be n distinct points which are situated in the interior of a closed contour C, and let $p(z)$ denote the polynomial

$$p(z) = (z - z_1)(z - z_2) \cdots (z - z_n).$$

If $f(z)$ is regular within and on C, show that

$$P(z) = \frac{1}{2\pi i} \int_C \frac{f(\zeta)}{p(\zeta)} \frac{p(\zeta) - p(z)}{\zeta - z} d\zeta$$

is a polynomial of order $n - 1$ and that $f(z_\nu) = P(z_\nu)$, $\nu = 1, \ldots, n$.

10. The Argument Principle. Let $f(z)$ be an analytic function which is single-valued in a domain D that is bounded by one or more closed contours C. We assume that, except for a finite number of poles in D, $f(z)$ is regular in $D + C$ and that, moreover, $f(z)$ does not vanish on C. By the residue theorem, the integral

$$\frac{1}{2\pi i} \int_C \frac{f'(z)}{f(z)} dz$$

is equal to the sum of the residues of the logarithmic derivative of $f(z)$ in D. Now the only possible singularities of $f'(z)/f(z)$ in D coincide with the zeros and poles of $f(z)$. In order to determine the residue of $f'(z)/f(z)$ at a zero of $f(z)$, we observe that, in the neighborhood of a zero a of the nth order, $f(z)$ has an expansion

$$f(z) = (z - a)^n[a_1 + a_2(z - a) + \cdots], \qquad a_1 \neq 0.$$

We therefore have $f(z) = (z - a)^n f_1(z)$, where $f_1(z) \neq 0$ in a certain neighborhood of $z = a$. Hence, $\log f(z) = n \log (z - a) + \log f_1(z)$ and

$$\frac{f'(z)}{f(z)} = \frac{n}{z - a} + \frac{f_1'(z)}{f_1(z)},$$

where the last term is regular at $z = a$. It follows that the residue of $f'(z)/f(z)$, or, as we also say, the *logarithmic residue* of $f(z)$, at $z = a$ is n, that is, it is equal to the order of the zero of $f(z)$ at $z = a$. If the zeros of $f(z)$ in D are counted with their multiplicities—a simple zero to be counted once, a double zero twice, etc.—the sum of the logarithmic residues of $f(z)$ at the zeros of $f(z)$ in D will therefore be equal to the number of these zeros. We now turn to the poles of $f(z)$ in D. If $z = b$ is a pole of order m, we have near $z = b$ an expansion

$$\begin{aligned} f'z) &= \frac{b_1}{(z - b)^m} + \cdots + \frac{b_m}{z - b} + b_{m+1} + \cdots \\ &= \frac{1}{(z - b)^m} [b_1 + b_2(z - b) + \cdots] = \frac{f_2(z)}{(z - b)^m}, \end{aligned}$$

where $f_2(z)$ is regular at $z = b$ and $f_2(b) \neq 0$. Hence,

$$\frac{f'(z)}{f(z)} = -\frac{m}{z - b} + \frac{f_2'(z)}{f_2(z)},$$

which shows that the logarithmic residue of $f(z)$ at a pole of $f(z)$ of order m is $-m$. If the poles of $f(z)$ in D are counted with their multiplicities, the sum of the logarithmic residues of $f(z)$ at the poles of $f(z)$ in D will therefore be equal to minus the number of these poles. Since $f'(z)/f(z)$ has no singularities in D except at the zeros and poles of $f(z)$, we have thus obtained the following result.

If the domain D is bounded by one or more closed contours C and if $f(z)$ is single-valued and regular in $D + C$ except for a finite number of poles in D and, moreover, $f(z) \neq 0$ on C, then

$$\frac{1}{2\pi i}\int_C \frac{f'(z)}{f(z)}\,dz = N_0 - N_\infty, \tag{46}$$

where N_0 and N_∞ are, respectively, the number of zeros and the number of poles of $f(z)$ in D. Both zeros and poles are to be counted with their multiplicities.

If, in (46), we replace $f(z)$ by $f(z) - a$, this formula will yield the difference between the number of zeros of $f(z) - a$ and the poles of $f(z) - a$. Since the latter are identical with the poles of $f(z)$, we thus find that

$$\frac{1}{2\pi i}\int_C \frac{f'(z)}{f(z) - a}\,dz = N_a - N_\infty,$$

where N_a indicates how often the value a is taken by $f(z)$ in D.

(46) can be brought into a different form in which the essentially geometric character of this identity becomes more apparent. If we write

$$\varphi = \arg\{f(z)\}, \qquad f(z) = |f(z)|e^{i\varphi},$$

we obtain

$$\begin{aligned}\frac{1}{2\pi i}\int_C \frac{f'(z)}{f(z)}\,dz &= \frac{1}{2\pi i}\int_C d\log f(z)\\ &= \frac{1}{2\pi i}\int_C [d\log|f(z)| + i\,d\varphi]\\ &= \frac{1}{2\pi i}\int_C d\log|f(z)| + \frac{1}{2\pi}\int_C d\varphi.\end{aligned}$$

We saw in Sec. 7, Chap. II, that $\log w$ is a many-valued function of w; if $\log w$ is continued along a closed curve which surrounds to origin, or, what amounts to the same thing, the point $w = \infty$, we shall not return to the value of $\log w$ with which we started. However, this many-valuedness was confined to $\operatorname{Im}\{\log w\} = \arg\{w\}$; $\operatorname{Re}\{\log w\} = \log|w|$ was single-valued. If we write $w = f(z)$, it follows therefore that

$$\int_C d\log|f(z)| = 0.$$

Indeed,

$$\int_{z_1}^{z_2} d \log |f(z)| = \log |f(z_2)| - \log |f(z_1)|,$$

and if the integration is performed over a closed contour, the terminals z_1 and z_2 of the integration coincide; because of the single-valuedness of $\log |f(z)|$, the value of the integral is therefore zero. Hence

$$\frac{1}{2\pi i} \int_C \frac{f'(z)}{f(z)} dz = \frac{1}{2\pi} \int_C d\varphi, \tag{47}$$

where $\varphi = \arg \{f(z)\}$. To interpret this formula, we observe that

$$\int_{z_1}^{z_2} d\varphi = \varphi(z_2) - \varphi(z_1) = \arg \{f(z_2)\} - \arg \{f(z_1)\}$$

is the change of the argument of $f(z)$, or, as we shall also say, the *variation of the argument* of $f(z)$, if z varies from z_1 to z_2. $\int_C d\varphi$ will therefore be the total variation of $\arg \{f(z)\}$ if z describes the entire boundary C of D. It is clear that the value of this integral must be an integral multiple of 2π. If z describes C, the point $f(z)$ describes a closed curve C', and if C' surrounds the origin m times in the positive sense, the increase of $\arg \{f(z)\}$ along C' will be $2\pi m$. In view of (46) and (47), we therefore obtain the following theorem.

Let the domain D be bounded by one or more closed contours C and let the analytic function $f(z)$ be single-valued and, apart from a finite number of poles, regular in $D + C$. If N_0 and N_∞ denote the number of zeros and poles of $f(z)$ in D, respectively, and $f(z) \neq 0$ on C, then

$$\frac{1}{2\pi} \Delta_C \arg \{f(z)\} = N_0 - N_\infty, \tag{48}$$

where Δ_C arg $\{f(z)\}$ denotes the total variation of arg $\{f(z)\}$ if z describes C.

This important result is known as the *argument principle.* It is worth pointing out that the argument principle essentially expresses a topological fact which is quite independent of the theory of analytic functions. We shall see later (Sec. 1, Chap. V), that the set of points w which corresponds to the points z of D by means of the relation $w = f(z)$ is also a domain, say D'. D' may be on ordinary plane domain such as D or it may be self-overlapping in a fashion reminiscent of the Riemann surfaces discussed in Sec. 7, Chap. II; the latter case will always happen if a value w_0 is taken at least twice by $w = f(z)$ in D. The number of zeros of $f(z)$ in D is simply the number of times the origin is covered by D', and the number of poles indicates how often $w = \infty$ is covered by D'. Suppose first that $f(z)$ is regular in D. In this case, D' is finite and it is intuitively clear that the number of times D' covers the origin is the same as the number of times the boundary C' of D' surrounds the origin. But this is identical with the statement of the argument principle in the case in which $f(z)$ is regular in D, as a glance at (48) shows. The case in which $f(z)$ has poles in D follows by observing that a closed curve which surrounds the origin in the negative sense surrounds at the same time the point at infinity in the positive sense; we only have to remember that "surrounding the point $w = \infty$" means "surrounding a domain containing the point $w = \infty$." Each covering of the point $w = \infty$ will therefore have the effect of diminishing by one unit the number of times C' surrounds the origin.

The argument principle can be extended to the case in which $f(z)$ has zeros or poles on the boundary C of D. Suppose that $f(z_0) = 0$, where z_0 is situated in the interior of a smooth section of C. $f(z)$ is regular at z_0, and we therefore have $f(z) = (z - z_0)^m f_1(z)$, $f_1(z_0) \neq 0$, if m is the multiplicity of the zero. In view of $\log f(z) = m \log (z - z_0) + \log f_1(z)$, it follows that $\arg \{f(z)\} = m \arg \{(z - z_0)\} + \arg \{f_1(z)\}$. At $z = z_0$, $f_1(z) \neq 0$ and $\log f_1(z)$ is regular. Hence, $\arg \{f_1(z)\}$ will vary continuously if z varies along C and passes through $z = z_0$. The expression $\arg \{(z - z_0)\}$, however, shows a different behavior. Since this is the angle between the parallel to the positive axis through z_0 and the linear segment drawn from z to z_0, it is clear that $\arg \{(z - z_0)\}$ jumps by the amount π if z_0 is passed. The contribution of this zero to $\arg \{f(z)\}$ will therefore be πm, that is, one-half of what it would have been if the zero were situated in the interior of D. If $z = z_0$ is a pole of order m, its contribution to $\arg \{f(z)\}$ will be $-\pi m$; this follows immediately from the fact that $[f(z)]^{-1}$ has a zero of order m at z_0 and that $\log \{[f(z)]^{-1}\} = -\log f(z)$. We therefore have the following extension of the argument principle:

The argument principle (48) *remains valid if $f(z)$ has poles and zeros on the boundary, provided that these poles and zeros are counted with half their multiplicities.*

As an application of the argument principle we prove the following result, known as *Rouché's theorem.*

If the functions $f(z)$ and $g(z)$ are regular and single-valued in a domain D and on its boundary C and if, on C, $|g(z)| < |f(z)|$, then the function $f(z) + g(z)$ has exactly as many zeros in D as the function $f(z)$.

We have

$$\log [f(z) + g(z)] = \log f(z) + \log \left[1 + \frac{g(z)}{f(z)} \right],$$

whence

$$\arg \{f(z) + g(z)\} = \arg \{f(z)\} + \arg \left\{ 1 + \frac{g(z)}{f(z)} \right\}. \tag{49}$$

On C, we have $\left|\frac{g(z)}{f(z)}\right| < 1$, and it follows therefore that the points

$$w = 1 + \frac{g(z)}{f(z)}, \qquad z \in C, \tag{50}$$

are all situated in the interior of the circle $|1 - w| < 1$. Since this circle does not contain the origin, the curve (50) cannot possibly surround that point. As a result, the total variation of the argument of (50) along C is zero. Hence, by (49),

$$\Delta_C \arg \{f(z) + g(z)\} = \Delta_C \arg \{f(z)\}.$$

Since both $f(z)$ and $f(z) + g(z)$ have no poles in D, it follows therefore from (48) that these two functions have the same number of zeros in F.

The application of Rouché's theorem is illustrated by the following short proof of the maximum principle. If $f(z)$ is regular in $D + C$ and there is a point z_0 in D such that, for $z \in C$, $|f(z)| < |f(z_0)|$, then it follows from Rouché's theorem that the functions $f(z_0) - f(z)$ and $f(z_0)$ have the same number of zeros in D. But the constant $f(z_0)$ $[f(z_0) \neq 0]$ has no zeros in D and the function $f(z) - f(z_0)$ has at least one zero there, namely, at $z = z_0$. The assumption $|f(z)| < |f(z_0)|$, $z \in C$, thus leads to a contradiction.

We close this section with two examples of how the argument principle is applied to problems in the theory of functions. We first show that an *analytic function which is regular in the closure of a domain D and takes only real values on the boundary C of D reduces to a constant.* Let $a = \alpha + i\beta$, $\beta \neq 0$, be a nonreal complex number and consider the values of

$$f(z) - a = f(z) - \alpha - i\beta$$

for $z \in C$. If $\beta > 0$, say, we shall have $\text{Im}\ \{f(z) - a\} = \beta > 0$ since $f(z)$ is real on C. The values of $f(z) - a$ are thus confined to the upper half-plane and the curve described by $f(z) - a$ if z describes C cannot surround the origin. Hence, $\Delta_C \arg \{f(z) - a\} = 0$; since $f(z) - a$ is regular in $D + C$, it follows therefore from the argument principle that $f(z) - a \neq 0$, that is, $f(z) \neq a$ in D. The same reasoning also applies to values a for which $\text{Im}\ \{a\} < 0$. We have thus proved that $f(z)$ does not take nonreal values in D. But this means that $f(z)$ reduces to a constant. Indeed, since $f(z)$ is regular in D, we have

$$f'(z) = \lim_{h \to 0} \frac{f(z + h) - f(z)}{h} = \lim_{h \to 0} \frac{f(z + ih) - f(z)}{ih}$$

if $h \to 0$ through positive values. Since $f(z)$ is real throughout D, the first limit is real and the second limit is imaginary. They can therefore be equal only if they are both zero. Since z was arbitrary, it follows that $f'(z) = 0$ throughout D; hence, $f(z) = \text{const.}$

The second application of the argument principle to be made here is concerned with the Green's function $g(z,\zeta)$ of a domain D which is bounded by n closed contours. We recall from Sec. 5, Chap. I, that $g(z,\zeta)(\zeta \in D)$ is harmonic in D except at the point ζ and that $g(z,\zeta) = 0$ if z is on C. Near $z = \zeta$, $g(z,\zeta)$ is of the form

$$g(z,\zeta) = -\log|z - \zeta| + g_1(z), \tag{51}$$

where $g_1(z)$ is harmonic at $z = \zeta$. A *critical point* of $g(z,\zeta)$ is defined as a point at which both partial derivatives $g_x(z,\zeta)$ and $g_y(z,\zeta)$ $(z = x + iy)$ are zero. We shall show that *the Green's function of a domain of connectivity n has exactly $n - 1$ critical points in D.* If $h(z,\zeta)$ denotes the harmonic conjugate of $g(z,\zeta)$ (see Sec. 8, Chap. I), then, as shown in Sec. 5, Chap. II, the function

$$p(z,\zeta) = g(z,\zeta) + ih(z,\zeta) \tag{52}$$

is an analytic function which is regular in D except, of course, at the point $z = \zeta$. Since $\log|z - \zeta|$ is the real part of $\log(z - \zeta)$, it follows that $p(z,\zeta)$ is of the form

$$p(z,\zeta) = -\log(z - \zeta) + p_1(z),$$

where $p_1(z)$ is regular in D. However, $p(z,\zeta)$ will not be single-valued in the multiply-connected domain D. It was shown in Sec. 10, Chap. I, that $h(z,\zeta)$ has constant additive periods which are associated with complete circuits around the "holes" of D; (52) shows that the same is true of $p(z,\zeta)$. However, since these periods are constant they will vanish if $p(z)$ is differentiated. Hence, the function

$$p'(z,\zeta) = -\frac{1}{z - \zeta} + p_1'(z) \tag{53}$$

is free of periods; since the period $2\pi i$ of the logarithm with respect to a closed circuit around $z = \zeta$ has also been removed by the differentiation, $p'(z,\zeta)$ is thus a single-valued function in D. We are now anticipating two results which will be proved later. The first of these is the fact (to be demonstrated in Sec. 2, Chap. VII) that it is sufficient to prove a result of the type we are concerned with for domains which are bounded by analytic curves (see Sec. 11, Chap. I); the second result (to be proved in Sec. 5, Chap. V) is that $p(z,\zeta)$ is regular on C if C consists of analytic curves. In this case, we shall have

$$p'(z,\zeta)\,dz = dp(z,\zeta) = \frac{\partial}{\partial s}\,p(z,\zeta)\,ds$$

if z is on C and s is the arc-length parameter on C. By (52), it follows that

$$p'(z,\zeta)\,dz = \frac{\partial}{\partial s}\,g(z,\zeta)\,ds + i\,\frac{\partial}{\partial s}\,h(z,\zeta)\,ds.$$

Since, on C, $g(z,\zeta) = 0$, this reduces to

$$p'(z,\zeta)\,dz = i\,\frac{\partial h}{\partial s}\,ds, \qquad h = h(z,\zeta),$$

or, if we wish to avoid the use of differentials,

$$\frac{1}{i}\,p'(z,\zeta)\,\frac{dz}{ds} = \frac{\partial h}{\partial s}. \tag{54}$$

As regards the subject of differentials, we remark here that in the theory of conformal mapping there are many occasions on which the use of the differential notation is extremely concise and convenient; whenever there is no danger of confusion we shall therefore not hesitate to avail ourselves of its advantages. Returning now to (54), we observe that h is the harmonic conjugate of $g = g(z,\zeta)$ and that therefore, by the Cauchy-Riemann equations (41) (Sec. 8, Chap. I),

$$\frac{\partial g}{\partial n} = \frac{\partial h}{\partial s},$$

where $\partial/\partial n$ denotes differentiation with respect to the outward pointing normal. In view of Exercise 3, Sec. 5, Chap. I, it follows therefore that $(\partial h/\partial s) \leq 0$. Hence, by (54),

$$\frac{1}{i}\,p'(z,\zeta)\,\frac{dz}{ds} \leq 0, \qquad z \in C.$$

This shows that, on C,

$$\arg\left\{p'(z,\zeta)\,\frac{dz}{ds}\right\} = \text{const.},$$

and therefore

$$\Delta_C \arg\left\{p'(z,\zeta)\,\frac{dz}{ds}\right\} = 0,$$

which can also be written

$$\Delta_C \arg\{p'(z,\zeta)\} + \Delta_C \arg\left\{\frac{dz}{ds}\right\} = 0. \tag{55}$$

Now arg $\{dz/ds\}$ is the angle which the tangent to C at the point z forms with the positive axis, as the reader will immediately verify. In order to find the total variation of this angle if z describes C, we observe that a finite domain of connectivity n has one "outer boundary," say C_1 and

$n - 1$ inner boundaries. If z describes C in the positive sense with respect to D, C_1 is also described in the positive sense with respect to the domain enclosed by it; the inner boundaries, however, will then be described in the negative sense with regard to the respective domains enclosed by them. If the point of contact of the tangent to C traverses, in turn, all the n separate boundary components in the positive direction with respect to D, the angle which the tangent makes with the positive axis will therefore increase by the amount 2π along C_1 and it will decrease by 2π along each of the inner boundaries. Hence,

$$\Delta_C \arg \left\{\frac{dz}{ds}\right\} = 2\pi - 2\pi(n - 1) = -2\pi(n - 2)$$

and, by (55),

$$\Delta_C \arg \{p'(z,\zeta)\} = 2\pi(n - 2).$$

Since, as shown by (53), $p'(z,\zeta)$ has a simple pole at $z = \zeta$ and is otherwise regular in $D + C$, it follows from the argument principle that the number of zeros of $p'(z,\zeta)$ in D is $(n - 2) + 1 = n - 1$. But this proves our result since, by (52) and the Cauchy-Riemann equations (14) (Sec. 5, Chap. II),

$$p'(z,\zeta) = \frac{\partial g}{\partial x} - i\frac{\partial g}{\partial y},$$

which shows that the zeros of $p'(z,\zeta)$ and the critical points of $g(z,\zeta)$ coincide.

EXERCISES

1. If $|z| = R$ and R is taken sufficiently large, show that $|a_n z^n| > |a_0 + a_1 z + \cdots + a_{n-1}z^{n-1}|$, where $a_n \neq 0$ and $a_0, \ldots, a_{n-1}$ are arbitrary complex numbers. Use this result and Rouché's theorem to give another proof of the fundamental theorem of algebra.

2. Show that all five roots of the algebraic equation $z^5 + 15z + 1$ must be situated in the interior of the circle $|z| < 2$, but that only one root of this equation is in the circle $|z| < \frac{3}{2}$.

3. Show that an analytic function which is regular and single-valued in a domain D except for m poles and which takes only real values on the boundary of D must have m zeros in D.

4. Show that the equation

$$ze^{a-z} = 1, \qquad a > 1,$$

has precisely one root in the circle $|z| \leq 1$; explain why this root is necessarily positive.

5. Show that a rational function which has m poles in the z-plane (the point $z = \infty$ included) takes every complex value exactly m times.

6. Show that, in $|z| < 1$, the function

$$f(z) = z + \frac{1}{z}$$

takes every nonreal value exactly once.

7. Let α_ν, $\nu = 1, 2, \ldots, m$ be complex numbers for which $|\alpha_\nu| < 1$. Verify that, for $|z| = 1$, $|\alpha_\nu - z| = |1 - \alpha_\nu z|$ and use this fact and Rouché's theorem in order to prove that the function

$$f(z) = \prod_{\nu=1}^{m} \left(\frac{\alpha_\nu - z}{1 - \bar{\alpha}_\nu z} \right)$$

takes every value a for which $|a| < 1$ exactly m times if $|z| < 1$; show further that for $|a| > 1$ the equation $f(z) = a$ has m roots in $|z| > 1$. *Hint:* Consider the function $f(z) - a$.

8. Let the function $f(z)$ be regular and single-valued in a domain D except for one simple pole, and let $|f(z)| = 1$ be satisfied at all points of the boundary of D. Prove that every value a for which $|a| > 1$ is taken by $f(z)$ in D once, and once only.

9. Use (53), (54), and the Cauchy-Riemann equations to give an alternative derivation of formula (16), Sec. 5, Chap. I, by means of the residue theorem.

CHAPTER IV

FAMILIES OF ANALYTIC FUNCTIONS

1. Equicontinuity and Uniform Boundedness. A great many proofs in various branches of mathematical analysis are based on the elementary limit point principle, also known as the Bolzano-Weierstrass theorem, according to which every infinite set of points which is contained in a finite domain must have at least one limit point. How useful this principle is becomes clear when we are dealing with a class of mathematical entities for which a corresponding principle does not hold. For instance, it is not true that, given an infinite set of functions $f(x)$ which are continuous in the interval $0 \leq x \leq 1$, we can always choose a subset which converges to a continuous function in $0 \leq x \leq 1$. This inability to extend the limit point principle to sets of continuous functions is largely responsible for the difficulty characteristic of many existence proofs in mathematical analysis.

It is now a fact of fundamental importance for the theory of analytic functions that there exists an analogue to the limit point principle for sets of analytic functions. However, there are certain limitations which are in the nature of things. We cannot expect it to be true that from an infinite set of analytic functions all of which are regular in the same domain D it should always be possible to select a subset which converges to an analytic function regular in D. To see this, we only have to consider the functions z, $2z$, $3z$, . . . , nz, ... All these functions are regular at all finite points of the plane but, nevertheless, any infinite subsequence of these functions obviously has the limit ∞ for all $z \neq 0$. It is therefore clear that additional restrictions have to be imposed on the class of analytic functions which are regular in a given domain, before an extension of the limit point principle to sets of such functions becomes feasible. However, before attacking the problem of finding out what these restrictions are, we shall introduce several concepts which refer to sets of functions or, as we also say, *classes of functions.*

The first of these concepts is that of *equicontinuity* of a class of functions in a domain D. Suppose S is an infinite set of functions $f(z)$ all of which are regular and single-valued in the same domain D. Since the functions $f(z)$ are regular in D, they are, of course, also continuous there; that is, for any z in D and for any positive ϵ there exists a δ such that $|f(z) - f(z_1)| < \epsilon$, provided $|z - z_1| < \delta = \delta(\epsilon,z)$. As in the case of real functions, and for the same reasons, a function $f(z)$ which is continuous in a closed domain is uniformly continuous there; this means that the number δ will only depend on ϵ and the closed domain but not on the individual point z. Thus, if A is a closed subdomain of D, we shall have $|f(z) - f(z_1)| < \epsilon$, provided $|z - z_1| < \delta = \delta(\epsilon,A)$, regardless of what particular point z in A we take. If we take another function of the class S, say $f_1(z)$, this function will again be uniformly continuous in A, and for a given ϵ we are again able to find a number δ_1 in the manner described above. However, it will generally not be true that the δ obtained from $f(z)$ will also serve in the case of the function $f_1(z)$; it may well be that δ_1 has to be taken considerably smaller than δ. In the latter case, δ_1 will serve for both functions. Extending this procedure to a finite number of functions $f_\nu(z)$, $\nu = 1, 2, \ldots, n$, we find that for a given ϵ there always exists a $\delta = \delta(\epsilon,A)$ such that $|f_\nu(z) - f_\nu(z_1)| < \epsilon$ if $|z - z_1| < \delta$ and $z \in A$. However, this argument breaks down if we are trying to find a δ which will serve an infinite number of functions. Although, in the finite case, $\delta > 0$, it is possible that δ decreases if more functions are added and finally tends to zero if the number of functions tends to infinity. In this case, there will not exist a positive δ which can serve all functions of the set. The functions of a set in which such an occurrence is impossible are called equicontinuous.

The functions $f(z)$ of a certain class are said to be equicontinuous in a domain D if for every given positive ϵ and for every closed subdomain A of D there exists a positive number $\delta = \delta(\epsilon,A)$ such that for $z \in A$, $z_1 \in A$ we have

$$|f(z) - f(z_1)| < \epsilon \qquad \text{if } |z - z_1| < \delta, \tag{1}$$

regardless of which function of the class we take.

Another concept which is pertinent in this connection is that of a class of *locally uniformly bounded* functions.

A class of functions $f(z)$ which are regular in a domain D is said to be locally uniformly bounded if for each $z_0 \in D$ there exist a positive number $M(z_0)$ and a neighborhood $N(z_0)$ of z_0 such that $|f(z)| < M(z_0)$ if z is in $N(z_0)$, regardless of which function $f(z)$ we take.

It is easy to see that a class of functions which is locally uniformly bounded in D is also uniformly bounded in any closed subdomain A of D. Indeed, if $z_0 \in A$, then there must exist a number $M(A)$ such that $M(z_0) < M(A)$. If this were not the case, there would exist a sequence of points $z_1, z_2, \ldots$ converging to a point z^* of the closed domain A such that $\lim_{n \to \infty} M(z_n) = \infty$. But this would mean that there does not exist a number $M(z^*)$ for which $|f(z)| < M(z^*)$ if z is in an arbitrarily small neighborhood of z^*.

If the functions $f(z)$ are locally uniformly bounded in D, the same is true of their derivatives $f'(z)$. To show this we draw a circle C about the point z such that C is entirely contained in D. There exists a constant M such that $|f(z)| < M$ on C. Hence, if z_0 is a point within C,

$$|f'(z_0)| = \left| \frac{1}{2\pi i} \int_C \frac{f(\zeta)\, d\zeta}{(\zeta - z_0)^2} \right| < \frac{MR}{d^2},$$

where R is the radius of C and d is the smallest distance of z_0 from C. If z is in a neighborhood of z_0 which is contained within C and whose distance from C is at least d_1, we have in this neighborhood $|f'(z)| < MR\, d_1^{-2}$, which shows that $f'(z)$ is indeed locally uniformly bounded.

We are now in a position to show that *a class of functions which is locally uniformly bounded in D is also equicontinuous there.* The proof is very simple. Since $f'(z)$ is also locally uniformly bounded, we have $|f'(z)| < M(A)$ if z is confined to any closed subdomain A of D. Hence, if $z \in D$ and z_1 is in the interior of a circle C entirely contained in D, we have

$$|f(z) - f(z_1)| = \left| \int_z^{z_1} f'(z)\, dz \right| \leq \int_z^{z_1} |f'(z)|\, |dz| \leq M(C)|z - z_0|.$$

This shows that (1) is satisfied within and on C if we take $\delta(\epsilon, C) \cdot M(C) = \epsilon$. The extension of this result from circles to arbitrary closed subdomains A of D now follows by the same type of argument as that used above to show that locally uniformly bounded functions are uniformly bounded in any A.

Conversely, *a class of functions which is equicontinuous in a domain D is also locally uniformly bounded there, provided it is uniformly bounded at one point of D.* If z_0 is the point at which the functions $f(z)$ are known to be uniformly bounded and z is an arbitrary point of D, we choose a closed subdomain A of D which contains z_0 and z and we connect z_0 and z by a contour C situated in the interior of D. Since the functions $f(z)$ are equicontinuous in D, (1) holds for any two points on C whose distance is less than δ. If $m > \frac{L}{\delta}$, where L is the length of C, we can therefore find points $z_1, z_2, \ldots, z_{m-1}$ on C such that $|f(z_1) - f(z_0)| < \epsilon$, $|f(z_2) - f(z_1)| < \epsilon$, $\cdots$ $|f(z) - f(z_{m-1})| < \epsilon$. It therefore follows that

$$|f(z) - f(z_0)| < m\epsilon,$$

whence

$$|f(z_0)| - m\epsilon < |f(z)| < |f(z_0)| + m\epsilon. \tag{2}$$

If $|f(z_0)| < M$, it follows therefore that $|f(z)| < M + m\epsilon$, which shows that the functions are uniformly bounded at every point of D. The uniform boundedness in a neighborhood of z follows by observing that, in a δ-neighborhood of z, we have $|f(z)| < M + (m + 1)\epsilon$. The left-hand inequality in (2) shows, incidentally, what may happen if we omit the assumption that the functions $f(z_0)$ be uniformly bounded at one point of D. If, for a sequence of functions $f_n(z)$, we have $\lim |f_n(z_0)| = \infty$, it follows from (2) that $|f_n(z)| \to \infty$. If the functions $f(z)$ are equicontinuous in D but not uniformly bounded at a point z_0 of D, they will also not be uniformly bounded at any other point of D.

EXERCISES

1. Let $\{f(z)\}$ be the class of functions

$$f(z) = \sum_{n=0}^{\infty} a_n z^n$$

which are regular for $|z| < 1$ and for which $|a_n| < M_n$, where $\lim\sup^n \sqrt{M_n} < \infty$ and the M_n are independent of the particular function $f(z)$. Show that the functions of $\{f(z)\}$ are equicontinuous in $|z| < 1$.

2. Prove the following converse of the result of the preceding exercise: If the functions of the class $\{f(z)\}$ are regular and equicontinuous in $|z| < 1$ and if $|f(0)| < M$, then there exist constants M_n, $n = 1, 2, \ldots$, such that $|a_n| < M_n$, where a_n is the coefficient of the expansion $f(z) = \sum_{n=0}^{\infty} a_n z^n$.

2. Normal Families of Analytic Functions. A class of functions for which an analogue of the limit point principle holds is called a *normal family* of functions.

G is called a normal family of regular functions $f(z)$ in a domain D if from any sequence $f_1(z), f_2(z), \ldots$ of functions of G it is possible to extract a subsequence

$$f_{n_1}(z), f_{n_2}(z), \ldots$$

which converges uniformly in any closed subdomain of D.

It is customary in this connection to extend the concept of uniform convergence so as to make it include the case in which a sequence of functions converges to "the constant ∞"; by this is meant that, for any positive M, we shall have $|f_n(z)| > M$ in any closed subdomain of D, provided n is large enough. Except in this case, the above uniformly convergent subsequence will converge to an analytic function which is regular in D; this follows from the theorem proved at the end of Sec. 3, Chap. III. However, it does not necessarily follow that this limit function is also included in the family G, just as the limit point of a subset of a given point set S does not always belong to S. This limit point will certainly belong to S if S is a closed set; it is precisely this property which is characteristic of a closed set. A family of analytic functions which is analogous to a closed set in this respect is said to be compact.

A normal family G of functions is said to be compact if the limits of all converging sequences of functions of G are also functions of G.

Once it is known that a particular class of analytic functions is a normal family, it is possible to draw many important conclusions regarding the functions of this class, as we shall see in the following section. It is therefore essential to find criteria which permit us to decide whether or not a given class of functions forms a normal family. The main result in this direction is the following theorem of Montel.

If the functions of a class G are regular and locally uniformly bounded in a domain D, then G is a normal family in D.

As a first step in the proof of Montel's theorem, we derive the following preliminary result.

Let $f_1(z), f_2(z), \ldots$ be a sequence of analytic functions which are regular and locally uniformly bounded in a domain D. If this sequence converges at all points of a point set E which is dense in D, then it converges uniformly in every closed subdomain of D.

If A is a closed subdomain of D, we partition A into a network of—complete or incomplete—equal squares as described in the proof of Cauchy's theorem (Sec. 2, Chap. III). The side of the square will be taken small enough in order to make sure that, for two points z and z_1 in each square, we have $|f_n(z) - f_n(z_1)| < \epsilon$, where ϵ is given in advance. This is possible since the functions $f_n(z)$ are locally uniformly bounded and therefore, as shown in the preceding section, equicontinuous. The set E in which the sequence $\{f_n(z)\}$ converges is dense in D; hence, there will be in each square s_p a point z_p at which $\{f_n(z)\}$ converges. Since there is only a finite number of squares s_p, we can therefore find a number N such that for $n > N$, $m > N$ and for all points z_p

$$|f_n(z_p) - f_m(z_p)| < \epsilon \tag{3}$$

if ϵ is arbitrarily given. If z is an arbitrary point of A, it will belong to one of the squares s_p. In this square we have

$$\begin{aligned}|f_n(z) - f_m(z)| &= |f_n(z) - f_n(z_p) + f_n(z_p) - f_m(z_p) + f_m(z_p) - f_m(z)| \\ &\leq |f_n(z) - f_n(z_p)| + |f_n(z_p) - f_m(z_p)| + |f_m(z_p) - f_m(z)|.\end{aligned}$$

The first and third terms on the right-hand side are each smaller than ϵ because of the equicontinuity of the functions $f_n(z)$. In view of (3) we therefore obtain

$$|f_n(z) - f_m(z)| < 3\epsilon, \qquad n, m > N = N(\epsilon).$$

Since ϵ was arbitrary, it follows that the sequence $\{f_n(z)\}$ converges uniformly in A.

We are now ready to prove Montel's theorem. We first choose a sequence of points $z_1, z_2, \ldots$ which are dense everywhere in the domain D. That this is possible follows from the well-known fact that the set of all points (a,b) with rational coordinates a, b is countable, that is, it may be arranged as a sequence of points bearing the subscripts 1, 2, 3, . . . ; the same is therefore true of those "rational points" which are in D. But this set of points is dense in D and it may therefore serve our purpose. Next, we take an arbitrary infinite sequence of functions $f_1(z), f_2(z), \ldots$ belonging to the class G. Since the functions $f_n(z)$ are locally uniformly bounded, there exists a number M such that $|f_n(z_1)| < M$. The points $w_n = f_n(z_1)$ are thus all situated in the interior of the circle $|w| < M$ and we can therefore select a subsequence $w_{n_1}{}^{(1)}, w_{n_2}{}^{(1)}, \ldots$ which converges to a point $w^{(1)}$ for which $|w^{(1)}| \leq M$. In other words, the subsequence

$$f_{n_1}{}^{(1)}(z), f_{n_2}{}^{(1)}(z), \ldots \tag{4}$$

of the original sequence will converge at the point $z = z_1$. Passing now to the point z_2, we again use the fact that the numbers $f_{n_\nu}{}^{(1)}(z_2)$, $\nu = 1, 2, \ldots$ are uniformly bounded. In the same way as before, we may therefore select a subsequence

$$f_{n_1}{}^{(2)}(z), f_{n_2}{}^{(2)}(z), \ldots \tag{5}$$

of (4) which converges at $z = z_2$. Being a subsequence of (4), (5) will, of course, also converge at $z = z_1$. We next select a subsequence $f_{n_\nu}{}^{(3)}(z)$ of (5) which converges at z_3; since this sequence is a subsequence of both (4) and (5), it will also converge at $z = z_1$ and $z = z_2$. Continuing in this fashion, we obtain after p steps a sequence

$$f_{n_1}{}^{(p)}(z), f_{n_2}{}^{(p)}(z), \ldots \tag{6}$$

which converges at the points $z_1, z_2, \ldots, z_p$. From (6), we again select a subsequence which converges also at z_{p+1}, and we continue this process indefinitely. We now apply what is known as the *diagonal process*. This consists in constructing a sequence of functions by taking first the first function of (4), then the second function of (5), etc. The pth function in this *diagonal sequence* will be the pth function of the sequence (6). It is easy to see that the diagonal sequence converges at all points $z_1, z_2, \ldots$. We only have to observe that all terms of this sequence, starting from the pth term, are also contained in the sequence (6). Since (6) converges at the point z_p, the same is therefore true of the diagonal sequence. But p is arbitrary, and this shows that the diagonal sequence converges at all points $z_1, z_2, \ldots$. Now the points $z_1, z_2, \ldots$ are dense in D. As shown above, it follows therefore that the diagonal sequence converges uniformly in every closed subdomain of D. We have thus proved that given any sequence of functions belonging to a locally uniformly bounded class, we can always select a subsequence which converges uniformly in any closed subdomain of D. In other words, the class of functions in question is a normal family. This completes the proof of Montel's theorem.

A particularly simple case arises when the class under consideration consists of uniformly bounded functions, that is, functions $f(z)$ for which there exists a positive constant M such that $|f(z)| \leq M$ if $z \in D$. Since this clearly implies the local uniform boundedness of the class, it follows from Montel's theorem that *the class of analytic functions which are regular and uniformly bounded in a domain D is a normal family.* This class is, moreover, compact. Indeed, if $\{f_n(z)\}$ is a uniformly converging sequence of such functions, it follows from $|f_n(z)| \leq M$ that the limit $f(z)$ of this sequence must also satisfy $|f(z)| \leq M$ at all points of D. But, as the limit of a uniformly converging sequence of functions which are regular in D, $f(z)$ is also an analytic function regular in this domain. $f(z)$ belongs therefore to the original class of functions, which has thus been shown to be compact. Another example of a family of functions which is normal and compact is the class of functions $f(z)$ which are regular and single-valued in a domain D and satisfy there $|f(z) - a| > A > 0$, where a is a constant. This class is easily transformed into a class of uniformly bounded functions. Indeed, if we write

$$g(z) = \frac{1}{f(z) - a},$$

then $|g(z)| < \frac{1}{A}$, and the normality and compactness of the family $\{f(z)\}$ is a consequence of the normality and compactness of the family $\{g(z)\}$.

EXERCISES

1. Let $\{f(z)\}$ be a normal family of functions which are regular and single-valued in a domain D. If the analytic function $F(w)$ is regular and single-valued for all values w taken by the functions $w = f(z)$ in D, show that the functions $g(z) = F[f(z)]$ also form a normal family.

2. Verify that any infinite subsequence of the sequence of functions $z, z^2, \ldots, z^n, \ldots$ converges to the constant 0 for $|z| < 1$ and to the constant ∞ for $|z| > 1$. Why is this possible?

3. Show that the family of functions $f(z)$ which are regular and single-valued in a domain D and satisfy there $\text{Re}\,\{f(z)\} \geq 0$ is normal and compact.

3. Extremal Problems. In this section, we shall prove an analogue to the well-known theorem of Weierstrass according to which a real function which is defined on a closed set attains its maximum at some point of the set. Suppose we are given a family $\{f(z)\}$ of analytic functions which are regular in a domain D, and that a *functional* $J(f)$ is defined with respect to all functions $f = f(z)$ of this family. By this is meant that, by a certain rule, a finite number is associated with each function of $\{f(z)\}$. For exampie, if $\{f(z)\}$ denotes the class of functions regular in $|z| < 1$, $J(f)$ may be a definite coefficient, say a_2, in the power series expansion $f(z) = \sum_{\nu=0}^{\infty} a_\nu z^\nu$. Another functional would be the value of $f(z)$ at a given point of D, and so forth. A functional $J(f)$ is said to be *continuous* if the convergence in D of the sequence $f_1(z), f_2(z), \ldots$ to the function $f(z)$ implies $J(f_n) \rightarrow J(f)$. The coefficient a_2, in our example, is a continuous functional. If $f_n(z) = \sum_{\nu=0}^{\infty} a\nu^{n}{}_{z}\nu$ and $f_n(z) \rightarrow f(z) = \sum_{\nu=1}^{\infty} a\nu z^{\nu}$, then $f_n(0) \rightarrow f(0)$, *i.e.*, $a_0^{(n)} \rightarrow a_0$. Hence $[f_n(z) - a_0^{(n)}]_z - 1 \rightarrow [f(z) - a_0]z^{-1}$; setting $z = 0$, we find that $a_1^{(n)} \rightarrow a_1$. It follows, in turn, that $z^{-2}[f_n(z) - a_0^{(n)} - a_1^{(n)} z] \rightarrow z^{-2}[f_n(z) - a_0 - a_1 z]$ and thus, for $z = 0$, $a_2^{(n)} \rightarrow a_2$. This shows that a_2 is indeed a continuous functional in the sense of our definition.

The analogue to the theorem of Weierstrass referred to above can now be stated as follows.

If $J(f)$ is a continuous functional defined in the normal and compact family $\{f(z)\}$, then the problem

$$|J(f)| = \max.$$

has a solution within the family $\{f(z)\}$.

In other words, there exists at least one function of the family $\{f(z)\}$, say $f_0(z)$, such that all functions $f(z)$ of $\{f(z)\}$ satisfy

$$|J(f)| \leq |J(f_0)|.$$

The proof is not difficult. The positive numbers $|J(f)|$ have a least upper bound A, which may be finite or infinite (we shall see presently that A is finite). In view of the definition of a least upper bound, there exists a sequence $f_1(z), f_2(z), \ldots$ of functions of $\{f(z)\}$ such that

$$\lim_{n \rightarrow \infty} |J(f_n)| = A.$$

Since $\{f(z)\}$ is normal and compact, we can select a subsequence $f_{n_1}, f_{n_2}, \ldots$ of this sequence which converges to a function $f_0(z)$ of $\{f(z)\}$. In view of the continuity of the functional $J(f)$, it follows therefore that

$$|J(f_0)| = \lim_{\nu \to \infty} |J(f_{n_\nu})| = A.$$

This shows that there exists a function $f_0(z)$ of $\{f(z)\}$ for which the upper limit A of the numbers $|J(f)|$ is attained; since the functional $J(f)$ takes only finite values, it also follows that A is finite. This completes the proof.

We remark that the same argument also shows that $|J(f)|$ has a minimum within the family; in the case in which $J(f) \neq 0$ for all members of the family, this minimum will be larger than zero. In the same way it follows that the problem Re $\{J(f)\}$—or, more generally, Re $\{e^{i\theta}J(f)\}$, $0 \leq \theta < 2\pi$, θ fixed—has a solution within the family.

This result is of fundamental importance in the treatment of *extremal problems*, that is, problems in which it is required to find the precise upper or lower limit of a functional defined with respect to a given class of functions. If this class can be shown to be a normal and compact family, then the existence of a solution of our extremal problem within the class is assured. If the extremum be a maximum, there will exist a function $f_0(z)$ of the class such that $|J(f)| \leq |J(f_0)|$ for all other functions of the class. In many cases, this inequality will yield a considerable amount of information with regard to the identity of the function $f_0(z)$, especially when the "functions of comparison" $f(z)$ are judiciously chosen. Examples of this procedure, which closely resembles that used in the calculus of variations, will be found in the following chapters.

We close this section with the proof of two results concerned with converging sequences of analytic functions. The first of these is the following.

Let $f_1(z)$, $f_2(z)$, . . . be a sequence of analytic functions which are regular and single-valued in a domain D and converge uniformly to a nonconstant function $f(z)$ regular in D. If $f(z_0) = 0$, where z_0 is a point of D, then, for any positive ϵ, there must be a zero of $f_n(z)$ within the circle $|z - z_0| < \epsilon$, provided n is taken sufficiently large.

We take ϵ small enough so that all points of the circle $|z - z_0| = \epsilon$ are in the interior of D and, moreover, $f(z)$ does not vanish in $|z - z_0| \leq \epsilon$ except at z_0. Since $f(z)$ is continuous on $|z - z_0| = \epsilon$, there exists a positive number m such that $|f(z)| > m$ on this circumference. The sequence $\{f_n(z)\}$ converges uniformly on $|z - z_0| = \epsilon$ and we shall therefore have $|f(z) - f_n(z)| < m$ for $|z - z_0| = \epsilon$, provided n is taken large enough. Hence,

$$|f(z) - f_n(z)| < m < |f(z)|, \qquad |z - z_0| = \epsilon.$$

By Rouché's theorem (Sec. 10, Chap. III), the function

$$f_n(z) = f(z) + [f_n(z) - f(z)]$$

will therefore have the same number of zeros in $|z - z_0| < \epsilon$ as the function $f(z)$. But $f(z)$ has one zero, at $z = z_0$, in this circle. Hence, the same is also true of $f_n(z)$, and our theorem is proved.

The second result refers to functions $f(z)$ which are regular and *univalent* in a domain D. A function $f(z)$ is said to be univalent in D if it does not take there the same value more than once, that is, we have $f(z_1) \neq f(z_2)$ if $z_1 \neq z_2$ and both z_1 and z_2 are in D.

If the functions of the sequence $f_1(z), f_2(z), \ldots$ *are regular and univalent in a domain* D *and converge in* D *to a regular nonconstant function* $f(z)$, *then* $f(z)$ *is also univalent in* D.

This is an easy consequence of the preceding result. Indeed, suppose $f(z_1) = f(z_2)$ $(z_1,z_2 \in D)$ and consider the sequence of functions

$$g_n(z) = f_n(z) - f_n(z_1).$$

Since $f_n(z)$ is univalent, we shall have $g_n(z) \neq 0$ except at $z = z_1$. The limit function $g(z) = f(z) - f(z_1)$ vanishes at $z = z_2$. By the preceding result $g_n(z)$ must therefore vanish within an arbitrarily small neighborhood of z_2, provided n is large enough. However, since $g_n(z)$ does not vanish in D except at z_1, this is impossible. Our assumption that $f(z_1) = f(z_2)$ thus leads to a contradiction and it follows that $f(z)$ is univalent in D.

EXERCISES

1. Let S denote the class of functions $f(z)$ which are regular and univalent in a domain D and satisfy $|f(z)| \leq 1$, $z \in D$. If $\zeta \in D$, show that there exists a function $f_0(z)$ of S such that

$$|f'(\zeta)| \leq |f_0'(\zeta)|,$$

where $f(z)$ is any other function of the class S.

2. Let $\{f_n(z)\}$ be a sequence of regular and single-valued functions in D which converges to a regular function $f(z)$ in D. If ϵ is arbitrarily small, show that the circle $|z - z_0| < \epsilon$ $(z_0 \in D)$ contains a point z_n such that $f_n(z_n) = f(z_0)$, provided n is taken large enough.

3. A p-valent function is defined as a function which takes no value more than p times in a given domain. Show that the (nonconstant) limit of a sequence of functions which are p-valent in D is also a p-valent function in D.

4. Show that

$$J(f) = \frac{f'(\zeta)}{1 - |f(\zeta)|^2}, \qquad \zeta \in D$$

is a continuous functional with respect to the class of functions $f(z)$ which are regular and single-valued in a domain D and satisfy there $|f(z)| < 1$.

5. If S is the family of functions defined in Exercise 1, show that there exists a function $f_0(z)$ of S such that

$$0 < |f_0(z_1) - f_0(z_2)| \leq |f(z_1) - f(z_2)|,$$

where z_1 and z_2 are two distinct points of D and $f(z)$ is any function of S.

6. Let $f(z)$ be regular and univalent in a domain D, and let z be a point of D. If $h_1, h_2, \ldots$ are complex numbers such that $z + h_n$ $(n = 1, 2, \ldots)$ is in D and $\lim_{n \to \infty} h_n \to 0$, show by considering the sequence of functions

$$\frac{f(z + h_n) - f(z)}{h_n}$$

that $f'(z)$ does not vanish in D.

CHAPTER V

CONFORMAL MAPPING OF SIMPLY-CONNECTED DOMAINS

1. Mapping Properties of Analytic Functions. A real function of a real variable can easily be visualized by means of a graph. If $y = f(x)$, we interpret x and y, respectively, as the abscissa and the ordinate of a point in a rectangular coordinate system, and the set of points obtained in this way if x is made to take all values of an interval $a \leq x \leq b$ "represents" the function $y = f(x)$ in this interval. This simple yet far-reaching idea which at one stroke, as it were, gives a geometrical meaning to all concepts and results of real analysis and which is the starting point of analytic geometry can, to a certain degree, be extended to the case of complex-valued functions of a complex variable. However, some changes are clearly called for. Since the values of both the function and the variable are complex numbers, a point representation of the function $w = f(z)$ would require four real numbers, thus necessitating the use of four-dimensional space. Since such a space is not accessible to our geometric visualization, but little would be gained by a geometric representation of this type.

A geometric representation of the function $w = f(z)$ which is free of this disadvantage is obtained by regarding z and w as points in two different planes—the z-plane and the w-plane—and by interpreting the function $w = f(z)$ as a *mapping* of points in the z-plane onto points in the w-plane. If $f(z)$ is regular and single-valued in a domain D, we shall say that the set of points in the w-plane which corresponds to the points of D by means of the mapping $w = f(z)$ is the *map* of D as given by this function; we shall also use the word *image* in this connection.

Some simple properties of these mappings are evident. If C is a continuous arc contained in a domain D in which $w = f(z)$ is regular and single-valued, its image will be an arc of the same nature. Indeed, if we write $w = u(x,y) + iv(x,y)$ $(z = x + iy)$, the image of the arc $x = x(t)$, $y = y(t)$, $t_1 \leq t \leq t_2$, is the arc $u = u[x(t),y(t)]$, $v = v[x(t),y(t)]$, and u and v are continuous functions of t if the same is true of x and y. Similarly, the image of a differentiable arc in D is again a differentiable arc. We leave it as an exercise to the reader to show that an analytic arc in D (see Sec. 11, Chap. I), is mapped onto an analytic arc. However, one property of these arcs is not necessarily preserved in such a mapping. The image of an arc that does not intersect itself may very well do so. In fact, if $f(z_1) = f(z_2)$, $z_1, z_2 \in D$, any nonintersecting continuous arc passing through z_1 and z_2 will obviously be mapped onto an arc that does intersect itself. If $f(z)$ is univalent in D such an occurrence is, of course, impossible.

The name *conformal mapping* for the mapping associated with a regular analytic function is derived from a property by which it is characterized. We say that a mapping is *conformal* if it preserves the angle between two differentiable arcs. To show that the mapping effected by a regular analytic function is indeed conformal, we proceed as follows. Suppose $f(z)$ is regular in the neighborhood of a point $z = z_0$ at which $f'(z_0) \neq 0$. The point $z = z_0$ is the terminal of two differentiable arcs $x_1 = x_1(t)$, $y_1 = y_1(t)$, $0 \leq t \leq 1$, and $x_2 = x_2(t)$, $y_2 = y_2(t)$, $0 \leq t \leq 1$, which intersect at $z = z_0$ under the angle α. In complex notation, the equations of these arcs are $z_1 = z_1(t)$ and $z_2 = z_2(t)$, respectively. If z_1 and z_2 are points on these curves which have the distance r from z_0, we have

$$z_1 - z_0 = re^{i\theta_1}, \qquad z_2 - z_0 = re^{i\theta_2}.$$

In view of

$$\frac{z_2 - z_0}{z_1 - z_0} = e^{i(\theta_2 - \theta_1)},$$

the angle formed by the linear segments connecting the points z_1 and z_0 and the points z_2 and z_0, respectively, is therefore $\theta_2 - \theta_1$. If $r \to 0$, this tends to the angle α. Hence

$$\alpha = \lim_{r \to 0} \arg \left\{ \frac{z_2 - z_0}{z_1 - z_0} \right\}. \tag{1}$$

It should be noted that, in this definition, α is measured starting from the arc $z_1 = z_1(t)$ and ending at the arc $z_2 = z_2(t)$. If w_1 and w_2 are the images of z_1 and z_2, respectively, then the images of the above two curves meet at $w_0 = f(z_0)$ under the angle

$$\beta = \lim_{r \to 0} \arg \left\{ \frac{w_2 - w_0}{w_1 - w_0} \right\}. \tag{2}$$

Hence,

$$\begin{aligned} \beta &= \lim_{r \to 0} \arg \left\{ \frac{f(z_2) - f(z_0)}{f(z_1) - f(z_0)} \right\}. \\ &= \lim_{r \to 0} \arg \left\{ \left(\frac{\dfrac{f(z_2) - f(z_0)}{z_2 - z_0}}{\dfrac{f(z_1) - f(z_0)}{z_1 - z_0}} \right) \left(\frac{z_2 - z_0}{z_1 - z_0} \right) \right\}. \end{aligned}$$

For $r \to 0$, both z_1 and z_2 tend to z_0. Hence,

$$\lim_{r\to 0} \frac{f(z_2) - f(z_0)}{z_2 - z_0} = \lim_{r\to 0} \frac{f(z_1) - f(z_0)}{z_1 - z_0} = f'(z_0).$$

Since $f'(z_0) \neq 0$, we obtain therefore

$$\beta = \lim_{r\to 0} \frac{z_2 - z_0}{z_1 - z_0} = \alpha.$$

The angle formed by the two arcs at $z = z_0$ is thus indeed identical with that formed by their images at $w_0 = f(z_0)$. A comparison of (1) and (2) shows that, moreover, the sense of the angle is preserved.

The preceding argument fails if $f'(z_0) = 0$, and it is easy to see that at points at which the derivative vanishes the mapping necessarily ceases to be conformal. If $f^{(m)}(z_0)$ is the first nonvanishing derivative of $f(z)$ at the point $z = z_0$, $f(z)$ can be expanded in the neighborhood of this point into a power series of the form

$$f(z) = f(z_0) + a_m(z - z_0)^m + a_{m+1}(z - z_0)^{m+1} + \cdots,$$

where $a_m \neq 0$. Hence, by (1) and (2),

$$\begin{aligned}\beta = \lim_{r\to 0} \arg \left\{\frac{f(z_2) - f(z_0)}{f(z_1) - f(z_0)}\right\} &= \lim_{r\to 0} \arg \left\{\left(\frac{z_2 - z_0}{z_1 - z_0}\right)^m\right\} \\ &= m \lim_{r\to 0} \arg \left\{\frac{z_2 - z_0}{z_1 - z_0}\right\} = m\alpha.\end{aligned}$$

This shows *that the angle between two differentiable curves which meet at a point* z_0 *at which* $f'(z_0) = \cdots = f^{(m-1)}(z_0) = 0$ *while* $f^{(m)}(z_0) \neq 0$ *is magnified* m *times by the mapping* $z \to f(z)$. The points z_0 at which $f'(z_0) = 0$ and at which, therefore, the mapping yielded by the function $w = f(z)$ ceases to be conformal will occasionally be referred to as critical points of the function $w = f(z)$.

In order to substantiate our previous statement that the conformality is a characteristic property of the mapping effected by regular analytic functions, we now have to show that no continuously differentiable mapping of one plane set onto another can be conformal unless it is the mapping associated with a regular analytic function. By the term "continuously differentiable mapping" we mean that the mapping is defined by a pair of real functions $u = u(x,y)$ and $v = v(x,y)$ which have continuous first partial derivatives with respect to x and y in the domain in question. We moreover suppose that the Jacobian $u_xv_y - u_yv_x$ of the transformation $(x,y) \to (u,v)$ does not vanish at the points considered. Let now $x = x_1(t)$, $y = y_1(t)$ and $x = x_2(t)$, $y = y_2(t)$ be two differentiable arcs which intersect at (x_0,y_0). The angles between the tangents to these arcs at (x_0,y_0) and the positive x-axis are given, respectively, by

$$\tan \alpha_1 = \frac{y_1'(t_0)}{x_1'(t_0)}, \qquad \tan \alpha_2 = \frac{y_2'(t_0)}{x_2'(t_0)}, \tag{3}$$

if t_0 denotes the value of the parameter corresponding to the point (x_0,y_0). For the sake of simplicity we shall restrict ourselves to arcs $x_1(t)$, $y_1(t)$ for which $y_1'(t_0) = 0$, that is, arcs whose tangents at (x_0,y_0) are parallel to the x-axis. We shall thus have $\alpha_1 = 0$ and the angle between the two arcs will be $\alpha = \alpha_2$. The images of the two arcs under the mapping $(x,y) \to (u,v)$ will be $u = u_\nu(t) = u[x_\nu(t),y_\nu(t)]$, $v = v_\nu(t) = v[x_\nu(t),y_\nu(t)]$, $\nu = 1, 2$. Since $u(x,y)$ and $v(x,y)$ have continuous first derivatives, we have $u_\nu' = u_x x_\nu' + u_y y_\nu'$, $v_\nu' = v_x x_\nu' + v_y y_\nu'$, $\nu = 1, 2$, where all functions are to be taken at the point (x_0,y_0). The angles between the two image arcs and the positive u-axis will therefore be given by

$$\tan \beta_1 = \frac{v_1'}{u_1'} = \frac{v_x x_1' + v_y y_1'}{u_x x_1' + u_y y_1'} = \frac{v_x}{u_x}, \qquad \tan \beta_2 = \frac{v_x x_2' + v_y y_2'}{u_x x_2' + u_y y_2'}.$$

If the mapping $(x,y) \to (u,v)$ is conformal at (x_0,y_0), the angle $\beta_2 - \beta_1$ must be equal to the angle $\alpha = \alpha_2$. In view of (3), we thus have

$$\frac{\dfrac{v_x + v_y \tan \alpha}{u_x + u_y \tan \alpha} - \dfrac{v_x}{u_x}}{1 + \dfrac{v_x}{u_x}\left(\dfrac{v_x + v_y \tan \alpha}{u_x + u_y \tan \alpha}\right)} = \tan \alpha,$$

whence

$$(u_x u_y + v_x v_y) \tan \alpha - u_x v_y + v_x u_y + u_x^2 + v_x^2 = 0.$$

Since this must be true for any $0 < \alpha < \frac{1}{2}\pi$, it follows that the two identities

$$u_x u_y + v_x v_y = 0, \qquad u_x v_y - v_x u_y = u_x^2 + v_x^2 \tag{4}$$

must hold. These identities can also be brought into the form

$$\begin{aligned} u_x(u_x - v_y) + v_x(v_x + u_y) &= 0, \\ -v_x(u_x - v_y) + u_x(v_x + u_y) &= 0. \end{aligned}$$

This is a system of two homogeneous linear equations for the unknowns $(u_x - v_y)$ and $(v_x + u_y)$, which can have nonzero solutions only if its determinant is zero. But the value of the determinant is $u_x^2 + v_y^2$; as shown by (4), this is precisely the value of the Jacobian of the transformation $(x,y) \to (u,v)$ which has been assumed not to vanish at (x_0,y_0). Hence, both solutions of the system of equations must be zero, that is, we have $u_x - v_y = 0$, $v_x + u_y = 0$. It follows that at a point at which the mapping $(x,y) \to (u,v)$ is conformal and its Jacobian does not vanish, the equations

$$u_x = v_y, \qquad u_y = -v_x$$

must hold. But, as a comparison with (14), Sec. 5, Chap. II, shows, these are the Cauchy-Riemann equations. By the results of Sec. 5, Chap. II, a pair of functions u and v which have continuous first derivatives at a point (x_0,y_0) and satisfy there the Cauchy-Riemann equations are the real part and the imaginary part of an analytic function $f(z) = u(x,y) + iv(x,y)$, $z = x + iy$, which is regular at $z = x_0 + iy_0$. We have thus proved that any continuously differentiable conformal mapping with a nonvanishing Jacobian is associated with a regular analytic function. Broadly speaking, the theory of analytic functions of a complex variable and the theory of conformal mapping are two different aspects of one and the same thing.

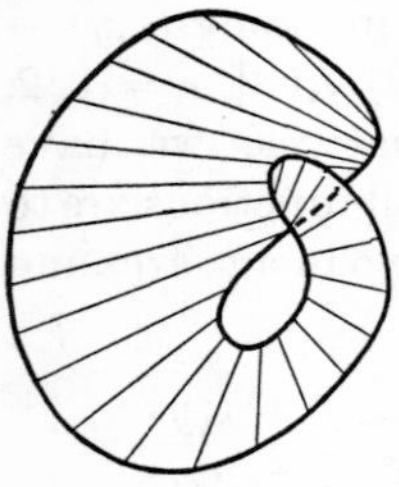

FIG. 8.

We shall now show that the conformal image of a domain D, given by an analytic function $w = f(z)$ which is regular and single-valued in D, is also a domain, provided the concept of domain is given a sufficiently general definition. That a more general definition is required becomes clear when we consider the case in which the function $w = f(z)$ takes the same value at two different points of D. Geometrically speaking, this means that parts of the conformal image of D "overlap." Figure 8 illustrates a particularly simple case of this type. A somewhat different case arises if the derivative of the mapping function vanishes at a point z_0 of D. We have seen before that the magnitude of an angle whose vertex is at z_0 is multiplied by the factor m if $f^{(m)}(z_0)$ is the first nonvanishing derivative of $f(z)$ at $z = z_0$. Thus, when a neighborhood of z_0 is swept out by arg $\{z - z_0\}$ growing from 0 to 2π, arg $\{f(z) - f(z_0)\}$ will increase by $2\pi m$. This shows that the conformal image of a neighborhood of z_0 is "wound" m times around the point $f(z_0)$. As an illustration, consider the function $w = f(z) = z^2$ in the unit circle $|z| < 1$. If $z = re^{i\theta}$, then $f(z) = r^2e^{2i\theta}$ which shows that the image of the circle $|z| = r$ is the doubly traversed circle $|w| = r^2$. All points of the circle $|w| < 1$ are taken twice in $|z| < 1$. At $z = 0$, we have $f'(0) = 0$ and the above-described phenomenon, with $m = 2$, occurs. The complete conformal map of $|z| < 1$ thus consists of two "sheets" of the circle $|w| < 1$ which have one point in common, namely, the point $w = 0$. This point is a *branch point of the first order* of this domain (compare the definition of a branch point in Sec. 7, Chap II). A branch point of order $m - 1$ will thus be a point w_0 such that in every neighborhood $|w - w_0| < \epsilon$ there will be points belonging to m sheets of the conformal image of D; the point w_0 itself belongs to all m sheets. Our definition of a domain will now be the same as before: a point set is a domain if it is open and connected. The only difference is that we now permit multiple coverings of the plane and the existence of isolated branch points which belong to more than one sheet at the same time.

After these preparations, we prove the following result: *The conformal map of a domain is again a domain.*

We note that this statement contains a slight inaccuracy, since the mapping ceases to be conformal at those points at which the derivative of the mapping function vanishes. It is, however, customary to speak of the "conformal map of a domain" even in the case in which there are critical points at which the mapping is not conformal; no confusion can arise from this practice. Let now $w = f(z)$ be regular in a domain D and let z_0 denote a point of D at which $f'(z_0) \neq 0$. The function $g(z) = f(z) - f(z_0)$ vanishes at $z = z_0$. Since $g(z)$ is regular in the neighborhood of z_0, this zero is isolated, and we can find a circle $|z - z_0| = \epsilon$ such that $g(z) \neq 0$ for $0 < |z - z_0| \leq \epsilon$. On $|z - z_0| = \epsilon$, $g(z)$ is continuous and there exists therefore a positive constant A such that $|g(z)| > A > 0$ for $|z - z_0| = \epsilon$. Let now ρ and φ be two constants such that $0 < \rho < A$ and $0 \leq \varphi < 2\pi$. Since, for $|z - z_0| = \epsilon$, $|g(z)| > \rho = |\rho e^{i\varphi}|$, it follows from Rouché's theorem (Sec. 10, Chap. III) that the functions $g(z)$ and $g(z) - \rho e^{i\varphi}$ have the same number of zeros in $|z - z_0| \leq \epsilon$. But $g(z)$ has precisely one zero there and the same is therefore true of

$$g(z) - \rho e^{i\varphi} = f(z) - f(z_0) - \rho e^{i\varphi}.$$

If we write $f(z_0) = w_0$ and $w = w_0 + \rho e^{i\varphi}$, it follows that, for $|z - z_0| \leq \epsilon$, the function $f(z)$ takes all values in the neighborhood $|w - w_0| < A$ and, moreover, it takes them exactly once. We have thus shown that if z_0 is not a critical point of the mapping function, then the conformal image D^* of D contains a neighborhood of the image point of z_0. If z_0 is a critical point, a slight modification of our argument is required. If the first nonvanishing derivative of $f(z)$ at $z = z_0$ is of the mth order, then $g(z) = f(z) - f(z_0)$ has a zero of the mth order at $z = z_0$. The rest of the argument runs as above, only that the values w for which $|w - w_0| < A$ will now be taken m times in $|z - z_0| \leq \epsilon$. It is therefore true of all points w_0 of D^* that they have neighborhoods which are entirely contained in D^*. In other words, D^* is an open set. In order to prove that D^* is a domain, we therefore only have to show that D^* is a connected set. But this is an immediate consequence of the fact that every continuous arc in D is mapped onto a continuous arc in D^*. This completes the proof.

We remark that the situation is particularly simple if the function $w = f(z)$ is univalent in D, that is, if it takes there no value more than once. In this case D^* contains no multiply covered points and D^* is therefore a plane domain in the usual sense. Such a domain which covers no point more than once is called a *schlicht* domain (some writers occasionally use the term *simple* as a substitute for the German word *schlicht*).

Another fact which is immediately apparent is that a conformal mapping preserves the connectivity of a domain. The conformal map of a simply-connected domain will thus be a simply-connected domain. Clearly, the connectivity of a domain is preserved under any continuous one-to-one transformation.

We close this section with the proof of an important theorem on the conformal correspondence of two domains.

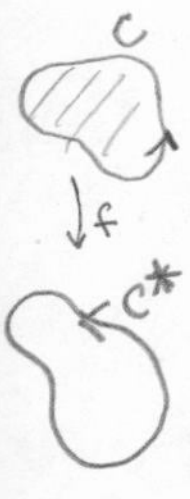

Let C and C^* be two simple closed contours in the z-plane and w-plane, respectively, and let $w = f(z)$ be regular on and within C. If $w = f(z)$ maps C onto C^* in such a way that C^* is traversed by w exactly once in the positive sense if z describes C in the positive sense, then $w = f(z)$ maps the domain bounded by C onto the domain bounded by C^*.

If the domains bounded by C and C^* are denoted by D and D^*, respectively, we have to prove that every point of D^* is taken exactly once if z is in D. By Sec. 10, Chap. III, the number of times a value w_0 is taken by $f(z)$ in D is

$$N = \frac{1}{2\pi i} \int_C \frac{f'(z)}{f(z) - w_0}\, dz.$$

With the substitution $w = f(z)$, $f'(z)\, dz = dw$, this can also be written

$$N = \frac{1}{2\pi i} \int_{C^*} \frac{dw}{w - w_0},$$

where the integration has now to be extended over the contour C^* into which C is transformed by $w = f(z)$. By the residue theorem, the value of this expression is 1 if w_0 is within C^*, and it is 0 if w_0 is outside C^*. This shows that every point in D^* is taken exactly once and that a value outside D^* is not taken at all. This proves our theorem.

EXERCISES

1. If the analytic function $w = f(z)$ $(z = x + iy,\ w = u + iv)$ maps a domain in the z-plane onto a domain in the w-plane, show that the value of the Jacobian of the transformation $(x,y) \to (u,v)$ at a point z_0 is $|f'(z_0)|^2$.

2. If the function $w = f(z)$ maps a domain D onto a schlicht domain D^*, show that the area A of D^* is

$$A = \iint_D |f'(z)|^2\, dx\, dy, \qquad z = x + iy.$$

Show that this formula also holds in the case in which D^* is not schlicht, provided the area of each sheet is counted separately.

3. If the function $f(z)$ is regular on a closed contour C and in the domain D enclosed by C, show that the length L of the boundary C^* of the domain onto which D is mapped by $w = f(z)$ is

$$L = \int_C |f'(z)||dz|.$$

4. Show that the last theorem of the text remains true if D is a domain of connectivity n whose boundary consists of n closed contours.

5. Find the families of curves C_1 and C_2 on which, respectively, $\operatorname{Re}\{z^2\} = \text{const.}$ and $\operatorname{Im}\{z^2\} = \text{const.}$; show that the two families C_1 and C_2 are orthogonal to each other. Which general principle does this illustrate?

6. If $w = e^z$, which curves in the w-plane correspond to the straight lines

$$\operatorname{Re}\{z\} = \text{const.}$$

and $\operatorname{Im}\{z\} = \text{const.}$ in the z-plane?

7. If the function $f(z)$ is regular and univalent in a domain D and satisfies there $|f(z)| \leq 1$, show that

$$\iint_D |f'(z)|^2\, dx\, dy \leq \pi.$$

8. If the function $f(z)$ is regular and $f'(z) \neq 0$ in a domain D, show that the mapping $w = \overline{f(z)}$ [the complex conjugate of $f(z)$] preserves the magnitude of the angles between two differentiable arcs in D but inverts their orientation.

9. Show that the function $w = e^z$ maps the interior of the rectangle $0 < x < 1$, $0 < y < 2\pi$ $(z = x + iy)$ onto the interior of the circular ring $1 < |w| < e$ which is cut along the positive axis. *Hint:* Study the mapping of the boundary, and use the last theorem of the text.

10. Let $w = f(z)$ be regular in $|z| \leq r$. If L denotes the length of the conformal image of the circle $|z| = r$ given by this mapping function, show that $L \geq 2\pi r|f'(0)|$. *Hint:* Express $f'(0)$ as a Cauchy integral.

11. Let $w = f(z)$ be regular in $|z| < 1$ and let A denote the area of the domain onto which the unit circle is mapped by $w = f(z)$. Use the result of the preceding exercise in order to show that $A \geq \pi|f'(0)|^2$. *Hint:* Use the Schwarz integral inequality.

12. If $U(x,y)$ is a harmonic function in a domain D and if $z = f(w)$ $(z = x + iy,$ $w = u + iv)$ maps a domain D^* in the w-plane onto D, show that

$$U^*(u,v) = U[x(u,v),y(u,v)]$$

$[f(w) = x(u,v) + iy(u,v)]$ is a harmonic function in D^*.

2. The Linear Transformation. A linear transformation is a transformation of the form

$$w = \frac{az + b}{cz + d}, \qquad ad - bc \neq 0, \tag{5}$$

where a, b, c, d are complex numbers and z is a complex variable. (5) is also known by the names of fractional linear transformation, bilinear transformation, and linear substitution. The condition $ad - bc \neq 0$ is necessary, as otherwise w will reduce to a constant. We first consider the particular cases of (5) in which $c = 0, d = 1$, and $a = 0, d = 0, b = c$, respectively, that is, the transformations

$$w = az + b \tag{6}$$

and

$$w = \frac{1}{z}. \tag{7}$$

It is easy to see what effect the transformation (6) has on any point in the z-plane. If, for the moment, we disregard the additive constant b and write $z = re^{i\theta}$, $a = |a|e^{i\varphi}$, we obtain $w = r|a|e^{i(\theta+\varphi)}$. Hence, the distances of all points z from the origin are multiplied by the same factor $|a|$ and, in addition, all points z are rotated by the angle α about the origin. The transformation $z \to az$ is thus equivalent to a magnification (or contraction, if $|a| < 1$) and a rotation of any geometric figure in the z-plane. In particular, the image of a circle under this transformation will again be a circle. The additive constant b in (6) will obviously cause the translation of a point $z_1 = x_1 + iy_1$ in the x- and y-directions by the amounts Re $\{b\}$ and Im $\{b\}$, respectively. It is therefore also true of the transformation (6) that it transforms circles into circles.

We now turn our attention to the transformation (7). Writing $z = re^{i\theta}$, we have

$$w = \frac{e^{-i\theta}}{r}, \tag{8}$$

which shows that w is in the exterior of the unit circle if z is in the interior, and vice versa. Moreover, arg $\{w\} = -$ arg $\{z\}$. Hence, although $|z| = 1$ is transformed into $|w| = 1$, the circumference $|w| = 1$ is described in the negative sense if $|z| = 1$ is described in the positive sense. (7), too, transforms circles into circles. In order to prove this property, we observe that all circles in the xy-plane have equations of the type $x^2 + y^2 + Ax + By + C = 0$, where A, B, C are real constants. In polar coordinates r, θ, this becomes $r^2 + r(A \cos\theta + B \sin\theta) + C = 0$. As shown by (8), the polar coordinates ρ, φ in the w-plane are connected with r, θ by the relations $\rho = 1/r$, $\varphi = -\theta$. Hence, the image of the above circle under the transformation (7) has the polar equation

$$\frac{1}{\rho^2} + \frac{1}{\rho}(A \cos\varphi - B \sin\varphi) + C = 0, \tag{9}$$

or, if $C \neq 0$,

$$\rho^2 + \rho\left(\frac{A}{C}\cos\varphi - \frac{B}{C}\sin\varphi\right) + \frac{1}{C} = 0.$$

This, obviously, is again the equation of a circle. If $C = 0$, it follows from (9) that

$$A\rho\cos\varphi - B\rho\sin\varphi + 1 = 0,$$

or, if $w = u + iv$, $Au - Bv + 1 = 0$. The image of the circle

$$x^2 + y^2 + Ax + By = 0,$$

which is the most general circle passing through the origin, is therefore a straight line. If we regard the straight lines as special cases of circles, namely, circles that pass through the point at infinity, we have thus shown that the transformation (7) indeed maps circles onto circles.

Returning now to (5), we remark that this transformation can also be written in the form

$$w = \frac{a}{c} + \frac{bc - ad}{c(cz + d)}. \tag{10}$$

This shows that (5) can be decomposed into the following three successive transformations:

$$z_1 = cz + d,$$

$$z_2 = \frac{1}{z_1},$$

$$w = \frac{a}{c} + \frac{bc - ad}{c} z_2.$$

The first and the third of these transformations are of the type (6) and the second coincides with (7). Since both (6) and (7) map circles onto circles, we have thus proved that *the general linear substitution* (5) *transforms all circles in the z-plane into circles in the w-plane.* Those circles in the w-plane which correspond to circles passing through the point $z = -(d/c)$ (whose image is $w = \infty$) degenerate into straight lines. Since, in (5), w is a regular analytic function for all values of z with the exception of $z = -(d/c)$ and since the derivative

$$\frac{dw}{dz} = \frac{ad - bc}{(cz + d)^2}$$

does not vanish at any finite point of the plane, the mapping effected by the linear transformation (5) is conformal at all finite points of the plane, with the possible exception of $z = -(d/c)$. However, the exceptional position of the points $z = \infty$ and $z = -(d/c)$ is only apparent. In view of (8), the transformation (7) will also preserve the angle between two differentiable arcs intersecting at the origin, although the latter is mapped onto $w = \infty$. If $f(z_0) = \infty$, the mapping effected by $w = f(z)$ at the point $z = z_0$ will therefore be conformal if the same is true of the mapping associated with the function $g(z) = [f(z)]^{-1}$. But this means that $g(z)$ must be regular at $z = z_0$ and $f'(z_0) \neq 0$. In other words, $g(z)$ has a simple zero at $z = z_0$, which shows that *the mapping by* $w = f(z)$ *is conformal at a point* $z = z_0$ *at which* $f(z_0) = \infty$ *if, and only if,* $f(z)$ *has a simple pole at* $z = z_0$. Hence, the mapping (5) is also conformal at $z = -(d/c)$. It follows in the same way that this mapping is conformal at $z = \infty$; we have only to replace z by $1/\zeta$ and to consider the point $\zeta = 0$.

In addition to yielding a conformal map of the whole z-plane, the function (5) is also distinguished by the fact that it is univalent in the entire z-plane. Indeed, if we write $w(z_\nu) = w_\nu (\nu = 1,2)$, it follows from (10) that

$$w_1 - w_2 = \frac{A(z_1 - z_2)}{(cz_1 + d)(cz_2 + d)}, \qquad A = ad - bc, \tag{11}$$

and this shows that $w_1 \neq w_2$ if $z_1 \neq z_2$. From the identity (11) we may draw another interesting conclusion. If z_1, z_2, z_3, z_4 are four distinct finite points in the z-plane and w_ν $(\nu = 1,2,3,4)$ their finite images in the w-plane as given by (5), it follows from (11) that

$$(w_1 - w_4)(w_3 - w_2) = \frac{A^2(z_1 - z_4)(z_3 - z_2)}{\prod_{\nu=1}^{4} (cz_\nu + d)}.$$

Similarly,

$$(w_1 - w_2)(w_3 - w_4) = \frac{A^2(z_1 - z_2)(z_3 - z_4)}{\prod_{\nu=1}^{4} (cz_\nu + d)}.$$

Hence,

$$\frac{(w_1 - w_4)(w_3 - w_2)}{(w_1 - w_2)(w_3 - w_4)} = \frac{(z_1 - z_4)(z_3 - z_2)}{(z_1 - z_2)(z_3 - z_4)}. \tag{12}$$

The expression

$$\frac{(z_1 - z_4)(z_3 - z_2)}{(z_1 - z_2)(z_3 - z_4)}$$

is called the *cross ratio* of the four points z_1, z_2, z_3, z_4. (12) shows that the cross ratio of four points is the same as the cross ratio of the images of these points as given by the transformation (5). In other words, *the cross ratio of four points is invariant under a linear transformation.* If one of the points w_ν, say w_1, is the point at infinity, the corresponding result is obtained by letting $w_1 \to \infty$ in (12). The left-hand side will then take the form

$$\frac{w_3 - w_2}{w_3 - w_4}.$$

This expression is therefore to be regarded as the cross ratio of the points ∞, w_2, w_3, w_4. A similar remark applies if one of the points z_ν is the point at infinity.

(12) is true for any four points z_ν $(\nu = 1,2,3,4)$ and the four points w_ν which correspond to them by the transformation (5). If z_4 is taken to be the variable z, w_4 will therefore be equal to w as defined by (5). As a consequence, it follows from (12) that the two variables z and w which are connected by (5) are also related by

$$\frac{(w_1 - w)(w_3 - w_2)}{(w_1 - w_2)(w_3 - w)} = \frac{(z_1 - z)(z_3 - z_2)}{(z_1 - z_2)(z_3 - z)}. \tag{13}$$

But it is easily seen that this relation between z and w is also of the form (5); hence, it is identical with it. (13) makes it possible to find a linear transformation which carries three arbitrarily given points z_1, z_2, z_3 into three preassigned points w_1, w_2, w_3. On the other hand, this formula also shows that a linear transformation is completely determined by three such correspondences. Since three points determine a circle and a linear transformation carries circles into circles, it follows therefore that, by a suitable linear transformation, any circle in the z-plane can be mapped onto any circle in the w-plane; we may, moreover, prescribe three points on the circumferences of each of these circles which are to correspond to each other. If this is done, the mapping is uniquely determined. If the circumference C_z is mapped onto the circumference C_w, it follows from the fact that the linear transformation is univalent in the entire plane and from the theorem at the end of the preceding section that the interior of C_z is mapped either onto the interior or onto the exterior of C_w. The latter case will happen if, and only if, the pole of the function (5) is situated in the interior of C_z. If C_w degenerates into a straight line, both the interior and the exterior of C_w become half-planes bounded by this straight line.

If we solve (5) for z, we obtain

$$z = \frac{b - dw}{-a + cw},$$

which shows that *the inverse of a linear transformation is also a linear transformation.* Another property which is immediately verified is the fact that *a linear transformation of a linear transformation is again a linear transformation.* Indeed, if we make the substitution

$$z = \frac{a_1 z_1 + b_1}{c_1 z_1 + d_1}$$

in (5), we find that the functional dependence of w with respect to z_1 is again of the form (5). In the language of algebra, these two properties can be expressed by saying that *the linear transformations* (5) *form a group.* The unit element of this group is, of course, the identical substitution $w = z$.

We have seen before that a linear transformation maps the interior of a circle onto the interior of a circle if the first circle does not contain the pole of the function (5). We shall now show that this property is characteristic of such transformations. In other words, we shall show that *an analytic function which maps the interior of a circle onto the interior of another circle is a linear transformation.** If the interior of the circle C_z is mapped onto the interior of the circle C_w, there is a point β within C_w which is the image of the center α of C_z. We proceed to show that, by a suitable linear transformation, C_w can be mapped onto a circle C_w' in such a way that β is carried into the center of C_w'. To this end, we construct the circle C_1 through β which intersects C_w at right angles and whose diameter through β passes through the center of C_w (Fig. 9). If γ denotes the other end of the diameter of C_1 through β, we consider the linear transformation

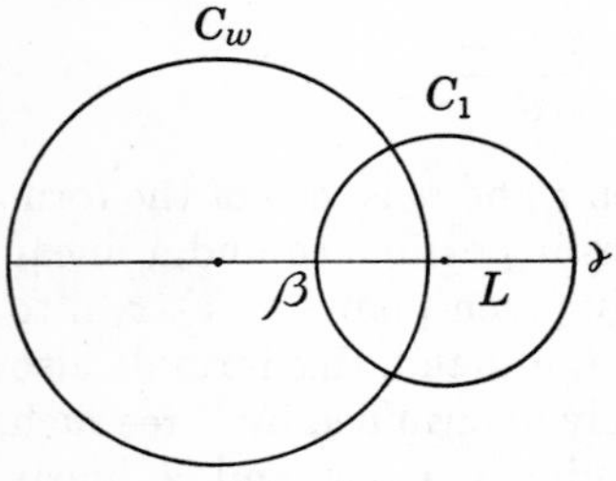

FIG. 9.

$$(14) \qquad w_1 = \frac{1}{w - \gamma}.$$

(14) transforms C_w into a circle C_w', while C_1 and the straight line L connecting β and γ are transformed into straight lines (both curves pass through $w = \gamma$). Since γ is outside C_w, the interior of C_w will be mapped onto the interior of C_w'. Since C_w and C_1 intersect at right angles, and since the mapping by (14) is conformal, the straight line onto which C_1 is mapped by (14) intersects C_w' at right angles. This shows that the image of C_1 is a diameter of C_w'. Similarly, the line L intersects C_w at right

* The reader should note that the phrase "a domain D is mapped *onto* a domain D'" implies a one-to-one correspondence between the points of D and those of D'. Hence, a case like that of the mapping $w = z^2$ which transforms the circle $|z| < 1$ into the doubly covered circle $|w| < 1$ is excluded from our present considerations.

angles and the same must therefore be true of the intersection of C_w' and the straight line which is the image of L. The latter is thus also a diameter of C_w'. Since β is situated at the point of intersection of C_1 and L within C_w, its image point δ must therefore lie on two different diameters of C_w'. Hence, δ is the center of C_w'.

Let now $w = f(z)$ be the function which maps the interior of C_z onto the interior of C_w. We may assume that the center α of C_z is at the origin; if this is not the case, we consider the function $f(z + \alpha)$ instead. If $f(0) = \beta$, then we define the point γ as above and consider the function

$$f_1(z) = \frac{1}{f(z) - \gamma}.$$

As shown before, $f_1(z)$ will map the interior of C_z onto the interior of a circle C_w' in such a way that the centers of the two circles are mapped into each other. The same is true of the function

$$g(z) = f_1(z) - \delta = \frac{1 + \gamma\delta - \delta f(z)}{f(z) - \gamma},$$

only that the image circle C_w'' of C_z has now its center at the origin and $g(0) = 0$. Consider now the function

$$h(z) = \frac{g(z)}{z}.$$

Since $g(0) = 0$ and the function $g(z)$ has, therefore, an expansion

$$g(z) = a_1 z + a_2 z^2 + \cdots = z(a_1 + a_2 z + \cdots)$$

near $z = 0$, $h(z)$ is regular at $z = 0$. Hence, $h(z)$ is regular throughout the interior of C_z. Clearly, $h(z)$ does not vanish in the interior of C_z. Indeed, the interior of C_w'' covers the origin only once, thus giving rise to one zero of $g(z)$ at $z = 0$; but this zero has been canceled by the factor z^{-1}. Moreover, $|h(z)|$ is constant on C_z. Indeed, the centers of both C_z and C_w'' are at the origin, and on C_z both $|z|$ and $|g(z)|$ are constant. The function $h(z)$ is thus free of zeros within C_z and has a constant absolute value on C_z. By Exercise 1, Sec. 8, Chap. III, it therefore reduces to a constant K. Hence,

$$g(z) = \frac{1 + \gamma\delta - \delta f(z)}{f(z) - \gamma} = Kz,$$

which shows that $f(z)$ is a linear transformation.

It is sometimes convenient to regard (5) not as a transformation of one plane onto another, but as a transformation of the plane onto itself. If we adopt this point of view, then the *fixed points* of the transformation become particularly interesting. By fixed points we mean those points of the plane which do not change their position if z is made subject to the transformation (5). Clearly, these points are the solutions of the equation

$$z = \frac{az + b}{cz + d},$$

or

$$cz^2 + (d - a)z - b = 0.$$

Except in the trivial case in which $a = d$, $b = c = 0$, the linear transformation will therefore have two fixed points; if $(d - a)^2 = -4bc$, these points coincide. If we exclude this case and denote the two fixed points by α and β, then it follows from (13) that the transformation can be written in the form

$$\left(\frac{w - \alpha}{w - \beta}\right)\left(\frac{w_2 - \beta}{w_2 - \alpha}\right) = \left(\frac{z - \alpha}{z - \beta}\right)\left(\frac{z_2 - \beta}{z_2 - \alpha}\right),$$

where z_2 is an arbitrary point of the z-plane and w_2 is the image of z_2. Since α, β, z_2, w_2 are constants, we may also write

$$\text{(15)} \qquad \frac{w - \alpha}{w - \beta} = \lambda\left(\frac{z - \alpha}{z - \beta}\right),$$

where λ is another constant. (15) represents the general form of a linear transformation if its two fixed points are given.

Two points P_1 and P_2 are called *inverse points* with respect to a circle of radius R and center P if P, P_1, P_2 lie, in that order, on the same straight line and if the distances $\overline{PP}_1$ and $\overline{PP}_2$ are related by

$$\overline{PP}_1 \cdot \overline{PP}_2 = R^2.$$

The linear transformation (5) has the important property that *the images of two points which are inverse with respect to a circle C_z are inverse with respect to the circle C_w into which C_z is transformed.* By a well-known theorem of elementary geometry, any circle C' which passes through two inverse points with respect to a circle C intersects C at right angles; conversely, if C' is orthogonal to C, any straight line passing through the center of C intersects C' at two points which are inverse with respect to C. The proof of this theorem follows from the easily established relation

$$\overline{PP}_1 \cdot \overline{PP}_2 = (\overline{PP}_3)^2$$

in Fig. 10, where PP_3 is the tangent to the circle and PP_2 is an arbitrary straight line intersecting the circle. Let now z_1 and z_2 be inverse points with respect to the circle C_z. All circles passing z_1 and z_2 are then orthogonal to C_z. The mapping (5), which transforms C_z into a circle C_w, transforms the circles passing through z_1 and z_2 into circles passing through the images w_1 and w_2 of z_1 and z_2. Since the mapping is conformal everywhere, these circles must intersect C_w at right angles. But this means that w_1 and w_2 are inverse points with respect to the circle C_w, and our statement is proved. We add that, in the case in which a circle degenerates into a straight line, a pair of inverse points becomes a pair of *symmetric points* with respect to the straight line, that is, points which are situated on a perpendicular to the line and are at equal distances from it. This is an immediate consequence of the fact that the line of symmetry passes through the centers of all circles which pass through the symmetric points.

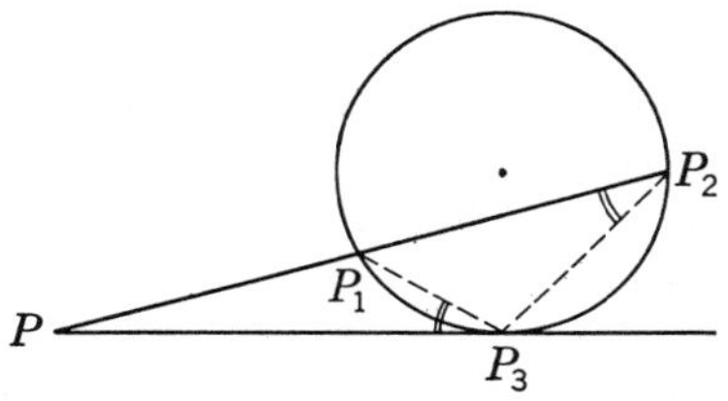

FIG. 10.

The fact that a linear substitution transforms inverse points into inverse points is of much help if it is desired to find a transformation which maps the interiors of two specified circles onto each other. As a first example, we determine the most general analytic function which maps the unit circle onto itself. As shown before, this must be a linear transformation. If $|z| < 1$ is mapped onto $|w| < 1$, there exists a point α ($|\alpha| < 1$) which is mapped onto the origin in the w-plane. The point inverse to α with respect to the circle $|z| < 1$ is clearly $1/\bar{\alpha}$. This point must be mapped onto the inverse to the origin with respect to $|w| = 1$, that is, onto the point $w = \infty$. The linear transformation (5) will therefore vanish for $z = \alpha$ and it will have its pole at $z = (\bar{\alpha})^{-1}$. Hence, it is necessarily of the form

$$w = k_1 \left(\frac{z - \alpha}{z - \frac{1}{\bar{\alpha}}} \right),$$

or

$$w = k \left(\frac{z - \alpha}{1 - \bar{\alpha} z} \right),$$

where k_1 and k are constants. For $|z| = 1$ we must have $|w| = 1$. Taking, in particular $z = 1$, we find that

$$w(1) = k \left(\frac{1 - \alpha}{1 - \bar{\alpha}} \right).$$

Since, obviously, $|1 - \alpha| = |1 - \bar{\alpha}|$, it follows that $|k| = 1$, or, what amounts to the same thing, $k = e^{it}, 0 \leq t < 2\pi$. Our transformation is therefore of the form

$$(16) \qquad w = e^{it}\left(\frac{z - \alpha}{1 - \bar{\alpha}z}\right), \qquad 0 \leq t < 2\pi, \ |\alpha| < 1.$$

Conversely, any transformation (16) maps the unit circle onto itself; this is an immediate consequence of the fact that, for $z = e^{i\theta}$,

$$|1 - \bar{\alpha}z| = |1 - \bar{\alpha}e^{i\theta}| = |e^{-i\theta}(1 - \bar{\alpha}e^{i\theta})| = |e^{-i\theta} - \bar{\alpha}|$$
$$= |e^{i\theta} - \alpha| = |z - \alpha|.$$

(16) is therefore the most general function mapping the unit circle onto itself. It contains three arbitrary real parameters, namely, t, Re $\{\alpha\}$, and Im $\{\alpha\}$. This corresponds to the fact, shown above, that on each of the circles $|z| = 1$ and $|w| = 1$ we can arbitrarily choose three points which are to be mapped onto each other and that, once this choice is made, the mapping is completely determined. (16) shows that the function mapping the unit circle onto itself is also completely determined if we prescribe the point α which is to correspond to $w = 0$ and the argument t of the derivative of the mapping function at the point $z = \alpha$.

As another example, we consider the mapping of the upper half-plane Im $\{z\} > 0$ onto the unit circle $|w| < 1$. If α is the point corresponding to $w = 0$, its inverse with respect to the real axis, that is, $\bar{\alpha}$, corresponds to the inverse of $w = 0$ with respect to $|w| = 1$, namely, the point $w = \infty$. The linear transformation in question must therefore be of the form

$$w = k\left(\frac{z - \alpha}{z - \bar{\alpha}}\right).$$

Since, for real z, $|z - \alpha| = |z - \bar{\alpha}|$, we have $|k| = 1$. Hence, the desired mapping is of the form

$$(17) \qquad w = e^{it}\left(\frac{z - \alpha}{z - \bar{\alpha}}\right), \qquad 0 \leq t < 2\pi.$$

EXERCISES

1. Show that the function

$$w = f(z) = \frac{z}{1 - z}$$

maps the unit circle $|z| < 1$ onto the half-plane Re $\{w\} > -\frac{1}{2}$.

2. Show that the most general transformation of the upper half-plane onto itself is of the form

$$w = \frac{az + b}{cz + d},$$

where a, b, c, d are real and $ad - bc > 0$.

3. If the transformation (13) is brought into the form (5), show that

$$ad - bc = (w_1 - w_2)(w_1 - w_3)(w_2 - w_3)(z_1 - z_2)(z_1 - z_3)(z_2 - z_3).$$

4. By the successive substitutions

$$w = w_1 = \frac{aw_2 + b}{cw_2 + d}, \quad w_2 = \frac{aw_3 + b}{cw_3 + d}, \quad \ldots, \quad w_{n-1} = \frac{aw_n + b}{cw_n + d}, \quad w_n = \frac{az + b}{cz + d},$$

we obtain a linear transformation

$$w = \frac{Az + B}{Cz + D},$$

which is called the "nth iterate" of the transformation $w = (az + b)(cz + d)^{-1}$. Show that the nth iterate will be the identical substitution $w = z$ if the original transformation can be brought into the form

$$\frac{w - \alpha}{w - \beta} = e^{\frac{2\pi i}{n}} \left(\frac{z - \alpha}{z - \beta}\right).$$

5. Show that the transformation $w = z^{-1}$ maps the common part of the two circles $|z - 1| < 1$ and $|z + i| < 1$ onto the quarter-plane Re $\{w\} > \frac{1}{2}$, Im $\{w\} > \frac{1}{2}$.

6. Let T be a curvilinear triangle bounded by three circular arcs. Show that there exists a linear transformation mapping T onto a rectilinear triangle if, and only if, the three circles bounding T have a common point.

7. Let $ad - bc = 1$ and denote by C the circle $|cz + d| = 1$. Show that the linear transformation

$$w = \frac{az + b}{cz + d}$$

increases lengths and areas within C and decreases lengths and areas outside C.

8. Show that the cross ratio of four distinct points is real if, and only if, the four points lie on the same circle.

9. If z_2 and z_4 are inverse points with respect to a circle C and z_1, z_3 are two arbitrary points of C, show that the cross ratio of the four points z_1, z_2, z_3, z_4 has the absolute value 1.

10. By different labeling of four given distinct points it is possible to obtain different values for their cross ratio. Show that there are altogether six different values for their cross ratio and that, if the value of one of them be denoted by λ, the values of the five others are

$$\frac{1}{\lambda}, \quad 1 - \lambda, \quad \frac{1}{1 - \lambda}, \quad \frac{\lambda}{\lambda - 1}, \quad \frac{\lambda - 1}{\lambda}.$$

11. Let C_1 and C_2 be two circles which intersect with the angle θ, and let L denote the straight line passing through the centers of C_1 and C_2. Show that one of the values of the cross ratio of the four points at which C_1 and C_2 intersect L is $\sin^2 \frac{1}{2}\theta$.

3. The Schwarz Lemma. The following very useful result is known by the name of the Schwarz lemma.

Let the analytic function $f(z)$ *be regular in the unit circle* $|z| < 1$ *and let* $f(0) = 0$. *If, in* $|z| < 1$, $|f(z)| \leq 1$, *then*

$$|f(z)| \leq |z|, \qquad |z| < 1, \tag{18}$$

where equality can hold only if $f(z) \equiv Kz$ *and* $|K| = 1$.

The proof is very simple. Since $f(0) = 0$, the function $f(z)/z$ is regular at $z = 0$ and therefore throughout the unit circle. If $0 < \rho < 1$, it follows from $|f(z)| \leq 1$ and the maximum principle that

$$\left|\frac{f(z)}{z}\right| \leq \frac{1}{\rho}, \qquad |z| < \rho.$$

Since ρ can be arbitrarily close to 1, we thus have

$$\left|\frac{f(z)}{z}\right| \leq 1, \qquad |z| < 1, \tag{19}$$

which is equivalent to (18). If, for a point z_0 in the unit circle, we have equality in (18), then the left-hand side of (19) takes the value 1 at $z = z_0$. By the maximum principle, it therefore must reduce to a constant K of absolute value 1. Hence, $f(z)$ is of the form Kz. This proves the Schwarz lemma. We remark that, in view of the power series expansion

$$f(z) = f'(0)z + \frac{f''(0)}{2!}z^2 + \cdots,$$

we also obtain from (19)

$$|f'(0)| \leq 1 \tag{20}$$

by setting $z = 0$. Equality in (20) will again only hold for $f(z) \equiv Kz$, $|K| = 1$.

A function $f(z)$ which is regular in a domain and satisfies there $|f(z)| \leq M$, where M is a positive constant, is called a *bounded function* in this domain. For the sake of concise formulation, we shall furthermore assume that $M = 1$ if we speak of a bounded function without referring to its bound; if $M \neq 1$, we have only to divide $f(z)$ by M in order to obtain a function of this type. The Schwarz lemma is valid for bounded functions in the unit circle which vanish at the origin. We shall now show that a similar result holds for functions which, although bounded in $|z| < 1$, do not vanish at $z = 0$. We first recall that the linear transformation (16) maps the unit circle onto itself. Since $|f(z)| \leq 1$ for $|z| < 1$, the points $w_1 = f(z)$ will be within or on the circle $|w_1| = 1$. It therefore follows from (16) that the points

$$w = g(z) = \frac{f(z) - \alpha}{1 - \bar{\alpha}f(z)}, \qquad |\alpha| < 1,$$

will satisfy $|w| \leq 1$. Hence $g(z)$ is also a regular and bounded analytic function in $|z| < 1$. Since $|f(0)| < 1$, we may take $\alpha = f(0)$. We thus obtain the bounded function

$$g(z) = \frac{f(z) - f(0)}{1 - \overline{f(0)}f(z)}, \tag{21}$$

which, obviously, vanishes for $z = 0$. To $g(z)$ we may apply the Schwarz lemma (18). Hence,

$$|g(z)| \leq |z|. \tag{22}$$

Solving (21) for $f(z)$, we obtain

$$f(z) = \frac{f(0) + g(z)}{1 + \overline{f(0)}g(z)}. \tag{23}$$

(22) shows that, if $|z| < \rho$, the values of $g(z)$ are contained in the circle of radius ρ about the origin. Since, by (23), $f(z)$ is obtained from $g(z)$ by the linear transformation

$$w_1 = \frac{f(0) + w}{1 + \overline{f(0)}w}, \tag{24}$$

the values of $f(z)$ must therefore be contained in the interior of the circle C onto which $|w| = \rho$ is mapped by (24). This is the Schwarz lemma for bounded functions $f(z)$ for which $f(0) \neq 0$. An estimate for the value of $|f(z)|$ can be obtained by means of the inequalities

$$\frac{|a| - |b|}{1 - |a||b|} \leq \frac{|a + b|}{|1 + \bar{a}b|} \leq \frac{|a| + |b|}{1 + |a||b|}, \qquad |a| < 1,\ |b| < 1, \tag{25}$$

the proof of which is left as an exercise to the reader. If $|w| < |f(0)|$, the circle C obviously does not contain the origin; hence, the distance from the origin of any point within C is larger than the distance from the origin of the point of C which is closest to the origin. It follows therefore from (23) and (25) that

$$\frac{|f(0)| - |z|}{1 - |f(0)||z|} \leq |f(z)| \leq \frac{|f(0)| + |z|}{1 + |f(0)||z|}. \tag{26}$$

The left-hand side of (26) has been proved for $|z| < |f(0)|$; for $|f(0)| \leq |z| < 1$ it holds trivially.

An estimate for the derivative of a bounded function $f(z)$ can be obtained in the following fashion. If $f(z)$ is bounded in $|z| < 1$, the same is true of the function

$$g(z) = \frac{f(z) - f(\zeta)}{1 - \overline{f(\zeta)}f(z)}, \qquad |\zeta| < 1.$$

$g(z)$ vanishes for $z = \zeta$. The function

$$(27) \qquad h(z) = \frac{g(z)}{\dfrac{z-\zeta}{1-\bar{\zeta}z}} = \left(\frac{f(z)-f(\zeta)}{z-\zeta}\right)\left(\frac{1-\bar{\zeta}z}{1-\overline{f(\zeta)}f(z)}\right)$$

is therefore regular at $z = \zeta$ and also at all other points of $|z| < 1$. Furthermore, $h(z)$ is bounded in $|z| < 1$. Indeed, $\overline{\lim\limits_{|z|\to 1}} \, |g(z)| \leq 1$ and

$$\left|\frac{z-\zeta}{1-\bar{\zeta}z}\right| = 1$$

for $|z| = 1$; hence, by the maximum principle, $|h(z)| \leq 1$ throughout $|z| < 1$. Setting $z = \zeta$, we thus find from (27) that

$$|f'(\zeta)| \left(\frac{1-|\zeta|^2}{1-|f(\zeta)|^2}\right) \leq 1,$$

whence

$$(28) \qquad |f'(z)| \leq \frac{1-|f(z)|^2}{1-|z|^2}.$$

If it is known that a bounded function $f(z)$ vanishes at the points $\alpha_1, \ldots, \alpha_n$ in the unit circle, it is possible to obtain more accurate information about the value of $|f(z)|$. If $f(\alpha_\nu) = 0$, $\nu = 1, \ldots, n$, and $|\alpha_\nu| < 1$, it follows by the argument used in showing that the function $h(z)$ is regular and bounded in $|z| < 1$ that the function

$$h(z) = \frac{f(z)}{\prod\limits_{\nu=1}^{n} \left(\dfrac{z-\alpha_\nu}{1-\bar{\alpha}_\nu z}\right)}$$

is regular and bounded in the unit circle. We thus have

$$(29) \qquad f(z) = h(z) \prod_{\nu=1}^{n} \left(\frac{z-\alpha_\nu}{1-\bar{\alpha}_\nu z}\right), \qquad |h(z)| \leq 1, \ |z| < 1.$$

It follows from (29) that

$$|f(z)| \leq \prod_{\nu=1}^{n} \left|\frac{z-\alpha_\nu}{1-\bar{\alpha}_\nu z}\right|$$

and in particular

$$(30) \qquad |f(0)| \leq \prod_{\nu=1}^{n} |\alpha_\nu|.$$

From the results on bounded functions it is easy to deduce corresponding results concerning *functions with a positive real part*, that is, functions $f(z)$ for which $\operatorname{Re}\{f(z)\} \geq 0$ in $|z| < 1$. The linear substitution

$$w_1 = \frac{w-1}{w+1}$$

transforms the right half-plane $\operatorname{Re}\{w\} > 0$ into the unit circle $|w_1| < 1$; indeed, the points $w = 0, i, \infty$ are transformed into the points $w_1 = -1, i, 1$, respectively, and $w_1(1) = 0$. If $\operatorname{Re}\{f(z)\} \geq 0$, it follows therefore that the function

$$g(z) = \frac{f(z)-1}{f(z)+1}$$

satisfies $|g(z)| \leq 1$. Hence, any function $f(z)$ with a positive real part can be written in the form

$$f(z) = \frac{1+g(z)}{1-g(z)}, \qquad |g(z)| \leq 1. \tag{31}$$

If, in particular, $f(0) = 1$, then $g(0) = 0$. In view of the obvious inequality

$$\left|\frac{1+a}{1-a}\right| \leq \frac{1+|a|}{1-|a|}, \qquad |a| < 1,$$

we thus obtain from (18) and (31) the inequality

$$|f(z)| \leq \frac{1+|z|}{1-|z|}. \tag{32}$$

The assumption $f(0) = 1$ is no restriction on the generality of this inequality. If $f(0) = \alpha + i\beta$ $(\alpha > 0)$, then $f_1(z) = \frac{1}{\alpha}[f(z) - i\beta]$ satisfies $\operatorname{Re}\{f_1(z)\} \geq 0$, $f_1(0) = 1$, and (31) applies to $f_1(z)$. A lower bound for $|f(z)|$ is obtained from the observation that the function $[f(z)]^{-1}$ also has a positive real part. Indeed,

$$\operatorname{Re}\left\{\frac{1}{f(z)}\right\} = \operatorname{Re}\left\{\frac{\overline{f(z)}}{|f(z)|^2}\right\} = \operatorname{Re}\left\{\frac{f(z)}{|f(z)|^2}\right\} \geq 0$$

[that $f(z) \neq 0$ for $|z| < 1$ is a trival consequence of $\operatorname{Re}\{f(z)\} \geq 0$]. It therefore follows from (32) that

$$\frac{1}{|f(z)|} \leq \frac{1+|z|}{1-|z|},$$

or

$$\frac{1-|z|}{1+|z|} \leq |f(z)| \leq \frac{1+|z|}{1-|z|}. \tag{33}$$

The example of the function

$$f_0(z) = \frac{1 + z}{1 - z} \tag{34}$$

shows that the inequalities (33) are the best possible. This function maps $|z| < 1$ onto the right half-plane and has therefore a positive real part in $|z| < 1$; for $z = -|z|$ and $z = |z|$, respectively, the sign of equality holds in the two inequalities (33). An inequality which is the "best possible" in this sense is also referred to as a *sharp* inequality.

Another set of sharp inequalities concerns the coefficients $a_1, a_2, \ldots$ of the power series expansion

$$f(z) = 1 + a_1 z + a_2 z^2 + \cdots + a_n z^n + \cdots \tag{35}$$

of a function $f(z)$ with a positive real part in $|z| < 1$. We have

$$|a_n| \leq 2, \qquad n = 1, 2, \ldots, \tag{36}$$

where equality holds for the function (34). For $n = 1$, (36) is an immediate consequence of (31) and (20). In order to prove (36) for general n, we assume that $f(z)$ is regular in $|z| \leq 1$ and consider the integral

$$I = \frac{1}{2\pi i} \int_{|z|=1} f(z) \left[2 - z^n - \frac{1}{z^n} \right] \frac{dz}{z}.$$

By the residue theorem and (35), we have

$$I = 2 - a_n.$$

Introducing polar coordinates $z = e^{i\theta}$ in the integral, we obtain

$$I = \frac{2}{\pi} \int_0^{2\pi} f(e^{i\theta}) \sin^2 \frac{\theta}{2} \, d\theta,$$

whence

$$\operatorname{Re} \{2 - a_n\} = \operatorname{Re} \{I\} = \frac{2}{\pi} \int_0^{2\pi} \operatorname{Re} \{f(e^{i\theta})\} \sin^2 \frac{\theta}{2} \, d\theta.$$

In view of $\operatorname{Re} \{f(e^{i\theta})\} \geq 0$, it follows therefore that $\operatorname{Re} \{2 - a_n\} \geq 0$, or

$$\operatorname{Re} \{a_n\} \leq 2. \tag{36$'$}$$

If $\operatorname{Re} \{f(z)\} \geq 0$, we obviously also have $\operatorname{Re} \{f(e^{it}z)\} \geq 0$ $(0 \leq t < 2\pi)$; the transformation $z \rightarrow e^{it}z$ amounts only to a rotation of the unit circle about the origin. The nth coefficient of the latter function is $e^{int}a_n$. Hence, by (36′),

$$\operatorname{Re} \{e^{int}a_n\} \leq 2, \qquad 0 \leq t < 2\pi.$$

Taking, in particular, a value of t such that $nt + \arg\{a_n\} = 0$, we obtain $e^{int}a_n = |a_n|$ and therefore

$$|a_n| \leq 2.$$

This proves (36) in the case in which $f(z)$ is regular in the closure of the unit circle. If $f(z)$ is regular only in $|z| < 1$, we observe that $f_1(z) = f(\rho z)$, $0 < \rho < 1$, is regular in $|z| < 1$ and has there a positive real part. Its nth coefficient is $\rho^n a_n$. Hence $\rho^n|a_n| \leq 2$, where ρ is any value between 0 and 1. Letting $\rho \to 1$, we obtain $|a_n| \leq 2$, which shows that the inequality (36) is also true in the general case. (36) is sharp since the function (34) has the expansion

$$f_0(z) = \frac{1+z}{1-z} = 1 + 2z + 2z^2 + \cdots + 2z^n + \cdots.$$

EXERCISES

1. If $f(z)$ is regular and bounded in $|z| < 1$ and if $|\zeta| < 1$, show that the function

$$g(z) = \frac{f\left(\frac{\zeta + z}{1 + \bar{\zeta}z}\right) - f(\zeta)}{1 - \overline{f(\zeta)}f\left(\frac{\zeta + z}{1 + \bar{\zeta}z}\right)}$$

is also regular and bounded and that $g(0) = 0$. Use this result and the inequality (20) in order to establish (28).

2. Using the identity

$$f(z_1) - f(z_2) = \int_{z_1}^{z_2} f'(z)\,dz$$

and integrating along the linear segment connecting z_1 and z_2, deduce from (28) that a bounded function $f(z)$ is subject to the inequality

$$\left|\frac{f(z_1) - f(z_2)}{z_1 - z_2}\right| \leq \frac{1}{1 - r^2}, \qquad |z_1| < r,\ |z_2| < r,\ r < 1.$$

3. If $f(z)$ is bounded and $f(z_1) = f(z_2) = \beta$, $|z_1| = |z_2| = \rho < 1$, $f(0) = 0$, show that

$$\frac{f(z) - \beta}{1 - \bar{\beta}f(z)} = \left(\frac{z - z_1}{1 - \bar{z}_1 z}\right)\left(\frac{z - z_2}{1 - \bar{z}_2 z}\right) h(z), \qquad |h(z)| \leq 1,\ |z| \leq 1,$$

and deduce that $|\beta| \leq \rho^2$.

4. If $f(z)$ is bounded in $|z| < 1$, $f(0) = 0$, and $|f'(0)| = \alpha$, show that

$$\rho(\alpha - \rho) \leq (1 - \alpha\rho)|f(\rho e^{i\theta})|, \qquad 0 \leq \theta < 2\pi.$$

Hint: Show that $z^{-1}f(z)$ is bounded and use (26).

5. Use the results of Exercises 3 and 4 to prove the following theorem.

If $f(z)$ is bounded in $|z| < 1$, $f(0) = 0$, and $|f'(0)| = \alpha$, then $f(z)$ is univalent in the interior of the circle

$$|z| < \rho_0 = \frac{\alpha}{1 + \sqrt{1 - \alpha^2}}.$$

Remark: In view of the last theorem in Sec. 1, it is sufficient to show that we cannot have $f(z_1) = f(z_2)$, $z_1 \neq z_2$, $|z_1| = |z_2| = \rho < \rho_0$.

6. By considering the bounded function

$$f_0(z) = z\left(\frac{\alpha - z}{1 - \bar{\alpha}z}\right), \qquad 0 < \alpha < 1,$$

show that the "radius of univalence" ρ_0 of the preceding exercise is the largest possible for the class of functions considered.

7. If ω is defined by

$$\omega = e^{\frac{2\pi i}{n}},$$

where n is an integer, show that

$$\sum_{\nu=1}^{n} \omega^{\nu m} = 0$$

if the integer m is neither 0 nor an integral multiple of n, and show that in these excluded cases the value of the sum is n.

8. If $f(z)$ is regular in $|z| < 1$ and has there the power series expansion

$$f(z) = a_0 + a_1 z + \cdots + a_\nu z^\nu + \cdots,$$

deduce from the result of the preceding exercise that

$$\begin{aligned} g(z) &= \frac{1}{n}[f(\omega z) + f(\omega^2 z) + \cdots + f(\omega^n z)] \\ &= a_0 + a_n z^n + a_{2n} z^{2n} + \cdots + a_{\nu n} z^{\nu n} + \cdots, \end{aligned}$$

and that the function $h(z) = g(\sqrt[n]{z})$ is also regular in $|z| < 1$.

9. Show that the function $h(z)$ of the preceding exercise is bounded in $|z| < 1$ if the same is true of the function $f(z)$ by means of which $h(z)$ has been defined. If

$$f(z) = a_0 + a_1 z + a_2 z^2 + \cdots,$$

use this result in order to prove the inequality

$$|a_n| \leq 1 - |a_0|^2, \qquad n = 1, 2, \ldots,$$

for the coefficients of a bounded function. *Hint:* Use the fact, following from (28), that $|f'(0)| \leq 1 - |f(0)|^2$ if $f(z)$ is bounded.

10. Use the result of the preceding exercise to prove that

$$\sum_{n=0}^{\infty} |a_n|\rho^n \leq 1, \qquad \rho \leq \tfrac{1}{3},$$

if $f(z) = \sum_{n=0}^{\infty} a_n z^n$ is bounded in $|z| < 1$. By considering the bounded function

$$f_0 = \frac{\alpha - z}{1 - \bar{\alpha}z}, \qquad |\alpha| < 1,$$

for suitable values of α, show that the constant $\frac{1}{3}$ is the largest possible.

11. Show that the linear substitution

$$w = \frac{1+z}{1-z}$$

transforms the circle $|z| < \rho$ into the circle

$$\left| w - \frac{1+\rho^2}{1-\rho^2} \right| < \frac{2\rho}{1-\rho^2}.$$

Use (31) and this result in order to give an alternative proof of the inequalities (33). Show that this method also yields the inequality

$$\text{Re } \{f(z)\} \geq \frac{1-|z|}{1+|z|}, \qquad |z| < 1,$$

for a function $f(z)$ with a positive real part. Why is this inequality generally better than the left-hand inequality (33)?

12. Let $f(z)$ be bounded in $|z| < 1$, and let $f(z)$ have an infinite number of zeros in $|z| < 1$. If $\alpha_1, \alpha_2, \ldots$ are these zeros [which cannot have a limit point in $|z| < 1$ unless $f(z)$ is identically zero], show by means of (30) that

$$\sum_{n=0}^{\infty} \log |\alpha_n|$$

converges.

13. If the non-Euclidean length L of an arc C within the unit circle is defined by

$$L = \int_C \frac{|dz|}{1-|z|^2},$$

show that the mapping $z \to w$ by a bounded function $w = f(z)$ has the effect of decreasing the non-Euclidean length of all arcs in $|z| < 1$.

14. Prove the inequalities (26) by applying the result of the preceding exercise to the linear segment connecting the point z with the origin. *Hint:* Use the fact that, on this segment,

$$\frac{d|f|}{d|z|} \leq \left|\frac{df}{dz}\right|.$$

4. The Riemann Mapping Theorem. We saw in Sec. 1 that the conformal map of a domain D is again a domain, say D^*. The shape of D^* depends, of course, on the particular function $w = f(z)$ by means of which the mapping is effected. While much insight into the nature of conformal mapping can be gained by investigating the mappings yielded by given functions $w = f(z)$ of given domains D, this is a piecemeal procedure which does not reveal the true potentialities inherent in the subject of conformal mapping. The decisive step, first taken by Bernhard Riemann (1826–1866), to which the theory of conformal mapping owes its modern development is a shift of emphasis from the function-theoretical to the geometric side of the problem. Instead of studying the mappings associated with a given analytic function, we assume that two domains D and D^* are given and we ask whether there exists an analytic function $w = f(z)$ by means of which D is mapped onto D^*. We may even avoid mentioning the function $f(z)$ altogether and ask whether, given two domains D and D^*, there always exists a continuously differentiable conformal mapping of D onto D^*.

Clearly, the answer to this question cannot be in the affirmative unless D and D^* are of the same order of connectivity. We shall here confine ourselves to the case of simply-connected domains D and D^*, postponing the discussion of the multiply-connected case to Chap. VII. But even two simply-connected domains cannot always be mapped conformally onto each other, or, as we shall also say, are not "conformally equivalent" to each other, as the following simple example shows. Let D be a schlicht domain which has only one boundary point, and let D^* be the unit circle. D thus consists of the whole plane with the exception of one point and it is clearly simply-connected. We may assume that the boundary point of D is the point at infinity, as this can always be achieved by a linear transformation. If D were conformally equivalent to D^*, there would exist an analytic function $w = f(z)$ which is regular at all finite points of the z-plane and which would take there values within the circle $|w| < 1$. Hence, $|f(z)| \leq 1$ for all finite z, and it follows by Liouville's theorem that $f(z)$ reduces to a constant, which is absurd.

Leaving aside the not very interesting case of domains whose boundary consists of only one point, we shall show that any two simply-connected schlicht domains whose boundaries consist of more than one point are conformally equivalent. In other words, given two arbitrary domains D and D^* whose boundaries consist of more than one point, there always exists an analytic function $w = f(z)$ which maps D onto D^*. In fact, there even exists an infinity of such mapping functions, depending on three real parameters. If these parameters are disposed of by the requirements that a given point z_0 of D should be mapped onto a given point of D^* and that $\arg \{f'(z_0)\}$ should have a given value, then the mapping function is uniquely determined. The fact that all simply-connected domains with more than one boundary point are conformally equivalent is known as the *Riemann mapping theorem* and is of fundamental importance in the theory of conformal mapping and in its applications. It is clearly sufficient to prove the Riemann mapping theorem for the case in which one of the two domains is the unit circle. If both D and D^* can be mapped conformally onto the unit circle, then D can be mapped onto D^* by first mapping D onto the unit circle and then mapping the unit circle onto D^*. The facts implied in this procedure, namely, that the inverse of a conformal map is also a conformal map and that the conformal map of a conformal map is again a conformal map, are self-evident. The Riemann mapping theorem is therefore equivalent to the following result.

Any simply-connected schlicht domain D whose boundary consists of more than one point can be mapped conformally onto the interior of the unit circle. It is, moreover, possible to make an arbitrary point of D and a direction through this point correspond, respectively, to the origin and the direction of the positive axis. If this is done, the mapping is unique.

As a first step toward the proof of this theorem we show that there exist analytic functions $f(z)$ which are regular and univalent in the domain D and which satisfy $|f(z)| \leq 1$ if z is in D. By hypothesis, there exist at least two distinct points, say a and b, on the boundary of D. Consider now the function

$$p(z) = \sqrt{\frac{z - a}{z - b}}.$$

The only singularities of this function are the points a and b. Since these points are not situated in the interior of D, $p(z)$ is thus regular at all points of D. In view of the fact that D is simply-connected, it follows therefore from the monodromy theorem (Sec. 5, Chap. III) that $p(z)$ is also single-valued in D. Furthermore, $p(z)$ is univalent in D. Indeed, if there were two distinct points z_1 and z_2 of D such that

$$\sqrt{\frac{z_1 - a}{z_1 - b}} = \sqrt{\frac{z_2 - a}{z_2 - b}},$$

it would follow that

$$\frac{z_1 - a}{z_1 - b} = \frac{z_2 - a}{z_2 - b},$$

which is only possible for $z_1 = z_2$. Another easily proved property of the function $p(z)$ is that it cannot take in D both the values α and $-\alpha$, where α is any complex number. From

$$\sqrt{\frac{z_1 - a}{z_1 - b}} = \alpha, \qquad \sqrt{\frac{z_2 - a}{z_2 - b}} = -\alpha$$

it would follow that

$$\alpha^2 = \frac{z_1 - a}{z_1 - b} = \frac{z_2 - a}{z_2 - b},$$

which again is only possible for $z_1 = z_2$. Now $w = p(z)$, being a regular function in D, maps D onto a domain D_1. If w_0 is one of the points of D_1, there exists therefore a neighborhood $|w - w_0| < \gamma$ which is contained in D_1. Since, as we have just shown, D_1 cannot contain both the points w

and $-w$, the neighborhood $|w + w_0| < \gamma$ will therefore be outside D_1. It follows that

$$|p(z) + w_0| \geq \gamma, \qquad z \in D.$$

As a result, the function

$$f_1(z) = \frac{\gamma}{p(z) + w_0}$$

satisfies $|f_1(z)| \leq 1$ at all points of D. Since $f_1(z)$ is obtained from $p(z)$ by a linear transformation and since $p(z)$ is univalent in D, it is also clear that $f_1(z)$ is univalent in D.

We now consider the class B of functions $f(z)$ which are regular and univalent in D and satisfy $|f(z)| \leq 1$ if z is in D. This class is not empty, as the example of the function $f_1(z)$ shows. Since the functions of B are uniformly bounded, it follows from the results of Sec. 2, Chap. IV, that B is a normal family of analytic functions in D. From any sequence of functions of B we can therefore select a subsequence which converges to a regular function in D. This limit function is again a function of B unless it reduces to a constant. Indeed, a limit of functions $f(z)$ such that $|f(z)| \leq 1$ is obviously also bounded by 1, and it was shown in Sec. 3, Chap. IV, that a limit of univalent functions in a domain D is again univalent in D unless it reduces to a constant. We now restrict B by the requirement $|f'(\zeta)| \geq |f_1'(\zeta)|$, where $f_1(z)$ is the function defined above and ζ is an arbitrary point of D. This restricted family, to be denoted by B_1, is not only normal but also compact. The compactness of B_1 follows from the fact that constant limit functions, which were possible in B, are now excluded. Indeed, the derivative of a constant limit function would vanish at ζ and this contradicts the fact that for all functions $f(z)$ of B_1 we have $|f'(\zeta)| \geq |f_1'(\zeta)|$ and that $f_1'(z)$, being the derivative of a univalent function, does not vanish in D.

We now consider the extremal problem

$$|f'(\zeta)| = \max., \qquad f(z) \in B_1. \tag{37}$$

The class B_1 is not empty and it is compact. Since $|f'(\zeta)|$ is a continuous functional in the sense of Sec. 3, Chap. IV, it follows from the results of Sec. 3, Chap. IV, that the problem (37) has a solution within the family B_1. In other words, there exists a function $f_0(z)$ which is univalent in D and satisfies there $|f_0(z)| \leq 1$, such that

$$|f'(\zeta)| \leq |f_0'(\zeta)|, \tag{38}$$

where $f(z)$ is any other function of B_1. Our aim is now to show that $w = f_0(z)$ maps D onto the interior of the unit circle $|w| < 1$. We shall do so by assuming that there is a point α in $|w| < 1$ which is not covered by the conformal map of D as given by $w = f(z)$, and then showing that this assumption leads to a contradiction.

We first remark that the function $f_0(z)$ solving the extremal problem (37) must vanish at $z = \zeta$. Indeed, since the linear substitution (16) maps the unit circle onto itself, the function

$$f^*(z) = \frac{f_0(z) - f_0(\zeta)}{1 - \overline{f_0(\zeta)}f(z)}$$

would also belong to the class B_1. But, in view of

$$|f_0^{*\prime}(\zeta)| = \frac{|f_0'(\zeta)|}{1 - |f_0(\zeta)|^2} > |f_0'(\zeta)|,$$

this would mean that $f_0(z)$ is not the solution of the problem (37). Suppose now that a value $\alpha(|\alpha| < 1)$ is not taken by $w = f(z)$ in D. The function

$$\varphi(z) = \sqrt{\frac{\alpha - f_0(z)}{1 - \bar{\alpha}f_0(z)}}, \tag{39}$$

where a definite branch of the square root has been taken, will then be regular in D; by the monodromy theorem, $\varphi(z)$ is therefore also single-valued in D. $\varphi(z)$ is furthermore univalent and satisfies $|\varphi(z)| \leq 1$ in D. The boundedness is again a consequence of the properties of the transformation (16). The univalence follows from the fact that, for $\varphi(z_1) = \varphi(z_2)$, we have

$$\frac{\alpha - f_0(z_1)}{1 - \bar{\alpha}f_0(z_1)} = \frac{\alpha - f_0(z_2)}{1 - \bar{\alpha}f_0(z_2)}$$

and, therefore, $f_0(z_1) = f_0(z_2)$. But this is impossible if z_1 and z_2 are two distinct points of D, since $f_0(z)$ is univalent. We next construct the function

$$g(z) = \frac{\varphi(z) - \varphi(\zeta)}{1 - \overline{\varphi(\zeta)}\varphi(z)}, \tag{40}$$

which evidently is also univalent and satisfies $|g(z)| \leq 1$ if z is in D. From (39) and (40), and using the fact that $f_0(\zeta) = 0$, we obtain

$$g'(\zeta) = -\frac{1 + |\alpha|}{2\sqrt{\alpha}} f_0'(\zeta).$$

Since

$$1 + |\alpha| = 2\sqrt{|\alpha|} + (1 - \sqrt{|\alpha|})^2 > 2\sqrt{|\alpha|}, \qquad |\alpha| < 1,$$

it follows therefore that

$$|g'(\zeta)| > |f_0'(\zeta)|.$$

But, since the function $g(z)$ belongs to the class B_1, this contradicts the fact that $f_0(z)$ solves the extremal problem (37). Our assumption that there exists a point α in $|w| < 1$ which is not covered by the conformal image of D as given by $w = f_0(z)$ has thus led to an absurdity. Since $f_0(z)$ is univalent in D, it follows therefore that $w = f_0(z)$ maps D onto the simply covered circular disk $|w| < 1$.

The function $w = f_0(z)$ which has thus been found to map D onto $|w| < 1$ and which lets the point $z = \zeta$ correspond to $w = 0$ is clearly not unique; the function $e^{i\theta}f_0(z)$, where θ is an arbitrary constant angle, has the same mapping properties. The angle θ may be made definite by the requirement that a given arc element dz at $z = \zeta$ be transformed into an arc element at $w = 0$ which has the direction of the positive axis, *i.e.*, by the condition $e^{i\theta}f_0'(\zeta)\,dz > 0$. If this is done, the mapping is uniquely determined. Indeed, if both $f_0(z)$ and $f_1(z)$ satisfy all these requirements, then the function

$$h(z) = \frac{f_0(z)}{f_1(z)}$$

is regular in D [the zeros of $f_0(z)$ and $f_1(z)$ cancel] and $|h(z)| \to 1$ if z approaches the boundary of D. By Exercise 1, Sec. 8, Chap. III, $h(z)$ therefore reduces to a constant which must have the absolute value 1. Since $e^{i\theta}f_\nu'(\zeta)\,dz > 0$, $\nu = 0, 1$, we have $h(\zeta) > 0$ and thus $h(z) \equiv 1$. Hence, $f_0(z)$ and $f_1(z)$ are identical. This completes the proof of the Riemann mapping theorem.

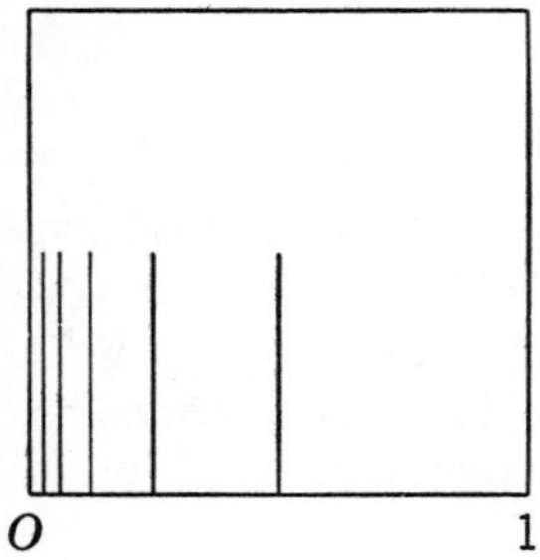

FIG. 11.

The mapping theorem establishes a one-to-one correspondence between the points of two simply-connected domains D and D^*, but it yields no information regarding the correspondence between the points of the boundaries C and C^* of these domains. In particular, the mapping theorem does not say that the mapping is continuous in the closure $D + C$ of D and that it establishes a one-to-one correspondence between $D + C$ and $D^* + C^*$. That this cannot be expected follows from the fact that our concept of a domain with more than one boundary point is very general and that it permits the occurrence of quite "pathological" cases. To illustrate what may happen, consider the domain D indicated in Fig. 11. D consists of the square $0 < x < 1$, $0 < y < 1$, from which the vertical lines of length $\frac{1}{2}$ and abscissas $\frac{1}{2}, \frac{1}{4}, \ldots, 2^{-n}, \ldots$ have been removed. Obviously, the origin cannot be connected with an interior point of D by a Jordan arc which is entirely in D; any such arc has points in common with the vertical lines which do not belong to D. Such a point is called an *inaccessible boundary point* of D. It is clear that the conformal mapping of the unit circle onto D cannot be continuous at all points of the circumference. If z_0 ($|z_0| = 1$) corresponds to the inaccessible point of D, then any arbitrarily small arc of $|z| = 1$ containing z_0 will contain points whose conformal images have a distance of more than $\frac{1}{2}$ from the origin.

In order to obtain continuity of the conformal map in the closure of a domain, it therefore becomes necessary to make restrictive assumptions which exclude such phenomena as the one described above. We shall show that *if D and D^* are two domains, bounded by simple closed contours C and C^*, respectively, then the conformal mapping $D \to D^*$ is continuous in $D + C$ and establishes a one-to-one correspondence between the points of C and C^*.* We remark that by a suitable refinement of our argument the same result can also be shown to hold in the case in which C and C^* are simple closed Jordan curves. Let z_0 be a point of C and let L_1 and L_2 be two different Jordan arcs in D which terminate at z_0; let further $w = f(z)$ denote the analytic function mapping D onto D^*. Since C is a contour, we can draw a small circle K_r of center z_0 and radius r such that K_r intersects C in two points only. We denote by D_r the domain within K_r which is cut off by K_r, and we denote by $z_{1,r}$ and $z_{2,r}$ the intersections of K_r with L_1 and L_2, respectively. We now show that

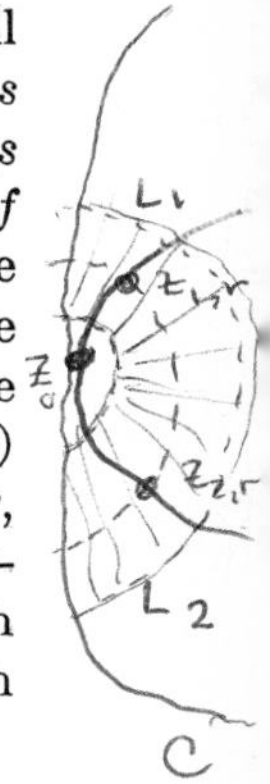

$$\lim_{r \to 0} |f(z_{1,r}) - f(z_{2,r})| = 0. \tag{41}$$

We have

$$d(r) = |f(z_{1,r}) - f(z_{2,r})| = \left| \int_{z_{1,r}}^{z_{2,r}} f'(z)\, dz \right|,$$

where the integration path may be taken to be the arc of K_r (contained in D) between the two points. It follows that

$$d(r) \le \int_{\theta_1}^{\theta_2} |f'(z)| r\, d\theta, \qquad z = z_0 + re^{i\theta},$$

whence, by the Schwarz integral inequality,

$$\frac{d^2(r)}{r} \le \int_{\theta_1}^{\theta_2} |f'(z)|^2 r\, d\theta \int_{\theta_1}^{\theta_2} d\theta \le 2\pi \int_{\theta_1}^{\theta_2} |f'(z)|^2 r\, d\theta.$$

Suppose now that $d^2(r) > d > 0$ for $R \ge r > \epsilon > 0$. Integrating the last inequality from ϵ to R, we obtain

$$d^2 \log \frac{R}{\epsilon} \le 2\pi \iint |f'(z)|^2 r\, dr\, d\theta.$$

This integral expresses the area of a subdomain of D^*. Its value is therefore not larger than the area A of D^*. Hence,

$$d^2 \leq \frac{2\pi A}{\log R - \log \epsilon}.$$

This tends to zero if $\epsilon \to 0$, which shows that $d = 0$. This proves (41).

If $\lim_{z \to z_0} f(z)$ exists for z tending to z_0 along an arc L_1, it follows from (41) that this limit exists and is the same if z_0 is approached along any other arc L_2 from within D. In order to show that the limit exists, we proceed as follows. We first remark that all limit points of sequences of points $f(z_1), f(z_2), \ldots$ such that $\lim_{n \to \infty} z_n = z_0$ must lie on C^*. They cannot lie outside the closure of D^*, since the points $f(z_n)$ are in D^*. On the other hand, they cannot lie in D^*, for the inverse mapping $D^* \to D$ is single-valued and conformal at all points of D^* and the images of sufficiently small neighborhoods of points of D^* do not contain boundary points of D. Now there are two possibilities. If z tends to z_0 along L_1, $\lim_{z \to z_0} f(z)$ may exist, or it may not. In order to show that the second case cannot happen, we first observe that the nonexistence of the limit implies that for two different sequences of points $z_1, z_2, \ldots$ of L_1 which tend to z_0 we can obtain two different limits, say α and β. In view of what was said before, both α and β must be points of C^*. Consider now the conformal image of the arc L_1 as given by $w = f(z)$. If we draw two mutually exclusive circles of sufficiently small radius ϵ about α and β, the conformal image of L_1 must oscillate an infinity of times between the interiors of these circles if $z \to z_0$. Since the two circles have a finite minimum distance from each other, we can find two points α_1 and β_1 and two continuous arcs T_1 and T_2 in D^* terminating at α_1 and β_1, respectively, such that the minimum distance between T_1 and T_2 is positive, say m, and both T_1 and T_2 intersect the image of L_1 an infinity of times. If we denote by S_1 and S_2 the arcs in D which correspond to T_1 and T_2, respectively, it follows therefore that both S_1 and S_2 intersect L_1 an infinity of times and that the points of intersection converge to z_0. Since both S_1 and S_2 are continuous—with the possible exception of their terminals on C—every circle of center z_0 and sufficiently small radius r must therefore intersect both S_1 and S_2. Since the distance between T_1 and T_2 is at least m, the absolute value of the difference of the values of $f(z)$ at these points is not smaller than m. But this does not agree with (41), and the assumption that $\lim_{z \to z_0} f(z)$ along L_1 does not exist has thus led to a contradiction. We have therefore proved that the limit of $f(z)$ for $z \to z_0$ exists and is independent of the arc in D along which z_0 is approached. Since we have also shown that the value of this limit is a point of C^*, and since the roles of D and D^* can be interchanged, it follows that the conformal mapping $D \to D^*$ is continuous in $D + C$ and that the points of C and C^* correspond to each other in a one-to-one fashion.

As an application of this result we prove that *the boundary value problem of the first kind for harmonic functions can be solved for a simply-connected domain D whose boundary is a closed contour C.* We recall from Sec. 5, Chap. I, that this problem consists in finding a function $U(x,y)$ of the two real variables x, y which is harmonic in D and takes prescribed boundary values on C; for the sake of simplicity, we confine ourselves to the case in which the boundary values $U(x',y')$ are a piecewise continuous function of the point (x',y') on C. We saw in Sec. 6, Chap. I, that this problem can be solved in the case in which C is a circle; in fact, the solution can be written down explicitly by means of the Poisson integral formula. If C is a general simple closed contour, it follows from the Riemann mapping theorem that there exists an analytic function $z = f(w)$ which maps the unit circle $|w| < 1$ onto D. If $U(x,y)$ is a harmonic function in D, the function $U^*(u,v) = U[x(u,v),y(u,v)][f(w) = x(u,v) + iy(u,v),\ w = u + iv]$ is a harmonic function in $|w| < 1$ (see Exercise 12, Sec. 1). Now we have seen that the conformal mapping $(u,v) \to (x,y)$ yielded by the function $z = f(w)$ also establishes a one-to-one correspondence between the points (x',y') of C and the points (u',v') of $|w| = 1$. The boundary value of the harmonic function $U(x,y)$ at the point (x',y') will therefore be the same as the boundary value of $U^*(u,v)$ at the corresponding point (u',v') on $|w| = 1$. The conformal mapping $(u,v) \to (x,y)$ thus reduces the boundary value problem for the domain D to a corresponding boundary value problem for the unit circle. Since the latter problem can be solved by means of the Poisson integral formula, it follows that the boundary value problem of the first kind can be solved for any simply-connected domain which is bounded by a simple closed contour.

It was shown in Sec. 5, Chap. I, that the Green's function $g(z,\zeta)$ of a domain D can be obtained by a particular boundary value problem. It can, however, also be expressed in terms of the function $w = F(z,\zeta)$ which maps D onto the unit circle and carries the point ζ of D into the origin. Consider the expression

$$g(z,\zeta) = -\log|F(z,\zeta)|, \tag{42}$$

which is the real part of the analytic function

$$p(z,\zeta) = -\log F(z,\zeta). \tag{43}$$

$p(z)$ is clearly regular at all points of D except at $z = \zeta$, where $F(z,\zeta)$ vanishes. In the neighborhood of $z = \zeta$, $F(z,\zeta)$ has a power series expansion

$$F(z,\zeta) = a_1(z - \zeta) + a_2(z - \zeta)^2 + \cdots, \qquad a_1 \neq 0.$$

Hence,

$$\begin{aligned} p(z,\zeta) &= -\log \left\{ a_1(z - \zeta)\left[1 + \frac{a_2}{a_1}(z - \zeta) + \cdots\right]\right\} \\ &= -\log (z - \zeta) - \log a_1 + \frac{a_2}{a_1}(z - \zeta) + \cdots \\ &= -\log (z - \zeta) + p_1(z,\zeta), \end{aligned}$$

where $p_1(z,\zeta)$ is regular at $z = \zeta$. By (42) and (43), the function $g(z,\zeta)$ is therefore harmonic in D except at $z = \zeta$; near $z = \zeta$, $g(z,\zeta)$ is of the form

$$\begin{aligned} g(z,\zeta) &= -\log |z - \zeta| + \operatorname{Re} \{p_1(z,\zeta)\} \\ &= -\log |z - \zeta| + g_1(z,\zeta), \end{aligned}$$

where $g_1(z,\zeta)$ is harmonic at all points of D. This shows that $g(z,\zeta)$ has the characteristic singularity of the Green's function at $z = \zeta$. If z approaches a point of the boundary C of D, then $w = F(z,\zeta)$ approaches the circle $|w| = 1$. Hence

$$\lim_{z \to C} g(z,\zeta) = \lim_{z \to C} \log |F(z,\zeta)| = 0.$$

Since there cannot be two different functions with these properties (their difference would be harmonic in D and zero on the boundary and, by the maximum principle for harmonic functions, it would therefore be zero throughout D), the function (42) must be the Green's function of D.

EXERCISES

1. If $w = f(z)$ is any function mapping D onto the unit circle and ζ is a point of D, show that the Green's function $g(z,\zeta)$ is given by

$$g(z,\zeta) = \log \left| \frac{1 - \overline{f(\zeta)}f(z)}{f(z) - f(\zeta)} \right|.$$

2. If the domain D contains the point at infinity and D is mapped onto the interior of the unit circle in such a way as to make the points $z = \infty$ and $w = 0$ correspond to each other, show that the mapping function $w = f(z)$ has the expansion

$$f(z) = \frac{a_1}{z} + \frac{a_2}{z} + \cdots + \frac{a_n}{z^n} + \cdots, \qquad a_1 \neq 0,$$

which converges in the exterior of a sufficiently large circle.

3. If D is the domain of the preceding exercise and D is mapped onto a domain which also contains the point at infinity in such a way that the points $z = \infty$ and $w = \infty$ correspond to each other, show that the mapping function $w = f(z)$ has an expansion

$$f(z) = bz + a_0 + \frac{a_1}{z} + \frac{a_2}{z^2} + \cdots, \qquad b \neq 0$$

in a neighborhood of $z = \infty$.

4. If D contains the point at infinity, show that the Green's function $g(z, \infty)$ is of the form

$$g(z, \infty) = \log |z| + \gamma + g_1(z), \qquad g_1(\infty) = 0,$$

where $g_1(z)$ is harmonic in D. γ is called the *Robin constant* of the complement $\bar{D}$ of D (that is, the set of all points in the z-plane which do not belong to D). The quantity $C = e^{-\gamma}$ is also known as the *capacity* or the *transfinite diameter* of $\bar{D}$.

5. If S_1 and S_2 are point sets in the plane whose complements are simply-connected domains and if S_1 contains S_2, prove that the capacity of S_2 cannot be larger than the capacity of S_1.

6. If S is a point set in the plane whose complement is a simply-connected domain and if S can be enclosed in a circle of radius R, show that the capacity of S is not larger than R. *Hint:* Use the result of the preceding exercise.

5. The Symmetry Principle. We saw in Sec. 5, Chap. III, that two function elements $f_1(z)$ and $f_2(z)$ which are defined in two, partly overlapping, domains D_1 and D_2 are analytic continuations of each other if $f_1(z)$ and $f_2(z)$ take the same values in the domain which belongs to both D_1 and D_2. We shall now show that the same result holds under much weaker assumptions.

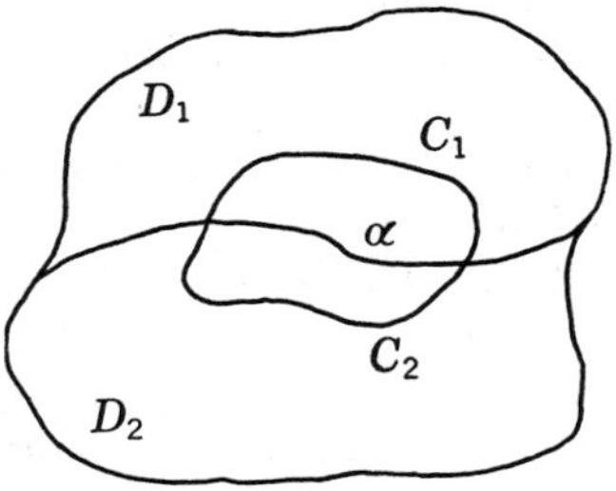

FIG. 12.

Let D_1 and D_2 be two adjacent domains whose common boundary is a smooth arc α. If the analytic functions $f_1(z)$ and $f_2(z)$ are regular in D_1 and D_2, respectively, and if the limits for $z \to \alpha$ of both functions coincide and are continuous on α, then $f_1(z)$ and $f_2(z)$ are direct analytic continuations of each other.

To prove this result, we construct a contour C which is divided by the arc α into two arcs C_1 and C_2, situated in D_1 and D_2, respectively (Fig. 12). We now define in the interior D of C a function $f(z)$ which is equal to $f_1(z)$ in those points of D which are also in D_1, and is equal to $f_2(z)$ in the common part of D and D_2. At the points of α which are within D, $f(z)$ is defined as the common limit of $f_1(z)$ and $f_2(z)$. Consider now the integral

$$\frac{1}{2\pi i} \int_{\Gamma_1} \frac{f(\zeta)}{\zeta - z}\, d\zeta,$$

taken along the boundary Γ_1 of the domain B_1 bounded by C_1 and α. By the Cauchy integral formula, the value of this integral is $f(z)$ if z is in B_1; if z is in the domain B_2 bounded by C_2 and α, the integral is zero. Similarly, the integral

$$\frac{1}{2\pi i}\int_{\Gamma_2}\frac{f(\zeta)}{\zeta - z}\,d\zeta,$$

taken along the boundary Γ_2 of B_2, has the value $f(z)$ if z is in B_2, and it is zero if z is in B_1. We now add the two integrals. Since both Γ_1 and Γ_2 are traversed in the positive direction with respect to the domains enclosed by them, the contributions of the arc α to these integrals cancel each other. Hence, the integral

$$\frac{1}{2\pi i}\int_{C}\frac{f(\zeta)}{\zeta - z}\,d\zeta$$

represents the function $f(z)$, irrespective of whether z is in B_1 or B_2. But, as shown in Sec. 3, Chap. III, this integral is an analytic function of z which is regular in the interior of C. $f_1(z)$ and $f_2(z)$ are therefore the values of one and the same function in the domains B_1 and B_2, respectively. Hence, they are analytic continuations of each other.

The most important application of this result is the *symmetry principle*, also called *reflection principle*, or *inversion principle*, due to Riemann and Schwarz. This principle, which is of fundamental importance both in the theory of conformal mapping and in the applications, may be stated as follows.

Let D_z and D_w be two domains and let the boundaries of D_z and D_w contain circular arcs α_z and α_w (which may also degenerate into linear segments), respectively. If $w = f(z)$ maps D_z onto D_w in such a way that α_w corresponds to α_z, then $f(z)$ can be continued analytically across α_z into the domain D_z^* obtained from D_z by inversion with respect to the circle C_z of which α_z forms a part. If $z \in D_z$ and $z^* \in D_z^*$ and if z and z^* are inverse points with respect to C_z, then the points $f(z)$ and $f(z^*)$ are inverse with respect to the circle C_w of which α_w forms a part.

We therefore not only know that $f(z)$ can be continued beyond α_z into D_z^*, but we can also find the values of $f(z)$ at all points of D_z^*. In particular, it follows that $w = f(z)$ maps the domain D_z^* onto the domain D_w^* which is obtained from D_w by inversion with respect to C_w. In order to prove the symmetry principle, it is sufficient to consider the case in which both α_z and α_w are linear segments. Indeed, as shown in Sec. 2, a linear substitution transforms two points which are inverse with respect to a circle into two points which are inverse with respect to the transformed circle. In particular, if the circle is transformed into a straight line, the inverse points become symmetric points with respect to that line. If, by suitable linear substitutions, C_z and C_w are both transformed into the real axis, our task is therefore reduced to the case in which both α_z and

α_w are parts of the real axis and in which the inverse points of z and $f(z)$ are their complex conjugates $\bar{z}$ and $\overline{f(z)}$, respectively.

Consider now the function $\overline{f(\bar{z})}$. This function is defined in D_z^* and it is, moreover, a regular analytic function in this domain. Indeed, the mapping $z \to \bar{z}$ does not change the magnitude of the angle between two curves; it only changes the orientation of the angle. Since the same is true of the mapping $f(z) \to \overline{f(z)}$, the composite mapping $z \to \overline{f(\bar{z})}$ restores the orientation of the angle and it is therefore conformal. But this means that $\overline{f(\bar{z})}$ is a regular analytic function in D_z^*. On α_z, we have $z = \bar{z}$ and therefore

$$\overline{f(\bar{z})} = f(z), \tag{44}$$

if $f(z)$ and $\overline{f(\bar{z})}$ are defined as the limits of these functions if $z \to \alpha_z$ from D_z. Since α_z is mapped by $f(z)$ onto part of the real axis [which provides the reason for putting the bar above the left-hand side of (44)], it follows from the results of the preceding section that the limits of both $f(z)$ and $\overline{f(\bar{z})}$ for $z \to \alpha_z$ are continuous on α_z. In view of the result proved further above, it follows therefore from (44) that $f(z)$ is regular on α_z and that $f(z)$ and $\overline{f(\bar{z})}$ are analytic continuations of each other. But it was shown in Sec. 5, Chap. III, that two analytic functions which are identical at the points of a small arc at which both are regular are identical everywhere. Hence, the identity (44), which so far has been shown to hold only at the points of α_z, holds for all analytic continuations of $f(z)$. Writing (44) in the form

$$f(\bar{z}) = \overline{f(z)},$$

we see therefore that the value of $f(z)$ at the point inverse to z with respect to α_z is the inverse of the point $f(z)$ with respect to α_w. This completes the proof of the symmetry principle.

As an example of the application of the symmetry principle, we give another demonstration of the fact, proved in Sec. 2, that a function $w = f(z)$ which maps the interior of a circle C onto the interior of another circle C^* is necessarily a linear transformation. The inverse of the interior of C with respect to C is obviously the exterior of C. It follows therefore from the symmetry principle that $w = f(z)$ can be continued analytically beyond all points of C and that this function also maps the exterior of C onto the exterior of C^*. Consequently, $w = f(z)$ is regular at all points of the z-plane, except at the point z_0 corresponding to $w = \infty$, and maps the entire z-plane onto the entire simply covered w-plane. Its only singularity is therefore a simple pole at $z = z_0$. $f(z)$ thus reduces to a rational function with only one pole, which is another way of saying that $w = f(z)$ is a linear transformation.

The most striking part of the symmetry principle, namely, the practical possibility of calculating the values of an analytic function in the exterior of the domain in which it was originally given, is due to the particular properties of the circle. There is, however, yet another aspect of the symmetry principle which is hardly less important and which can be generalized to a much wider class of functions. This is the fact that an analytic function which maps a circular boundary arc onto a circular boundary arc is regular on that arc. Before we generalize this property, we recall the definition of an *analytic arc*. A continuous arc $z = x + iy = x(t) + iy(t)$, $t_1 \leq t \leq t_2$, is said to be an analytic arc if, in some neighborhood $|t - t_0| < \epsilon$ of every $t_1 < t_0 < t_2$, both $x(t)$ and $y(t)$ can be expanded into converging power series $x(t) = \sum_{n=0}^{\infty} a_n(t - t_0)^n$, $y(t) = \sum_{n=0}^{\infty} b_n(t - t_0)^n$; we further assume that $x'(t)$ and $y'(t)$ do not vanish at the same time. We now prove the following generalization of the symmetry principle.

If $w = f(z)$ maps a domain D_z onto a domain D_w and if, in this mapping, an analytic boundary arc α of D_z corresponds to an analytic boundary arc β of D_w, then $f(z)$ is regular at the points of α.

We may also state the conclusion of the theorem by saying that $f(z)$ can be continued analytically beyond the boundary arc α. To prove the statement, we observe that the parametric representation $x = x(t)$, $y = y(t)$ of the analytic arc α defines, in the neighborhood of t_0, a regular analytic function of t if t is taken to be a complex variable. This follows from the fact that $z(t) = x(t) + iy(t)$ can be expressed as a power series which converges for $t_0 - \epsilon < t < t_0 + \epsilon$ and therefore in the whole circle $|t - t_0| < \epsilon$. For real values of t, the point $z(t)$ describes a piece α_0 of the arc α. Hence, the function $z = z(t)$ maps the circle $|t - t_0| < \epsilon$ onto a domain containing α_0 in such a way that a piece of the real axis is mapped onto α_0. Suppose now that the analytic arc β corresponding to α by the mapping $w = f(z)$ is a linear segment. The function $f_1(t) = f[z(t)]$ will then map a piece γ of the real axis onto a linear segment. By the inversion principle, $f_1(t)$ is therefore regular at the points of γ. But since $z(t)$ is regular there, this means that $f(z)$ must be regular at the points of α_0. The same argument can be applied to a neighborhood of any point of α, and it thus follows that $f(z)$ is regular at all points of α. If β is not a linear segment but a general analytic arc, we observe that the function $f_1(t) = f[z(t)]$ maps $t_0 - \epsilon < t < t_0 + \epsilon$ onto a piece of the analytic arc β. By what we have just proved, $f_1(t)$ must therefore be regular for these values of t. But since $z(t)$ is also regular there, this again implies that $f(z)$ is regular at α_0, and thus at all points of α.

In the case in which β is an algebraic curve whose equation is $F(u,v) = 0$, where $F(u,v)$ is a polynomial in u and v, it is possible to obtain explicit formulas for the analytic continuation of a function which maps a piece γ of the real axis onto β. Since, on γ, $z = \bar{z}$ and

$$\text{Re}\,\{f(z)\} = \frac{1}{2}[f(z) + \overline{f(z)}], \qquad \text{Im}\,\{f(z)\} = \frac{1}{2i}[f(z) - \overline{f(z)}],$$

we obtain the identity

$$F\left\{\frac{1}{2}[f(z) + \overline{f(\bar{z})}],\ \frac{1}{2i}[f(z) - \overline{f(\bar{z})}]\right\} = 0, \tag{45}$$

if z is a point of γ. Since an algebraic curve is analytic, it follows that $f(z)$ is regular on γ. In view of the principle of the permanence of a functional equation (Sec. 5, Chap. III), the identity (45) connecting the analytic functions $f(z)$ and $\overline{f(\bar{z})}$ persists therefore for all values of z to which these functions can be continued analytically. As an example consider the function $w = f(z)$ which maps the upper half-plane Im $\{z\} > 0$ onto the interior of the ellipse $b^2u^2 + a^2v^2 < a^2b^2$, $w = u + iv$. By (45), its analytic continuation across the real axis is given by

$$\frac{b^2}{4}[f(z) + \overline{f(\bar{z})}]^2 - \frac{a^2}{4}[f(z) - \overline{f(\bar{z})}]^2 = a^2b^2.$$

This formula enables us to compute the values of $f(z)$ in the lower half-plane if its values in the upper half-plane are known.

Our result concerning the mapping of analytic boundary curves enables us to prove a fact which was anticipated at the end of Sec. 10, Chap. III. We used there, without proof, the fact that *the Green's function $g(z,\zeta)$ of a domain D which is bounded by one or more closed analytic curves C is harmonic at the points of C.* This is equivalent to saying that the analytic function

$$p(z,\zeta) = g(z,\zeta) + ih(z,\zeta)$$

whose real part is $g(z,\zeta)$ is regular at the points of C. The proof is now very simple. Since $g(z,\zeta) = 0$ if z tends to any point of C, it follows that the function $w = p(z,\zeta)$ maps a neighborhood of a point z_0 on C onto a domain in such a way that a piece of C containing z_0 corresponds to a section of the imaginary axis. Since a straight line is an analytic curve, it is seen that $p(z,\zeta)$ must be regular at all points of the analytic curves C.

EXERCISES

1. Let $w = f(z)$ map a domain D_z onto a domain D_w such that a circular boundary arc α_z corresponds to a circular boundary arc α_w, and let a, b and r, R be the centers and radii of α_z and α_w, respectively. By the theorem proved at the beginning of this section, show that the analytic continuation of $f(z)$ beyond α_z is given by

$$[f(z) - b]\left[\overline{f\left(a + \frac{r^2}{\bar{z} - \bar{a}}\right)} - \bar{b}\right] = R^2,$$

and explain why this relation is equivalent to the symmetry principle. *Hint:* Observe that a point z on α_z satisfies $r^2 = |z - a|^2 = (z - a)(\bar{z} - \bar{a})$ and that two points z_1 and z_2 which are inverse with respect to α_z are connected by the relation

$$(z_1 - a)(\bar{z}_2 - \bar{a}) = r^2.$$

2. Show that an analytic function which maps the circle $|z| < 1$ onto the n-times covered circle $|w| < 1$ must be a rational function with n poles.

3. Show that an analytic function $w = f(z)$ which maps the unit circle $|z| < 1$ onto the full w-plane from which the ray $-\infty \leq w \leq -\frac{1}{4}$ has been removed can be continued analytically beyond $|z| = 1$ by the relation

$$f(z) = \overline{f\left(\frac{1}{\bar{z}}\right)}.$$

Show further that $w = f(z)$ maps the full z-plane onto the doubly covered full w-plane and that, therefore, $f(z)$ must be a rational function with two poles. Verify that, with the additional conditions $f(0) = 0$, $f'(0) > 0$, $f(z)$ is of the form

$$f(z) = \frac{z}{(1 - z)^2}.$$

4. If $w = f(z)$ maps the ring $0 < \rho < |z| < 1$ onto the circle $|w| < 1$ from which the linear segment $-\alpha \leq w \leq \alpha$ $(0 < \alpha < 1)$ has been removed, show that $w = f(z)$ maps the ring $\rho < |z| < \rho^{-1}$ onto the full w-plane from which the linear segment $-\alpha \leq w \leq \alpha$ and the rays $-\infty \leq w \leq -\alpha^{-1}$, $\alpha^{-1} \leq w \leq \infty$ have been removed. Show further that $f(z)$ can be continued analytically beyond $|z| = 1$ by the relation

$$f(z)\overline{f\left(\frac{1}{\bar{z}}\right)} = 1,$$

while its continuation beyond $|z| = \rho$ is given by

$$f(z) = \overline{f\left(\frac{\rho^2}{\bar{z}}\right)}.$$

5. If $w = f(z)$ maps the rectangle with the corners $-a, a, a + bi, -a + bi$ $(a, b > 0)$ onto the half-plane $\text{Im}\,\{w\} > 0$ and if $f(-a) = \alpha$, $f(a) = \beta$, $\alpha < \beta$, show that $w = f(z)$ maps the rectangle with the corners $-a - bi$, $a - bi$, $a + bi$, $-a + bi$ onto the full w-plane from which the rays $-\infty \leq w \leq \alpha$, $\beta \leq w \leq \infty$ have been removed. If $\alpha > \beta$, show that the map of the large rectangle is the full w-plane from which the linear segment $\beta \leq w \leq \alpha$ has been removed.

6. Let the function $f(z)$ be regular in a domain D whose boundary includes an analytic arc α. If the limits for $z \to \alpha$ of either of the expressions

$$\text{Re}\,\{f(z)\}, \qquad \text{Im}\,\{f(z)\}, \qquad |f(z) - \gamma|,$$

where γ is a constant, are the same at all points of α, show that $f(z)$ is regular on α.

7. If the domain D is bounded by n closed analytic curves $C_1, \ldots, C_n$, and if $\omega_\nu(z)$ denotes the harmonic measure of C_ν $(\nu = 1, \ldots, n)$ defined in Sec. 10, Chap. I, show that $\omega_\nu(z)$ is harmonic at all points of the boundary curves $C_1, \ldots, C_n$.

6. The Schwarz-Christoffel Formula. While the Riemann mapping theorem assures us that any two simply-connected domains with more than one boundary point can be mapped conformally upon each other, it is not of much help when we are faced with the practical problem of finding the mapping function which transforms two given domains into each other. The necessity thus arises of developing special techniques which will help us in the treatment of a given mapping problem. It is obviously sufficient to adapt these techniques to the case in which one of the two domains is a circle; if we can map two domains onto the same circle, we can also map them onto each other. The choice of the circle as the *standard domain*, or the *canonical domain*, in the simply-connected case has the advantage of leading to comparatively simple formulas. We shall also employ a half-plane in the capacity of a canonical domain if this results in even greater simplification. Since a circle and a half-plane are transformed into each other by a linear substitution, the mapping formulas involving these two domains can be easily transformed into each other.

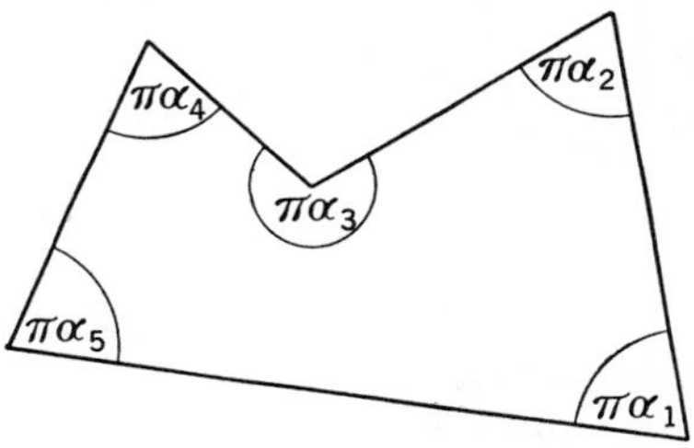

FIG. 13.

While it is in the nature of things that a simple solution of the general mapping problem cannot be expected, there are many important cases in which the mapping functions can be found by means of comparatively simple devices. In this section we shall treat the conformal mapping of a general polygon onto a circle or a half-plane.

Let then D be a polygon in the w-plane and let $\pi\alpha_1, \pi\alpha_2, \ldots, \pi\alpha_n$ denote its interior angles (see Fig. 13 for the case $n = 5$). In the formulas, it is more convenient to use the exterior angles $\pi\mu_1, \ldots, \pi\mu_n$, defined by $\pi\alpha_\nu + \pi\mu_\nu = \pi$, $\nu = 1, \ldots, n$. It is clear that $\mu_\nu > 0$ corresponds to a projecting corner and that $\mu_\nu < 0$ corresponds to the opposite case. If the polygon is convex, then all numbers μ_ν are positive. By a theorem of elementary geometry (which is almost self-evident), the sum of all exterior angles of a closed polygon is 2π. Hence, the quantities μ_ν introduced above are connected by the relation

$$\sum_{\nu=1}^{n} \mu_\nu = 2. \tag{46}$$

Let now $w = f(z)$ be an analytic function that maps the upper half-plane $\text{Im}\,\{z\} > 0$ onto the interior of the polygon D and let the points a_1,

$a_2, \ldots, a_n$ in the z-plane correspond to the vertices of the polygon whose exterior angles are $\pi\mu_1, \pi\mu_2, \ldots, \pi\mu_n$.

The points $a_1, \ldots, a_n$ divide the real axis into n parts each of which is mapped by $w = f(z)$ onto a linear segment. By the symmetry principle, $f(z)$ is therefore regular at all points of the real axis except at the points $a_1, \ldots, a_n$, and across each of the intervals bounded by these points $f(z)$ can be continued analytically by simple symmetry. The mirror image D' of D with respect to one of its sides will thus be the conformal map of the lower half-plane $\text{Im}\,\{z\} < 0$. Applying the symmetry principle again, this time with respect to one of the sides of D', we find that $w = f(z)$ maps the upper half-plane onto a figure D'' that is congruent with D but has a different location in the w-plane (see Fig. 14). These two inversions return a point z into its original position, while a point w is made subject to a translation and a rotation about the origin. The value $f_1(z)$ of the function $f(z)$ with which we return is therefore of the form

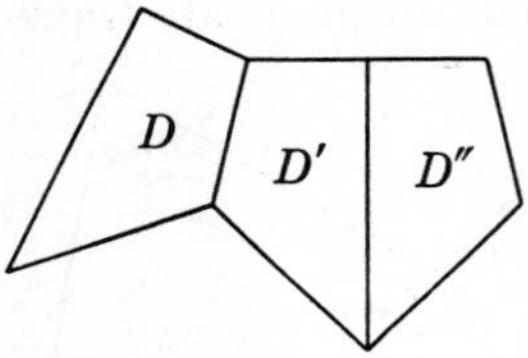

FIG. 14.

$$f_1(z) = Af(z) + B, \tag{46'}$$

where A and B are constants. It should be noted that the singularities of $f_1(z)$ are also situated at the points $a_1, \ldots, a_n$; obviously, these points are not affected by inversions with respect to the real axis. From (46′), we obtain

$$\frac{f_1''(z)}{f_1'(z)} = \frac{f''(z)}{f'(z)}.$$

This shows that the function

$$g(z) = \frac{f''(z)}{f'(z)} \tag{47}$$

returns to its initial value if z returns to its initial position by means of two inversions with respect to the real axis. Since, clearly, all possible branches of $f(z)$ are obtained from the initial branch by an even number of inversions (all images of the upper half-plane are polygons congruent to D and all images of the lower half-plane are congruent to the mirror image of D), it follows that the function $g(z)$ defined in (47) is single-valued in the whole z-plane. The singularities of $g(z)$ can only be located at the points $a_1, \ldots, a_n$.

In order to determine the character of these singularities, we consider the behavior of $f(z)$ in the neighborhood of the point a_ν; for the time being, we assume that none of the points $a_1, \ldots, a_n$ coincide with $z = \infty$. A small section of the real axis containing a_ν is mapped by $w = f(z)$ onto two linear segments which intersect at $w = f(a_\nu)$ with the angle $\pi\alpha_\nu$. Consider now the function

$$h(z) = [f(z) - f(a_\nu)]^{\frac{1}{\alpha_\nu}} \tag{48}$$

near the point $z = a_\nu$. Since the mapping $z \rightarrow z^\gamma$ $(\gamma > 0)$ transforms rays emanating from the origin into rays emanating from the origin but multiplies all angles with vertex at the origin by the factor γ, it follows that $h(z)$ maps a small section of the real axis containing a_ν into two linear segments forming the angle π. In other words, the two segments belong to the same straight line. Thus, the function maps a linear segment containing a_ν onto a linear segment; by the symmetry principle, it is therefore regular at $z = a_\nu$. With $\alpha_\nu = 1 - \mu_\nu$, it thus follows from (48) that $f(z)$ is of the form

$$f(z) = f(a_\nu) + [h(z)]^{1-\mu_\nu},$$

where $h(z)$ is regular at $z = a_\nu$ and $h(a_\nu) = 0$, $h'(a_\nu) \neq 0$. $h(z)$ may also be written in the form $h(z) = (z - a_\nu)h_1(z)$, where $h_1(z)$ is regular at $z = a_\nu$ and $h_1(a_\nu) \neq 0$. Using this, we obtain

$$f(z) = f(a_\nu) + (z - a_\nu)^{1-\mu_\nu}[h_1(z)]^{1-\mu_\nu},$$

whence

$$\frac{f''(z)}{f'(z)} = -\frac{\mu_\nu}{(z - a_\nu)} + k(z),$$

where $k(z)$ is regular at $z = a_\nu$ and the fact that $h_1(a_\nu) \neq 0$ has been used. Comparison with (47) shows that the function

$$g(z) + \frac{\mu_\nu}{z - a_\nu}$$

is regular at the point $z = a_\nu$. Carrying through the same procedure at all points $a_1, \ldots, a_n$, we thus find that the function

$$g_1(z) = g(z) + \sum_{\nu=1}^{n} \frac{\mu_\nu}{z - a_\nu} \tag{49}$$

is regular at all points $a_1, \ldots, a_n$. But these were the only singular points of the single-valued function $g(z)$. Hence, $g(z)$ is regular and single-valued in the entire plane (including $z = \infty$); by Liouville's theorem, it therefore reduces to a constant. Moreover, this constant must be zero. Indeed, $f(z)$ is regular at $z = \infty$ and has therefore an expansion $f(z) = f(\infty) + c_1z^{-1} + c_2z^{-2} + \cdots$ near $z = \infty$. Differentiating, we find that $f'(z)$ has a zero of second order at $z = \infty$ while $f''(z)$ has a zero of third order there. It therefore follows from (47) that $g(z)$

vanishes at $z = \infty$. Since $(z - a_\nu)^{-1}$ also vanishes there, we conclude from (49) that $g_1(\infty) = 0$. But $g_1(z)$ was shown to be a constant and it is therefore zero everywhere. Combining (47) and (49), we therefore obtain

$$\frac{f''(z)}{f'(z)} = -\sum_{\nu=1}^{n} \frac{\mu_\nu}{z - a_\nu}.$$

By integrating this expression, the desired mapping function is finally found to be

$$f(z) = \alpha \int_0^z \frac{dz}{(z - a_1)^{\mu_1}(z - a_2)^{\mu_2} \cdots (z - a_n)^{\mu_n}} + \beta, \tag{50}$$

where α and β are integration constants determining the position and size of the polygon.

The formula (50) holds if none of the points a_ν coincide with the point at infinity. However, this restriction can easily be removed by means of a linear transformation. If, for instance, we transform the point a_n into the point at infinity by the linear substitution $z = a_n - (1/\zeta)$, we obtain

$$f(\zeta) = \alpha \int_0^\zeta \left(a_n - a_1 - \frac{1}{\zeta}\right)^{-\mu_1} \cdots \left(a_n - a_{n-1} - \frac{1}{\zeta}\right)^{-\mu_{n-1}} \left(-\frac{1}{\zeta}\right)^{-\mu_n} \frac{d\zeta}{\zeta^2} + \beta_1,$$

whence, in view of (46),

$$f(z) = \alpha_1 \int_0^z \frac{dz}{(z - a_1')^{\mu_1} \cdots (z - a_{n-1}')^{\mu_{n-1}}} + \beta_1, \tag{51}$$

where $a_1', \ldots, a_{n-1}'$ are constants. Hence, the effect on (50) of one of the points a_ν coinciding with the point at infinity simply consists in the corresponding term being left out of the formula.

By the linear transformation

$$z = i\left(\frac{1 + \zeta}{1 - \zeta}\right), \qquad \zeta = \frac{z - i}{z + i}$$

which maps $|\zeta| < 1$ onto $\operatorname{Im}\{z\} > 0$, we can also obtain from (50) a formula for the conformal map of the unit circle onto the polygon D. We have

$$\begin{aligned}(z - a_\nu)^{\mu_\nu} &= \left[i\left(\frac{1 + \zeta}{1 - \zeta}\right) - a_\nu\right]^{\mu_\nu} \\ &= \left(\frac{a_\nu + i}{1 - \zeta}\right)^{\mu_\nu}\left[\zeta - \frac{a_\nu - i}{a_\nu + i}\right]^{\mu_\nu}\end{aligned}$$

and

$$dz = \frac{2i\,d\zeta}{(1 - \zeta)^2}.$$

If we denote by b_ν the point

$$b_\nu = \frac{a_\nu - i}{a_\nu + i}$$

on the unit circle which is mapped onto the vertex of index ν, it follows therefore from (46) and (50) that

$$f(z) = \alpha_2 \int_0^z \frac{dz}{(b_1 - z)^{\mu_1} \cdots (b_n - z)^{\mu_n}} + \beta_2, \tag{52}$$

where the variable has again been denoted by z, and α_2, β_2 are constants.

By the same procedure we can also find the mapping function of the exterior of the polygon D. Since the angles $\pi\alpha_\nu$ are now replaced by $2\pi - \pi\alpha_\nu = \pi(2 - \alpha_\nu)$, the quantities $\mu_\nu = 1 - \alpha_\nu$ have to be replaced by $1 - (2 - \alpha_\nu) = -(1 - \alpha_\nu) = -\mu_\nu$. In analogy to (49), the function

$$g_2(z) = \frac{f''(z)}{f'(z)} - \sum_{\nu=1}^{n} \frac{\mu_\nu}{z - a_\nu} \tag{53}$$

will therefore be regular at the points $a_1, \ldots, a_n$. However, since the conformal map now contains the point at infinity, we cannot conclude without further investigation that $g_2(z)$ is a constant. Confining ourselves to the case in which the original domain is the unit circle, and assuming that $f(0) = \infty$, we thus have to study the behavior of the function f''/f' at $z = 0$. Since the mapping by $w = f(z)$ is conformal at $z = 0$, the singularity of $f(z)$ at this point is a simple pole. Hence, $f(z)$ must be of the form

$$f(z) = \frac{f_1(z)}{z},$$

where $f_1(z)$ is regular at $z = 0$ and $f_1(0) \neq 0$. It follows that

$$\frac{f''(z)}{f'(z)} = -\frac{2}{z} + \frac{zf_1''(z)}{zf_1'(z) - f_1(z)}.$$

This shows that

$$\frac{f''(z)}{f'(z)} + \frac{2}{z}$$

is regular at $z = 0$ and vanishes there. Combining this with the properties of the function $g_2(z)$ of (53), we find that the function

$$\frac{f''(z)}{f'(z)} - \sum_{\nu=1}^{n} \frac{\mu_\nu}{z - a_\nu} + \frac{2}{z} - \lambda, \qquad \lambda = \sum_{\nu=1}^{n} \frac{\mu_\nu}{d_\nu}$$

is regular and single-valued at all points of the plane and has the value 0 at $z = 0$. Hence, it reduces to the constant zero. Setting $z = \infty$, we find that we must have $\lambda = 0$. We thus obtain the formula

$$(54) \qquad f(z) = \alpha \int_{z_0}^{z} (z - a_1)^{\mu_1} \cdots (z - a_n)^{\mu_n} \frac{dz}{z^2}, \qquad z_0 \neq 0,$$

for the analytic function mapping $|z| < 1$ onto the exterior of the given polygon. In using this formula it should be noted that μ_ν is the exterior angle with respect to the interior of the polygon. The actual angle of the conformal map of $|z| < 1$ at the point $f(a_\nu)$ is $\pi(1 + \mu_\nu)$ (see Fig. 15).

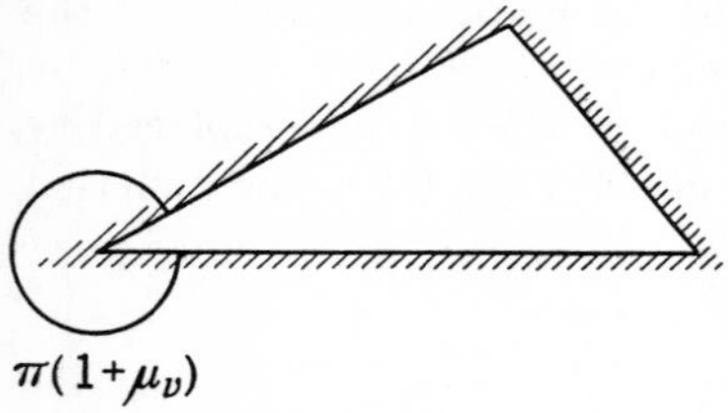

FIG. 15.

Formulas (50), (51), (52), and (54) are referred to as the *Schwarz-Christoffel formula.*

As an example for the use of the Schwarz-Christoffel formula we construct an analytic function $w = f(z)$ which maps the upper half-plane $\operatorname{Im}\{z\} > 0$ onto the interior of a triangle of angles $\pi\alpha$, $\pi\beta$, $\pi\gamma$. As shown in the proof of the Riemann mapping theorem, the correspondence of three points on the boundaries of two simply-connected domains can be arbitrarily prescribed. We shall thus ask for the three vertices of the triangle to correspond to the points $z = 0$, $z = 1$, $z = \infty$. Under these conditions it follows from (51) that

$$(55) \qquad f(z) = C_1 \int_0^z z^{\alpha-1}(1 - z)^{\beta-1}\, dz + C_2.$$

From (55) we can easily find the lengths of the sides of the triangle. If a, b, c denote the sides of the triangle opposite the angles $\pi\alpha$, $\pi\beta$, $\pi\gamma$, respectively, and we set $C_1 = 1$, we have

$$c = \int_0^1 |f'(z)\, dz| = \int_0^1 |z^{\alpha-1}(1 - z)^{\beta-1}\, dz|$$

$$= \int_0^1 \rho^{\alpha-1}(1 - \rho)^{\beta-1}\, d\rho = \frac{\Gamma(\alpha)\Gamma(\beta)}{\Gamma(\alpha + \beta)},$$

where $\Gamma(x)$ is the gamma function. In view of $\alpha + \beta + \gamma = 1$ and $\Gamma(x)\Gamma(1 - x) = \pi[\sin \pi x]^{-1}$, this can be brought into the form

$$c = \frac{1}{\pi} \sin \pi\gamma\, \Gamma(\alpha)\Gamma(\beta)\Gamma(\gamma).$$

Instead of evaluating a and b in a similar fashion, we can also observe that, by elementary trigonometry,

$$\frac{a}{\sin \pi\alpha} = \frac{b}{\sin \pi\beta} = \frac{c}{\sin \pi\gamma}.$$

Hence,

$$a = \frac{1}{\pi} \sin \pi\alpha\Gamma(\alpha)\Gamma(\beta)(\Gamma\gamma), \qquad b = \frac{1}{\pi} \sin \pi\beta\Gamma(\alpha)\Gamma(\beta)\Gamma(\gamma).$$

The determination of the vertices of the triangle for $C_1 = 1$, $C_2 = 0$ is left as an exercise to the reader.

The Schwarz-Christoffel formula remains correct if one of the corners of the polygon coincides with the point at infinity. As an example, we consider the function $w = f(z)$ which maps the upper half-plane $\text{Im }\{z\} > 0$ onto the interior of the "half-strip"

$$-\tfrac{1}{2}\pi < \text{Re }\{w\} < \tfrac{1}{2}\pi, \text{ Im }\{w\} > 0$$

(see Fig. 16) in such a way that the points $z = -1, 1, \infty$ and $w = -\frac{1}{2}\pi, \frac{1}{2}\pi, \infty$ correspond to each other in this order. Since the three angles of this "triangle" are $\frac{1}{2}\pi$, $\frac{1}{2}\pi$, 0, we obtain from (51)

$$f(z) = \alpha_1 \int_0^z \frac{dz}{\sqrt{1 - z^2}} + \beta_1$$
$$= \alpha_1 \sin^{-1} z + \beta_1.$$

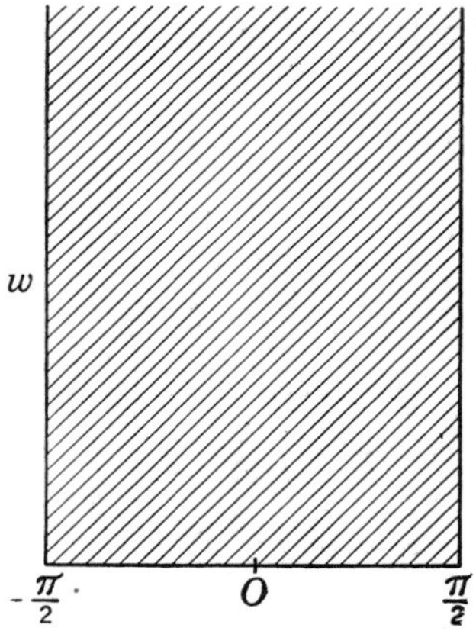

FIG. 16.

The constants α_1 and β_1 are determined by the conditions

$$-\tfrac{1}{2}\pi = f(-1) = -\tfrac{1}{2}\pi\alpha_1 + \beta_1, \qquad \tfrac{1}{2}\pi = f(1) = \tfrac{1}{2}\pi\alpha_1 + \beta_1.$$

It follows that $\beta_1 = 0$, $\alpha_1 = 1$, whence

$$w = f(z) = \sin^{-1} z.$$

We thus have the result that the inverse function $z = \sin w$ maps the half-strip of Fig. 16 onto the upper half-plane $\text{Im }\{z\} > 0$.

Finally, we consider the mapping $w = f(z)$ of the unit circle $|z| < 1$ onto the infinite strip $-\frac{1}{4}\pi < \text{Re }\{w\} < \frac{1}{4}\pi$ under the conditions

$$f(i) = f(-i) = \infty.$$

This infinite strip can be considered as a polygon with two sides which, at

$w = \infty$, form the angles 0. By (52), we obtain

$$f(z) = \alpha_2 \int_0^z \frac{dz}{(z - i)(z + i)} + \beta_2 = \alpha_2 \int_0^z \frac{dz}{1 + z^2} + \beta_2.$$

Hence, in view of $\tan^{-1}(\pm 1) = \pm\frac{1}{4}\pi$, it follows that $\alpha_2 = 1$, $\beta_2 = 0$ and, therefore,

$$f(z) = \tan^{-1} w.$$

The inverse function $w = \tan z$ thus maps the infinite strip $-\frac{1}{4}\pi < \text{Re}\{w\} < \frac{1}{4}\pi$ onto the unit circle $|z| < 1$. It should be pointed out that the use of the Schwarz-Christoffel formula in this degenerate case requires special justification and that it is easier to verify the result directly. The reader is recommended to do so by means of the known properties of the function $\tan w$.

EXERCISES

1. Show that

$$w = \int_0^z \frac{dz}{\sqrt{z(1 - z^2)}}$$

maps the upper half-plane $\text{Im}\{z\} > 0$ onto the interior of a square of side length

$$\frac{1}{2\sqrt{2\pi}} \Gamma^2\left(\frac{1}{4}\right).$$

2. Show that

$$w = \int_0^z \frac{dz}{\sqrt{1 - z^4}}$$

maps $|z| < 1$ onto the interior of a square of diagonal $[2\sqrt{2\pi}]^{-1}\Gamma^2(\frac{1}{4})$.

3. Show that

$$w = \int_{z_0}^z \frac{\sqrt{1 - z^4}}{z^2}\, dz, \qquad z_0 \neq 0,$$

maps $|z| < 1$ onto the exterior of a square.

4. Show that

$$w = \int_0^z \frac{dz}{(1 - z^n)^{\frac{2}{n}}}$$

maps $|z| < 1$ onto the inside of a regular polygon of order n whose side length is

$$\frac{1}{n} 2^{1-\frac{4}{n}} \frac{\Gamma^2\left(\frac{1}{2} - \frac{1}{n}\right)}{\Gamma\left(1 - \frac{2}{n}\right)}.$$

5. Show that

$$w = \int_0^z \frac{(1 - z^5)^{\frac{2}{5}}}{(1 + z^5)^{\frac{4}{5}}}\, dz$$

maps the unit circle $|z| < 1$ onto the pentagram of Fig. 17. If R is the radius of the circumscribed circle, show that

$$R = \int_{-1}^{0} (1 - t^5)^{\frac{2}{5}}(1 + t^5)^{-\frac{4}{5}}\, dt,$$

whence, by the substitution

$$u = \left(\frac{1 + t^5}{1 - t^5}\right)^2,$$

$$R = \frac{1}{5 \cdot 2^{\frac{2}{5}}} \int_0^1 u^{-\frac{9}{10}}(1 - u)^{-\frac{4}{5}}\, du$$

and thus

$$R = \frac{\Gamma(\frac{1}{10})\Gamma(\frac{1}{5})}{2^{\frac{2}{5}} \cdot 5\Gamma(\frac{3}{10})}.$$

6. Show that the function

$$w = \int_0^z \frac{dz}{(1 - z^4)\sqrt{1 + z^4}}$$

FIG. 17.

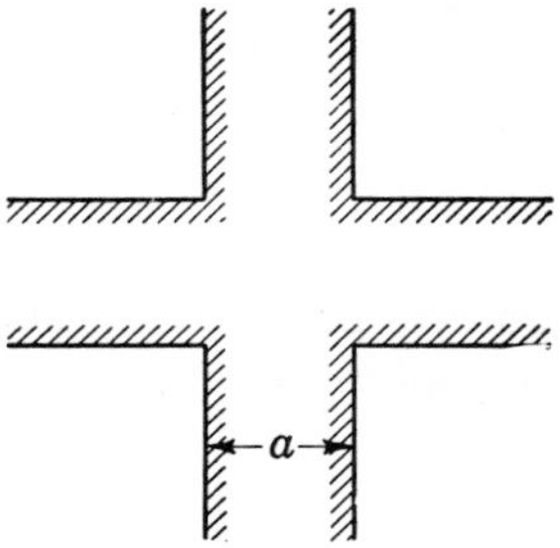

FIG. 18.

maps $|z| < 1$ onto the domain indicated in Fig. 18, and show that the width a is

$$a = \frac{1}{\sqrt{2}} \int_{-1}^{1} \frac{dx}{(1 + x^4)\sqrt{1 - x^4}}$$
$$= \tfrac{1}{8}[\pi\sqrt{2} + \pi^{-\frac{1}{2}}\Gamma^2(\tfrac{1}{4})].$$

7. Show that the function

$$w = \int_0^z \frac{1 - 2\cos 2\alpha z^2 + z^4}{(1 - z^2)(1 + z^2)^2}\, dz$$
$$= \int_0^z \left[\frac{\sin^2\alpha}{1 - z^2} + \frac{(1 - z^2)\cos^2\alpha}{(1 + z^2)^2}\right] dz$$
$$= \frac{1}{2}\sin^2\alpha \log\left(\frac{1 + z}{1 - z}\right) + \frac{z\cos^2\alpha}{1 + z^2}$$

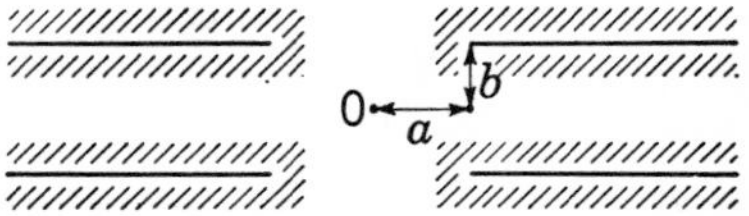

FIG. 19.

maps $|z| < 1$ onto the full w-plane which has been cut as indicated in Fig. 19. Show that the distances a and b are given by

$$a + ib = f(e^{i\alpha}) = \tfrac{1}{2}\sin^2\alpha \log\cot\tfrac{1}{2}\alpha + \tfrac{1}{2}\cos\tfrac{1}{2}\alpha + \tfrac{1}{2}\pi i \sin^2\alpha.$$

8. Using the symmetry principle, show that the function $w = \sin z$ maps the infinite strip $-\frac{1}{2}\pi < \text{Re}\{z\} < \frac{1}{2}\pi$ onto the full w-plane which has been cut along the two rays $-\infty \le w \le -1, 1 \le w \le \infty$.

9. Show that an analytic function $w = f(z)$ which maps the unit circle onto a convex polygon and satisfies $f(0) = 0, f'(0) = 1$ must be of the form

$$f(z) = \int_0^z \frac{dz}{\prod_{\nu=1}^{n} (1 - e^{i\theta_\nu}z)^{\mu_\nu}}, \qquad \mu_\nu > 0, 0 \le \theta_\nu < 2\pi,$$

and deduce that $f(z)$ is subject to the inequalities

$$|f'(z)| \le \frac{1}{(1-\rho)^2}, \qquad |f(z)| \le \frac{\rho}{1-\rho}, \qquad |z| = \rho < 1.$$

10. Show that the entire w-plane which has been cut along the ray $-\infty \le w \le -\frac{1}{4}$ can be regarded as a polygon with the two vertices $w = -\frac{1}{4}$ and $w = \infty$ and the corresponding exterior angles $-\pi$ and 3π. Using the Schwarz-Christoffel formula, show then that the function $w = f(z)$ which maps $|z| < 1$ onto this cut plane and satisfies $f(-1) = -\frac{1}{4}, f(1) = \infty, f(0) = 0$, is of the form

$$w = f(z) = \frac{z}{(1-z)^2}.$$

7. Domains Bounded by Circular Arcs. In this section we shall consider the conformal mapping of domains which are bounded by a finite number of circular arcs. For greater brevity, such a domain will be referred to as a *curvilinear polygon* (Fig. 20). Our aim is to find the function $w = f(z)$ which maps the upper half-plane Re $\{z\} > 0$ onto the interior of this figure. In the similar problem of the preceding section, the crucial step was the introduction of the differential operator w''/w' which is not affected if the function w is replaced by $aw + b$, where a and b are arbitrary constants. To put it differently, this operator is invariant under a linear substitution which transforms any straight line into any other straight line. In the present problem, the domain in question is not bounded by linear segments but by circular arcs, and it may therefore be expected that a fundamental role will be played by a differential operator which is not susceptible to transformations carrying circles into circles, *i.e.*, general linear transformations.

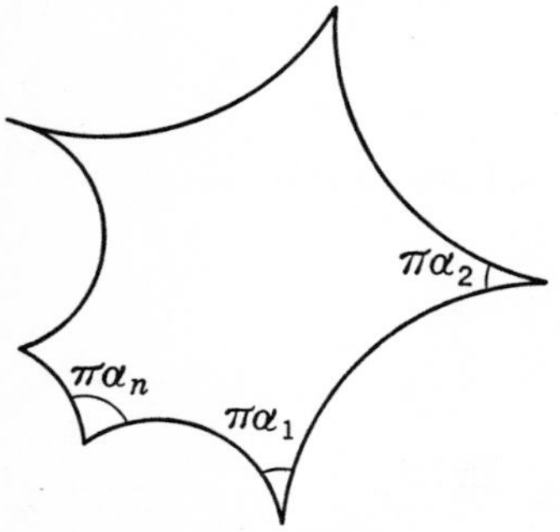

FIG. 20.

We shall show that the differential operator

$$\{w,z\} = \left(\frac{w''}{w'}\right)' - \frac{1}{2}\left(\frac{w''}{w'}\right)^2, \qquad w' = \frac{dw}{dz}, \tag{56}$$

which is known by the name of the *Schwarzian derivative* or the *Schwarzian differential parameter*, has precisely this property. In the notation (56), this amounts to showing that

$$\{W,z\} = \{w,z\}, \qquad W = \frac{az+b}{cz+d}, \qquad ad - bc \neq 0, \tag{57}$$

where a, b, c, d are constants. The identity (57) can be confirmed by a formal computation. We have

$$W' = \frac{(ad - bc)}{(cw + d)^2}\, w',$$

whence, by logarithmic differentiation,

$$\frac{W''}{W'} = \frac{w''}{w'} - \frac{2cw'}{cw + d}.$$

Hence,

$$\left(\frac{W''}{W'}\right)' = \left(\frac{w''}{w''}\right)' + \frac{2c^2w'^2}{(cw + d)^2} - \frac{2cw''}{cw + d},$$

and

$$\left(\frac{W''}{W'}\right)^2 = \left(\frac{w''}{w'}\right)^2 + \frac{4c^2w'^2}{(cw + d)^2} - \frac{4cw''}{cw + d},$$

and therefore

$$\left(\frac{W''}{W'}\right)' - \frac{1}{2}\left(\frac{W''}{W'}\right)^2 = \left(\frac{w''}{w'}\right)' - \frac{1}{2}\left(\frac{w''}{w'}\right)^2,$$

which, in view of (56), proves (57).

Returning now to the problem of mapping the upper half-plane onto the curvilinear polygon, we first observe that, in view of the symmetry principle, the mapping function $w = f(z)$ must be regular at all points of the real axis except at the points $a_1, a_2, \ldots, a_n$ which correspond to the vertices of the polygon. In view of the fact that the mapping $z \rightarrow f(z)$ is conformal at all points of $\text{Im}\,\{z\} > 0$ and at the points of the real axis other than a_ν, $\nu = 1, \ldots, n$, the derivative $f'(z)$ does not vanish there. It follows that the expression $\{w,z\}$ $[w = f(z)]$ is regular in the closed half-plane $\text{Im}\,\{z\} \geq 0$, with the exception of the points a_ν. We now use the invariance property (57) of the Schwarzian derivative. By a suitable linear transformation, any one of the circular arcs bounding our polygon can be mapped onto part of the real axis. If $w \rightarrow W$ is this linear transformation, it follows from (57) that $\{w,z\} = \{W,z\}$. Now the transformation $z \rightarrow W$ maps a part of the real axis—the linear segment $a_\nu < z < a_{\nu+1}$, say—onto another part of the real axis. For these values of z, the function $W = W(z)$ is a real function, and the same is therefore true of its derivatives. In view of the definition (56) of the Schwarzian derivative $\{W,z\}$, it follows that $\{W,z\}$ takes real values for $a_\nu < z < a_{\nu+1}$. Hence, $\{w,z\}$ is also real for $a_\nu < z < a_{\nu+1}$. We thus have proved that the Schwarzian derivative $\{w,z\}$ of the mapping function $w = f(z)$ is real at all points of the real axis except at the points $a_1, \ldots, a_n$.

At the points $a_1, \ldots, a_n$, the function $w = f(z)$—and therefore also its Schwarzian derivative—will have singularities. In order to study the nature of these singularities of $\{w,z\}$, we use again the invariance of $\{w,z\}$ with respect to an arbitrary linear transformation. Considering, in particular, the vertex corresponding to $z = a_\nu$, we can perform a linear transformation which carries this vertex into the origin and transforms the two circles meeting at this vertex with the angle $\pi\alpha_\nu$ (see Fig. 20) into two straight lines. Since the mapping is conformal, these two straight lines will meet at the origin with the same angle $\pi\alpha_\nu$. Now $\{w,z\}$ was not affected by this linear transformation; hence, the singularity of the function $\{w,z\}$ at $z = a_\nu$ can be obtained from the assumption that the function $w = f(z)$ maps a piece of the real axis containing $z = a_\nu$ onto two linear segments meeting at the origin with the angle $\pi\alpha_\nu$. As shown in the preceding section, such a function $f(z)$ is of the form

$$f(z) = (z - a_\nu)^{\alpha_\nu} f_1(z),$$

where $f_1(z)$ is regular at $z = a_\nu$, $f_1(a_\nu) \neq 0$, and $f_1(z)$ is a real function if z is real. Using this representation, we obtain, by an elementary computation,

$$\{w,z\} = \left[\frac{f''(z)}{f'(z)}\right]' - \frac{1}{2}\left[\frac{f''(z)}{f'(z)}\right]^2 = \frac{1}{2}\frac{1 - \alpha_\nu^2}{(z - a_\nu)^2} + \frac{\beta_\nu}{z - a_\nu} + f_2(z),$$

where $f_2(z)$ is regular at $z = a_\nu$ and

$$\beta_\nu = \frac{1 - \alpha_\nu^2}{\alpha_\nu} \frac{f_1'(a_\nu)}{f_1(a_\nu)}$$

is real. By applying the same procedure to all the points $a_1, \ldots, a_n$, we thus find that the expression

$$\{w,z\} - \frac{1}{2}\sum_{\nu=1}^{n} \frac{1 - \alpha_\nu^2}{(z - a_\nu)^2} - \sum_{\nu=1}^{n} \frac{\beta_\nu}{z - a_\nu}$$

is regular at the points $a_1, \ldots, a_n$ and, therefore, at all points of the real axis. This expression is, moreover, real at all points of the real axis. Indeed, $\{w,z\}$ was shown above to be real for real z, and the reality of the terms involving $(z - a_\nu)^{-1}$ and $(z - a_\nu)^{-2}$ follows from the fact that the constants α_ν, β_ν, a_ν are real. But, as shown in Sec. 10, Chap. III, an analytic function which is regular in the closure of a domain and takes real values on the boundary reduces to a constant. Hence, *the function* $w = f(z)$ *mapping* $\operatorname{Im}\{z\} > 0$ *onto a curvilinear polygon with angles* $\pi\alpha_1, \ldots, \pi\alpha_n$ *satisfies the differential equation*

$$\{w,z\} = \frac{1}{2}\sum_{\nu=1}^{n} \frac{1 - \alpha_\nu^2}{(z - a_\nu)^2} + \sum_{\nu=1}^{n} \frac{\beta_\nu}{z - a_\nu} + \gamma, \tag{58}$$

where $\beta_1, \ldots, \beta_n, \gamma$ *are real constants.*

The constants γ and β_ν are, however, not entirely independent of each other. If none of the points $a_1, \ldots, a_n$ coincide with the point at infinity, $w = f(z)$ must be regular at $z = \infty$. Hence, there is an expansion

$$f(z) = c_0 + \frac{c_1}{z} + \frac{c_2}{z^2} + \cdots$$

which converges near $z = \infty$. Inserting this in (56), we find by a formal computation that the expansion of $\{w,z\}$ near $z = \infty$ starts with the term in z^{-4}. Since the first terms of the corresponding expansion of the right-hand side of (58) are

$$\gamma + \frac{1}{z}\sum_{\nu=1}^{n} \beta_\nu + \frac{1}{z^2}\sum_{\nu=1}^{n}\left[a_\nu\beta_\nu + \frac{1}{2}(1 - \alpha_\nu^2)\right] + \frac{1}{z^3}\sum_{\nu=1}^{n}[\beta_\nu a_\nu^2 + a_\nu(1 - \alpha_\nu^2)] + \cdots,$$

it follows that the conditions

$$\gamma = 0, \qquad \sum_{\nu=1}^{n} \beta_\nu = 0, \qquad \sum_{\nu=1}^{n} [2a_\nu\beta_\nu + 1 - \alpha_\nu^2] = 0, \qquad \sum_{\nu=1}^{n} [\beta_\nu a_\nu^2 + a_\nu(1 - \alpha_\nu^2)] = 0 \tag{59}$$

must be satisfied. The reader will confirm without difficulty that the first two conditions (59) also hold if one of the vertices of the polygon corresponds to $z = \infty$ and that, in this case, the expansion of $\{w,z\}$ near

$z = \infty$ starts with the term $\frac{1}{2}(1 - \alpha^2)z^{-2}$, if $\pi\alpha$ is the corresponding angle of the polygon.

It is easy to see that the four identities (59) are the only general relations which can exist between the constants entering the equation (58). The solutions of (58) must be able to represent the most general curvilinear polygon P_n whose sides are n circular arcs. Since a circle is determined by $3n$ real parameters (radius and coordinates of the center), there is a $3n$-parameter family of such polygons. From these we have to deduct six real parameters since the equation (58) determines P_n only up to an arbitrary linear transformation depending on six real parameters (three arbitrary points can be made to correspond to three given points). This leaves $3n - 6$ independent real parameters determining a polygon P_n. In (58), there appear explicitly $3n + 1$ parameters, namely, γ and the constants α_ν, a_ν, β_ν. Deducting from these the four relations (59), we still have a balance of $3n - 3$, three more than we need. However, this excess of parameters is only apparent. By a linear transformation of the upper half-plane onto itself, three of the points $a_1, \ldots, a_n$ can be brought into prescribed positions on the real axis, thus eliminating another triple of constants entering (58). The number of these constants is thus reduced to $3n - 6$. Since this is precisely the number of parameters characterizing a polygon P_n, it follows that no more relations between the constants in (58) can be expected.

The difficulty in constructing the mapping function of a given curvilinear polygon from the differential equation (58) is due not so much to the fact that we have to integrate a differential equation of the third order; we shall see presently that our task can be reduced to the integration of a comparatively simple linear differential equation of the second order. The real difficulty is caused by the fact that the connection between the constants entering (58)—excepting, of course, the α_ν which are given by the angles—and the geometric configuration of the polygon P_n is extremely unobvious. $n - 3$ of these constants can be determined by "non-Euclidean" conditions, namely, by prescribing the points a_ν on the real z-axis which are to correspond by the mapping $w = f(z)$ to the vertices of P_n (it has been mentioned before that three of the points a_ν are arbitrary). Deducting further the n constants α_ν which are given by the angles of P_n, we are thus left with $n - 3$ constants, the so-called *accessory parameters*, whose determination by means of geometric conditions, whether Euclidean or non-Euclidean, is an extremely difficult task. Except for the case $n = 2$ (which can be treated by much more elementary means), the only case which is free of accessory parameters is that of a curvilinear triangle.

In this case, all constants entering the equation (58) can be expressed in terms of the given quantities. If we write $\alpha_1 = \alpha$, $\alpha_2 = \beta$, $\alpha_3 = \gamma$,

$a_1 = a$, $a_2 = b$, $a_3 = c$, it follows from (59) and an elementary computation that the equation (58) reduces in this case to

$$(60)\quad \{w,z\} = \frac{1}{(z-a)(z-b)(z-c)}\left[\frac{1-\alpha^2}{2}\frac{(a-b)(a-c)}{z-a} + \frac{1-\beta^2}{2}\frac{(b-a)(b-c)}{z-b} + \frac{1-\gamma^2}{2}\frac{(c-a)(c-b)}{z-c}\right].$$

This expression can be simplified by identifying a, b, c with the points $z = 0$, $z = \infty$, $z = 1$, respectively. If we let $b \to \infty$ in (60), the expressions

$$\frac{a-b}{z-b},\qquad \frac{(b-a)(b-c)}{(z-b)^2},\qquad \frac{c-b}{z-b}$$

tend to 1 and we obtain

$$\{w,z\} = \frac{1}{(z-a)(z-c)}\left[\frac{1-\alpha^2}{2}\frac{a-c}{z-a} + \frac{1-\beta^2}{2} + \frac{1-\gamma^2}{2}\frac{c-a}{z-c}\right].$$

Hence, for $a = 0$, $c = 1$,

$$\{w,z\} = \frac{1}{z(z-1)}\left[-\frac{1-\alpha^2}{2z} + \frac{1-\beta^2}{2} + \frac{1-\gamma^2}{2(z-1)}\right],$$

which can also be brought into the form

$$(61)\quad \{w,z\} = \frac{1-\alpha^2}{2z^2} + \frac{1-\gamma^2}{2(z-1)^2} + \frac{\alpha^2+\gamma^2-\beta^2-1}{2z(z-1)}.$$

However, before we enter into a further discussion of (61), we have to examine the differential equation (58) in greater detail. If w is a solution of (58), the same is true of the function W defined in (57) which contains three arbitrary constants (one of the four constants a, b, c, d can be made equal to 1 without altering the value of W). Since, on the other hand, (58) is a differential equation of the third order whose general solution cannot contain more than three independent arbitrary constants, it is thus sufficient to find one solution of (58); the general solution will then follow by an arbitrary linear transformation. The finding of one particular solution is further facilitated by a connection existing between an equation of the type (58) and a linear differential equation of the second order. The result to which we are referring is the following. *If u_1 and u_2 are two linearly independent solutions of the linear differential equation*

$$u''(z) + p(z)u(z) = 0,$$

then

$$w(z) = \frac{u_1(z)}{u_2(z)}$$

is a solution of the equation

$$\{w,z\} = 2p(z). \tag{62}$$

The truth of this result is easily confirmed. Substituting u_2w for u_1 in the equation $u_1'' + pu_1 = 0$, we obtain

$$u_2w'' + 2u_2'w' + w(u_2'' + pu_2) = 0$$

and therefore, in view of $u_2'' + pu_2 = 0$, $u_2w'' + 2u_2'w' = 0$. Hence,

$$\frac{w''}{w'} = -2\frac{u_2'}{u_2},$$

and thus

$$\left(\frac{w''}{w'}\right)' - \frac{1}{2}\left(\frac{w''}{w'}\right)^2 = -2\left(\frac{u_2'}{u_2}\right)' - 2\left(\frac{u_2'}{u_2}\right)^2 = -\frac{2u_2''}{u_2}.$$

In view of $u_2'' + pu_2 = 0$ and (56), this is equivalent to (62).

Combining (62) with (58) and (59), we thus obtain the following result.

If $w = f(z)$ *maps the upper half-plane* Im $\{z\} > 0$ *onto a curvilinear polygon composed of n circular arcs and if the point* $z = a_\nu$ *on the real axis corresponds to a vertex of angle* $\pi\alpha_\nu$, *then*

$$w = f(z) = \frac{u_1(z)}{u_2(z)}, \tag{63}$$

where $u_1(z)$ *and* $u_2(z)$ *are two linearly independent solutions of the linear differential equation*

$$u''(z) + \left[\frac{1}{4}\sum_{\nu=1}^{n}\frac{1-\alpha_\nu^2}{(z-a_\nu)^2} + \frac{1}{2}\sum_{\nu=1}^{n}\frac{\beta_\nu}{z-a_\nu}\right]u(z) = 0 \tag{64}$$

and the real constants β_ν *are subject to the relations*

$$\sum_{\nu=1}^{n}\beta_\nu = 0, \qquad \sum_{\nu=1}^{n}[2a_\nu\beta_\nu + 1 - \alpha_\nu^2] = 0, \tag{65}$$

$$\sum_{\nu=1}^{n}[\beta_\nu a_\nu^2 + a_\nu(1 - \alpha_\nu^2)] = 0.$$

Since a differential equation of the type (64) is easily solved in terms of power series expansions, our mapping problem is therefore to be regarded as solved if the constants β_ν are known. However, as already pointed out, the determination of the $n - 3$ independent constants β_ν in terms of the geometrical configuration of the curvilinear polygon is an exceedingly difficult task.

A complete treatment is possible in the case of a curvilinear triangle. If $\pi\alpha$, $\pi\beta$, $\pi\gamma$ are the angles of the triangle and $z = 0, \infty, 1$ the points corresponding to the vertices, it follows from (61) that the differential equation (64) takes the form

$$(66) \qquad u'' + \frac{1}{4}\left[\frac{1-\alpha^2}{z^2} + \frac{1-\gamma^2}{(z-1)^2} + \frac{\alpha^2+\gamma^2-\beta^2-1}{z(z-1)}\right]u = 0.$$

Since we are interested not in the individual solutions of this equation but in the quotient (63) of two solutions, we may replace (66) by a differential equation of the type

$$(67) \qquad y'' + P(z)y' + Q(z)y = 0$$

whose solutions are related to those of (66) by the identity

$$(68) \qquad y(z) = \sigma(z)u(z),$$

where $\sigma(z)$ is a given function. If $y_1(z)$ and $y_2(z)$ are two linearly independent solutions of (67), we clearly have

$$\frac{y_1(z)}{y_2(z)} = \frac{u_1(z)}{u_2(z)},$$

where $u_1(z)$ and $u_2(z)$ are linearly independent solutions of (66). Taking the function $\sigma(z)$ in (68) to be of the form

$$\sigma(z) = e^{-\frac{1}{2}\int^z P(z)\,dz},$$

we find by an elementary computation that the equation (67) is equivalent to the equation

$$u'' + [Q - \tfrac{1}{4}P^2 - \tfrac{1}{2}P']u = 0$$

for the function $u = u(z)$ defined in (68). A comparison with (66) shows therefore that the equations (66) and (67) will be equivalent (for our purposes) if the relation

$$(69) \quad Q - \frac{1}{4}P^2 - \frac{1}{2}P' = \frac{1}{4}\left[\frac{1-\alpha^2}{z^2} + \frac{1-\gamma^2}{(z-1)^2} + \frac{\alpha^2+\gamma^2-\beta^2-1}{z(z-1)}\right]$$

is satisfied. We leave it as an exercise to the reader to show that (69) holds if

$$P(z) = \frac{c - (a + b + 1)z}{z(1 - z)}, \qquad Q(z) = -\frac{ab}{z(1 - z)},$$

where the constants a, b, c are defined by

$$(70) \quad a = \tfrac{1}{2}(1 + \beta - \alpha - \gamma), \qquad b = \tfrac{1}{2}(1 - \alpha - \beta - \gamma), \qquad c = 1 - \alpha.$$

With these values of $P(z)$ and $Q(z)$, the equation (67) can be brought into the form

$$(71) \qquad z(1 - z)y'' + [c - (a + b + 1)z]y' - aby = 0.$$

This differential equation is known as the *hypergeometric equation* and plays an important part in many branches of pure and applied mathematical analysis. The properties of its solutions have been thoroughly investigated and have been made the subject of an extensive literature. Those properties of the solutions of (71) which are relevant from our present point of view will be found in the chapter on the hypergeometric function in Whittaker-Watson's "Modern Analysis," to which the reader is referred.

Summing up, we thus have the following result: *The function* $w = f(z)$ *which maps the upper half-plane* Im $\{z\} > 0$ *onto the interior of a curvilinear triangle with the angles* $\pi\alpha$, $\pi\beta$, $\pi\gamma$ *is of the form*

$$f(z) = \frac{y_1(z)}{y_2(z)},$$

where $y_1(z)$ *and* $y_2(z)$ *are two linearly independent solutions of the hypergeometric equation* (71) *and the constants* a, b, c *in* (71) *are related to* α, β, γ *by* (70).

The equation (71) is solved by the hypergeometric series

$$(72) \quad F(a,b,c;z) = 1 + \frac{ab}{c}z + \frac{a(a+1)b(b+1)}{c(c+1)2!}z^2 + \frac{a(a+1)(a+2)b(b+1)(b+2)}{c(c+1)(c+2)3!}z^3 + \cdots, \qquad |z| < 1,$$

as the reader will verify without difficulty. The function $F(a,b,c;z)$ can also be represented in the form of a definite integral. We have

$$(73) \qquad F(a,b,c;z) = \frac{\Gamma(c)}{\Gamma(b)\Gamma(c - b)} \int_0^1 t^{b-1}(1 - t)^{c-b-1}(1 - zt)^{-a}\,dt,$$

where the conditions $b > 0$, $c > b$ are necessary for the existence of the integral. The identity of (73) and the series (72) is easily established by expanding $(1 - zt)^{-a}$ by powers of z and integrating term by term. In view of (70), the conditions $b > 0, c > b$ are equivalent to $\alpha + \beta + \gamma < 1$, $\alpha < 1 + \beta + \gamma$; both will therefore be satisfied if the sum of the three angles of the curvilinear triangle is smaller than π. The integral representation (73) has many advantages over the infinite series (72). While the use of (72) is restricted to values of z such that $|z| < 1$, no such restriction applies to (73). (73) may therefore be used for all values of z in the upper half-plane. Besides, the convergence of the series (72) is very slow unless $|z|$ is small, while the value of the integral (73) can be easily computed with great accuracy.

For the solution of our mapping problem we need yet another solution of the equation (71). Such a solution is easily obtained from the observation that the substitution of $1 - z$ for z transforms (71) into

$$z(1 - z)y'' + [a + b - c + 1 - (a + b + 1)z]y' - aby = 0,$$

which is another hypergeometric equation. The parameters of this equation are $a_1 = a$, $b_1 = b$, $c_1 = a + b - c + 1$. A glance at (73) shows that this equation is solved by

$$y = \int_0^1 t^{b-1}(1 - t)^{a-c}(1 - zt)^{-a}\,dt, \tag{74}$$

where the conditions $b > 0$ and $a > c - 1$ are required for the existence of the integral. These conditions are identical with $\alpha + \beta + \gamma < 1$, $\gamma - \beta - \alpha < 1$, and are therefore satisfied if the sum of the angles is smaller than π. If, in (74), z is again replaced by $1 - z$, we obtain a solution of the equation (71); the confirmation that (74), after this substitution, is not a constant multiple of (73) is left as an exercise to the reader. Our result concerning the mapping function of the curvilinear triangle takes thus the following explicit form.

The function

$$w = f(z) = \frac{\int_0^1 t^{-\frac{1}{2}(1+\alpha+\beta+\gamma)}(1 - t)^{-\frac{1}{2}(1+\alpha-\beta-\gamma)}(1 - zt)^{-\frac{1}{2}(1-\alpha+\beta-\gamma)}\,dt}{\int_0^1 t^{-\frac{1}{2}(1+\alpha+\beta+\gamma)}(1 - t)^{-\frac{1}{2}(1-\alpha-\beta+\gamma)}(1 - t + zt)^{-\frac{1}{2}(1-\alpha+\beta-\gamma)}\,dt} \tag{75}$$

maps the upper half-plane Im $\{z\} > 0$ *onto a curvilinear triangle with the angles* $\pi\alpha, \pi\beta, \pi\gamma$, *provided the sum of the angles is smaller than* π.

If $\alpha + \beta + \gamma = 1$, the triangle can be made rectilinear by a suitable linear transformation and the mapping function can be constructed by means of the Schwarz-Christoffel formula. If $\alpha + \beta + \gamma > 1$, then (73) and (74) have to be replaced by integral representations of the hypergeometric function which converge for these values of the parameters. The interested reader will find integral representations of this type in the book of Whittaker and Watson mentioned above.

A further discussion of the function $f(z)$ defined in (75) will be found in Sec. 5, Chap. VI.

EXERCISES

1. Using (75) and the identities

$$\int_0^1 t^{r-1}(1-t)^{s-1}\,dt = \frac{\Gamma(r)\Gamma(s)}{\Gamma(r+s)}, \qquad r > 0,\ s > 0,$$

$$\Gamma(r)\Gamma(1-r) = \frac{\pi}{\sin \pi r},$$

show that the vertices with the angles $\pi\alpha$ and $\pi\gamma$ (corresponding to $z = 0$ and $z = 1$, respectively) are situated at the points

$$w = \frac{\sin \pi\alpha}{\cos \frac{\pi}{2}(\alpha - \beta - \gamma)}$$

and

$$w = \frac{\cos \frac{\pi}{2}(\alpha + \beta - \gamma)}{\sin \pi\gamma},$$

respectively.

2. If $\alpha = \gamma$, show that the function (75) satisfies the relation $f(z)f(1-z) = 1$. Use your result to prove that $w = f(z)$ maps the straight line $\text{Re}\,\{z\} = \frac{1}{2}$ onto part of the circumference $|w| = 1$. *Hint:* Use the fact that—for suitable determinations of the powers under the integral signs—$f(z)$ is real for $0 < z < 1$, and apply the symmetry principle.

3. Show that in the case in which the three circles forming the curvilinear triangle are tangent to each other (Fig. 21), the mapping function (75) takes the form

$$f(z) = \frac{T(z)}{T(1-z)},$$

where

$$T(z) = \int_0^1 \frac{dt}{\sqrt{t(1-t)(1-zt)}}.$$

Show further that this is equivalent to

$$f(z) = \frac{K(z)}{K(1-z)},$$

where

$$K(z) = \int_0^1 \frac{dt}{\sqrt{(1-t^2)(1-zt^2)}}.$$

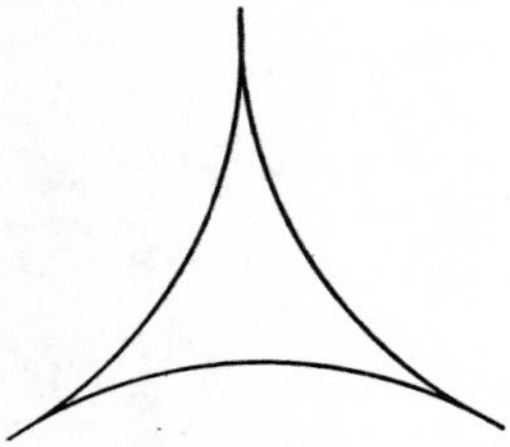

FIG. 21.

4. Show that the equation (58) for the function $w = f(z)$ which maps $\text{Im}\,\{z\} > 0$ onto the interior of a crescent-shaped (or lense-shaped, as the case may be) figure of angle α (see Fig. 22) is

$$\{w,z\} = \frac{(1-\alpha^2)(a-b)^2}{2(z-a)^2(z-b)^2},$$

where a and b are the points on the real axis corresponding to the vertices. Show further that the associated linear differential equation

$$u'' + \frac{(1 - \alpha^2)(a - b)^2}{4(z - a)^2(z - b)^2} u = 0$$

is equivalent to the differential equation

$$y'' + (1 - \alpha)\left[\frac{1}{z - a} + \frac{1}{z - b}\right] y' - \frac{\alpha(1 - \alpha)}{(z - a)(z - b)} y = 0.$$

Verify that the latter equation has the solutions $y = (z - a)^\alpha$ and $y = (z - b)^\alpha$ and deduce that the mapping function is of the form

$$f(z) = \frac{A(z - a)^\alpha + B(z - b)^\alpha}{C(z - a)^\alpha + D(z - b)^\alpha},$$

where A, B, C, D are complex constants for which $AD - BC \neq 0$.

FIG. 22.

8. Univalent Functions. Of especial importance from the point of view of conformal mapping are those analytic functions $f(z)$ which are univalent in a given domain D. We recall that a univalent function in D is characterized by the fact that it takes in D no value more than once and that, consequently, it maps D onto a schlicht domain, *i.e.*, a domain which is not self-overlapping and contains no branch points. For the latter reason, univalent functions are also often referred to as *schlicht functions*. In the present section, we shall investigate some of the properties of analytic functions which are univalent in a given simply-connected domain D. We may, without an essential restriction of the generality of our considerations, confine ourselves to the case in which D is the unit circle. Indeed, by the Riemann mapping theorem, any simply-connected domain can be mapped onto the unit circle; accordingly, any univalent function in D is associated with a univalent function in the unit circle and the properties of the latter function can be easily translated into properties of the original function if the function mapping D onto the unit circle is known. The choice of the unit circle as the domain of definition of a univalent function has the advantage of simplifying the computations and of leading to short and elegant formulas.

A function $f(z)$ which is regular and univalent in the unit circle may further be normalized by the conditions $f(0) = 0$, $f'(0) = 1$. Indeed, if $f(z)$ is univalent, so is the function

$$f_1(z) = \frac{f(z) - f(0)}{f'(0)},$$

and any property of the function $f_1(z)$ is immediately translated into a corresponding property of $f(z)$; we add that the division by $f'(0)$ is permissible since the derivative of a univalent function does not vanish. The class of analytic functions which are regular and univalent in the unit circle and which are normalized by the condition $f(0) = 0, f'(0) = 1$, will be denoted by S. There also exist functions which are univalent but not regular in the unit circle. For example, the function

$$g(z) = \frac{af(z) + b}{cf(z) + d}, \qquad ad - bc \neq 0, f(z) \in S$$

is obviously univalent in $|z| < 1$ but it will have a simple pole if $-(d/c)$ is one of the values taken by $f(z)$ in the unit circle. On the other hand, a univalent function can have no other singularities but one simple pole, as otherwise the value ∞ would be taken more than once. The class of univalent functions with one simple pole in $|z| < 1$ will be normalized by the requirement that the pole be located at the origin. If this condition is not satisfied and the function $f(z)$ has its pole at $z = \alpha(|\alpha| < 1)$, we consider instead of $f(z)$ the function

$$g(z) = f\left(\frac{z + \alpha}{1 + \bar{\alpha}z}\right).$$

Since the transformation $z \to (z + \alpha)(1 + \bar{\alpha}z)^{-1}$ maps the schlicht unit circle onto itself, $g(z)$ is also univalent in $|z| < 1$ and the pole of $g(z)$ is clearly situated at $z = 0$. The univalent functions with a pole at the origin will further be subjected to the normalization condition that the residue of the pole have the value 1; this can always be achieved by multiplying by a suitable constant. The class of these functions, *i.e.*, the univalent functions in $|z| < 1$ which have a Laurent expansion

$$(76) \qquad f(z) = \frac{1}{z} + b_0 + b_1 z + b_2 z^2 + \cdots, \qquad |z| < 1,$$

will be denoted by P.

The coefficients $b_1, b_2, \ldots$ of the Laurent expansion (76) of a function of P are subject to the important inequality

$$(77) \qquad \sum_{n=1}^{\infty} n|b_n|^2 \leq 1,$$

which is known as the *area theorem*. This name is due to the circumstance that the inequality (77) is but the analytic expression of the geometrically evident fact that the complement E (see Fig. 23) of the domain D upon which $|z| < 1$ is mapped by $w = f(z)$ has an area which is either positive or zero. To prove (77), we first remark that the common boundary C of

D and E is traversed in the negative direction with respect to D if it is described in the positive direction with respect to E. To avoid difficulties which might arise from the possible irregular character of the boundary of the domain upon which $w = f(z)$ maps the unit circle, we take D in Fig. 23 to be the domain which is the image of $|z| = r$ $(r < 1)$ by means of the mapping $z \to f(z)$. The curve C will then be analytic. If we introduce polar coordinates R, ϕ in the w-plane, then, by elementary calculus, the area of E will be

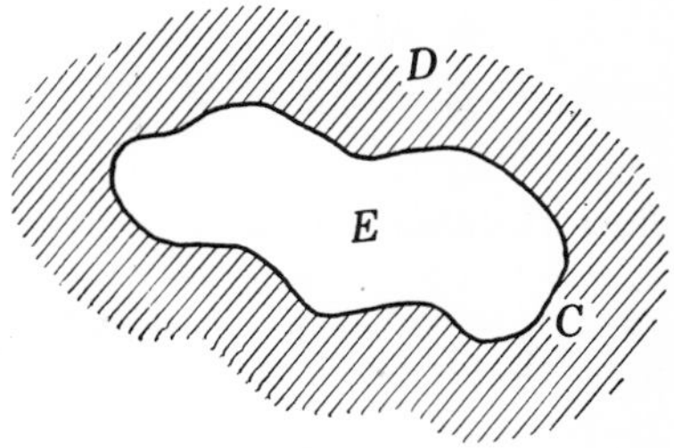

FIG. 23.

$$A = \frac{1}{2}\int_C R^2\,d\phi = -\frac{1}{2}\int_0^{2\pi} R^2\frac{\partial\phi}{\partial\theta}\,d\theta,$$

where r, θ are polar coordinates on the circle $|z| = r$ whose image is the curve C. The negative sign is due to the fact pointed out before that the positive direction with respect to E is the negative direction with respect to D, and therefore also with respect to $|z| = r$. Since, by the Cauchy-Riemann equations,

$$\frac{\partial\phi}{\partial\theta} = \frac{r}{R}\frac{\partial R}{\partial r},$$

it follows that

$$A = -\frac{r}{2}\int_0^{2\pi} R\frac{\partial R}{\partial r}\,d\theta = -\frac{r}{4}\frac{\partial}{\partial r}\left[\int_0^{2\pi} R^2\,d\theta\right],$$

whence

$$A = -\frac{r}{4}\frac{\partial}{\partial r}\left[\int_0^{2\pi} |f(re^{i\theta})|^2\,d\theta\right].$$

On the other hand,

$$\begin{aligned}\int_0^{2\pi} |f(re^{i\theta})|^2\,d\theta &= \int_0^{2\pi} f(re^{i\theta})\overline{f(re^{i\theta})}\,d\theta \\ &= \int_0^{2\pi}\left[\frac{1}{re^{i\theta}} + \sum_{n=0}^{\infty} b_n r^n e^{in\theta}\right]\left[\frac{1}{re^{-i\theta}} + \sum_{n=0}^{\infty} \bar{b}_n r^n e^{-in\theta}\right]d\theta \\ &= 2\pi\left[\frac{1}{r^2} + \sum_{n=0}^{\infty} |b_n|^2 r^{2n}\right],\end{aligned}$$

all other terms being zero; the term-by-term integration is permissible because of the uniform convergence. Hence,

$$A = -\frac{\pi r}{2}\frac{\partial}{\partial r}\left[\frac{1}{r^2} + \sum_{n=0}^{\infty} |b_n{}^2| r^{2n}\right],$$

and thus

$$(\pi^{-1})A = \frac{1}{r^2} - \sum_{n=1}^{\infty} n|b_n|^2 r^{2n}.$$

Since the area A cannot be negative, it follows that

$$\sum_{n=1}^{\infty} n|b_n|^2 r^{2n} \leq \frac{1}{r^2}.$$

This inequality holds for all values of r between 0 and 1. It therefore must also be true for $r \to 1$. This proves (77).

An immediate consequence of (77) is the inequality

$$|b_1| \leq 1. \tag{78}$$

In order to see whether this inequality is "sharp," *i.e.*, whether there exists a function of P for which $|b_1| = 1$, we observe from (77) that this is only possible if $b_2 = b_3 = \cdots = 0$. The function (76) will in this case reduce to

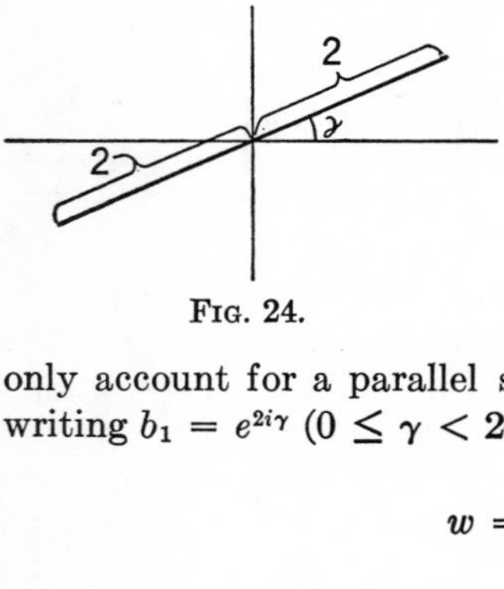

FIG. 24.

$$f_0(z) = \frac{1}{z} + b_0 + b_1 z, \qquad |b_1| = 1.$$

Disregarding the constant b_0, which can only account for a parallel shift of the conformal map of $|z| < 1$, and writing $b_1 = e^{2i\gamma}$ $(0 \leq \gamma < 2\pi)$, we have

$$w = f_0(z) = \frac{1}{z} + e^{2i\gamma}z.$$

This shows that $f_0(e^{i\theta}) = 2e^{i\gamma}\cos(\theta + \gamma)$. Hence, if z describes the unit circle, its conformal image w describes "both sides" of the linear segment indicated in Fig. 24. It follows that $w = f_0(z)$ is univalent in $|z| < 1$ and that this function maps $|z| < 1$ onto the full w-plane which has been cut along the linear segment indicated in Fig. 24. Hence, the inequality (78) is sharp and equality in (78) can hold only for functions mapping $|z| < 1$ onto the full w-plane which has a rectilinear "slit" of length 4.

We now turn our attention to functions of the class S, that is, univalent functions which have a Taylor expansion

$$f(z) = z + a_2z^2 + a_3z^3 + \cdots, \qquad |z| < 1, \tag{79}$$

in the unit circle. Since the reciprocal of $f(z)$ is a function of P, we can use (78) in order to obtain information regarding the coefficients of the expansion (79). We have

$$\frac{1}{f(z)} = \frac{1}{z(1 + a_2z + a_3z^2 + \cdots)} = \frac{1}{z}\{1 - (a_2z + a_3z^2 + \cdots) + (a_2z + a_3z^2 + \cdots)^2 - \cdots\},$$

whence,

$$\frac{1}{f(z)} = \frac{1}{z} - a_2 + (a_2^2 - a_3)z + \cdots,$$

where the omitted terms contain higher powers of z. Using (78), we thus obtain

$$|a_2^2 - a_3| \leq 1. \tag{80}$$

From (80), it is possible to arrive at an inequality which refers to the coefficient a_2 only. To this end, we observe that the function

$$g(z) = \sqrt{f(z^2)} = z\sqrt{1 + a_2z^2 + a_3z^4 + \cdots} = z + \frac{a_2}{2}z^3 + \cdots \tag{81}$$

is regular and univalent in $|z| < 1$ if the same is true of the function (79). Indeed, from $g(z_1) = g(z_2)$ ($z_1 \neq z_2$, $|z_1| < 1$, $|z_2| < 1$) it would follow by (81) that $f(z_1^2) = f(z_2^2)$. Since $f(z)$ in univalent in $|z| < 1$ and both z_1^2 and z_2^2 are points of the unit circle, this is only possible if $z_1^2 = z_2^2$. In view of $z_1 \neq z_2$, we are left with the possibility $z_2 = -z_1$. But we cannot have $g(z_1) = g(-z_1)$ for $z_1 \neq 0$ since, by (81), $g(z)$ is an *odd function* of z, that is, $g(-z) = -g(z)$. The second and third coefficient of the univalent function $g(z)$ are 0 and $\frac{1}{2}a_2$, respectively. Applying the inequality (80), we thus find that

$$|a_2| \leq 2. \tag{82}$$

This is again a sharp inequality. To see this, we only have to consider the function

$$w = f_0(z) = \frac{z}{(1 - z)^2} = z + 2z^2 + 3z^2 + \cdots + nz^n + \cdots \tag{83}$$

which, as the reader will confirm without difficulty, is univalent in $|z| < 1$ and maps $|z| < 1$ onto the full w-plane that is furnished with a slit along the negative axis from $-\frac{1}{4}$ to $-\infty$. Another useful exercise for the reader will be to retrace the steps leading to (82) and to show that the sign of equality in (82) can occur only if $f(z)$ is of the form $f(z) = e^{-i\gamma}f_0(e^{i\gamma}z)$, $0 \leq \gamma < 2\pi$, where $f_0(z)$ is defined in (83).

From (82) we can derive an important item of information concerning the conformal map of the unit circle which is yielded by the univalent function (79). Suppose ζ is a value not taken by the function $w = f(z)$ in $|z| < 1$. The function

$$f_1(z) = \frac{f(z)}{1 - \dfrac{f(z)}{\zeta}} = z + \left(a_2 + \frac{1}{\zeta}\right) z^2 + \cdots \tag{84}$$

will then be regular in $|z| < 1$, and it will also be univalent in $|z| < 1$ since the linear transformation of a univalent function leads again to a univalent function. Applying (82) to the second coefficient of this function, we obtain

$$\left| a_2 + \frac{1}{\zeta} \right| \leq 2,$$

whence

$$\frac{1}{|\zeta|} \leq 2 + |a_2| \leq 4,$$

the last inequality following by another application of (82). We thus find that

$$|\zeta| \geq \tfrac{1}{4}.$$

This shows that any point of the w-plane which is not taken by $w = f(z)$ in $|z| < 1$ has at least the distance $\frac{1}{4}$ from the origin. In other words, *the conformal map of the unit circle as yielded by the univalent function* (79) *contains all points of the circle* $|w| < \frac{1}{4}$. This result is sharp, *i.e.*, the constant $\frac{1}{4}$ cannot be replaced by any larger number. That this is so is shown by the function $f_0(z)$ in (83) which does not take the value $w = -\frac{1}{4}$ in $|z| < 1$.

The last result is an example of a group of theorems which are known by the name of *distortion theorems*. The idea underlying this appellation is that a conformal map of a domain D onto a domain D^* can be regarded as a "distortion" of the shape of D and of all subsets of D, resulting in the shape of D^* and in corresponding subsets of D^*. Any result giving limitations for the amount of this distortion is thus called a distortion theorem. The above "$\frac{1}{4}$-theorem" limits the distortion of the boundary of the unit circle in the conformal mapping by means of the univalent function (79); this boundary cannot be distorted so far as to come within a distance of less than $\frac{1}{4}$ of the origin.

By a repeated use of the transformation (84) we can obtain more detailed information regarding the distortion of the boundary. If ζ and η are two values not taken by $f(z)$ in $|z| < 1$, then the function $f_1(z)$ defined in (84) is regular and univalent in $|z| < 1$ and does not take there the value

$$\frac{\eta}{1 - (\eta/\zeta)}.$$

By the theorem just proved, the modulus of this number must therefore be at least $\frac{1}{4}$. Hence, by an obvious manipulation,

$$\left| \frac{1}{\zeta} - \frac{1}{\eta} \right| \leq 4.$$

Suppose now that ζ and η lie on the same straight line through the origin, so that the origin separates ζ and η. We then have $\zeta = |\zeta|e^{i\alpha}$ and $\eta = -|\eta|e^{i\alpha}$. Hence,

$$\frac{1}{|\zeta|} + \frac{1}{|\eta|} \leq 4.$$

This shows that at least one of the numbers $|\zeta|$ and $|\eta|$ must be larger than $\frac{1}{2}$, as otherwise the sum of their reciprocals would be larger than 4. The only exception is the case in which both $|\zeta|$ and $|\eta|$ are equal to $\frac{1}{2}$. We thus have the following result. *If ζ and η are two values not taken by a function of S and if the straight line connecting ζ and η passes through the origin, then the distance of at least one of the points ζ, η, from the origin is larger than $\frac{1}{2}$, or else both distances are equal to $\frac{1}{2}$.* We leave it as an exercise to the reader to show by means of the univalent function

$$w = \frac{z}{1 - z^2}$$

that the constant $\frac{1}{2}$ in this theorem is the largest possible.

As already pointed out above, the function

$$f\left(\frac{z + \zeta}{1 + \bar{\zeta}z}\right), \qquad |\zeta| < 1$$

is univalent in $|z| < 1$ if the same is true of the function $f(z)$. The normalized function

$$(85) \qquad g(z) = \frac{f\left(\dfrac{z + \zeta}{1 + \bar{\zeta}z}\right) - f(\zeta)}{f'(\zeta)(1 - |\zeta|^2)}$$

clearly satisfies the conditions $g(0) = 0$, $g'(0) = 1$ and is therefore a function of S. A formal computation shows that the second coefficient of its

Taylor expansion is

$$\frac{1}{2}\left\{\frac{f''(\zeta)(1-|\zeta|^2)}{f'(\zeta)} - 2\bar{\zeta}\right\}.$$

Hence, by (82),

$$\frac{1}{2}\left|\frac{f''(\zeta)(1-|\zeta|^2)}{f'(\zeta)} - 2\bar{\zeta}\right| \leq 2.$$

If we replace the variable ζ by z, we obtain, by an obvious manipulation,

$$\left|\frac{zf''(z)}{f'(z)} - \frac{2|z|^2}{1-|z|^2}\right| \leq \frac{4|z|}{1-|z|^2},$$

whence

$$(86) \qquad \frac{2\rho^2 - 4\rho}{1-\rho^2} \leq \text{Re}\left\{\frac{zf''(z)}{f'(z)}\right\} \leq \frac{4\rho + 2\rho^2}{1-\rho^2}, \qquad |z| = \rho < 1.$$

In view of

$$\text{Re}\left\{\frac{zf''(z)}{f'(z)}\right\} = \text{Re}\left\{\frac{\partial \log f'(z)}{\partial \log z}\right\} = |z|\frac{\partial}{\partial |z|}\text{Re}\,\{\log f'(z)\} = \rho\frac{\partial}{\partial\rho}\log|f'(z)|,$$

(86) may also be written in the form

$$\frac{2\rho - 4}{1-\rho^2} \leq \frac{\partial}{\partial\rho}\log|f'(z)| \leq \frac{4+2\rho}{1-\rho^2}.$$

Integrating these inequalities from 0 to ρ, we obtain

$$\log(1-\rho) - 3\log(1+\rho) \leq \log|f'(z)| \leq \log(1+\rho) - 3\log(1-\rho),$$

whence, by taking exponentials,

$$(87) \qquad \frac{1-\rho}{(1+\rho)^3} \leq |f'(z)| \leq \frac{1+\rho}{(1-\rho)^3}, \qquad |z| = \rho < 1.$$

(87) is again a distortion theorem; it shows that the "local distortion" $|f'(z)|$ of the conformal map $z \to f(z)$ must remain between certain bounds. The function $f_0(z)$ of (83) shows that both inequalities (87) are sharp.

From (87) we can obtain bounds for the function $f(z)$ itself. If we integrate along the linear segment connecting the origin and z, we have

$$|f(z)| = \left|\int_0^z f'(z)\,dz\right| \leq \int_0^\rho |f'(z)|\,d\rho \leq \int_0^\rho \frac{1+\rho}{(1-\rho)^3}\,d\rho = \frac{\rho}{(1-\rho)^2}.$$

In order to obtain a lower bound for $|f(z)|$, we first assume that $|f(z)| < \frac{1}{4}$. By the distortion theorem proved further above, the linear segment connecting the origin with the point $f(z)$ will then be entirely covered by the values of $f(z)$ in $|z| < 1$. If L is the arc in $|z| < 1$ which is mapped by

$w = f(z)$ onto this linear segment, we shall have $\arg \{dw\} = \arg \{f'(z) dz\} = \text{const.}$ along L. Hence,

$$|f(z)| = \left| \int_L f'(z)\, dz \right| \geq \int_L |f'(z)|\, d\rho \geq \int_0^\rho \frac{1-\rho}{(1+\rho)^3}\, d\rho = \frac{\rho}{(1+\rho)^2}.$$

Since $\rho(1+\rho)^{-2} < \frac{1}{4}$ for $0 \leq \rho < 1$, this inequality will be satisfied trivially if $|f(z)| \geq \frac{1}{4}$; it therefore holds in all cases. Collecting our results, we thus have

$$\frac{\rho}{(1+\rho)^2} \leq |f(z)| \leq \frac{\rho}{(1-\rho)^2}, \qquad |z| = \rho < 1. \tag{88}$$

From (88) we can draw the interesting conclusion that *the family S of functions* (79) *which are regular and univalent in* $|z| < 1$ *is normal and compact.* By the right-hand inequality (88), the family S is locally uniformly bounded at all points of the unit circle; it follows therefore from the results of Sec. 2, Chap. IV, that S is normal. Moreover, it was shown in Sec. 3, Chap. IV, that a converging sequence of univalent functions converges either to a univalent function or to a constant. Since all functions of S satisfy $f'(0) = 1$, the latter possibility is excluded. Hence, S is compact. This result is not restricted to functions which are regular and univalent in the unit circle. By an elementary transformation, the unit circle may be carried into any other circle. Since a family of functions is normal in a domain D if it is normal in all circles contained in D, it therefore follows that *the family of functions which are univalent in a domain D is normal.* Any subfamily of this family which is so chosen that constant limits are excluded is compact. For example, if ζ is a point of D, then the functions $f(z)$ which are univalent in D and satisfy $|f'(\zeta)| > C > 0$, where C is a constant, are a normal and compact family.

Since the family S is compact, it follows from Sec. 3, Chap. IV, that there must exist functions of S for which the moduli of the coefficients a_2, a_3, . . . in the expansion (79) attain their largest possible values (which are necessarily finite). (82) and (83) show that, in the case of the coefficient a_2, this problem is solved by the function $f_0(z)$ defined in (83). It can also be shown that the extremal function of the problem $|a_3| = \max.$ is again the function $f_0(z)$, the largest value of $|a_3|$ thus being 3. These and other considerations, partly to be discussed later, confer a certain degree of plausibility to the conjecture that the function $f_0(z)$ solves all the problems $|a_n| = \text{maximum}$, $n = 2, 3, \ldots$ and that, therefore, the coefficient d_n of a function of S is subject to the inequality $|a_n| \leq n$. So far, however, this conjecture has neither been proved nor disproved. The following weaker result is easily demonstrated.

The coefficients a_n of the univalent function (79) *are subject to the inequality*

$$|a_n| < e \cdot n, \tag{89}$$

where e is the base of the natural logarithms.

To prove (89), we start from the remark that the function $g(z)$ defined in (81) is univalent in $|z| < 1$ if $f(z)$ is in S. Hence the area of the conformal map of the circle $|z| < \rho$ $(\rho < 1)$ as yielded by $w = g(z)$ cannot be larger than that of the circle whose center is at the origin and whose radius is $\max_{|z|=\rho} |g(z)|$. By (88),

$$|g(z)| = |\sqrt{f(z^2)}| \leq \left[\frac{\rho^2}{(1-\rho^2)^2}\right]^{\frac{1}{2}} = \frac{\rho}{1-\rho^2}, \qquad |z| = \rho,$$

and therefore

$$\iint_{|z|<\rho} |g'(z)|^2 \, dx \, dy \leq \frac{\pi \rho^2}{(1-\rho^2)^2}.$$

On the other hand, if

$$g(z) = \sum_{n=1}^{\infty} b_n z^n, \qquad |z| < 1$$

is the Taylor expansion of $g(z)$, we have

$$\iint_{|z|<\rho} |g'(z)|^2 \, dx \, dy = \int_0^\rho \rho \, d\rho \int_0^{2\pi} |g'(\rho e^{i\theta})|^2 \, d\theta$$

$$= \int_0^\rho \rho \, d\rho \left[2\pi \sum_{n=1}^{\infty} |nb_n|^2 \rho^{2n-2}\right] = \pi \sum_{n=1}^{\infty} n|b_n|^2 \rho^{2n},$$

where Parseval's theorem (Sec. 4, Chap. III) has been used. Hence,

$$\sum_{n=1}^{\infty} n|b_n|^2 \rho^{2n-1} \leq \frac{\rho}{(1-\rho^2)^2}.$$

Integrating this inequality from 0 to ρ, we obtain

$$\sum_{n=0}^{\infty} |b_n|^2 \rho^{2n} \leq \frac{\rho^2}{1-\rho^2}.$$

In view of Parseval's theorem, this is equivalent to

$$\frac{1}{2\pi} \int_0^{2\pi} |g(\rho e^{i\theta})|^2 \, d\theta \leq \frac{\rho^2}{1-\rho^2}.$$

Since $g(z) = \sqrt{f(z^2)}$, it follows that

$$\frac{1}{2\pi}\int_0^{2\pi} |f(\rho^2 e^{2i\theta})|\, d\theta \leq \frac{\rho^2}{1-\rho^2}$$

or, with $\rho^2 = r$ and replacing 2θ by θ,

$$\frac{1}{2\pi}\int_0^{2\pi} |f(re^{i\theta})|\, d\theta \leq \frac{r}{1-r}.$$

Now the coefficient a_n is given by

$$a_n = \frac{1}{2\pi i}\int_{|z|=r} \frac{f(z)}{z^{n+1}}\, dz = \frac{1}{2\pi}\int_0^{2\pi} e^{-in\theta} r^{-n} f(re^{i\theta})\, d\theta,$$

whence

$$|a_n| r^n \leq \frac{1}{2\pi}\int_0^{2\pi} |f(re^{i\theta})|\, d\theta.$$

Combining this with the preceding inequality, we obtain

$$|a_n| \leq \frac{1}{r^{n-1}(1-r)},$$

where r may be any number between 0 and 1. Taking, in particular, $r = 1 - (1/n)$, we have

$$|a_n| \leq n\left(1 + \frac{1}{n-1}\right)^{n-1},$$

and this proves (89).

Although it is not known whether the inequality $|a_n| \leq n$ holds for all functions of S, there are some important subclasses of S for which it can easily be shown to be true. We shall prove first that *if the coefficients a_n of the univalent function* (79) *are real, then* $|a_n| \leq n$. If $f(z)$ is univalent in $|z| < 1$, then $f(z_1) - f(z_2) \neq 0$ for any two distinct points z_1 and z_2 in the unit circle. Taking, in particular, $z_1 = re^{i\theta}$, $z_2 = re^{-i\theta}$, $r < 1$, $0 < \theta < \pi$, then

$$\begin{aligned} f(z_1) - f(z_2) &= \sum_{n=1}^{\infty} a_n r^n (e^{in\theta} - e^{-in\theta}) \\ &= 2i \sum_{n=1}^{\infty} a_n r^n \sin n\theta \neq 0, \qquad 0 < \theta < \pi. \end{aligned}$$

Since $\sin\theta \neq 0$ for $0 < \theta < \pi$, we thus have

$$p(\theta) = \sum_{n=1}^{\infty} a_n r^n \sin n\theta \sin\theta \neq 0, \qquad 0 < \theta < \pi.$$

The coefficients a_n are real. Hence, $p(\theta)$ is real in the interval $0 < \theta < \pi$; since $p(\theta)$ is continuous and does not vanish there, it is of constant sign throughout the interval. But $\sin(-\theta) = -\sin\theta$, and therefore $p(-\theta) = p(\theta)$, which shows that either $p(\theta) \geq 0$ or $p(\theta) \leq 0$ for all values of θ. By the addition theorem of the cosine function, we have

$$p(\theta) = \frac{1}{2}\sum_{n=1}^{\infty} a_n r^n[\cos(n-1)\theta - \cos(n+1)\theta]$$

$$= \frac{r}{2}\left[1 + a_2 r\cos\theta + \sum_{n=2}^{\infty}\left(a_{n+1} - \frac{a_{n-1}}{r^2}\right) r^n \cos n\theta\right].$$

This shows that $\int_0^{2\pi} p(\theta)\, d\, d\theta = \pi r$ and that, therefore, $p(\theta) \geq 0$ for $0 \leq \theta < 2\pi$. In view of $1 \pm \cos n\theta \geq 0$, we thus obtain

$$0 \leq \frac{2}{\pi r}\int_0^{2\pi} p(\theta)(1 \pm \cos n\theta)\, d\theta = 2 \pm \left(a_{n+1} - \frac{a_{n-1}}{r^2}\right) r^n, \qquad n > 2,$$

whence

$$\left| a_{n+1} - \frac{a_{n-1}}{r^2}\right| r^n \leq 2.$$

Since this holds for all r between 0 and 1, it follows that

$$| a_{n+1} - a_{n-1} | \leq 2.$$

For $n = 2$, we obtain $|a_2| \leq 2$. Since $a_1 = 1$ and the difference of two terms whose subscripts differ by 2 is not larger than 2, it follows that

$$|a_n| \leq n, \qquad n = 2, 3, \ldots . \tag{90}$$

We next show that *the inequality* (90) *also holds for the coefficients of a univalent function* (79) *which maps the unit circle onto a star-shaped domain.* Here, a star-shaped domain is defined as a domain which is intersected by any straight line passing through the origin in a single linear segment. A more graphic definition would be that of a domain all of whose points can be "seen" from the origin. We first show that if $w = f(z)$ maps $|z| < 1$ onto a star-shaped domain D, then every circle $|z| = \rho < 1$ is also mapped onto a domain of this type, say D_ρ. To this end, we observe that D is star-shaped if, and only if, all points $tf(z)$ ($|z| < 1$) are in D if $0 < t < 1$. Since, by the mapping $w = f(z)$, these points correspond to points in $|z| < 1$, this can also be expressed by saying that $tf(z) = f[\omega(z)]$, where $\omega(z)$ is regular in $|z| < 1$ and $|\omega(z)| \leq 1$.

Furthermore we have, clearly, $\omega(0) = 0$. It follows therefore from the Schwarz lemma that

(91) $$tf(z) = f[\omega(z)], \qquad |\omega(z)| \leq |z|, \qquad 0 < t < 1.$$

This is a necessary and sufficient condition for $w = f(z)$ mapping $|z| < 1$ onto a starlike domain. Let now ρ be between 0 and 1, and consider the function $g(z) = f(\rho z)$. By (91), we have

$$tg(z) = tf(\rho z) = f[\omega(\rho z)] = f\left[\rho\,\frac{\omega(\rho z)}{\rho}\right].$$

In view of $|\omega(\rho z)| \leq \rho|z|$, the function

$$\omega_1(z) = \frac{\omega(\rho z)}{\rho}$$

satisfies $|\omega_1(z)| \leq |z|$. Hence,

$$tg(z) = g[\omega_1(z)], \qquad |\omega_1(z)| \leq |z|,\ 0 < t < 1,$$

and a comparison with (91) shows that $g(z)$ also maps $|z| < 1$ onto a star-shaped domain. But, in view of $g(z) = f(\rho z)$, this domain is identical with the conformal image D_ρ of $|z| < \rho$ as given by $w = f(z)$.

With the help of this result, we show that *the necessary and sufficient condition for the function* $w = f(z)$ *to map* $|z| < 1$ *onto a star-shaped domain is*

(92) $$\operatorname{Re}\left\{\frac{zf'(z)}{f(z)}\right\} \geq 0, \qquad |z| < 1.$$

We have seen that the conformal image of the circle $|z| = \rho$, $\rho < 1$, is a star-shaped curve. Hence, if the point $z = \rho e^{i\theta}$ describes this circle in the positive sense, the argument ϕ of the image point $f(z) = Re^{i\phi}$ must vary always in the same direction. If this were not the case, there would be rays emanating from $w = 0$ which intersect this curve more than once. Since this direction is necessarily the positive one, we have

(93) $$\frac{\partial}{\partial\theta}\arg\{f(z)\} = \frac{\partial\phi}{\partial\theta} \geq 0.$$

Conversely, if this condition is satisfied, the curve in question is clearly star-shaped. Now $\log f(z) = \log R + i\phi$; the last condition can therefore be written in the form

$$\operatorname{Im}\left\{\frac{\partial \log f(z)}{\partial\theta}\right\} \geq 0,$$

or, in view of

$$\frac{\partial}{\partial\theta} = i\rho e^{i\theta}\frac{d}{dz} = iz\frac{d}{dz}, \qquad z = \rho e^{i\theta}, \qquad \rho = \text{const.},$$

$$\text{Im}\left\{iz\frac{d}{dz}\log f(z)\right\} = \text{Re}\left\{\frac{zf'(z)}{f(z)}\right\} \geq 0.$$

This proves (92).

The inequality (90) is an easy consequence of (92). If $b_1, b_2, \ldots$ are the coefficients of the power series expansion

$$\frac{zf'(z)}{f(z)} = 1 + b_1z + b_2z^2 + \cdots,$$

then, by (92) and (36), $|b_n| \leq 2$, $n = 1, 2, \ldots$. Hence, in view of

$$\sum_{n=1}^{\infty} na_nz^n = zf'(z) = \frac{zf'(z)}{f(z)}f(z) = \left(1 + \sum_{n=1}^{\infty} b_nz^n\right)\left(\sum_{n=1}^{\infty} a_nz^n\right)$$

$$= \sum_{n=1}^{\infty} c_nz^n, \qquad c_n = a_n + b_1a_{n-1} + \cdots + b_{n-1},$$

we obtain

$$(n-1)|a_n| = |b_1a_{n-1} + \cdots + b_{n-1}| \leq 2(|a_{n-1}| + \cdots + |a_2| + 1).$$

For $n = 2$, this yields $|a_2| \leq 2$. Using this, we have $2|a_3| \leq 2(1 + |a_2|) \leq 6$, whence $|a_3| \leq 3$, etc. If (90) has been proved for $n = 2, 3, \ldots, m$, we obtain

$$m|a_{m+1}| \leq 2(1 + 2 + \cdots + m) = m(m+1).$$

Hence $|a_{m+1}| \leq m + 1$, and (90) is proved by complete induction. We add that the inequalities (90) are sharp since the function $f_0(z)$ defined in (83) maps $|z| < 1$ onto a star-shaped domain.

Exact bounds can also be obtained for the coefficients of a function

$$f(z) = z + a_2z^2 + a_3z^3 + \cdots, \qquad |z| < 1,$$

which is univalent in $|z| < 1$ and maps $|z| < 1$ onto a convex domain. To find these bounds, we first show that the function $w = f(z)$ maps all circles $|z| = \rho < 1$ onto convex curves. Clearly, the map yielded by $w = f(z)$ is convex if, and only if, the function $w_1 = f(z) - f(\zeta)$ maps $|z| < 1$ onto a star-shaped domain for any $|\zeta| < 1$. If the map of $w = f(z)$ is convex, it follows therefore that $w_1 = f(z) - f(\zeta)$ maps the circles $|z| = \rho < 1$ onto star-shaped curves, provided $|\zeta| < \rho$. Since ζ is otherwise arbitrary, the image of $|z| = \rho$ by $w = f(z)$ is thus a convex curve. The angle between the tangent to this curve at the point w and the positive axis is

$$\arg\{dw\} = \arg\{f'(z)\,dz\} = \arg\{izf'(z)\} \qquad z = \rho e^{i\theta}.$$

The curve is convex if, and only if, this angle grows monotonically if z describes the circle $|z| = \rho$, that is, if

$$\frac{\partial}{\partial\theta}\arg\{izf'(z)\} \geq 0.$$

Comparing this with (93), we find that $w = f(z)$ *maps the circle* $|z| < 1$ *onto a convex domain if, and only if,* $zf'(z)$ *maps* $|z| < 1$ *onto a star-shaped domain.* By (90), the nth coefficient of the function

$$zf'(z) = z + 2a_2z^2 + 3a_3z^3 + \cdots$$

is therefore smaller than, or equal to, n. Hence,

$$|a_n| \leq 1, \qquad n = 2, 3, \ldots .$$

These inequalities are sharp. This is shown by the function

$$w = f(z) = \frac{z}{1 - z} = z + z^2 + z^3 + \cdots,$$

which maps $|z| < 1$ onto the half-plane $\operatorname{Re}\{w\} > -\frac{1}{2}$.

In the case of a convex mapping, the $\frac{1}{4}$-theorem can be improved upon. It is easy to show that *if the function* $w = f(z)$ *with the normalization* (79) *maps* $|z| < 1$ *onto a convex domain* D, *then* D *contains the circle* $|w| < \frac{1}{2}$. By the residue theorem, we have

$$\frac{1}{2\pi i}\int_{|\zeta|=\rho} f(\zeta)\left[1 + \frac{\zeta}{2\rho z} + \frac{\rho z}{2\zeta}\right]\frac{d\zeta}{\zeta} = \frac{1}{2}\,\rho z.$$

In polar coordinates $\zeta = \rho e^{i\theta}$, $z = e^{i\varphi}$, this reads

$$\frac{1}{2}\,\rho^2 e^{i\varphi} = \frac{1}{\pi}\int_0^{2\pi} f(\rho e^{i\theta})\cos^2\frac{\theta - \varphi}{2}\,d\theta.$$

In view of

$$\frac{1}{\pi}\int_0^{2\pi}\cos^2\frac{\theta - \varphi}{2}\,d\theta = 1,$$

the right-hand side may be regarded as the center of gravity of a "mass distribution" of density $\pi^{-1}\cos^2\dfrac{\theta - \varphi}{2}$ at the points of the curve $w = f(\rho e^{i\theta})$, $0 \leq \theta < 2\pi$. Since the curve is convex, this center of gravity must be in its interior. Hence, the point $\frac{1}{2}\rho e^{i\varphi}$ must be one of the points of the map of $|z| \leq \rho$ by means of $w = f(z)$. Since φ is arbitrary, this map must therefore contain the circle $|w| < \frac{1}{2}\rho$. Letting $\rho \to 1$, we obtain our result. The function $w = z(1 - z)^{-1}$ shows that the constant $\frac{1}{2}$ is the largest possible.

EXERCISES

1. Show that the function

$$w = \frac{z}{(1 - z)^3}$$

in univalent for $|z| < \frac{1}{2}$, but not in a larger circle about the origin.

2. By applying (88) to the univalent function (85) and setting $z = -\zeta$ in the result, prove that a function $f(z)$ of S is subject to the inequalities

$$\frac{1}{\rho}\left(\frac{1 - \rho}{1 + \rho}\right) \leq \left|\frac{f'(z)}{f(z)}\right| \leq \frac{1}{\rho}\left(\frac{1 + \rho}{1 - \rho}\right), \qquad |z| = \rho < 1.$$

Show that these inequalities are sharp.

3. If $f(z)$ is in S and $|z_1| = |z_2| = \rho < 1$, show that

$$\left(\frac{1 - \rho}{1 + \rho}\right)^4 \leq \left|\frac{f'(z_1)}{f'(z_2)}\right| \leq \left(\frac{1 + \rho}{1 - \rho}\right)^4.$$

4. If the function $f(z)$ of S is bounded by the number M, that is, $|f(z)| \leq M$ for $|z| < 1$, show that (82) can be replaced by the better inequality

$$|a_2| \leq 2\left(1 - \frac{1}{M}\right).$$

Hint: Show that the function

$$f_1(z) = \frac{f(z)}{\left[1 + \dfrac{e^{i\gamma}}{M} f(z)\right]^2}, \qquad 0 \leq \gamma < 2\pi,$$

is univalent in $|z| < 1$, and apply (82) to $f_1(z)$. Show that this inequality is sharp and that the extremal function maps $|z| < 1$ onto the circle $|w| < M$ to which a rectilinear cut, pointing at the origin, has been applied (Fig. 25). If λ is the distance of the free end of the cut from the origin, show that

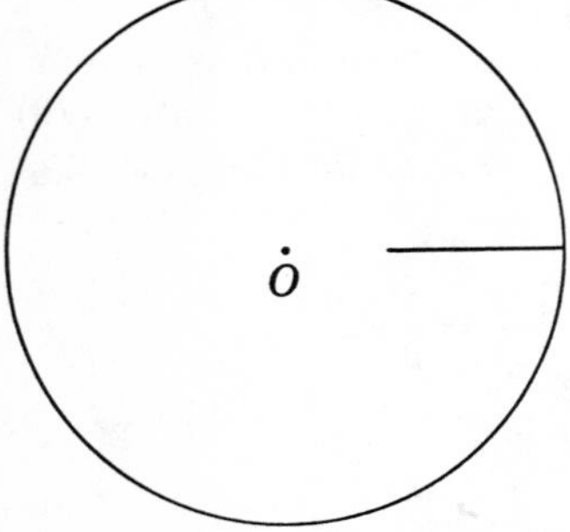

FIG. 25.

$$4\lambda M^2 = (\lambda + M)^2.$$

5. Show that a necessary and sufficient condition for a function $w = f(z)$ of S to map $|z| < 1$ onto a convex domain is

$$\text{Re}\left\{1 + \frac{zf''(z)}{f'(z)}\right\} \geq 0, \qquad |z| < 1.$$

6. Using (86) and the result of the preceding exercise, prove that a function $w = f(z)$ of S maps the circle $|z| < 2 - \sqrt{3}$ onto a convex domain. Show, by means of the function $f_0(z)$ of (83), that this bound for the "radius of convexity" of $f(z)$ is sharp.

7. If $w = f(z)$ maps $|z| < 1$ onto a convex domain, show that

$$f_1(z) = \frac{f(ze^{i\theta}) - f(ze^{-i\theta})}{e^{i\theta} - e^{-i\theta}} = \sum_{n=1}^{\infty} a_n \frac{\sin n\theta}{\sin \theta} z^n$$

maps $|z| < 1$ onto a star-shaped domain. *Hint:* Use the fact that the chord of a convex curve turns monotonically if the terminals of the chord describe the convex curve in the same sense without overtaking each other.

8. Use the result of the preceding exercise to show that if the function $w = f(z)$ of S maps $|z| < 1$ onto a convex domain, then

$$|f(e^{i\theta_1}) - f(e^{i\theta_2})| \geq \cot \frac{|\theta_1 - \theta_2|}{4}.$$

Hint: Deduce from (84) that a value ζ not taken by a function

$$w = f(z) = z + b_2 z^2 + \cdots$$

of S satisfies $|\zeta| \geq (2 + |b_2|)^{-1}$, and apply this result to the function $f_1(z)$ of the preceding exercise. Show, by means of the function $f(z) = z(1 - z)^{-1}$, that the above inequality is sharp.

9. An analytic function $f(z) = z + a_2 z^2 + \cdots$ which is regular in $|z| < 1$ and is real for real values of z but for no other values in $|z| < 1$ is called "typically real." Show that the inequalities $|a_n| \leq n$ proved in the text for univalent functions with real coefficients are also true for typically real functions which are not univalent. *Hint:* Show that $\text{Im } \{f(z)\} > 0$ if $\text{Im } \{z\} > 0$ and $\text{Im } \{f(z)\} < 0$ if $\text{Im } \{z\} < 0$, and consider the function $\sin \theta \text{ Im } \{f(\rho e^{i\theta})\}$ for $0 \leq \theta < 2\pi$.

10. Show that the conformal image of $|z| < 1$ as yielded by a typically real function $w = f(z)$ covers the linear segment $-\frac{1}{4} < w < \frac{1}{4}$. *Hint:* Use the result of the preceding exercise for $n = 2$ and employ a procedure similar to that used in the proof of the $\frac{1}{4}$-theorem.

11. Show that, for functions $w = f(z)$ of S which map $|z| < 1$ onto convex domains, the inequalities (87) and (88) can be improved to

$$\frac{1}{(1 + \rho)^2} \leq |f'(z)| \leq \frac{1}{(1 - \rho)^2}$$

and

$$\frac{\rho}{1 + \rho} \leq |f(z)| \leq \frac{\rho}{1 - \rho},$$

respectively.

12. Show that an odd univalent function, *i.e.*, a function of S whose expansion (79) contains only odd powers of z, maps $|z| < 1$ onto a domain covering the entire circle $|w| < \frac{1}{2}$.

13. Show that the univalent function

$$g(z) = z + c_{n+1} z^{n+1} + c_{2n+1} z^{2n+1} + c_{3n+1} z^{3n+1} + \cdots$$

can be written in the form

$$g(z) = \sqrt[n]{f(z^n)},$$

where $f(z)$ is a function of S, and conclude that the inequalities

$$|c_{n+1}| \leq \frac{2}{n}$$

and

$$\frac{\rho}{(1+\rho^n)^{2/n}} \leq |g(z)| \leq \frac{\rho}{(1-\rho^n)^{2/n}}$$

hold.

14. The two univalent functions

$$(94)\quad f(z) = z + a_2z^2 + \cdots, \qquad g(z) = 1 + b_1z + b_2z^2 + \cdots, \qquad |z| < 1,$$

will be called "unrelated" if they have no values in common, that is, if $f(z_1) \neq g(z_2)$, where z_1 and z_2 are any two points in $|z| < 1$. If

$$\frac{1}{f(z)} = \frac{1}{z} + \sum_{n=0}^{\infty} \alpha_n z^n, \qquad \frac{1}{g(z)} = 1 + \sum_{n=1}^{\infty} \beta_n z^n,$$

show by a suitable generalization of the proof of the area theorem (77) that

$$\sum_{n=1}^{\infty} n(|\alpha_n|^2 + |\beta_n|^2) \leq 1.$$

15. If $f(z)$ and $g(z)$ are univalent and unrelated, show that the same is true of the functions $\sqrt{f(z^2)}$ and $\sqrt{g(z)}$. Use this fact and the result of the preceding exercise in order to show that the coefficients a_2 and b_1 in (94) are connected by the inequality

$$|a_2|^2 + |b_1|^2 \leq 4.$$

16. If $f(z)$ is univalent in $|z| \leq 1$, show that the function $g(z)$ defined by

$$g(z) = \frac{f'(\zeta)(1-|\zeta|^2)}{f\left(\frac{z+\zeta}{1+\bar{\zeta}z}\right) - f(\zeta)}, \qquad |\zeta| < 1,$$

is also univalent in $|z| < 1$ and that it has the expansion

$$g(z) = \frac{1}{z} + \bar{\zeta} - \frac{1}{2}\frac{f''(\zeta)}{f'(\zeta)}(1-|\zeta|^2) - \frac{1}{6}(1-|\zeta|^2)^2\{w,\zeta\}z + \cdots,$$

where $w = f(\zeta)$ and $\{w,\zeta\}$ is the Schwarzian derivative defined in (56). Deduce that any univalent function $w = f(z)$, regardless of normalization, satisfies the sharp inequality

$$|\{w,z\}| \leq \frac{6}{(1-|z|^2)^2}, \qquad |z| < 1.$$

9. Subordination. The *principle of subordination* to be discussed in this section is but a simple extension of the Schwarz lemma. This principle enables us to derive a considerable amount of information about an analytic function $w = f(z)$ which is regular in the unit circle, if certain geometric details of the conformal map associated with this function are known.

Let $w = F(z)$ be an analytic function which is regular and univalent in $|z| < 1$, and let D denote the schlicht domain in the w-plane onto which the unit circle is mapped by $w = F(z)$. Consider now an analytic function $w = f(z)$ which is regular, but not necessarily univalent, in $|z| < 1$, and whose conformal map D' of $|z| < 1$ is entirely contained in D. In other words, let $w = f(z)$ be such that all its values in $|z| < 1$ are also values taken by $w = F(z)$ in $|z| < 1$. The inverse $z = F^{-1}(w)$ of $w = F(z)$ is regular in D and maps D onto $|z| < 1$. Since all the values taken by $w = f(z)$ in $|z| < 1$ are in D, it follows therefore that the function

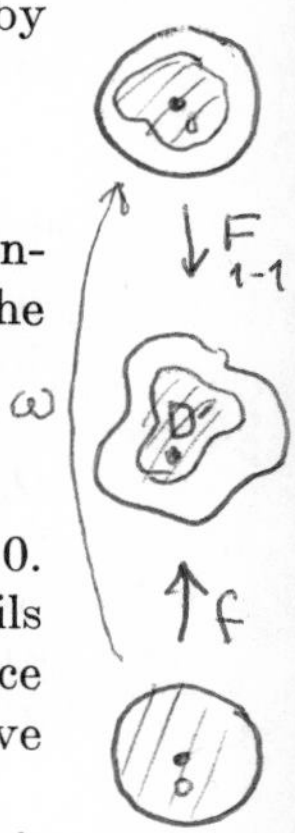

(95) $$\omega(z) = F^{-1}[f(z)]$$

is regular in $|z| < 1$ and that its values must lie in the interior of the conformal map of D as yielded by $z = F^{-1}(w)$, that is, in the interior of the unit circle. Solving (95) for $f(z)$, we thus obtain

$$f(z) = F[\omega(z)], \qquad |\omega(z)| \leq 1,\ |z| < 1.$$

If it is known, furthermore, that $f(0) = F(0)$, then we must have $\omega(0) = 0$. Indeed, $f(0) = F[\omega(0)] = F(0)$ and, since $F(z)$ is univalent, this entails $\omega(0) = 0$. In view of the Schwarz lemma, we may in this case replace $|\omega(z)| \leq 1$ by the stronger inequality $|\omega(z)| \leq |z|$. Summing up, we have the following result.

Let the function $f(z)$ and $F(z)$ *be regular in* $|z| < 1$ *and let* $F(z)$ *be univalent there. Let further* D' *and* D *denote the domains onto which the unit circle is mapped by* $w = f(z)$ *and* $w = F(z)$, *respectively. If* $f(0) = F(0)$ *and if* D' *is contained in* D, *then*

(96) $$f(z) = F[\omega(z)],$$

where $\omega(z)$ *is regular in* $|z| < 1$ and

(97) $$|\omega(z)| \leq |z|.$$

The sign of equality in (97) *is possible only if the domains* D *and* D' *coincide.*

The latter remark follows by considering the case in which the sign of equality holds in the Schwarz lemma. If the functions $f(z)$ and $F(z)$ are related by (96), we say that $f(z)$ is *subordinate* to $F(z)$. This relation is, of course, also possible in the case in which $F(z)$ is not univalent in $|z| < 1$, although the geometric relation between the conformal maps associated with $f(z)$ and $F(z)$ will then be somewhat less obvious. Some cases of this kind will be discussed later.

Immediate consequences of (96) and (97) are the inequalities

(98) $$|f(z)| \leq \max_{|z|=\rho} |F(z)|, \qquad |z| = \rho < 1,$$

and

$$|f'(0)| \leq |F'(0)|. \tag{99}$$

(98) follows from the maximum principle since, in view of (97), the values of $F[\omega(z)]$ for $|z| = \rho < 1$ are a subset of the values of $F(z)$ for $|z| < \zeta$. The inequality (99) follows by noting that

$$f'(0) = \omega'(0)F'(0)$$

and therefore, in view of $|\omega'(0)| \leq 1$,

$$|f'(0)| = |\omega'(0)||F'(0)| \leq |F'(0)|.$$

The reader will verify that, in both (98) and (99), the sign of equality can occur only if $f(z) = F(e^{i\gamma}z)$, $0 \leq \gamma < 2\pi$, that is, in the case in which the domains D and D' coincide.

(98) can be supplemented by the more general statement that, *if* D_ρ *and* D_ρ' *denote the conformal images of* $|z| < \rho$ *by means of* $w = F(z)$ *and* $w = f(z)$, *respectively, then* D_ρ' *is contained in* D_ρ. This is an obvious consequence of the fact, expressed by (97), that the conformal image of $|z| < \rho$ by means of the mapping $z \to \omega(z)$ is contained in the circle $|z| < \rho$.

The simplest illustrations of the principle of subordination are afforded by the cases in which the domain D is the unit circle $|w| < 1$ or the right half-plane $Re\,\{w\} > 0$. The subordinate functions are then the bounded functions and the functions with a positive real part, respectively. In the case of the family of bounded functions $f(z)$, the "superordinate" function $F(z)$ may be taken to be

$$F_B(z) = \frac{f(0) + z}{1 + \overline{f(0)}z},$$

while in the case of functions with a positive real part an appropriate superordinate is

$$F_P(z) = a\left(\frac{1 + z}{1 - z}\right) + ib, \qquad f(0) = a + ib.$$

The reader will easily verify that the functions $F_B(z)$ and $F_P(z)$ map $|z| < 1$ onto the unit circle and the right half-plane, respectively. These functions are, of course, not unique; instead of $F(z)$ we might as well have taken the function $F(e^{i\gamma}z)$, where γ is an arbitrary angle. Most of the results on bounded functions and on functions with a positive real part which were derived in Sec. 3 can be re-proved with the help of the principle of subordination and of the special forms of the functions $F_B(z)$ and $F_P(z)$. For example, if Re $\{f(z)\} \geq 0$ and $f(0) = 1$, then $F_P(z) = (1 + z)(1 - z)^{-1}$

and $F_P'(0) = 2$. Hence, $|f'(0)| \leq 2$, in accordance with the results of Sec. 3.

Evidently, the successful application of the subordination method depends on the explicit knowledge of the mapping functions for as large a number of domains as possible. On the other hand, the explicit knowledge of each particular mapping function automatically yields results of the type (98) and (99) for the class of functions which are subordinate to it. For example, consider the function

$$F(z) = \frac{z}{(1 - z)^2},$$

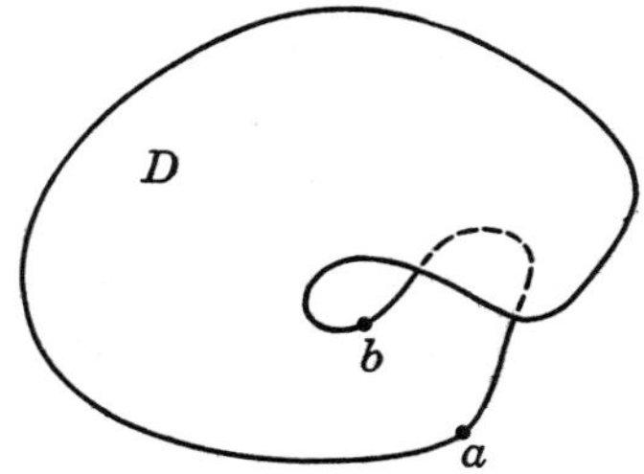

FIG. 26.

which played a fundamental role in the preceding section. As shown before, $w = F(z)$ maps $|z| < 1$ onto the full w-plane which has been furnished with a slit along the ray $-\infty \leq w \leq -\frac{1}{4}$. Clearly, any function $f(z)$ which is regular in $|z| < 1$ and does not take there values which are situated on this slit is subordinate to $F(z)$. In view of (98) and (99), we thus have the result that *any function* $w = f(z)$ *such that* $f(0) = 0$ *which is regular in* $|z| < 1$ *and leaves out the values* $-\infty \leq w \leq -\frac{1}{4}$ *is subject to the inequalities*

$$|f(z)| \leq \frac{\rho}{(1 - \rho)^2}, \qquad |z| = \rho < 1, \qquad |f'(0)| \leq 1.$$

Before we continue to discuss other examples we derive a general result, concerning functions $f(z)$ regular in $|z| < 1$, by combining the subordination principle with the $\frac{1}{4}$-theorem of the preceding section. This result refers to the "outer boundary" of the domain D onto which the unit circle is mapped by a function $w = f(z)$. A point a on the boundary of D is said to belong to the outer boundary of D if it is possible to connect a with the point at infinity by a continuous curve C none of whose points is in D. In Fig. 26, the point a belongs to the outer boundary of D while the point b does not. We now prove that *any point on the outer boundary of a domain* D *onto which the unit circle is mapped by a function*

$$w = f(z) = z + a_2z^2 + \cdots, \qquad |z| < 1,$$

has at least the distance $\frac{1}{4}$ *from the origin.* If a is a point of the outer boundary of D, a can be connected with $w = \infty$ by a continuous curve C. Consider now the analytic function $w = F(z)$ which maps $|z| < 1$ onto the full w-plane which is furnished with a slit along the curve C; this function exists by virtue of the Riemann mapping theorem. $F(z)$ is univalent in

$|z| < 1$ and does not take the value a. By the $\frac{1}{4}$-theorem, we have therefore

$$|a| \geq \tfrac{1}{4}|F'(0)|.$$

On the other hand, $f(z)$ is clearly subordinate to $F(z)$ and it follows therefore from (99) and from $f'(0) = 1$ that

$$|a| \geq \tfrac{1}{4}|F'(0)| \geq \tfrac{1}{4}|f'(0)| = \tfrac{1}{4}.$$

This completes the proof.

Continuing with examples for the application of the principle of subordination, we now consider the class of analytic functions $f(z)$ which are regular and have a bounded real part in $|z| < 1$, that is, $-1 \leq \text{Re}\,\{f(z)\} \leq 1$. These functions are subordinate to the function $w = F(z)$ which maps $|z| < 1$ onto the infinite strip $-1 < \text{Re}\,\{w\} < 1$ and satisfies $F(0) = f(0)$. As shown in Sec. 6, the desired mapping is effected by the function

$$w = F(z) = \frac{4}{\pi} \tan^{-1} z.$$

By (99), we thus find that *if* $f(z)$ *is regular and* $-1 < \text{Re}\,\{f(z)\} < 1$ *in* $|z| < 1$ *and if* $f(0) = 0$, *then*

$$|f'(0)| \leq \frac{4}{\pi},$$

where equality is possible only in the case in which $f(z) = (4/\pi) \tan^{-1} (e^{i\gamma} z)$, $0 \leq \gamma < 2\pi$. To apply (98), we have to find the maximum of the extremal function for $|z| = \rho$. In view of

$$|\tan^{-1} z| = \left| \frac{1}{2i} \log \frac{1 + iz}{1 - iz} \right| \leq \frac{1}{2} \log \frac{1 + \rho}{1 - \rho},$$

it therefore follows that

$$|f(z)| \leq \frac{2}{\pi} \log \frac{1 + \rho}{1 - \rho}, \qquad |z| = \rho < 1,$$

where this inequality is again sharp.

The application of the principle of subordination is not necessarily restricted to functions which are regular in $|z| < 1$. Since the special position of the point at infinity is easily abolished by means of a linear transformation which transforms $w = \infty$ into a finite point, both the functions $f(z)$ and $F(z)$ may be permitted to have poles in $|z| < 1$; if, as so far has been the case, $F(z)$ is univalent in $|z| < 1$, this function can, of course, have only one simple pole in the unit circle. As an example, we consider the function $w = F(z)$ which maps $|z| < 1$ onto the full w-plane to which a slit along the linear segment $\alpha \leq w \leq \beta$ $(\alpha > 0)$ has been applied. The function $F(z)$ is easily seen to be of the form

$$F(z) = \frac{\delta z}{(\gamma + z)(1 + \gamma z)},$$

where δ and γ are suitable positive constants and $\gamma < 1$. Indeed, we have

$$F(e^{i\theta}) = \frac{\delta}{|\gamma + e^{i\theta}|^2},$$

which shows that $w = F(z)$ maps $|z| = 1$ onto the doubly traversed linear segment

$$\frac{\delta}{(1 + \gamma)^2} \leq w \leq \frac{\delta}{(1 - \gamma)^2}.$$

Identifying the two terminals of this segment with α and β, we obtain by an elementary computation

$$\gamma = \frac{\sqrt{\beta} - \sqrt{\alpha}}{\sqrt{\beta} + \sqrt{\alpha}}, \qquad \delta = \frac{4\alpha\beta}{(\sqrt{\alpha} + \sqrt{\beta})^2},$$

whence

$$F'(0) = \frac{4\alpha\beta}{\beta - \alpha}.$$

We thus have the following result: *If $f(0) = 0$ and $f(z)$ is regular in $|z| < 1$ with the possible exception of a finite number of poles and if, furthermore, $w = f(z)$ does not take in $|z| < 1$ values w for which $\alpha \leq w \leq \beta$ $(\alpha > 0)$, then*

$$|f'(0)| \leq \frac{4\alpha\beta}{\beta - \alpha}.$$

This bound for $|f'(0)|$ is the smallest possible.

We mentioned before that the principle of subordination can also be applied in cases in which the superordinate function $F(z)$ is not univalent in $|z| < 1$. The domain D upon which $|z| < 1$ is mapped by $w = F(z)$ will then be of the general character of a Riemann surface and it may be self-overlapping and have branch points. In this case there will be no simple geometric answer to the question of when another Riemann domain D' is "contained" in D. The principle of subordination will therefore, in the case of nonunivalent functions $F(z)$, render useful services only if we can show directly that the function (95) is regular in $|z| < 1$ and that its values lie within the unit circle. An example typical of the cases which can be treated in this fashion is that of the class of functions $f(z)$ for which

$f(0) = 0$ and $|f(z)| \leq 1$ in $|z| < 1$ and which do not take a certain value α ($|\alpha| < 1$). That there exists a function $F(z)$ which is superordinate to all functions $f(z)$ with these properties can be shown as follows. We take an infinite number of replicas of the unit circle $|w| < 1$ to all of which a cut along the linear segment $0 < \alpha \leq w < 1$ has been administered (obviously, the assumption $\alpha > 0$ does not restrict the generality of the above class of functions). We now place these replicas of the slit circle upon each other and connect each circle with its two neighbors in the following manner. The lower edge of the slit is connected with the upper edge of the slit in the circle immediately above, and the upper edge is connected with the lower edge of the slit in the circle immediately below. If we continue in this manner ad infinitum, there will remain no free edges of the slits $\alpha \leq w < 1$. We thus obtain a surface D which covers the unit circle infinitely often and which has, within $|w| < 1$, the only boundary point $w = \alpha$. The procedure used in the construction of the surface D is reminiscent of that employed in the construction of the Riemann surface of the function $\log z$ in Sec. 7, Chap. II. In fact, D is identical with that part of the surface of $\log (w - \alpha)$ which is contained in the circle $|w| < 1$, as the reader will verify without difficulty.

Before we explicitly determine the function $w = F(z)$ which maps $|z| < 1$ onto this surface D and satisfies $F(0) = 0$, we show that $F(z)$ is the superordinate function required for our purposes. Let $f(0) = 0$, $|f(z)| \leq 1$ and $f(z) \neq \alpha$ in $|z| < 1$, and consider the function

$$\omega(z) = F^{-1}[f(z)], \qquad |z| < 1.$$

Since $w = F(z)$ maps $|z| < 1$ onto the surface D which has no boundary points in $|w| < 1$ except the point $w = \alpha$ and, moreover, since D has no branch points, $\omega(z)$ cannot have singularities. Indeed, such singularities can arise only if $f(z)$ takes values w beyond which the function $F^{-1}[w]$ cannot be continued or, at least, cannot be continued in a single-valued fashion. Both cases cannot occur. The boundary of D consists of the circle $|w| = 1$ and the point $w = \alpha$, that is, of points not taken by $w = f(z)$ in $|z| < 1$, and D is free of branch points. The function $\omega(z)$ is therefore regular in $|z| < 1$. Moreover, all its values must lie in the unit circle since the values of $w = f(z)$ did not overstep the boundary of D. Hence, $|\omega(z)| \leq |z|$, and $f|z|$ is indeed subordinate to $F(z)$.

Our next task is to construct an analytic expression for $F(z)$. As a first step, we transform D into a similar surface D_1 whose "hole" is at the origin instead of the point $w = \alpha$. This is achieved by the linear substitution

$$F_1(z) = \frac{\alpha - F(z)}{1 - \alpha F(z)}$$

which transforms the unit circle $|w| < 1$ into itself. Consider now the function

$$g(z) = \log F_1(z).$$

$g(z)$ is regular in $|z| < 1$ since $F_1(z)$ does not vanish in $|z| < 1$. Writing $F_1(z) = Re^{i\phi}$, we have

$$g(z) = \log R + i\phi.$$

This shows that all values of $g(z)$ in $|z| < 1$ are situated in the left half-plane $\text{Re}\,\{g(z)\} < 0$ and that all finite boundary points of the domain filled with the values of $g(z)$ are points of the imaginary axis. $g(z)$ is further univalent in $|z| < 1$. Indeed, from $g(z_1) = g(z_2)$ it follows that $R(z_1) = R(z_2)$ and $\phi(z_1) = \phi(z_2)$; but this is impossible since two points $w_1 = F_1(z_1)$ and $w_2 = F_1(z_2)$ such that $w_1 = w_2$, $z_1 \neq z_2$, lie in different sheets of D, and their arguments thus differ by a nonvanishing integral multiple of 2π. Hence, $g(z)$ maps $|z| < 1$ onto the simply covered left half-plane. Since, moreover, $g(0) = \log F_1(0) = \log \alpha$, it follows that

$$g(z) = \log \alpha \left(\frac{1 + z}{1 - z}\right),$$

and, by a simple computation,

$$F(z) = \frac{\alpha - e^{\log \alpha\left(\frac{1+z}{1-z}\right)}}{1 - \alpha e^{\log \alpha\left(\frac{1+z}{1-z}\right)}},$$

or

$$F(z) = \alpha \frac{1 - \alpha^{\frac{2z}{1-z}}}{1 - \alpha^{\frac{2}{1-z}}}.$$

In particular, we obtain

$$F'(0) = \frac{2\alpha \log (1/\alpha)}{1 - \alpha^2}.$$

This implies the following result.

If $f(0) = 0$ *and* $|f(z)| \leq 1$ *in* $|z| < 1$ *and* $f(z)$ *leaves out a value* α *such that* $|\alpha| < 1$, *then*

$$|f'(0)| \leq \frac{2|\alpha| \log \frac{1}{|\alpha|}}{1 - |\alpha|^2}. \tag{100}$$

This bound for $|f'(0)|$ *is the smallest possible.*

We have worked this example in great detail in order to bring out more clearly the geometric background of the problem. The inequality (100) may, however, be derived in a much shorter way by using known results on functions with a positive real part. If $|f(z)| \leq 1$, $f(0) = 0$, $f(z) \neq \alpha$, then

$$f_1(z) = \frac{\alpha - f(z)}{1 - \bar{\alpha}f(z)}$$

satisfies $|f_1(z)| \leq 1$, $f_1(0) = \alpha$, $f_1(z) \neq 0$. Hence, the function

$$h(z) = -\log f_1(z)$$

is regular in $|z| < 1$ and, in view of Re $\{\log w\} = \log |w|$, has a positive real part. Hence, by (36),

$$\frac{1 - |\alpha|^2}{|\alpha|} |f'(0)| = |h'(0)| \leq 2 \text{ Re } \{h(0)\} = 2 \log \frac{1}{|\alpha|},$$

and this is equivalent to (100).

We close this section with a result which may be regarded as a generalization of the $\frac{1}{4}$-theorem to a class of functions in which the condition of univalence has been replaced by a much weaker assumption. Instead of requiring the whole conformal map D of $|z| < 1$ as yielded by $w = f(z)$ to be schlicht, we merely suppose that there exists a boundary point α of D which can be connected with the origin by a "schlicht arc" C. By this is meant that the equation $w = f(z)$ has only one solution in $|z| < 1$ if w is a point of C. We now state our result.

Let

$$w = f(z) = z + a_2z^2 + \cdots, \qquad |z| < 1,$$

map $|z| < 1$ *onto a domain* D *and let* α *be a point of the boundary of* D *which can be connected with* $w = 0$ *by a schlicht arc. Then*

$$|\alpha| \geq \frac{1}{\pi^2}.$$

This inequality is sharp.

The function $w = g(z) = \alpha^{-1}f(z)$ maps $|z| < 1$ onto a domain D' which contains a schlicht arc C connecting $w = 0$ and $w = 1$, where $w = 1$ is a boundary point of D'. Clearly, $g(z) \neq 1$ in $|z| < 1$; if there were a point z_1 ($|z_1| < 1$) such that $g(z_1) = 1$, then there would be points near z_1 which are mapped on points of C and C would not be a schlicht arc. Consider now the function

$$h(z) = \sqrt{g(z^2)} = \sqrt{\frac{f(z^2)}{\alpha}} = \frac{z}{\sqrt{\alpha}} \sqrt{1 + a_2z^2 + \cdots}.$$

Since the arc C terminates at $w = 0$, $g(z)$ does not vanish in $|z| < 1$ except at $z = 0$ and $h(z)$ is therefore regular in $|z| < 1$. Moreover, If D''

denotes the map given by $w_1 = h(z)$, then the points $w_1 = 1$ and $w_1 = -1$ can be connected with $w_1 = 0$ by arcs C' and C'' which are covered by D'' only once. These arcs correspond to the arc C by means of the mapping $w_1^2 = w$. Since we can only have $h(z_1) = h(z_2)$ ($z_1 \in C', C''$, $z_2 \in C', C''$, $z_1 \neq z_2$) if $g(z_1^2) = g(z_2^2)$ and $z_1^2 \in C$, $z_2^2 \in C$, and since C is a schlicht arc of the map of $g(z)$, it follows that $h(z_1) = h(z_2)$ entails $z_1 = -z_2$. But, in view of $h(-z) = -h(z)$, this means that one of these two points is on C' and the other is on C''. Hence, both C' and C'' are schlicht arcs. We now form the function

$$F(z) = \frac{2}{\pi} \sin^{-1} h(z) = \frac{2}{\pi\sqrt{\alpha}} z + \cdots .$$

Clearly, $F(z)$ is regular in $|z| < 1$. Indeed,

$$\sin^{-1} w = \int_0^w \frac{dw}{\sqrt{1 - w^2}}$$

has its only singularities at $w = \pm 1$ and these are precisely the values which $h(z)$ does not take in $|z| < 1$. The two arcs C' and C'' of D'' are carried into two arcs γ' and γ'' of the map yielded by $w_2 = F(z)$. Obviously, γ' and γ'' connect $w_2 = 0$ with $w_2 = 1$ and $w_2 = -1$, respectively, and γ' and γ'' are symmetric to each other with respect to the origin. Let now γ_n denote the arc congruent to $\gamma' + \gamma''$ which connects $w_2 = 2n - 1$ with $w_2 = 2n + 1$, where n is a nonzero integer. We shall show that none of the points of the closed arcs γ_n are contained in the domain Δ onto which $|z| < 1$ is mapped by $w_2 = F(z)$. Indeed, if w^* is an interior point of γ_n, then $w^* - 2n$ is a point of γ' or γ''. Hence, if $w^* = F(z_1)$ then $h(z_1) = \sin \frac{1}{2}\pi w^* = \pm \sin \frac{1}{2}\pi(w^* - 2n) = h(z_2)$, where $h(z_2)$ is a point on one of the arcs C', C'' in the w_1-plane. But C' and C'' are schlicht arcs and we cannot have $h(z_1) = h(z_2)$ for two different points z_1, z_2 in the unit circle. This proves that the interior points of the arcs γ_n do not belong to Δ. The same is also true of their end points, *i.e.*, the points $w_2 = 2n + 1$, since $F(z_1) = 2n + 1$ would entail $h(z_1) = \sin (\pi/2)(2n + 1) = \pm 1$. Now the chain of arcs $\gamma_1, \gamma_2, \ldots$ obviously forms a continuous curve δ' connecting $w_2 = 1$ with $w_2 = \infty$; similarly, the arcs $\gamma_{-1}, \gamma_{-2}, \ldots$ form a continuous curve δ'' between $w_2 = -1$ and $w_2 = \infty$ which is symmetric to δ' with respect to the origin.

We now make one more transformation. The function

$$w_3 = G(z) = [F(\sqrt{z})]^2 = \frac{4}{\pi^2\alpha} z + \cdots \tag{101}$$

is regular in $|z| < 1$. Indeed, $F(z)$ is an odd function, that is, it has an expansion

$$F(z) = c_1 z + c_3 z^3 + \cdots + c_{n+1} z^{2n+1} + \cdots, \qquad |z| < 1.$$

Hence,

$$G(z) = z(c_1 + c_3 z + \cdots + c_{2n+1} z^n + \cdots)^2,$$

and this is regular in $|z| < 1$. Since δ' and δ'' are symmetric to each other with respect to the origin both of these curves are carried into the same curve η in the w_3-plane which connects the points $w_3 = 1$ and $w_3 = \infty$. Since none of the points of η are values taken by $w_3 = G(z)$ in $|z| < 1$, we see therefore that $w_3 = 1$ is a point of the outer boundary of the domain onto which $|z| < 1$ is mapped by $w_3 = G(z)$. Now we have shown above that, in the case of a function $\varphi(z)$ with the normalization $\varphi(0) = 0$, $\varphi'(0) = 1$, the points of the outer boundary have a distance of at least $\frac{1}{4}$ from the origin. The function

$$\frac{G(z)}{G'(0)}$$

has this normalization, and the outer boundary of its mapping contains the point $[G'(0)]^{-1}$. Hence $|G'(0)|^{-1} \geq \frac{1}{4}$, or

$$|G'(0)| \leq 4. \tag{102}$$

In view of (101), we thus have

$$\frac{4}{\pi^2|\alpha|} \leq 4,$$

whence

$$|\alpha| \geq \frac{1}{\pi^2}.$$

This completes the proof. The extremal function of the problem can be found by observing that we have equality in (102) if

$$G(z) = \frac{4z}{(1 + z)^2}.$$

Performing the various transformations used in reversed order, we find that the extremal function is of the form

$$w = f_0(z) = \frac{1}{\pi^2} \sin^2 \left[\frac{\pi \sqrt{z}}{1 + z}\right].$$

It will be a useful exercise for the reader to verify that the map associated with this function contains the schlicht segment $0 \leq w < \pi^{-2}$ and that it does not contain the point $w = \pi^{-2}$.

EXERCISES

1. Show that the function

$$w = \left(\frac{1+z}{1-z}\right)^{\frac{2\alpha}{\pi}}, \qquad 0 < \alpha < \pi.$$

maps $|z| < 1$ onto the angular region $|\arg \{w\}| < \alpha$. With the help of this mapping, prove that a function $f(z)$ which is regular in $|z| < 1$ and satisfies $f(0) = 1$, $|\arg \{f(z)\}| < \alpha$, is subject to the inequality

$$|f'(0)| \leq \frac{4\alpha}{\pi}.$$

2. Show that a function $w = f(z)$ which is regular in $|z| < 1$, vanishes at the origin, and does not take in $|z| < 1$ values w such that $-\infty \leq w \leq -\alpha$ and $\alpha \leq w \leq \infty$ $(\alpha > 0)$, satisfies the inequality

$$|f'(0)| \leq 2\alpha.$$

3. Show that the linear transformation

$$w = a - \frac{ar(\bar{a} - rz)}{\bar{a}(r - az)}, \qquad r < |a|,$$

maps $|z| < 1$ onto the exterior of the circle of radius r and center a in such a way that the points $z = 0$ and $w = 0$ correspond to each other. Use this mapping in order to prove the following result: Let $f(z)$ be regular in $|z| < 1$ except for a possible finite number of poles and let $f(0) = 0$. If the values taken by $f(z)$ in $|z| < 1$ leave out a circle of radius r which is seen from the origin under the angle α, then

$$|f'(0)| \leq r \cot^2 \tfrac{1}{2}\alpha.$$

4. Show that the result concerning functions with a bounded real part which was proved in the text can be generalized in the following manner.

If $f(z)$ is regular and $-1 \leq \operatorname{Re} \{f(z)\} \leq 1$ in $|z| < 1$, then

$$|f'(0)| \leq \frac{4}{\pi} \cos\left[\frac{\pi}{2} \operatorname{Re} \{f(0)\}\right].$$

Hint: Note that an imaginary constant may be added to $f(z)$ without violating the hypotheses and that it is therefore sufficient to consider functions $f(z)$ for which $f(0)$ is real. Then, show that the appropriate superordinate function $F_1(z)$ is obtained from the superordinate function $F(z)$ used in the text by means of the relation

$$F_1(z) = F\left(\frac{z + \zeta}{1 + \zeta z}\right),$$

where the real constant ζ is determined by $F(\zeta) = f(0)$.

5. Show that the function

$$w = \frac{4z}{2z - c(1 - z^2)}$$

maps $|z| < 1$ onto the full w-plane which is slit along an arc of the circle $|w - 1| = 1$, and find the extremities of this slit. Use your result to prove the following theorem:

Let $w = f(z)$ be regular with the possible exception of a finite number of poles in $|z| < 1$, and let $f(0) = 0$. If C is a circular circumference which passes through $w = 0$ and if the values of $w = f(z)$ in $|z| < 1$ leave out an arc of C which is seen from the origin under the angle θ, then

$$|f'(0)| \leq 4r \cot \tfrac{1}{2}\theta,$$

where r is the radius of C.

6. Let $F(0) = 0$, $F'(0) = 1$, and let $w = F(z)$ map $|z| < 1$ onto a convex domain D. If

$$f(z) = a_1 z + a_2 z^2 + a_3 z^3 + \cdots$$

is subordinate to $F(z)$, show that $|a_n| \leq 1$, $n = 1, 2, \ldots$. *Hint:* Use the result of Exercise 8, Sec. 3, and the fact that the center of gravity of n mass points in the convex domain D cannot be outside D.

7. Let $F(0) = 0$, $F'(0) = 1$, and let $w = F(z)$ map $|z| < 1$ onto a convex domain D. If $f(z)$ is subordinate to $F(z)$ and, moreover, $f(z)$ is an odd function of z, that is, $f(-z) = -f(z)$, show that

$$|f'(z)| \leq \frac{1}{1 - |z|^2}.$$

Hint: Use the fact that the value

$$\frac{1}{2}\left[f\left(\frac{z+\alpha}{1+\bar{\alpha}z}\right) + f\left(\frac{z-\alpha}{1-\bar{\alpha}z}\right)\right], \qquad |\alpha| < 1,$$

is in D.

8. Let the univalent function $F(z)$ have an expansion

$$F(z) = \frac{1}{z} + a_0 + a_1 z + a_2 z^2 + \cdots$$

in $|z| < 1$ and let $w = F(z)$ map $|z| < 1$ onto a domain D. If

$$f(z) = \frac{\alpha}{z} + b_0 + b_1 z + \cdots$$

is regular in $0 < |z| < 1$ and if all the values which $w = f(z)$ takes in $|z| < 1$ lie in D, show that

$$|\alpha| \geq 1.$$

9. If

$$w = f(z) = \frac{\alpha}{z} + b_0 + b_1 z + \cdots$$

is regular in $|z| < 1$ except at $z = 0$ and does not take values w such that $a < w < b$ (a,b real) show that

$$|\alpha| \geq \frac{b-a}{4}.$$

10. Show that the function

$$w = \rho z + \frac{1}{\rho z}, \qquad \rho > 0,$$

maps $|z| < 1$ onto the exterior of an ellipse whose semiaxes are $\rho^{-1} + \rho$ and $|\rho^{-1} - \rho|$, respectively. Use this mapping in order to prove the following result: If

$$w = f(z) = \frac{\alpha}{z} + b_0 + b_1 z + b_2 z^2 + \cdots$$

is regular in $|z| < 1$ except at $z = 0$ and if its values in $|z| < 1$ leave out the points of an ellipse of semiaxes a and b, then

$$|\alpha| \geq \frac{a+b}{2}.$$

11. In Sec. 8, it was shown that if $w = f(z) = z + a_2 z^2 + \cdots$ maps $|z| < 1$ onto a convex domain D, then D contains the circle $|w| < \frac{1}{2}$. Give an alternative proof of this result, based on the principle of subordination and on the fact that

$$w = \frac{cz}{1-z}, \qquad c = \text{const.},$$

maps $|z| < 1$ onto a certain half-plane.

12. If $f(z) = b_1 z + b_2 z^2 + \cdots$ is subordinate to $F(z) = a_1 z + a_2 z^2 + \cdots$, show that $|b_2|$ is smaller than, or equal to, the larger of the two numbers $|a_1|$, $|a_2|$. *Hint:* If $\omega(z) = c_1 z + c_2 z^2 + \cdots$ is the bounded function defined by $f(z) = F[\omega(z)]$, use the fact that $|c_2| \leq 1 - |c_1|^2$.

10. The Kernel Function. By the methods described in Secs. 6 and 7, we can find the analytic function which maps a polygon, whose sides are either linear segments or circular arcs, onto the interior of a circle. A large number of other mappings can be found by a detailed investigation of known analytic functions and by suitable combination of such functions. However, it is in the nature of things that mappings which are found in this way are distinguished by the simple geometric character of the domains mapped onto the unit circle; the simplicity of the associated domains is an expression of the simplicity of the properties which characterize a "known" function. Neither the methods mentioned above nor a "catalogue" of mappings by known functions will therefore be of much help if we are faced by the problem of mapping an irregularly shaped simply-connected domain onto the unit circle. It is the objective of this section to describe a method for the construction of the mapping function which can be used for a simply-connected domain of arbitrary shape.

A few preparations are necessary before we can attack our problem. We introduce the differential operators $\partial/\partial z$ and $\partial/\partial \bar{z}$ by the definitions

$$(103) \qquad \frac{\partial}{\partial z} = \frac{1}{2}\left(\frac{\partial}{\partial x} - i\frac{\partial}{\partial y}\right), \qquad \frac{\partial}{\partial \bar{z}} = \frac{1}{2}\left(\frac{\partial}{\partial x} + i\frac{\partial}{\partial y}\right), \qquad z = x + iy.$$

These operators can be applied to any function of x and y which has partial derivatives of the first order. If the operator $\partial/\partial z$ is applied to an analytic function $f(z) = u + iv$, we obtain

$$\frac{\partial f(z)}{\partial z} = \frac{1}{2}(u_x + iv_x - iu_y + v_y) = u_x + iv_x = f'(z),$$

in view of the Cauchy-Riemann equations. Hence, application of the operator $\partial/\partial z$ to an analytic function $f(z)$ results in ordinary differentiation with respect to z. On the other hand, it follows from the Cauchy-Riemann equations that

$$\frac{\partial f(z)}{\partial \bar{z}} = \frac{1}{2}(u_x + iv_x + iu_y - v_y) = 0.$$

This shows that with respect to the differential operator $\partial/\partial \bar{z}$ all analytic functions $f(z)$ play the role of constants. The reader will easily verify that, similarly,

$$\frac{\partial \overline{f(z)}}{\partial z} = 0, \qquad \frac{\partial \overline{f(z)}}{\partial \bar{z}} = \overline{f'(z)}.$$

We now express Green's formula in terms of the differential operators (103). If $r(x,y)$ is a function which has continuous first partial derivatives in a domain D (not necessarily simply-connected) and on its piecewise smooth boundary C, it follows from (5) (Sec. 3, Chap. I) that

$$\iint_D \frac{\partial r}{\partial \bar{z}}\, dx\, dy = \frac{1}{2}\iint_D (r_x + ir_y)\, dx\, dy = \frac{1}{2}\int_C (r\, dy - ir\, dx)$$
$$= \frac{1}{2i}\int_C r(dx + i\, dy) = \frac{1}{2i}\int_C r\, dz.$$

A more general formula is obtained by writing $r(x,y) = p(x,y) \cdot q(x,y)$, where both p and q have continuous first derivatives in $D + C$. In view of

$$\frac{\partial}{\partial \bar{z}}(pq) = p\frac{\partial q}{\partial \bar{z}} + q\frac{\partial p}{\partial \bar{z}},$$

we find that

$$\text{(104)} \qquad \iint_D p\frac{\partial q}{\partial \bar{z}}\, dx\, dy = \frac{1}{2i}\int_C pq\, dz - \iint_D q\frac{\partial p}{\partial \bar{z}}\, dx\, dy.$$

The proof of the analogous formula

$$\text{(105)} \qquad \iint_D p\frac{\partial q}{\partial z}\, dx\, dy = -\frac{1}{2i}\int_C pq\, d\bar{z} - \iint_D q\frac{\partial p}{\partial z}\, dx\, dy$$

is left as an exercise to the reader. (104) takes a particularly simple form if $p(x,y) = f(z)$ and $q(x,y) = \overline{g(z)}$, where $f(z)$ and $g(z)$ are regular analytic functions in $D + C$. In view of

$$\frac{\partial \overline{g(z)}}{\partial \bar{z}} = \overline{g'(z)}, \qquad \frac{\partial f(z)}{\partial \bar{z}} = 0,$$

we obtain

$$\iint_D f(z)\overline{g'(z)}\, dx\, dy = \frac{1}{2i} \int_C f(z)\overline{g(z)}\, dz. \tag{106}$$

We now turn to the consideration of orthogonal sets of regular analytic functions. Let D be a schlicht domain in the z-plane and let $L^2 = L^2(D)$ denote the class of analytic functions $f(z)$ which are regular and single-valued in D and for which the integral

$$(f,f) = \iint_D |f(z)|^2\, dx\, dy \tag{107}$$

has a finite value. If $f(z)$ is not continuous in the closure of D, the integral (107) is defined by

$$\iint_D |f(z)|^2\, dx\, dy = \lim_{n\to\infty} \iint_{D_n} |f(z)|^2\, dx\, dy,$$

where $\{D_n\}$ is a sequence of domains such that $D_n \in D_{n+1}$ and $D_n \to D$ if $n \to \infty$. Since the value of the integral (107) increases with the domain over which it is extended, this limit always exists and is either a finite positive number or $+\infty$. (107) is called the *norm* of the function $f(z)$.

If $f(z)$ and $g(z)$ are functions of L^2, their *inner product* (f,g) is defined by

$$(f,g) = \iint_D f(z)\overline{g(z)}\, dx\, dy. \tag{108}$$

The existence of the inner product is assured by the fact that, in view of the Schwarz integral inequality,

$$\left| \iint_D f(z)\overline{g(z)}\, dx\, dy \right|^2 \le \left(\iint_D |f(z)||g(z)|\, dx\, dy \right)^2$$
$$\le \iint_D |f(z)|^2\, dx\, dy \iint_D |g(z)|^2\, dx\, dy.$$

This also shows that

$$|(f,g)|^2 \le (f,f)(g,g). \tag{109}$$

Another property of the inner product is the fact, immediately visible from (108), that

$$(g,f) = \overline{(f,g)}. \tag{110}$$

(110) is also expressed by saying that the inner product (108) is *Hermitian.* If $(g,f) = 0$, that is, the two functions $f(z)$ and $g(z)$ have a vanishing inner product, we say that the functions $f(z)$ are *orthogonal* to each other. Obviously, a function cannot be orthogonal to itself unless it is identically zero.

A set of functions $f_1(z), f_2(z), \ldots, f_n(z)$ of L^2 is said to be *linearly independent* if there exists no identical relation of the type

$$a_1f_1(z) + a_2f_2(z) + \cdots + a_nf_n(z) = 0, \tag{111}$$

where $a_1, a_2, \ldots, a_n$ are constants not all of which are zero. If the functions $f_1(z), \ldots, f_n(z)$ are all orthogonal to each other, that is, if $(f_\nu,f_\mu) = 0$ for $\nu \neq \mu$, then they certainly form a linearly independent set. Indeed, suppose there exists a relation of the type (111). Multiplying (111) by $\overline{f_\nu(z)}$ and integrating over D, we obtain by (108)

$$a_1(f_1,f_\nu) + a_2(f_2,f_\nu) + \cdots + a_n(f_n,f_\nu) = 0.$$

Since all inner products of different functions vanish, it follows that $a_\nu(f_\nu,f_\nu) = 0$. But, as remarked above, $(f_\nu,f_\nu) > 0$, whence $a_\nu = 0$. This shows that a relation (111) is only possible in the trivial case in which all coefficients a_ν are zero. Hence, the orthogonal set $f_1, \ldots, f_n$ is linearly independent. Conversely, given a linearly independent set of functions $f_1, \ldots, f_n$, it is always possible to form n linear combinations of the type

$$u_\nu(z) = \sum_{\mu=1}^{n} a_{\nu\mu}f_\mu(z), \qquad \nu = 1, \ldots, n,$$

which are orthogonal to each other. This can be done by the following *orthogonalization process* which is also known as the *Schmidt process.* We set

$$u_1(z) = f_1(z).$$

Next, we determine the constant α_{11} by the requirement that the function

$$u_2(z) = f_2(z) + \alpha_{11}f_1(z) \tag{112}$$

be orthogonal to $u_1(z)$. In view of $u_1(z) = f_1(z)$, we must have

$$(u_2,u_1) = (f_2,f_1) + \alpha_{11}(f_1,f_1) = 0.$$

Since $(f_1,f_1) > 0$, we therefore obtain

$$\alpha_{11} = -\frac{(f_2,f_1)}{(f_1,f_1)}.$$

We now write

$$u_3(z) = f_3(z) + \alpha_{22}u_2(z) + \alpha_{21}u_1(z) \tag{113}$$

and try to determine the constants α_{22} and α_{21} by the condition that u_3 be orthogonal to both u_1 and u_2. Since u_1 and u_2 are orthogonal to each other, this yields

$$\begin{aligned}(u_3,u_1) &= (f_3,u_1) + \alpha_{21}(u_1,u_1) = 0,\\ (u_3,u_2) &= (f_3,u_2) + \alpha_{22}(u_2,u_2) = 0.\end{aligned}$$

In view of $(u_1,u_1) > 0$, $(u_2,u_2) > 0$, this determines α_{21} and α_{22}. It is important to note that the function $u_3(z)$ cannot be identically zero. Indeed, it follows from (112) and (113) that $u_3(z)$ is a linear combination of the type (111), and the identical vanishing of u_3 would therefore mean that the functions $f_1, f_2, \ldots, f_n$ are not linearly independent.

The Schmidt process can be continued until we obtain a set n functions $u_1, \ldots, u_n$ which are orthogonal to each other. To show that this is true, suppose we have already found linear combinations $u_1, u_2, \ldots, u_m$ $(m < n)$, involving the functions $f_1, \ldots, f_m$, which are mutually orthogonal. We now set

$$u_{m+1}(z) = f_{m+1}(z) + \alpha_{mm}u_m(z) + \cdots + \alpha_{m2}u_2(z) + \alpha_{m1}u_1(z).$$

The requirement that u_{m+1} be orthogonal to the functions $u_1, u_2, \ldots, u_m$ yields the conditions

$$(u_{m+1},u_\nu) = \alpha_{m\nu}(u_\nu,u_\nu), \qquad \nu = 1, 2, \ldots, m.$$

This determines the constants $\alpha_{m1}, \ldots, \alpha_{mm}$. Moreover, $u_{m+1}(z)$ is not identically zero. Clearly, u_{m+1} is a linear combination of the type (111) and this cannot vanish identically since the functions $f_1, \ldots, f_n$ are linearly independent. We thus have proved that the Schmidt orthogonalization process, if applied to n linearly independent functions $f_1, \ldots, f_n$, yields n mutually orthogonal functions $u_1, \ldots, u_n$, none of which is identically zero. It is hardly necessary to add that $u_1, \ldots, u_n$ are functions of L^2 if the same is true of $f_1, \ldots, f_n$.

It is convenient to normalize the orthogonal functions thus obtained by the additional requirement that their norms be equal to 1. Since $(u_\nu,u_\nu) > 0$, $\nu = 1, \ldots, n$ (none of the functions u_ν vanish identically), the functions

$$v_\nu(z) = \frac{u_\nu(z)}{\sqrt{(u_\nu,u_\nu)}}, \qquad \nu = 1, \ldots, n,$$

will clearly have the required normalization. A set of functions $v_1, \ldots, v_n$ which is both orthogonal and normalized is called an *orthonormal set.*

Using the symbol $\delta_{\nu\mu}$ which is defined by

$$\delta_{\nu\mu} = 0, \qquad \mu \neq \nu, \qquad \delta_{\nu\nu} = 1, \tag{114}$$

we may characterize an orthonormal set $v_1, \ldots, v_n$ by the conditions

$$(v_\nu, v_\mu) = \delta_{\nu\mu}. \tag{115}$$

The restriction to a finite number of functions f_ν in the preceding developments is not essential. An infinite set of functions $f_1, f_2, \ldots$ is said to be linearly independent if no relation of the type

$$a_1 f_{n_1}(z) + a_2 f_{n_2}(z) + \cdots + a_m f_{n_m}(z) = 0$$

exists. Obviously, the Schmidt process can also be applied in this case and it will lead to an infinite sequence of functions $v_1, v_2, \ldots$ which are orthonormalized by the conditions (115).

If $f(z)$ is a function of L^2 and if $v_1, v_2, \ldots$ is an orthonormal set of functions of L^2, then the numbers

$$a_n = (f, v_n) = \iint_D f(z)\overline{v_n(z)}\, dx\, dy \tag{116}$$

are called the *Fourier coefficients* of the function $f(z)$ with respect to the orthonormal set $\{v_n\}$. They are subject to the important inequality

$$\sum_{n=1}^{\infty} |a_n|^2 \leq (f, f), \tag{117}$$

whose proof follows from the observation that the integral

$$I = \iint_D \left| f(z) - \sum_{n=1}^{m} a_n v_n(z) \right|^2 dx\, dy \tag{118}$$

is nonnegative. We have

$$I = \left(f - \sum_{n=1}^{m} a_n v_n,\, f - \sum_{n=1}^{m} a_n v_n\right)$$
$$= (f,f) - \sum_{n=1}^{m} \bar{a}_n (f, v_n) - \sum_{n=1}^{m} a_n (v_n, f) + \sum_{n=1}^{m} \sum_{\nu=1}^{m} a_n \bar{a}_\nu (v_n, v_\nu).$$

Hence, by (115) and (116),

$$I = (f,f) - \sum_{n=1}^{m} |a_n|^2 - \sum_{n=1}^{m} |a_n|^2 + \sum_{n=1}^{m} |a_n|^2$$
$$= (f,f) - \sum_{n=1}^{m} |a_n|^2.$$

Since I is nonnegative and m may now be taken arbitrarily large, (117) follows. (117) is known as the *Bessel inequality*.

The integral (118) provides a measure of how close the functions f and $\sum_{n=1}^{m} a_n v_n$ are, or, as we also say, how well the function f is approximated by the linear combination $\sum_{n=1}^{m} a_n v_n$. Generally speaking, we say that a function f is *approximated in the mean* by a sequence of functions $g_1, g_2, \ldots,$ if

$$\lim_{n \to \infty} \iint_D |f(z) - g_n(z)|^2 \, dx \, dy = 0.$$

In view of the value of the integral I, it follows therefore from (118) that a function f of L^2 can be approximated in the mean by linear combinations $\sum_{n=1}^{m} a_n v_n$ if

$$\sum_{n=1}^{\infty} |a_n|^2 = (f,f). \tag{119}$$

If the identity (119) is satisfied for all functions f of L^2, we say that the orthonormal set $\{v_n\}$ is *complete*, or *complete in* L^2. If a complete set $\{v_n\}$ is given, then any function of L^2 can be approximated in the mean by linear combinations of the functions v_n. If we choose this latter property as the definition of a complete set, then this definition also extends to the case of a set of linearly independent functions which is not orthonormal. We leave it as an exercise to the reader to show that the completeness of a set of functions is not affected by the Schmidt process.

We shall now show that, given an arbitrary simply-connected domain D, there always exist sets of functions $\{v_n\}$ which are complete with respect to the class $L^2(D)$. Consider first the particularly simple case in which D is the unit circle. Any function $f(z)$ which is regular in $|z| < 1$ can be expanded there into a power series

$$f(z) = b_0 + b_1 z + b_2 z^2 + \cdots \tag{120}$$

which converges uniformly in any circle $|z| \leq \rho < 1$. This shows that in a circle $|z| \leq \rho < 1$ the function $f(z)$ can be approximated by linear combinations of the set of functions

$$1, z, z^2, z^3, \ldots .$$

These functions also constitute an orthogonal set. Indeed, we have by (106)

$$(z^n,z^m) = \iint\limits_{|z|<1} z^n\bar{z}^m\,dx\,dy = \frac{1}{2i(m+1)}\int_{|z|=1} z^n\bar{z}^{m+1}\,dz$$

$$= \frac{1}{2(m+1)}\int_0^{2\pi} e^{i(n-m)\theta}\,d\theta,$$

and this is zero unless $m = n$. For $n = m$, we obtain $(z^n,z^n) = \pi(n+1)^{-1}$. Hence, the orthonormalized set is

$$v_n(z) = \sqrt{\frac{n+1}{\pi}}\,z^n, \qquad n = 0, 1, 2, \ldots . \tag{121}$$

Next, we compute the Fourier coefficients (116). We have

$$a_n = \sqrt{\frac{n+1}{\pi}}\iint\limits_{|z|<1} f(z)\bar{z}^n\,dx\,dy = \lim_{\rho\to1}\sqrt{\frac{n+1}{\pi}}\iint\limits_{|z|<\rho} f(z)\bar{z}^n\,dx\,dy$$

whence, by (106),

$$a_n = \lim_{\rho\to1}\frac{1}{2i}\sqrt{\frac{n+1}{\pi}}\int_{|z|=\rho} f(z)\,\frac{\bar{z}^{n+1}}{n+1}\,dz.$$

Since $|z|^2 = \rho^2$, we have $\bar{z} = \rho^2 z^{-1}$. Hence

$$a_n = \lim_{\rho\to1}\frac{1}{\sqrt{\pi(n+1)}}\,\frac{\rho^{2n+2}}{2i}\int_{|z|=\rho}\frac{f(z)}{z^{n+1}}\,dz$$

$$= \lim_{\rho\to1}\frac{1}{\sqrt{\pi(n+1)}}\,\pi\rho^{2n+2}b_n = \sqrt{\frac{\pi}{n+1}}\,b_n,$$

where b_n is the nth coefficient in the expansion (120). It follows that

$$\sum_{n=0}^{\infty}|a_n|^2 = \pi\sum_{n=0}^{\infty}\frac{|b_n|^2}{n+1}. \tag{122}$$

On the other hand, we have

$$\iint\limits_{|z|<1}|f(z)|^2\,dx\,dy = \int_0^1\rho\,d\rho\int_0^{2\pi}|f(\rho e^{i\theta})|^2\,d\theta \tag{123}$$

$$= 2\pi\int_0^1\rho\,d\rho\left(\sum_{n=0}^{\infty}|b_n|^2\rho^{2n}\right)$$

$$= \pi\sum_{n=0}^{\infty}\frac{|b_n|^2}{n+1},$$

where Parseval's theorem (Sec. 4, Chap. III) has been used. Comparing the last result with (122), we find that the relation (119) is indeed satisfied for any function of L^2. Hence, the set of functions (121) is complete in the unit circle.

By means of the Riemann mapping theorem, this result is easily generalized to arbitrary simply-connected domains. If D is a finite domain which is simply-connected and has at least two boundary points, then there exists a function $w = w(z)$ which maps D onto the unit circle $|w| < 1$. We now assert that the functions

$$\varphi_n(z) = \sqrt{\frac{n+1}{\pi}}\,[w(z)]^n w'(z), \qquad n = 0, 1, \ldots$$

form a complete orthonormal set in D. We have

$$(\varphi_n,\varphi_m) = \iint_D \varphi_n(z)\overline{\varphi_m(z)}\,dx\,dy = \frac{n+1}{\pi}\iint_D w^n\bar{w}^m|w'(z)|^2\,dx\,dy.$$

If $w = u + iv$, then the Jacobian of the transformation $u, v \to x, y$ is $|w'(z)|^2$. Hence

$$(\varphi_n,\varphi_m) = \frac{n+1}{\pi}\iint_{|w|<1} w^n\bar{w}^m\,du\,dv,$$

and, as shown before, this is 0 or 1, according as $n \neq m$ or $n = m$. This proves the orthonormality. As to the completeness, we observe that if $z = p(w)$ is the inverse to $w = w(z)$, then with each function $f(z)$ in D we can associate a function $g(w)$ by the relation $g(w) = p'(w)f[p(w)]$. Since the powers $1, w, w^2, \ldots$ are complete in $|w| < 1$, we can find suitable coefficients $c_1, c_2, \ldots$ such that

$$\iint_{|w|<1}\left|g(w) - \sum_{\nu=1}^{m} c_\nu w^\nu\right|^2 du\,dv < \epsilon, \qquad w = u + iv,$$

provided $m > m_0(\epsilon)$. Passing to the variable z, we obtain

$$\iint_D\left|p'(w)f(z) - \sum_{\nu=1}^{m} c_\nu[w(z)]^\nu\right|^2 |w'(z)|^2\,dx\,dy < \epsilon.$$

In view of $p'(w)\,w'(z) = 1$ and the definition of $\varphi_n(z)$, this is equivalent to

$$\iint_D\left|f(z) - \sum_{\nu=1}^{m} c_\nu'\varphi_\nu(z)\right|^2 dx\,dy < \epsilon,$$

where

$$c_\nu' = \sqrt{\frac{\pi}{\nu + 1}}\, c_\nu.$$

This shows that the functions $\varphi_n(z)$ form indeed a complete set in D.

Proceeding with the preparations for the main result of this section, we now show that the family of functions $f(z)$ of $L^2(D)$ which satisfy $(f,f) \leq M$, where M is a positive constant, is normal and compact. If ρ is the radius of the largest circle about ζ ($\zeta \in D$) which is entirely within D, we have an expansion

$$f(z) = C_0 + C_1(z - \zeta) + C_2(z - \zeta)^2 + \cdots, \qquad |z - \zeta| < \rho.$$

Making a few obvious changes in (123), we obtain

$$\iint\limits_{|z-\zeta|<\rho} |f(z)|^2\, dx\, dy = \pi \sum_{n=0}^{\infty} \frac{|C_n|^2 \rho^{2n+2}}{n + 1}. \tag{124}$$

If $|z - \zeta| \leq r < \rho$, it follows therefore that

$$|f(z)|^2 = \left| \sum_{n=0}^{\infty} C_n (z - \zeta)^n \right|^2 \leq \left(\sum_{n=0}^{\infty} |C_n| r^n \right)^2$$

$$= \left(\sum_{n=0}^{\infty} \frac{|C_n|}{\sqrt{n+1}} \rho^{n+1} \frac{\sqrt{n+1}\, r^n}{\rho^{n+1}} \right)^2 \leq \sum_{n=0}^{\infty} \frac{|C_n|^2 \rho^{2n+2}}{n+1} \sum_{n=0}^{\infty} \frac{1}{\rho^2} (n+1) \left(\frac{r}{\rho}\right)^{2n}$$

$$= \frac{\rho^2}{\pi(\rho^2 - r^2)^2} \iint\limits_{|z-\zeta|<\rho} |f(z)|^2\, dx\, dy \leq \frac{\rho^2}{\pi(\rho^2 - r^2)^2} (f,f).$$

Since $(f,f) < M$, we obtain

$$|f(z)| \leq \frac{\rho \pi^{-\frac{1}{2}}}{\rho^2 - r^2} \sqrt{M}.$$

This shows the family of functions in question is locally uniformly bounded. By the results of Sec. 2, Chap. IV, it is therefore normal. The compactness, *i.e.*, the fact that the limit $f(z)$ of a sequence $\{f_n(z)\}$ for which $(f_n,f_n) \leq M$ also satisfies $(f,f) \leq M$, is obvious.

We now pose the following extremal problem. From among all functions $f(z)$ of $L^2(D)$ which satisfy $f(\zeta) = 1$, where ζ is a point of D, to find the particular function, or functions, $f_0(z)$ for which (f,f) is a minimum. We first have to ascertain whether such a function $f_0(z)$ exists. To this end we observe that we can confine ourselves to such functions of L^2 for which $f(\zeta) = 1$ and $(f,f) \leq A$, where A is the area of D. Indeed, the function $f(z) \equiv 1$ satisfies the conditions of our problem and we have, for this function, $(f,f) = \iint_D dx\,dy = A$. Since we are interested in the function minimizing (f,f), it is therefore sufficient to investigate functions for which $(f,f) \leq A$. As shown above, these functions form a normal and compact family. Hence, by the results of Sec. 3, Chap. IV, our extremal problem has a solution and there exists a function $f_0(z)$ such that

$$(125) \qquad (f_0,f_0) \leq (f,f), \qquad f_0 \in L^2, \qquad f \in L^2, \qquad f_0(\zeta) = f(\zeta) = 1.$$

We further observe that, since $f_0(\zeta) = 1$, $f_0(z)$ cannot be identically zero.

Let now $g(z)$ be a function of L^2 for which $g(\zeta) = 0$. The function $f^*(z) = f_0(z) + \epsilon e^{i\theta}g(z)$, where $0 \leq \theta < 2\pi$ and ϵ is a small positive parameter, will then also be in L^2. Since, moreover, $f^*(\zeta) = 1$, it follows from (125) that $(f^*,f^*) \geq (f_0,f_0)$. Written in full, this means that

$$\iint_D |f_0|^2\,dx\,dy \leq \iint_D |f_0 + \epsilon e^{i\theta}g|^2\,dx\,dy$$
$$= \iint_D |f_0|^2\,dx\,dy + 2\epsilon \operatorname{Re}\left\{e^{i\theta}\iint_D \bar{f}_0 g\,dx\,dy\right\} + \epsilon^2 \iint_D |g|^2\,dx\,dy.$$

Hence, in view of $\epsilon > 0$,

$$2 \operatorname{Re}\left\{e^{i\theta}\iint_D \bar{f}_0 g\,dx\,dy\right\} + \epsilon \iint_D |g|^2\,dx\,dy \geq 0.$$

This is true for any positive ϵ. Letting $\epsilon \to 0$, we thus obtain

$$(126) \qquad \operatorname{Re}\left\{e^{i\theta}\iint_D \bar{f}_0 g\,dx\,dy\right\} \geq 0.$$

But this is impossible unless

$$(127) \qquad \iint_D \bar{f}_0 g\,dx\,dy = 0.$$

Indeed, if this were not the case, we could by a suitable choice of the arbitrary argument θ make the left-hand side of (126) negative.

(127) has been proved for every function $g(z)$ of L^2 which satisfies $g(\zeta) = 0$. Given an arbitrary function $f(z)$ of L^2, we may therefore set $g(z) = f(z) - f(\zeta)$ in (127). This yields

$$\iint_D \overline{f_0(z)}f(z)\,dx\,dy = f(\zeta)\iint_D \overline{f_0(z)}\,dx\,dy.$$

Taking, in particular, $f(z) = f_0(z)$, we obtain

$$\iint_D |f_0(z)|^2\,dx\,dy = \iint_D \overline{f_0(z)}\,dx\,dy,$$

whence

$$\iint_D \overline{f_0(z)} f(z)\, dx\, dy = f(\zeta) \iint_D |f_0(z)|^2\, dx\, dy.$$

With the notation

$$\frac{f_0(z)}{(f_0,f_0)} = K(z,\zeta),$$

this can be written

$$f(\zeta) = \iint_D \overline{K(z,\zeta)} f(z)\, dx\, dy. \tag{128}$$

The functions $K(z,\zeta)$ and $f_0(z)$ differ only by a constant factor. Since $f_0(\zeta) = 1$, we therefore have

$$f_0(z) = \frac{K(z,\zeta)}{K(\zeta,\zeta)}. \tag{129}$$

We note that the denominator in (129) is different from zero. Indeed, by (128),

$$K(\zeta,\zeta) = \iint_D |K(z,\zeta)|^2\, dx\, dy, \tag{130}$$

and this is positive.

$K(z,\zeta)$ is known as the *kernel function* or *Bergman kernel function* of D. $K(z,\zeta)$ is completely characterized by the *reproducing property* (128), where this appellation is derived from the fact that any function $f(z)$ of L^2 is "reproduced" by multiplication with $\overline{K(z,\zeta)}$ and integration over D. That $K(z,\zeta)$ is the only function of L^2 with this property is seen in the following manner. Suppose there exists a function $K_1(z,\zeta)$ of L^2 for which also

$$f(\zeta) = \iint_D \overline{K_1(z,\zeta)} f(z)\, dx\, dy, \qquad f(z) \in L^2.$$

Subtracting this from (128), we obtain

$$\iint_D \overline{(K - K_1)} f\, dx\, dy = 0.$$

This holds for any $f(z)$ in L^2. Taking, in particular, $f(z) = K(z,\zeta) - K_1(z,\zeta)$, we have

$$\iint_D |K - K_1|^2\, dx\, dy = 0.$$

But this is only possible if $K(z,\zeta) \equiv K_1(z,\zeta)$.

Let now $\{v_n(z)\}$ be a complete orthonormal set of functions of L^2, and compute the Fourier coefficients (116) of the kernel function $K(z,\zeta)$ with respect to the set $\{v_n\}$. By (116) and (128), we obtain

$$a_n = \iint_D K(z,\zeta)\overline{v_n(z)}\,dx\,dy$$
$$= \overline{\iint_D \overline{K(z,\zeta)}v_n(z)\,dx\,dy} = \overline{v_n(\zeta)}.$$

Since the set $\{v_n\}$ is complete, it therefore follows from (119) and (130) that

$$K(\zeta,\zeta) = \sum_{n=1}^{\infty} |v_n(\zeta)|^2. \tag{131}$$

With the help of this identity, we shall now show that *the kernel function $K(z,\zeta)$ can be expanded into the infinite series*

$$K(z,\zeta) = \sum_{n=1}^{\infty} v_n(z)\overline{v_n(\zeta)} \tag{132}$$

which converges absolutely and uniformly in any closed domain which is entirely within D. To prove (132), consider the expression

$$S = K(\eta,\zeta) - \sum_{n=1}^{m} v_n(\eta)\overline{v_n(\zeta)}, \qquad \eta \in D,\ \zeta \in D.$$

By (128), we have

$$S = \iint_D \overline{K(z,\eta)}\left[K(z,\zeta) - \sum_{n=1}^{m} v_n(z)\overline{v_n(\zeta)}\right] dx\,dy.$$

Hence, in view of the Schwarz integral inequality,

$$|S|^2 \le \iint_D |K(z,\eta)|^2\,dx\,dy \iint_D \left|K(z,\zeta) - \sum_{n=1}^{m} v_n(z)\overline{v_n(\zeta)}\right|^2 dx\,dy.$$

If we use (128), (130), and the fact that the set $\{v_n\}$ is orthonormal, we find after a short computation that this is equivalent to

$$|S|^2 \le K(\eta,\eta)\left[K(\zeta,\zeta) - \sum_{n=1}^{m} |v_n(\zeta)|^2\right],$$

or, in view of (131),

$$|S|^2 \le K(\eta,\eta) \sum_{n=m+1}^{\infty} |v_n(\zeta)|^2.$$

The sum on the right-hand side is the remainder of the converging series (131); it can therefore be made smaller than an arbitrary positive ϵ by taking m large enough. Changing the variable η into z and taking into account the definition of S, we thus obtain

$$\left| K(z,\zeta) - \sum_{n=1}^{m} v_n(z)\overline{v_n(\zeta)} \right|^2 \leq \epsilon K(z,z).$$

This proves (132). That the convergence is absolute follows from

$$\left| \sum_{m+1}^{\infty} v_n(z)\overline{v_n(\zeta)} \right|^2 \leq \sum_{m+1}^{\infty} |v_n(z)|^2 \sum_{m+1}^{\infty} |v_n(\zeta)|^2$$
$$\leq K(z,z) \sum_{m+1}^{\infty} |v_n(\zeta)|^2 \leq \epsilon K(z,z).$$

This also shows that the convergence is uniform in any closed subdomain D_1 of D. Indeed, ϵ depends only on ζ and we may employ the (finite) maximum of $K(z,z)$ in D_1.

The bilinear expansion (132) of the kernel function $K(z,\zeta)$ makes it possible to compute $K(z,\zeta)$ if a complete set of functions of $L^2(D)$ is known. If the set is not orthonormal, we orthonormalize it by the Schmidt process and then set up the expansion (132). As an example, consider the case in which D is the unit circle $|z| < 1$. As seen before, the functions (121) constitute a complete orthonormal set in $|z| < 1$. Hence,

$$K(z,\zeta) = \frac{1}{\pi} \sum_{n=0}^{\infty} (n+1)(\bar{\zeta}z)^n,$$

or

$$K(z,\zeta) = \frac{1}{\pi} \frac{1}{(1 - \bar{\zeta}z)^2}. \tag{133}$$

We are now ready to prove the main result of this section.

Let D be a finite simply-connected domain with more than one boundary point and let $w = f(z) = f(z,\zeta)$ be the analytic function which maps D onto the unit circle in such a way that $f(\zeta) = 0$ ($\zeta \in D$) and $f'(\zeta) > 0$. Then

$$f'(z) = \sqrt{\frac{\pi}{K(\zeta,\zeta)}}\, K(z,\zeta), \tag{134}$$

where $K(z,\zeta)$ is the Bergman kernel function of D. $K(z,\zeta)$ can be computed in terms of an arbitrary complete orthonormal set $\{v_n\}$ of functions of $L^2(D)$ by means of the bilinear expansion (132).

Let D_ρ denote the subdomain of D which is mapped by $w = f(z)$ onto the circle $|w| < \rho$ ($\rho < 1$), and let C_p be the boundary of D_ρ. Obviously, $f(z)$ is regular at the points of C_p and C_p is an analytic curve. If $g(z)$ is an arbitrary function of $L^2(D)$, we may therefore conclude from the residue theorem that

$$\frac{g(\zeta)}{f'(\zeta)} = \frac{1}{2\pi i} \int_{C_\rho} \frac{g(z)}{f(z)}\, dz. \tag{135}$$

Indeed, $f(z)$ has its only zero within C_ρ at the point $z = \zeta$ and

$$\lim_{z \to \zeta} \frac{(z - \zeta)g(z)}{f(z)} = \frac{g(\zeta)}{f'(\zeta)}.$$

Since, on C_ρ, $|f(z)|^2 = \rho^2$, we may replace $f(z)$ by $\rho^2[\overline{f(z)}]^{-1}$ in (135). This yields

$$\frac{g(\zeta)}{f'(\zeta)} = \frac{1}{2\pi i \rho^2} \int_{C_\rho} \overline{f(z)} g(z)\, dz.$$

With the help of Green's formula (106), this can be cast into the form

$$\frac{g(\zeta)}{f'(\zeta)} = \frac{1}{\pi \rho^2} \iint_{D_\rho} \overline{f'(z)} g(z)\, dx\, dy.$$

It is now clearly permissible to let ρ tend to 1. We obtain

$$g(\zeta) = \iint_D \left[\frac{\overline{f'(z) f'(\zeta)}}{\pi} \right] g(z)\, dx\, dy.$$

This shows that the function

$$\frac{f'(z) \overline{f'(\zeta)}}{\pi}$$

has the characteristic reproducing property (128) with respect to any function $g(z)$ of $L^2(D)$. Since we have shown before that the only function with this property is the kernel function $K(z,\zeta)$, it follows therefore that

$$\frac{f'(z) \overline{f'(\zeta)}}{\pi} = K(z,\zeta).$$

For $z = \zeta$, we obtain

$$[f'(\zeta)]^2 = \pi K(\zeta,\zeta).$$

If we use this to eliminate $f'(\zeta)$ from the preceding formula, we arrive at the identity (134). This completes the proof.

As an example, consider the case in which D is the unit circle. The kernel function will then have the form (133) and it thus follows from (134) that

$$f'(z) = \frac{1 - |\zeta|^2}{(1 - \bar{\zeta} z)^2}.$$

The function $f(z)$ mapping the unit circle onto itself in such a way that $f(\zeta) = 0, f'(\zeta) > 0$ is therefore of the form

$$f(z) = \frac{z - \zeta}{1 - \bar{\zeta}z},$$

in accordance with previous results.

(132) and (134) show that the problem of mapping a simply-connected domain D conformally onto the unit circle is solved if we know a complete orthonormal set of functions of $L^2(D)$. Such a set can always be obtained by means of the Schmidt orthogonalization process if a complete set of functions of $L^2(D)$ is known. Obviously, the practical applicability of the theory developed in this section depends largely on the a priori knowledge of such a set. For a wide class of domains, this knowledge is provided by the following theorem.

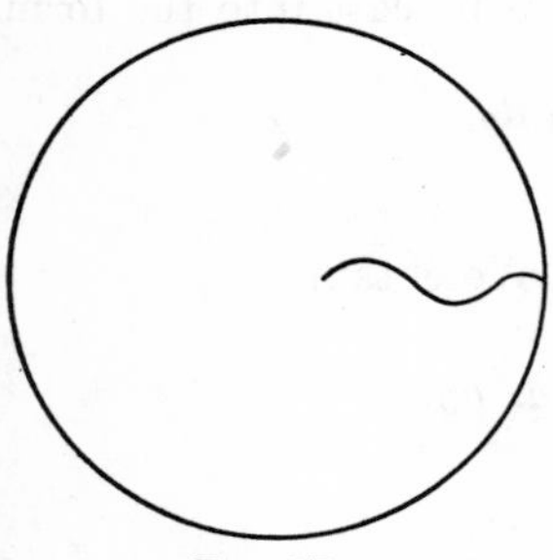

FIG. 27.

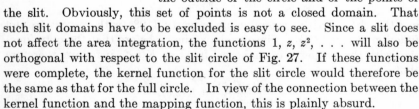

If D is a finite domain whose complement is a closed domain (i.e., the closure of a domain), then the powers

$$1, z, z^2, z^3, \ldots, z^n, \ldots$$

form a complete set in $L^2(D)$.

The condition that the complement of D should be a closed domain excludes slit domains of the type indicated in Fig. 27. The complement D' of the domain of Fig. 27 consists of the outside of the circle and of the points of the slit. Obviously, this set of points is not a closed domain. That such slit domains have to be excluded is easy to see. Since a slit does not affect the area integration, the functions $1, z, z^2, \ldots$ will also be orthogonal with respect to the slit circle of Fig. 27. If these functions were complete, the kernel function for the slit circle would therefore be the same as that for the full circle. In view of the connection between the kernel function and the mapping function, this is plainly absurd.

To prove the theorem, we orthonormalize the functions $1, z, z^2, \ldots$ by the Schmidt process and obtain a set of orthonormal polynomials $P_0(z)$, $P_1(z), P_2(z), \ldots$. With these polynomials, we set up a function $K_1(z,\zeta)$ defined by

$$K_1(z,\zeta) = \sum_{n=0}^{\infty} P_n(z)\overline{P_n(\zeta)}. \tag{136}$$

As shown before, this series converges uniformly in any closed subdomain of D and its sum is therefore regular in D [this does not depend on the completeness of the $P_n(z)$]. Our aim is to show that $K_1(z,\zeta)$ and the kernel function $K(z,\zeta)$ are identical.

The function (136) clearly has the reproducing property (128) with respect to such functions $f(z)$ which are either polynomials or can be approximated by polynomials. In particular, we have

$$\zeta^n = \iint_D \overline{K_1(z,\zeta)} z^n \, dx \, dy. \tag{137}$$

We shall show that, as a consequence of (137), we also have

$$\frac{1}{\eta - \zeta} = \iint_D \overline{K_1(z,\zeta)} \left(\frac{1}{\eta - z}\right) dx \, dy, \tag{138}$$

if η is a point which is not in D. Suppose first that η is situated in the outside of a circle about the origin which contains D; since D is finite, such a circle exists. The expansion

$$\frac{1}{\eta - z} = \frac{1}{\eta} + \frac{z}{\eta^2} + \frac{z^2}{\eta^3} + \cdots$$

will then converge uniformly for all z in D. Hence

$$\iint_D \overline{K_1(z,\zeta)} \left(\frac{1}{\eta - z}\right) dx \, dy = \sum_{n=0}^{\infty} \eta^{-n-1} \iint_D \overline{K_1(z,\zeta)} \, z^n dx \, dy.$$

By (137), this is equal to

$$\sum_{n=0}^{\infty} \eta^{-n-1}\zeta^n = \frac{1}{\eta - \zeta}.$$

This shows that (138) is true for values η outside the above-mentioned circle. Now both sides of the relation (138) are clearly regular analytic functions of η in the complement of D. In view of the principle of permanence, the identity (138) persists therefore for all such values of η.

Let now $f(z)$ be an otherwise arbitrary function which is regular in the closure of D. By the definition of regularity at a point, $f(z)$ will then be regular in a domain D^* which includes both D and its boundary. We now draw a closed contour C which is contained in D^* but has no points in common with either D or its boundary. If ζ is a point of D, we have

$$f(\zeta) = \frac{1}{2\pi i} \int_C \frac{f(\eta)}{\eta - \zeta} \, d\eta.$$

Hence, by (138),

$$f(\zeta) = \frac{1}{2\pi i} \int_C f(\eta) \left[\iint_D \overline{K_1(z,\zeta)} \left(\frac{1}{\eta - z} \right) dx\, dy \right] d\eta.$$

Changing the order of integration, we obtain

$$f(\zeta) = \iint_D \overline{K_1(z,\zeta)} \left[\frac{1}{2\pi i} \int_C \frac{f(\eta)}{\eta - z}\, d\eta \right] dx\, dy,$$

whence

$$f(\zeta) = \iint_D \overline{K_1(z,\zeta)} f(z)\, dx\, dy.$$

This shows that $\overline{K_1(z,\zeta)}$ has the reproducing property (128) with respect to functions $f(z)$ which are regular in the closure of D, and therefore also with respect to such functions of $L^2(D)$ which can be approximated by functions regular in the closure of D. If we can show that any function of $L^2(D)$ can be approximated by functions regular in the closure of D, it will thus follow that $K_1(z,\zeta)$ has the reproducing property (128) with respect to all functions of $L^2(D)$. Since this property is characteristic of the kernel function $K(z,\zeta)$, the identity of $K(z,\zeta)$ and $K_1(z,\zeta)$ will then be established. The reader will also confirm without difficulty that this identity is equivalent to the fact that the functions $P_n(z)$ of (136), and therefore also the powers of z, form a complete set in $L^2(D)$.

To show that any function of $L^2(D)$, or, more generally, any function $f(z)$ which is regular in D, can be approximated by functions regular in the closure of D, we proceed as follows: Since the complement of D is a closed domain, it is possible to find a sequence of domains $D_1, D_2, \ldots$ such that $D_1 \supset D_2 \supset D_3 \supset \cdots$ and $\lim_{n\to\infty} D_n = D$, where the boundary of D has no point in common with the boundaries of the domains D_n. If $w = p_n(z)$ denotes the function, whose existence is guaranteed by the Riemann mapping theorem, which maps D_n onto D in such a way that $p_n(\zeta) = \zeta$, $p_n'(\zeta) > 0$, then $\lim_{n\to\infty} p_n(z) = z$. Clearly, $p_n(z)$ is regular in the closure $D + C$ of D and maps $D + C$ onto a closed domain which is entirely contained in D. Hence, the function $f_n(z) = f[p_n(z)]$ is regular in $D + C$. Since, for $n \to \infty$, we have $p_n(z) \to z$, we also have $f_n(z) \to f(z)$, which proves that $f_n(z)$ can be approximated by functions regular in $D + C$. This completes the proof.

Summing up our results, we have thus arrived at the following procedure for the construction of the mapping function.

Let D be a simply-connected domain whose complement is a closed domain and let $P_0(z)$, $P_1(z)$, $P_2(z)$, . . . be the polynomials obtained by orthonor-

malizing the powers 1, z, z^2, . . . *by the Schmidt process with respect to* D. *The series*

$$K(z,\zeta) = \sum_{n=0}^{\infty} P_n(z)\overline{P_n(\zeta)}, \qquad \zeta \in D,$$

converges absolutely and uniformly in any closed subdomain of D *and represents there a regular function. If* $w = f(z)$ *is the function mapping* D *onto the unit circle such that* $f(\zeta) = 0$, $f'(\zeta) > 0$, *then*

$$f'(z) = \sqrt{\frac{\pi}{K(\zeta,\zeta)}}\, K(z,\zeta).$$

In a practical application, the series for $K(z,\zeta)$ has of course to be broken off after a finite number of steps. If the last orthonormal polynomial taken is $P_n(z)$, we thus obtain a polynomial

$$K_n(z,\zeta) = \sum_{\nu=0}^{n} P_\nu(z)\overline{P_\nu(\zeta)}$$

of order n which approximates $K(z,\zeta)$. $K_n(z,\zeta)$ can also be characterized as the polynomial of order n which yields the best approximation in the mean of $K(z,\zeta)$. Indeed, since $P_\nu(z)$ is a polynomial of degree ν, any polynomial $R_n(z)$ of degree n can be written in the form

$$R_n(z) = \sum_{\nu=0}^{n} a_\nu P_\nu(z).$$

In view of (128) and the orthonormality of the $P_\nu(z)$, we have

$$\begin{aligned}
\iint_D \Big| K(z,\zeta) - \sum_{\nu=1}^{n} a_\nu P_\nu(z) \Big|^2 dx\,dy \\
&= K(\zeta,\zeta) - 2\,\mathrm{Re}\left\{ \sum_{\nu=0}^{n} a_\nu P_\nu(\zeta)\right\} + \sum_{\nu=0}^{n} |a_\nu|^2 \\
&= K(\zeta,\zeta) - \sum_{\nu=1}^{n} |P_\nu(\zeta)|^2 + \sum_{\nu=1}^{n} |a_\nu - \overline{P_\nu(\zeta)}|^2 \\
&= \iint_D \Big| K(z,\zeta) - \sum_{\nu=1}^{n} P_n(z)\overline{P_n(\zeta)} \Big|^2 dx\,dy + \sum_{\nu=1}^{n} |a_\nu - \overline{P_\nu(\zeta)}|^2.
\end{aligned}$$

Hence,

$$\iint_D |K(z,\zeta) - R_n(z)|^2\, dx\,dy > \iint_D |K(z,\zeta) - K_n(z,\zeta)|^2\, dx\,dy,$$

unless $R_n(z)$ and $K_n(z,\zeta)$ are identical.

The actual carrying out of the Schmidt process involves two types of steps, namely, the evaluation of the integrals

$$I_{nm} = \iint_D z^n \bar{z}^m \, dx \, dy$$

and a certain amount of elementary algebraic manipulation. Unless the domain D is fairly complicated, it is generally the latter which is the more tiresome. If D is smoothly bounded, the evaluation of I_{nm} is further facilitated by the use of Green's formula (106) which transforms I_{nm} into the line integral

$$I_{nm} = \frac{1}{2i(m+1)} \int_C z^n \bar{z}^{m+1} \, dz,$$

where C is the boundary of D.

We close this section with the discussion of the case in which D is the ellipse $b^2x^2 + a^2y^2 < a^2b^2$ and in which the orthogonal polynomials $P_n(z)$ can be easily determined. We assume that the foci of the ellipse are situated at $z = \pm 1$, that is, $a^2 - b^2 = 1$; obviously, this does not restrict the generality of our procedure. The function

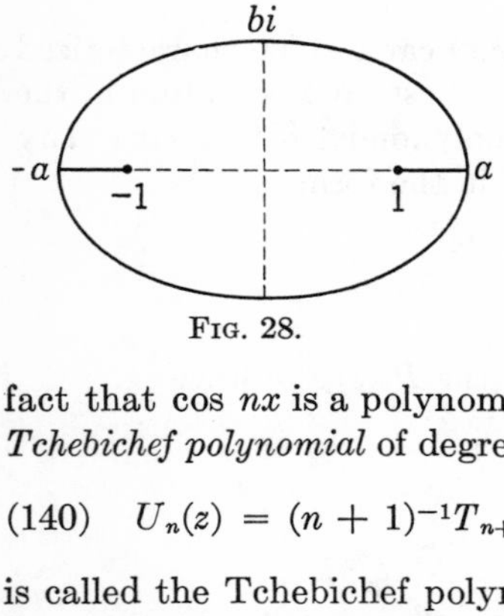

FIG. 28.

$$(139) \qquad T_n(z) = \cos(n \cos^{-1} z)$$

is a polynomial of degree n; this is an immediate consequence of the easily proved fact that $\cos nx$ is a polynomial of degree n in $\cos x$. $T_n(z)$ is called the *Tchebichef polynomial* of degree n. The polynomial of degree n

$$(140) \quad U_n(z) = (n+1)^{-1}T_{n+1}'(z) = (1-z^2)^{-\frac{1}{2}} \sin[(n+1)\cos^{-1} z]$$

is called the Tchebichef polynomial of the second kind. We shall show that the polynomials $U_n(z)$ are orthogonal to each other with respect to the above-mentioned ellipse.

To this end, we apply to the ellipse two cuts, from $-a$ to -1 and from 1 to a, respectively (see Fig. 28). Obviously, these cuts do not affect the value of the area integral

$$(141) \qquad A_{nm} = \iint_D U_n(z)\overline{U_m(z)} \, dx \, dy.$$

We now use the fact to be proved later (Sec. 2, Chap. VI) that by the conformal mapping $z = \cos w$ the cut ellipse of Fig. 28 is transformed into the rectangle $(-ci, -ci + \pi, ci + \pi, ci)$, where $a = \cosh c$, $b = \sinh c$ $(c > 0)$.

Since the Jacobian of the transformation is

$$\frac{\partial(x,y)}{\partial(u,v)} = \left|\frac{dz}{dw}\right|^2 = |1 - z^2|, \qquad w = u + iv,$$

it follows from (140) and (141) that

$$A_{nm} = \iint_R \sin(n+1)w \, \overline{\sin(m+1)w} \, du \, dv,$$

where the integration is now to be extended over the above rectangle. We now evaluate this integral. We have

$$\begin{aligned} A_{n-1,m-1} &= \int_{-c}^{c} \int_0^{\pi} \sin(nu + inv) \sin(mu - imv) \, du \, dv \\ &= \int_{-c}^{c} \cosh nv \cosh mv \, dv \int_0^{\pi} \sin nu \sin mu \, du \\ &\quad + \int_{-c}^{c} \sinh nv \sinh mv \, dv \int_0^{\pi} \cos nu \cos mu \, du \\ &\quad + i \int_{-c}^{c} \sinh nv \cosh mv \, dv \int_0^{\pi} \cos nu \sin mu \, du \\ &\quad - i \int_{-c}^{c} \cosh nv \sinh mv \, dv \int_0^{\pi} \sin nu \cos mu \, du. \end{aligned}$$

The last two integrals over v vanish because the integrands are odd; the first two trigonometric integrals obviously vanish if $n \neq m$. Hence, $A_{nm} = 0$ if $n \neq m$ and the polynomials (140) are indeed orthogonal with respect to integration over the ellipse. For $n = m$, we obtain

$$A_{nn} = \frac{\pi}{2(n+1)} \sinh 2(n+1)c,$$

or, in view of $a + b = e^c$,

$$A_{nn} = \frac{\pi}{4(n+1)} (\rho^{n+1} - \rho^{-n-1}), \qquad (a+b)^2 = \rho.$$

The orthonormalized polynomials are therefore

$$P_n(z) = 2\sqrt{\frac{n+1}{\pi}} (\rho^{n+1} - \rho^{-n-1})^{-\frac{1}{2}} U_n(z), \qquad n = 0, 1, \ldots .$$

The set $\{P_n(z)\}$ contains polynomials of all degrees and is therefore complete. Hence, the kernel function of our ellipse is of the form

$$K(z,\zeta) = \frac{4}{\pi} \sum_{n=0}^{\infty} \frac{(n+1) U_n(z) \overline{U_n(\zeta)}}{\rho^{n+1} - \rho^{-n-1}}, \qquad \rho = (a+b)^2.$$

In view of (140), this can also be written

$$K(z,\zeta) = \frac{4}{\pi} \sum_{n=0}^{\infty} \frac{T_{n+1}'(z)\overline{U_n(\zeta)}}{\rho^{n+1} - \rho^{-n-1}}$$

where $T_n(z)$ is the Tchebichef polynomial (139). If $f(z,\zeta)$ is the function mapping the ellipse onto the unit circle, satisfying $f(\zeta) = 0, f'(\zeta) > 0$, we have, in view of (134),

$$f(z,\zeta) = \sqrt{\frac{\pi}{K(\zeta,\zeta)}} \sum_{n=0}^{\infty} \frac{[T_{n+1}(z) - T_{n+1}(\zeta)]\overline{U_n(\zeta)}}{\rho^{n+1} - \rho^{-n-1}},$$

where

$$K(\zeta,\zeta) = \frac{4}{\pi} \sum_{n=0}^{\infty} \frac{(n+1)|U_n(\zeta)|^2}{\rho^{n+1} - \rho^{-n-1}}.$$

If we make the substitutions $z = \cos w$, $\zeta = \cos \eta$ and use (139) and (140), we obtain

$$f(\cos w, \cos \eta) = \frac{\pi\alpha|\sin \eta|}{2 \sin \eta} \sum_{n=0}^{\infty} \frac{[\cos (n+1)w - \cos (n+1)\eta] \sin (n+1)\eta}{\rho^{n+1} - \rho^{-n-1}},$$

where

$$\alpha^{-2} = \sum_{n=0}^{\infty} \frac{(n+1)|\sin^2 (n+1)\eta|}{\rho^{n+1} - \rho^{-n-1}} = \sum_{n=1}^{\infty} \frac{n|\sin n\eta|^2}{\rho^n - \rho^{-n}}.$$

This result becomes particularly simple if the center of the ellipse is to be mapped onto the center of the circle. In this case, we have $\zeta = 0$ and therefore $\eta = \frac{1}{2}\pi$. With $f(z,0) = f(z)$, it follows that

$$f(\cos w) = \frac{\pi\alpha}{2} \sum_{n=0}^{\infty} \frac{(-1)^n \cos (2n+1)w}{\rho^{2n+1} - \rho^{-2n-1}}, \tag{142}$$

where

$$\alpha^{-2} = \sum_{n=0}^{\infty} \frac{2n+1}{\rho^{2n+1} - \rho^{-2n-1}}, \qquad \rho = (a+b)^2. \tag{142'}$$

EXERCISES

1. Let $f(z)$ be regular and single-valued in a domain D (not necessarily simply-connected) and on the closed contour or contours C by which D is bounded. If ζ is a point of D, show that the integral

$$I = \iint_D \frac{f'(z)}{z - \zeta}\, dx\, dy$$

exists and that

$$I = \frac{1}{2i} \int_C \frac{\overline{f(z)}}{z - \zeta} \, dz - \pi \overline{f(\zeta)}.$$

Hint: Apply Green's formula (106) to the domain obtained by deleting from D a small circle of center ζ.

2. If $\{v_n(z)\}$ is a complete orthonormal set in $L^2(D)$, show that any function $f(z)$ of $L^2(D)$ can be expanded into a series

$$f(z) = \sum_{n=1}^{\infty} a_n v_n(z), \qquad a_n = (f, v_n),$$

which converges absolutely and uniformly in any closed subdomain of D. *Hint:* Consider the mth remainder of the series and apply the inequality

$$\left| \sum_n p_n q_n \right|^2 \leq \sum_n |p_n|^2 \sum_n |q_n|^2.$$

3. If $f(z)$ is a function of $L^2(D)$, use the reproducing property (128) of the kernel function $K(z,\zeta)$ to show that

$$|f(\zeta)|^2 \leq K(\zeta,\zeta) \iint_D |f(z)|^2 \, dx \, dy.$$

Use this result to show that the function

$$f_0(z) = \frac{K(z,\zeta)}{K(\zeta,\zeta)}$$

has the extremal property (125).

4. Let $f(z)$ be regular in $|z| < 1$ and let $M(f)$ be the mean value defined by

$$M^2(f) = \frac{1}{\pi} \iint_{|z|<1} |f(z)|^2 \, dx \, dy.$$

If $M(f) \leq 1$, show that

$$|f(z)| \leq \frac{1}{(1 - |z|^2)^2}, \qquad |z| < 1.$$

5. Let D_1 be a domain contained in D and let $K_1(z,\zeta)$ and $K(z,\zeta)$ denote the kernel functions of D_1 and D, respectively. If ζ is a point of D_1, show that

$$K(\zeta,\zeta) \leq K_1(\zeta,\zeta).$$

Hint: Consider the extremal problem $(f,f) = \text{min.}$, $f(\zeta) = 1$ in both the domains D and D_1, and use the fact that the solution of a minimum problem cannot become larger if the class of competing functions is enlarged.

6. Show that the Hermitian form

$$H(x,x) = \sum_{\nu=1}^{n} \sum_{\mu=1}^{n} \bar{x}_\nu x_\mu K(\zeta_\nu,\zeta_\mu), \qquad \zeta_\nu \in D, \ \nu = 1, \ldots, n,$$

is positive-definite. *Hint:* Consider the integral

$$\iint_D \left| \sum_{\nu=1}^{n} x_\nu K(z_\nu, \zeta_\nu) \right|^2 dx \, dy.$$

7. Let D be a simply-connected domain which contains the circle $|z| < r$ and is contained in the circle $|z| < R$. If $K(z,\zeta)$ is the kernel function of D and $|\zeta| < r$, show that

$$\frac{R^2}{(R^2 - |\zeta|^2)^2} \leq \pi K(\zeta,\zeta) \leq \frac{r^2}{(r^2 - |\zeta|^2)^2}.$$

Hint: Use the result of Exercise 5.

8. Let $w = f(z)$ be regular in a simply-connected domain D and let A be the area of the domain onto which D is mapped by $w = f(z)$, that is,

$$A = \iint_D |f'(z)|^2 \, dx \, dy.$$

If $f'(\zeta) = 1$, where ζ is a point of D, show that A takes its smallest possible value for the function $w = f_0(z)$ $[f_0'(\zeta) = 1]$ which maps D onto the interior of a circle about the origin. *Hint:* If D' is the domain onto which D is mapped by the extremal function $w = f_0(z)$, show that D' is independent (except for a magnification) of the domain D, and conclude that it is sufficient to solve the above problem in the case in which D is the unit circle. In the latter case, use the relation

$$\iint_{|z|<1} |f'(z)|^2 \, dx \, dy = \pi \sum_{n=1}^{\infty} n|b_n|^2,$$

valid for a function

$$f(z) = \sum_{n=0}^{\infty} b_n z^n$$

which is regular in $|z| < 1$.

9. Use the result of the preceding exercise to identify the kernel function $K(z,\zeta)$ of a simply-connected domain D with a constant multiple of $f'(z)$, where $w = f(z)$ maps D onto the unit circle and $f(\zeta) = 0$.

10. Prove that

$$|K(\zeta,\eta)|^2 \leq K(\zeta,\zeta)K(\eta,\eta).$$

11. Let ζ and η be two distinct points of a domain D and let $K_1(z,\zeta)$ be defined by

$$K_1(z,\zeta) = K(z,\zeta) - \alpha K(z,\eta),$$

where

$$\alpha = \frac{K(\eta,\zeta)}{K(\eta,\eta)}.$$

Show that $K_1(z,\zeta)$ has the reproducing property

$$f(\zeta) = \iint_D \overline{K_1(z,\zeta)} f(z) \, dx \, dy$$

with respect to all functions $f(z)$ of $L^2(D)$, including $K_1(z,\zeta)$ itself, for which $f(\eta) = 0$. Deduce that the function

$$f_0(z) = \frac{K_1(z,\zeta)}{K_1(\zeta,\zeta)}$$

solves the extremal problem

$$\iint_D |f(z)|^2 \, dx \, dy = \text{min.}, \qquad f(\zeta) = 1, \qquad f(\eta) = 0, \qquad f(z) \in L^2(D),$$

and show that the value of the minimum is $[K_1(\zeta,\zeta)]^{-1}$.

12. If $u(x,y)$ is a real harmonic function and if $\partial/\partial z$ and $\partial/\partial \bar{z}$ are the differential operators defined in (103), show that $\partial u/\partial z$ is an analytic function of z and that $\partial u/\partial \bar{z}$ is the conjugate of an analytic function of z.

13. If $v(x,y)$ is a complex-valued function of x and y which possesses continuous second derivatives, show that

$$4 \frac{\partial^2 v}{\partial z \, \partial \bar{z}} = \Delta v,$$

where Δ is the Laplace operator.

14. Let $f(z)$ be regular in the closure of a smoothly bounded domain D, and let the function $\tau(z)$ be defined by

$$\tau(z) = \iint_D \overline{f'(\zeta)} \log |z - \zeta| d\xi \, d\eta, \qquad \zeta = \xi + i\eta$$

Show that $\tau(z)$ satisfies the differential equation

$$\Delta \tau(z) = 4\pi \overline{f'(z)}.$$

Hint: Use the results of Exercises 1 and 12.

11. Conformal Mapping of Nearly Circular Domains. We end this chapter with a few remarks on the conformal mapping of nearly circular domains. To be specific, let D be a simply-connected domain in the z-plane whose boundary has the polar equation $r = 1 + \epsilon p(\theta)$, where $p(\theta)$ is bounded and piecewise continuous and ϵ is a small positive parameter. Our aim is to find an expression for the analytic function $f(z)$ mapping D onto the unit circle, which is correct up to quantities of the first order of magnitude in ϵ. In other words, we wish to represent $f(z)$ in the form

$$f(z) = f_0(z) + \epsilon p(z) + \epsilon q(z),$$

where $f_0(z)$ and $p(z)$ are independent of ϵ and $q(z) \to 0$ if $\epsilon \to 0$. This may also be expressed in the form

$$f(z) = f_0(z) + \epsilon p(z) + o(\epsilon).$$

If ϵ is very small, then the error committed by neglecting the term $o(\epsilon)$ becomes insignificant and the expression $f_0(z) + \epsilon p(z)$ yields a good approximation for the mapping function $f(z)$. We add that the conformal mapping of nearly circular domains is of importance in a number of practical applications. What is required in these cases is to find the mapping function of a simply-connected domain Δ^* if the mapping function of a domain Δ, which is very close to Δ^*, is known. Since the domains Δ considered in practical applications are generally bounded by analytic arcs, on which, in view of the symmetry principle, the mapping function is regular, the function mapping Δ onto the unit circle will map Δ^* onto a nearly circular domain. Hence, it is sufficient to consider domains of the latter type.

The problem of mapping a nearly circular domain D onto the unit circle is easily solved by means of the Hadamard variation formula for the Green's function of a domain, developed in Sec. 11, Chap. I. According to the last formula of Sec. 11, Chap. I, the Green's function $g(z,0)$ of the nearly circular domain D considered above is

$$g(z,0) = -\log|z| + \frac{\epsilon}{2\pi}\int_0^{2\pi}\frac{(1-\rho^2)p(\theta)\,d\theta}{1-2\rho\cos(\theta-\varphi)+\rho^2} + o(\epsilon),$$

where $z = \rho e^{i\varphi}$. In view of

$$\text{Re}\left\{\frac{e^{i\theta}+z}{e^{i\theta}-z}\right\} = \text{Re}\left\{\frac{e^{i\theta}+\rho e^{i\varphi}}{e^{i\theta}-\rho e^{i\varphi}}\right\} = \frac{1-\rho^2}{1-2\rho\cos(\theta-\varphi)+\rho^2}$$

and $\log|z| = \text{Re}\,\{\log z\}$, this can also be written

$$(143)\qquad g(z,0) = \text{Re}\left\{-\log z + \frac{\epsilon}{2\pi}\int_0^{2\pi}\frac{e^{i\theta}+z}{e^{i\theta}-z}\,p(\theta)\,d\theta\right\} + o(\epsilon).$$

As shown in Sec. 4 of this chapter, the Green's function $g(z,0)$ of D and the analytic function $f(z)$ mapping D onto the unit circle and satisfying $f(0) = 0$ are related by the identity

$$(144)\qquad g(z,0) = -\log|f(z)| = -\text{Re}\,\{\log f(z)\}.$$

Now the expression enclosed in braces in (143) is an analytic function of z in D. Since the real parts of two analytic functions can be equal throughout a domain only if the difference of the two functions is an imaginary constant iC, it follows therefore from (143) and (144) that

$$\log f(z) = \log z - \frac{\epsilon}{2\pi}\int_0^{2\pi}\frac{e^{i\theta}+z}{e^{i\theta}-z}\,p(\theta)\,d\theta + iC + o(\epsilon),$$

where $o(\epsilon)$ is an analytic function of z such that $\epsilon^{-1}o(\epsilon) \to 0$ if $\epsilon \to 0$. Taking exponentials and observing that

$$e^{-\epsilon q} = 1 - \epsilon q + \frac{\epsilon^2}{2}q^2 - \cdots = 1 - \epsilon q + o(\epsilon),$$

we obtain, for $C = 0$,

$$(145)\qquad f(z) = z - \frac{\epsilon z}{2\pi}\int_0^{2\pi}\frac{e^{i\theta}+z}{e^{i\theta}-z}\,p(\theta)\,d\theta + o(\epsilon).$$

The function (145) *yields the conformal mapping onto the unit circle of the nearly circular domain D whose boundary has the polar equation $r = 1 + \epsilon p(\theta)$, where $p(\theta)$ is bounded and piecewise continuous and ϵ is a small positive parameter.* If $C \neq 0$, there appears a factor e^{iC} which causes a rotation of the unit circle about the origin.

If it is desired to find the function $z = F(w)$ inverse to $w = f(z)$, we proceed as follows. Clearly, $f(w)$ must be of the form

$$F(w) = w + \epsilon q(w) + o(\epsilon).$$

Hence

$$\begin{aligned} w = f(z) &= f[F(w)] = f[w + \epsilon q(w) + o(\epsilon)] \\ &= f(w) + \epsilon q(w)f'(w) + o(\epsilon). \end{aligned}$$

Using, for a moment, the abbreviation $f(z) = z - \epsilon t(z) + o(\epsilon)$ for (145), we thus find that

$$\begin{aligned} w &= w - \epsilon t(w) + \epsilon q(w)[1 - \epsilon t'(w) + o(\epsilon)] + o(\epsilon) \\ &= w - \epsilon[t(w) - q(w)] + o(\epsilon), \end{aligned}$$

whence

$$q(w) = t(w) + \epsilon^{-1}o(\epsilon).$$

Since $q(w)$ and $t(w)$ are independent of ϵ, this means that $q(w) = t(w)$. Using the definition of $t(w)$ and replacing the variable w by z, we thus have the following result.

The function

$$F(z) = z + \frac{\epsilon z}{2\pi} \int_0^{2\pi} \frac{e^{i\theta} + z}{e^{i\theta} - z} p(\theta)\, d\theta + o(\epsilon) \tag{146}$$

maps $|z| < 1$ *onto the nearly circular domain whose boundary has the polar equation* $r = 1 + \epsilon p(\theta)$, *where* $p(\theta)$ *is bounded and piecewise continuous and* ϵ *is a small positive parameter.*

EXERCISES

1. If the function $p(\theta)$ in (146) has the form

$$p(\theta) = a_0 + \sum_{n=1}^{m} (a_n \cos n\theta + b_n \sin n\theta),$$

show that the associated mapping function is

$$f(z) = z + \epsilon z \left[a_0 + \sum_{n=1}^{m} (a_n - ib_n)z^n \right] + o(\epsilon).$$

Hint: Substitute ζ for $e^{i\theta}$, and evaluate the integral by means of the residue theorem.

2. Show that the function mapping $|z| < 1$ onto the nearly circular ellipse $b^2u^2 + a^2v^2 < a^2b^2$, where $b = 1$, $a = 1 + \epsilon$, is of the form

$$f(z) = z + \frac{\epsilon}{2} z(1 + z^2) + o(\epsilon).$$

CHAPTER VI

MAPPING PROPERTIES OF SPECIAL FUNCTIONS

1. The Rational Function of the Second Degree. The most general rational function of the second degree is of the form

$$w = f(z) = \frac{az^2 + bz + c}{a'z^2 + b'z + c'}. \tag{1}$$

Since $f(z)$ remains unaltered if both the numerator and the denominator are multiplied by the same constant, $f(z)$ depends on only five arbitrary constants. In view of

$$f(z) - w_0 = \frac{az^2 + bz + c - w_0(a'z^2 + b'z + c')}{a'z^2 + b'z + c'},$$

the equation $f(z) - w_0 = 0$ is of the second degree in z, which shows that every value w_0 is taken by $w = f(z)$ exactly twice. It follows that $w = f(z)$ maps the full z-plane onto the doubly covered w-plane. We may also say that $w = f(z)$ maps the z-plane onto a two-sheeted Riemann surface R and that both sheets of R cover the entire w-plane. The branch points of R, that is, those points w which are common to both sheets of R, correspond to those points z at which either $f'(z) = 0$ or $f(z)$ has a double pole. The reader will easily verify from (1) that there are precisely two such branch points.

We shall now show that the general transformation (1) can be built up from two linear substitutions and the transformation $z \to z^2$. To this end, consider first the case in which the two branch points of the Riemann surface associated with (1) are situated at $w = 0$ and $w = \infty$. If $f(\alpha) = 0$ and $f(\beta) = \infty$, the expansions of $f(z)$ near $z = \alpha$ and $z = \beta$ will then be of the forms

$$f(z) = \gamma_2(z - \alpha)^2 + \gamma_3(z - \alpha)^3 + \cdots, \qquad \gamma_2 \neq 0,$$

and

$$f(z) = \frac{\delta_{-2}}{(z - \beta)^2} + \frac{\delta_{-1}}{(z - \beta)} + \delta_0 + \delta_1(z - \beta) + \cdots, \qquad \delta_{-2} \neq 0.$$

Since $f(z)$ takes no value more than twice, $f(z)$ will be finite and different from zero at all points other than α and β. The function $g(z) = \sqrt{f(z)}$ can therefore have singularities only at the points α and β. However,

$g(z)$ is clearly regular at $z = \alpha$ and has a simple pole at $z = \beta$. Indeed, in view of the above expansions, we have

$$g(z) = \sqrt{f(z)} = (z - \alpha)\sqrt{\gamma_2}\left[1 + \frac{\gamma_3}{2\gamma_2}(z - \alpha) + \cdots\right]$$

and

$$g(z) = \sqrt{f(z)} = \frac{\sqrt{\delta_{-2}}}{z - \beta}\left[1 + \frac{\delta_{-1}}{2\delta_{-2}}(z - \beta) + \cdots\right]$$

near $z = \alpha$ and $z = \beta$, respectively. The only singularity of $g(z)$ in the entire z-plane is therefore a simple pole at $z = \beta$. Hence, $g(z)$ must be a linear substitution of the form

$$g(z) = \frac{Az + B}{Cz + D}.$$

Since $f(z) = [g(z)]^2$, it follows that

$$w = f(z) = \left(\frac{Az + B}{Cz + D}\right)^2. \tag{2}$$

The case of the general transformation (1) can be reduced to this special case by an additional linear substitution which transforms the two branch points of the Riemann surface associated with (1) into the points 0 and ∞. We thus have the following result.

The transformation

$$w = \frac{az^2 + bz + c}{a'z^2 + b'z + c'}$$

can be decomposed into three successive transformations of the type

$$z_1 = \frac{Az + B}{Cz + D},$$
$$z_2 = z_1^2,$$
$$w = \frac{A'z_2 + B'}{C'z_2 + D'}.$$

Another important decomposition of the transformation (1) can be obtained in the case in which both branch points of the Riemann surface of $f(z)$ are situated at finite points, say at $w = \alpha$ and $w = \beta$. By the transformation

$$w_1 = g(z) = \frac{2f(z) - \alpha - \beta}{\alpha - \beta}$$

we then obtain a function $g(z)$ which maps the z-plane onto a two-sheeted Riemann surface with branch points at the points $w_1 = \pm 1$. By the same reasoning as above, the function

$$h(z) = g(z) + \sqrt{g^2(z) - 1}$$

will therefore be regular at the points z at which $g(z) = \pm 1$. The only other possible singularities of $h(z)$ are the points z_1 and z_2 at which $g(z)$ has simple poles. At these points, $h(z)$ is either regular or it has a simple pole. Indeed, if

$$g(z) = \frac{\eta}{z - z_1} + a_0 + a_1(z - z_1) + \cdots,$$

then

$$\begin{aligned} h(z) &= g(z) + \sqrt{g^2(z) - 1} \\ &= \frac{\eta}{z - z_1} + a_0 + a_1(z - z_1) + \cdots \\ &\qquad + \frac{\eta}{z - z_1}\sqrt{1 + \frac{2a_0}{\eta}(z - z_1) + \cdots} \\ &= \frac{\eta}{z - z_1} + a_0 + \cdots \pm \left(\frac{\eta}{z - z_1} + a_0 + \cdots\right). \end{aligned}$$

Hence, there corresponds a simple pole of $h(z)$ to the positive branch of the square root, while the negative branch of the square root leads to a regular point of $h(z)$. We thus find that the function $h(z)$ is regular for all values of z, with the possible exception of the points z_1 and z_2 at which $h(z)$ may have simple poles. It is clear that $h(z)$ must have at least one pole as otherwise it would reduce to a constant; $h(z)$ is therefore a rational function of either the first or the second degree. To determine the degree of $h(z)$, it is sufficient to find the number of times an arbitrary value w_0 is taken by $h(z)$. For $w_0 = 1$, we obtain $g + \sqrt{g^2 - 1} = 1$, which is possible only for $g(z) = 1$. But the equation $g(z) = 1$ has only one solution, say z_0, since $g'(z_0) = 0$ and the two solutions of the equation $g(z) = w_0$ coincide in this case. Hence, $h(z)$ is a rational function of the first degree, *i.e.*, a linear substitution. Since $w_1 = w + \sqrt{w^2 - 1}$ entails

$$w = \frac{1}{2}\left(w_1 + \frac{1}{w_1}\right),$$

we thus have the following result.

The transformation

$$w = \frac{az^2 + bz + c}{a'z^2 + b'z + c'}$$

can be decomposed either into the three successive transformations

$$z_1 = \frac{Az + B}{Cz + D},$$

$$z_2 = \frac{1}{2}\left(z_1 + \frac{1}{z_1}\right),$$

$$w = A'z_2 + B',$$

or into the two successive transformations

$$z_1 = \frac{Az + B}{Cz + D},$$
$$w = z_1^2 + A'.$$

The latter possibility corresponds to the case in which one of the two branch points of the associated Riemann surface is situated at infinity. If A' is the location of the second branch point, then $f(z) - A'$ has its branch points at zero and infinity and must therefore be of the form (2).

Apart from a trivial translation and rotation, any transformation (1) can thus be decomposed into one linear substitution and either the transformation $w = z^2$ or the transformation $w = \frac{1}{2}(z + z^{-1})$. Since the mapping properties of linear substitutions are very well known, the study of the transformation reduces therefore essentially to that of the two transformations

$$w = z^2 \tag{3}$$

and

$$w = \frac{1}{2}\left(z + \frac{1}{z}\right). \tag{4}$$

The mapping properties of the transformation (3) are easily obtained. If $w = u + iv$, then (3) is equivalent to

$$u = x^2 - y^2, \qquad v = 2xy.$$

Given any curve in the z-plane or w-plane we thus can immediately determine its conformal image in the w-plane or z-plane, respectively. Taking, for instance, the straight lines $u =$ const. and $v =$ const., we find that their conformal images are the equilateral hyperbolas $x^2 - y^2 =$ const. and $2xy =$ const. It is clear that these two families of equilateral hyperbolas must be orthogonal to each other; indeed, the lines $u =$ const. and $v =$ const. are orthogonal to each other, and the mapping is conformal. If, on the other hand, we set $x = c$ or $y = c$, we obtain

$$v^2 = 4c^2(c^2 - u) \qquad \text{or} \qquad v^2 = 4c^2(c^2 + u).$$

The conformal images of the straight lines $x =$ const. and $y =$ const. are therefore two mutually orthogonal families of confocal parabolas.

It is also of interest to consider the curves in the z-plane which correspond to circles in the w-plane. If α ($\alpha \neq 0$) is the center of a circle and R its radius, then the conformal image of this circle by means of the transformation (3) is $|z^2 - \alpha| = R$. With $\alpha = \beta^2$, this can be written $|z - \beta||z + \beta| = R$. Hence, the images of circles are the loci of points whose distances from two fixed points have a constant product. These curves are known as *Cassinians*. The reader will readily verify that for $R > |\beta|^2$ we obtain one closed curve, while for $R < |\beta|^2$ the Cassinian splits into two separate closed curves. For $R = |\beta|^2$, we obtain the ordinary lemniscate; indeed, with $\beta = \rho e^{i\gamma}$, we have

$$|z^2 - \beta^2| = |r^2 e^{2i\theta} - \rho^2 e^{2i\gamma}| = \rho^2,$$

whence $r^4 - 2r^2\rho^2 \cos 2(\theta - \gamma) + \rho^4 = \rho^4$, and therefore

$$r^2 = 2\rho^2 \cos 2(\theta - \gamma).$$

The orthogonal trajectories of a family of confocal Cassinians (*i.e.*, belonging to the same β) must be the conformal images of the orthogonal trajectories of the corresponding family of concentric circles. Since the latter trajectories are the rays $\arg \{w - \beta^2\} = \text{const.}$, the desired curves are found to have the equations

$$\frac{2xy - \operatorname{Im}\{\beta^2\}}{x^2 - y^2 - \operatorname{Re}\{\beta^2\}} = C.$$

The reader will verify that this is a family of equilateral hyperbolas.

We now turn to the transformation (4). Setting $z = re^{i\theta}$, $w = u + iv$, we obtain

$$(5) \qquad u = \frac{1}{2}\left(r + \frac{1}{r}\right)\cos\theta, \qquad v = \frac{1}{2}\left(r - \frac{1}{r}\right)\sin\theta,$$

whence

$$\frac{u^2}{\left[\frac{1}{2}\left(r + \frac{1}{r}\right)\right]^2} + \frac{v^2}{\left[\frac{1}{2}\left(r - \frac{1}{r}\right)\right]^2} = 1.$$

This shows that the transformation (4) maps the circles $r = \text{const.}$ onto the ellipses of semiaxes $\frac{1}{2}(r + r^{-1})$ and $\frac{1}{2}|r - r^{-1}|$. In view of

$$\frac{1}{4}\left(r + \frac{1}{r}\right)^2 - \frac{1}{4}\left(r - \frac{1}{r}\right)^2 = 1,$$

these ellipses have the foci ± 1. The circles $r = c$ and $r = c^{-1}$ obviously yield the same ellipse, and for $r = 1$ the ellipse degenerates into the linear segment connecting $w = 1$ and $w = -1$ (see Fig. 29). Figure 29 also indicates the family of confocal hyperbolas which is orthogonal to these ellipses. Being the conformal images of the rays $\theta = \text{const.}$, these hyperbolas have, by (5), the equations

$$\frac{u^2}{\cos^2\theta} - \frac{v^2}{\sin^2\theta} = 1.$$

The transformation (4) maps both the interior and the exterior of the unit circle onto the full w-plane which has a rectilinear slit from $w = -1$ to $w = 1$.

The transformation (4) is of importance in certain aerodynamical applications. This is due to the fact that (4) maps the outside of certain circles onto the outside of curves which have the general character of airplane wing profiles. Consider a circle C in the z-plane which passes through the point $z = 1$ and contains the point $z = -1$ in its interior. Since the derivative $w'(z)$ vanishes at $z = 1$, $z = 1$ is a critical point of the transformation and the angles whose vertices are at $z = 1$ are doubled. Hence, if D denotes the exterior of C and D^* is the conformal map of D, the boundary of D^* will have an angle of 2π at the point corresponding to $z = 1$. If C is symmetrical with respect to the real axis, C contains the circle $|z| = 1$. Since the latter is mapped onto the w-plane with a slit from $w = -1$ to $w = 1$, the map of the domain D bounded by such a circle C which is fairly close to $|z| = 1$ will therefore be of the type indicated in Fig. 30. If C is not symmetrical with respect to the real axis, more general profiles can be obtained in this fashion. An example is indicated in Fig. 31. The shapes obtainable in

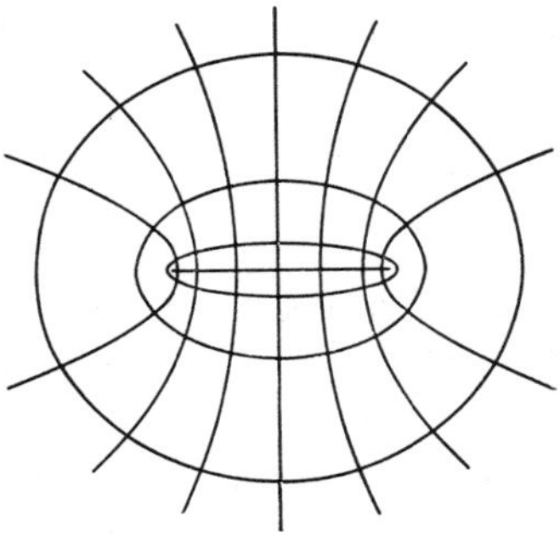

FIG. 29.

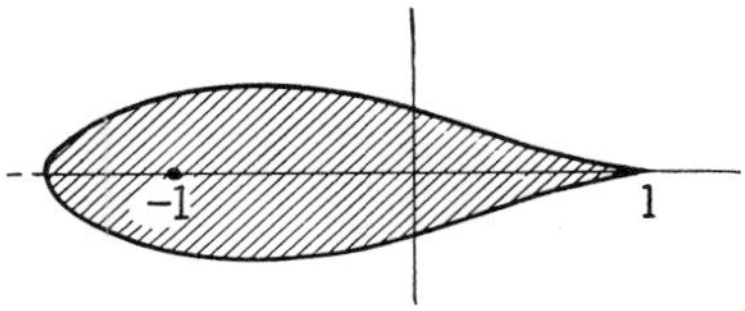

FIG. 30.

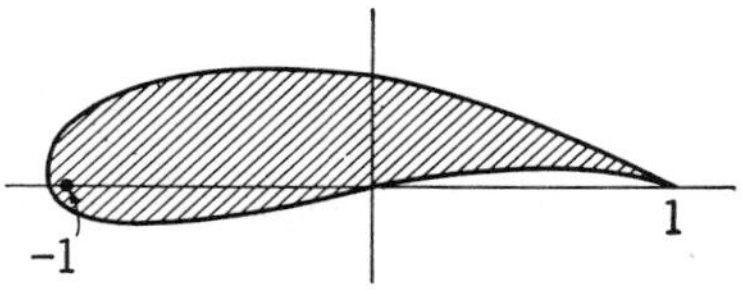

FIG. 31.

this way are known as *Joukowski profiles.* If the circle C is replaced by a nearly circular curve, the transformation (4) will yield a shape which is close to a Joukowski profile. With the help of the mappings of Sec. 11, Chap. V, it is therefore possible to map a circle onto a wide variety of wing profiles.

EXERCISES

1. Show that the transformation (3) maps the straight line $Au + Bv = C$ (A,B,C real, $A \neq 0$, $w = u + iv$) onto an equilateral hyperbola whose center is at the origin and whose half axis is $CA^{-2}\sqrt{A^2 + B^2}$.

2. Show that the transformation (3) maps the circle $|z - a| < |a|$ onto the interior of the cardioid whose polar equation is

$$\rho = 2|a|[1 + \cos(\theta + \theta_0)], \qquad \theta_0 = \arg\{a\}.$$

3. Show that the transformation

$$w = \frac{4a}{(z+1)^2}, \qquad a > 0,$$

maps the unit circle $|z| < 1$ onto the exterior of the parabola $v^2 = 4a(a - u)$, where the exterior of a parabola is defined as that region which does not contain the focus.

4. Show that the transformation

$$w = \frac{z^2 - c^2 + 2cz}{z^2 - c^2 - 2cz}, \qquad c > 0,$$

maps the half circle $|z| < c$, Re $\{z\} > 0$ onto the unit circle $|w| < 1$.

5. Show that the transformation

$$w = \sqrt{1 - z^2}$$

maps the hyperbola $2x^2 - 2y^2 = 1$ onto itself.

6. Show that the mapping

$$w = z\sqrt{1 - a^2} + a\sqrt{1 - z^2}, \qquad 0 \le a \le 1,$$

transforms any ellipse of foci ± 1 into itself.

7. Show that the mapping

$$w = \sqrt{1 - z^2}$$

transforms the hyperbola

$$\frac{x^2}{\cos^2\theta} - \frac{y^2}{\sin^2\theta} = 1, \qquad 0 < \theta < \frac{\pi}{2},$$

into the hyperbola

$$\frac{x^2}{\sin^2\theta} - \frac{y^2}{\cos^2\theta} = 1.$$

8. Show that the transformation

$$w = \frac{z}{\sqrt{z^2 + 1}}$$

maps the upper half-plane Im $\{z\} > 0$ which is cut along the imaginary axis from $z = i$ to $z = i\infty$ onto the upper half-plane Im $\{w\} > 0$.

9. Show that the transformation

$$w = \sqrt{\frac{z-1}{z+1}}$$

maps the entire z-plane which is cut along the rays $-\infty \le z \le -1$ and $1 \le z \le \infty$ onto the upper half-plane Im $\{w\} > 0$.

10. Show that there exists a one-parameter family of circles which are mapped by the transformation (4) onto doubly covered circles. *Hint:* Use the fact that the transformation (4) can be written in the form

$$\frac{w+1}{w-1} = \left(\frac{z+1}{z-1}\right)^2.$$

11. A lemniscate of degree n is defined as the locus of those points whose distances from n given points (the "foci" of the lemniscate) form a constant product. If

$$w = a_0 + a_1 z + a_2 z^2 + \cdots + a_n z^n, \qquad a_n \neq 0,$$

show that the curves in the z-plane corresponding to circles in the w-plane are lemniscates of degree n.

12. Let T be an orthogonal trajectory of a family of confocal lemniscates whose foci are at the points $P_1, \ldots, P_n$ and let θ_k denote the angle between the positive axis and the linear segment PP_k, where P is a point of T. Show that

$$\theta_1 + \theta_2 + \cdots + \theta_n = \text{const.},$$

if P describes T.

2. Exponential and Trigonometric Functions. The fundamental mapping properties of the transformation

$$w = e^z \tag{6}$$

are an immediate consequence of the decomposition of the exponential function into its real and imaginary parts. With $w = u + iv$, $z = x + iy$, we have

$$u + iv = e^{x+iy} = e^x(\cos y + i \sin y),$$

whence

$$u = e^x \cos y, \qquad v = e^x \sin y.$$

It follows that

$$u^2 + v^2 = e^{2x}, \qquad \frac{v}{u} = \tan y.$$

This shows that *the mapping* (6) *transforms parallels to the imaginary axis into circles around the origin, and parallels to the real axis into rays emanating from the origin.*

For a more detailed investigation, it is more convenient to consider the transformation inverse to (6), *i.e.*, the mapping

$$z = \log w.$$

If $w = Re^{i\phi}$, we have

$$x + iy = \log R + i\phi,$$

and therefore

$$x = \log R, \qquad y = \phi. \tag{7}$$

(7) shows that the transformation (6) maps the boundary of any rectangle in the z-plane whose sides are parallel to the axes onto a closed contour consisting of two circular arcs about the origin and two linear segments pointing at the origin. Hence, (6) maps the rectangle of Fig. 32a onto the domain indicated in Fig. 32b. If the rectangle degenerates into the infinite strip $\theta_1 < \text{Im}\,\{z\} < \theta_2$, then its conformal image becomes the entire angular space $\theta_1 < \arg\,\{w\} < \theta_2$. If the width of the rectangle or of the strip exceeds 2π, its conformal image is clearly self-overlapping. On the other hand, any schlicht domain in the z-plane which is contained

in an infinite strip parallel to the x-axis of width less than 2π is mapped by (6) onto a schlicht domain.

We now turn to the trigonometric functions $w = \sin z$ and $w = \cos z$. In view of

$$\cos z = \frac{1}{2}(e^{iz} + e^{-iz}), \qquad \sin z = \frac{1}{2i}(e^{iz} - e^{-iz}), \tag{8}$$

the mapping properties of these function can be obtained by combining those of the transformations (4) and (6). It is sufficient to consider the function $w = \sin z$. If its mapping properties are known, those of

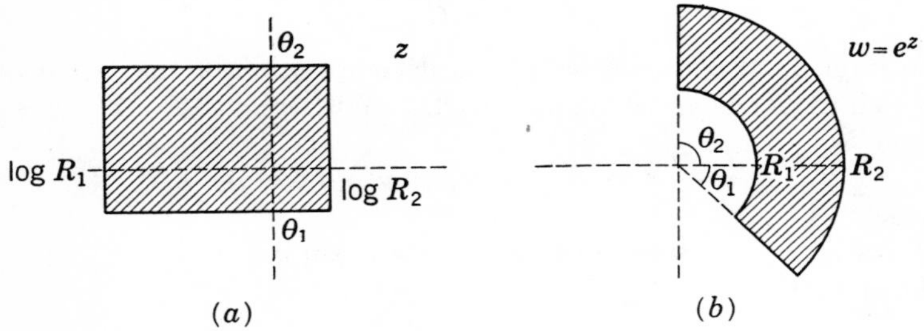

FIG. 32.

$w = \cos z$ follow trivially by means of the relation $\cos z = \sin(z + \frac{1}{2}\pi)$. In view of (8), the transformation $w = \sin z$ can be built up from the following four successive transformations:

$$z_1 = iz, \tag{9a}$$

$$z_2 = e^{z_1}, \tag{9b}$$

$$z_3 = -iz_2, \tag{9c}$$

$$w = \frac{1}{2}\left(z_3 + \frac{1}{z_3}\right). \tag{9d}$$

In order to find the conformal images of the straight lines $y = \text{const.}$ and $x = \text{const.}$, we observe that (9a) transforms $y = c$ into $x = -c$ and $x = c$ into $y = c$; (9b) transforms these lines into circles about the origin and rays emanating from the origin, respectively. This is, obviously, not changed by the transformation (9c). Finally, as shown in the preceding section, (9d) transforms circles about the origin and rays emanating from the origin into confocal ellipses and hyperbolas, respectively, whose common foci are at the points $w = \pm 1$. Hence, $w = \sin z$ *transforms the lines* $y = $ *const. and* $x = $ *const. into confocal ellipses and hyperbolas whose common foci are at* $w = \pm 1$.

This result can also be obtained with the help of the addition theorem of the function $\sin z$. We have

$$u + iv = w = \sin z = \sin (x + iy)$$
$$= \sin x \cos iy + \cos x \sin iy.$$

Since $\sin it = i \sinh t$ and $\cos it = \cosh t$, where the hyperbolic functions $\sinh t$ and $\cosh t$ are real for real values of t, it follows that

$$u + iv = \sin x \cosh y + i \cos x \sinh y,$$

whence

(10) $$u = \sin x \cosh y, \qquad v = \cos x \sinh y.$$

Hence,

(11) $$\frac{u^2}{(\cosh y)^2} + \frac{v^2}{(\sinh y)^2} = 1$$

and, in view of $\cosh^2 y - \sinh^2 y = \cos^2 iy + \sin^2 iy = 1$,

(12) $$\frac{u^2}{(\sin x)^2} - \frac{v^2}{(\cos x)^2} = 1.$$

(11) shows that the line $y = c$ is transformed into an ellipse of half axes $\cosh c$ and $\sinh c$, respectively. In view of $\cosh^2 y - \sinh^2 y = 1$, the foci of the ellipse are at ± 1. It should be noted that this ellipse is traversed an infinity of times if the point z describes the entire straight line $y = c$. Indeed, in view of (10), this ellipse is given by the parametric representation

$$u = \cosh c \sin x, \qquad v = \sinh c \cos x,$$

and this shows that a segment of length 2π of the line $y = c$ corresponds to the complete perimeter of the ellipse.

(12) shows that the line $x = c$ is transformed into an hyperbola of half axes $\sin c$ and $\cos c$. Since $\sin^2 c + \cos^2 c = 1$, the foci of this hyperbola are at ± 1. (12) can also be written in the form

$$u^2 \cos^2 x - v^2 \sin^2 x = \sin^2 x \cos^2 x,$$

which shows that for $x = \pm \frac{1}{2}\pi$ and $x = 0$ the hyperbolas degenerate into the real axis and the imaginary axis, respectively. A similar degeneration is shown by the ellipses (11). Writing (11) in the form

$$u^2 \sinh^2 y + v^2 \cosh^2 y = \sinh^2 y \cosh^2 y,$$

we find that for $y = 0$ the ellipse degenerates into a part of the real axis which is necessarily the linear segment connecting the points $w = -1$ and $w = 1$. By considering these degenerate cases, we arrive at an interesting result. The total boundary of the infinite half-strip indicated in Fig. 33*a* corresponds to the entire real axis in Fig. 33*b*. The reader will easily verify that if the boundary of the infinite half-strip is described in the positive sense with respect to the half-strip, the real axis in the w-plane is traversed monotonically from $-\infty$ to ∞. It follows that *the transformation* $w = \sin z$ *maps the infinite half-strip* $-\frac{1}{2}\pi < \text{Re}\,\{z\} < \frac{1}{2}\pi$, $\text{Im}\,\{z\} > 0$ *onto the upper half-plane* $\text{Im}\,\{w\} > 0$.

Since both domains have rectilinear boundaries, the function $w = \sin z$ can be continued beyond the boundary of the half-strip by means of the symmetry principle of Sec. 5, Chap. V. The linear segments $-\frac{1}{2}\pi < z < \frac{1}{2}\pi$ and $-1 < w < 1$ correspond to each other; hence, if $w = \sin z$ is continued beyond $-\frac{1}{2}\pi < z < \frac{1}{2}\pi$, its values in points symmetric with respect to the real axis must themselves be symmetric with respect to the

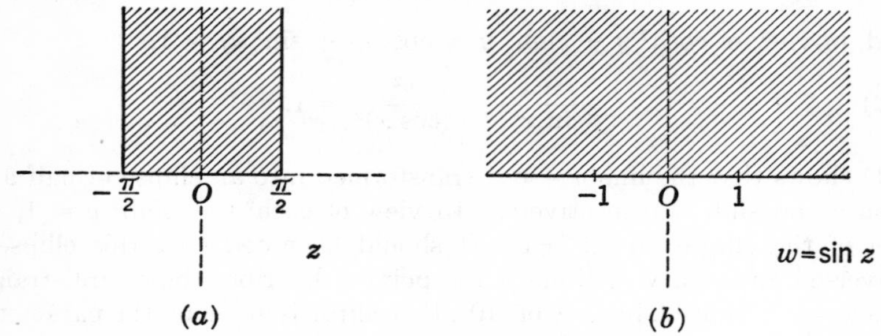

FIG. 33.

real axis. It follows that $w = \sin z$ maps the lower half-strip $-\frac{1}{2}\pi < \text{Re}\,\{z\} < \frac{1}{2}\pi$, $\text{Im}\,\{z\} < 0$ onto the lower half-plane $\text{Im}\,\{w\} < 0$. Taking into account the fact that the latter mapping was obtained from the former by analytic continuation beyond the linear segment $-\frac{1}{2}\pi < z < \frac{1}{2}\pi$, we find that $w = \sin z$ *maps the infinite strip* $-\frac{1}{2}\pi < \text{Re}\,\{z\} < \frac{1}{2}\pi$ *onto the full* w*-plane which is cut along the rays* $-\infty \leq w \leq -1$ *and* $1 \leq w \leq \infty$.

Applying the symmetry principle to the analytic continuation of $w = \sin z$ beyond the ray $z = \frac{1}{2}\pi + it$, $0 < t < \infty$, we find that $w = \sin z$ maps the infinite half-strip $\frac{1}{2}\pi < \text{Re}\,\{z\} < \frac{3}{2}\pi$, $\text{Im}\,\{z\} > 0$ onto the lower half-plane $\text{Im}\,\{w\} < 0$; furthermore, it follows that the half-strip $-\frac{1}{2}\pi < \text{Re}\,\{z\} < \frac{3}{2}\pi$, $\text{Im}\,\{z\} > 0$ is mapped onto the full w-plane which is cut along the ray $-\infty \leq w \leq 1$. The mapping of Fig. 33 can also be used, together with the symmetry principle, in order to deduce the periodicity of the function $w = \sin z$. If z and z_1 are symmetric with respect to the line $\text{Re}\,\{z\} = \frac{1}{2}\pi$, we obviously have $z_1 + \bar{z} = \pi$. Similarly, if z_2 and z are symmetric with respect to the line $\text{Re}\,\{z\} = -\frac{1}{2}\pi$, we have $z_2 + \bar{z} = -\pi$. Since, in the mapping $w = \sin z$, symmetry with respect to these lines corresponds to symmetry with respect to the real axis, it follows that

$$\sin(\pi - \bar{z}) = \overline{\sin z}, \qquad \sin(-\pi - \bar{z}) = \overline{\sin z}.$$

Hence,

$$\sin(z + 2\pi) = \sin[\pi - (-\pi - z)] = \overline{\sin(-\pi - \bar{z})} = \sin z,$$

which is the required periodicity property.

The conformal mapping properties of the function $w = \tan z$ follow from the definition

$$w = \tan z = \frac{\sin z}{\cos z} = \frac{1}{i}\frac{e^{iz} - e^{-iz}}{e^{iz} + e^{-iz}} = \frac{1}{i}\frac{e^{2iz} - 1}{e^{2iz} + 1},$$

which shows that the transformation $w = \tan z$ can be built up from the three successive transformations

$$z_1 = 2iz,$$
$$z_2 = e^{z_1},$$
$$w = \frac{1}{i}\frac{z_2 - 1}{z_2 + 1}.$$

By considering the properties of these transformations, the reader will conclude without difficulty that $w = \tan z$ transforms the straight lines $\operatorname{Re}\{z\} = \text{const.}$ into the family of circular arcs which terminate at the points $w = i$ and $w = -i$; the lines $\operatorname{Im}\{z\} = \text{const.}$ are transformed into a family of circles orthogonal to these circular arcs. In particular, the lines $\operatorname{Re}\{z\} = -\frac{1}{4}\pi$ and $\operatorname{Re}\{z\} = \frac{1}{4}\pi$ are transformed into the two halves of the circle $|w| = 1$ between the points $w = -i$ and $w = i$. The reader will verify that if the point z travels "down" the line $\operatorname{Re}\{z\} = -\frac{1}{4}\pi$ and, subsequently, "up" the line $\operatorname{Re}\{z\} = \frac{1}{4}\pi$, the point w describes the full circumference $|w| = 1$ in the positive sense. It follows that *the transformation $w = \tan z$ maps the infinite strip $-\frac{1}{4}\pi < \operatorname{Re}\{z\} < \frac{1}{4}\pi$ onto the unit circle $|w| < 1$.*

As an application of this mapping, we derive the addition theorem of the function $\tan z$. Since the strip $-\frac{1}{4}\pi < \operatorname{Re}\{z\} < \frac{1}{4}\pi$ remains unchanged if it is shifted vertically by the amount t (t real), it follows that the function $w_1 = \tan(z + it)$ likewise maps this strip onto the unit circle. Hence, the transformation $w \to w_1$ maps the unit circle onto itself and it must therefore be of the form

$$w_1 = e^{i\gamma}\left(\frac{w + \beta}{1 + \bar{\beta}w}\right), \qquad 0 \leq \gamma < 2\pi.$$

We thus obtain

$$\tan(z + it) = e^{i\gamma}\left(\frac{\tan z + \beta}{1 + \bar{\beta}\tan z}\right)$$

Since $\tan 0 = 0$ and $\tan(-z) = -\tan z$ it follows, by setting $z = 0$ and $z = -it$, that

$$\tan(it) = e^{i\gamma}\beta, \qquad \beta = -\tan(-it) = \tan it.$$

Hence $e^{i\gamma} = 1$ and therefore

$$\tan(z + it) = \frac{\tan z + \tan(it)}{1 + \overline{\tan(it)} \tan z}.$$

$\tan z$ is real for real z. By the symmetry principle, we thus have

$$\overline{\tan z} = \tan \bar{z},$$

and therefore $\overline{\tan(it)} = \tan(-it) = -\tan(it)$. This yields

$$\tan(z + it) = \frac{\tan z + \tan(it)}{1 - \tan(it) \tan z},$$

or, writing $it = \zeta$,

$$\tan(z + \zeta) = \frac{\tan z + \tan \zeta}{1 - \tan \zeta \tan z}.$$

This identity has so far only been proved for values of ζ which are pure imaginary and for values of z for which $-\frac{1}{4}\pi < \operatorname{Re}\{z\} < \frac{1}{2}\pi$. However, since the two sides of this identity are analytic functions of both z and ζ, it follows from the principle of permanence that the identity persists for all values of z and ζ for which both its sides are regular. This proves the addition theorem.

EXERCISES

1. The polar equation of a logarithmic spiral is $\rho = Ae^{B\theta}$ (A, B real). Show that $w = e^z$ transforms all straight lines in the z-plane into logarithmic spirals.

2. Show that

$$w = e^{-a\left(\frac{1+z}{1-z}\right)}, \qquad a > 0,$$

maps the unit circle $|z| < 1$ onto a domain D which covers the entire unit circle $|w| < 1$ an infinite number of times, with the exception of the point $w = 0$ which is not covered at all; show that all boundary points of D satisfy either $|w| = 1$ or $w = 0$.

3. Derive the addition theorem $e^{z+\zeta} = e^z e^\zeta$ from the fact that $w = e^z$ maps an infinite strip parallel to the real axis onto an angular space D whose vertex is at the origin. *Hint:* Use the fact that the mapping $z \to z + a$ (a real) leaves the strip unchanged and that the only conformal mapping of D onto itself which preserves the points 0 and ∞ is $w \to bw$, $b > 0$.

4. Show that the analytic function $w = f(z)$ which maps the domain of Fig. 32a onto the domain of Fig. 32b must be such that

$$\frac{f'(z)}{f(z)}$$

is real on the boundary. Deduce that $f(z)$ is necessarily of the form $f(z) = ae^{bz}$, where a and b are constants.

5. Show that the transformation

$$w = \log \frac{1+z}{1-z}$$

maps $|z| < 1$ onto the infinite strip $-\frac{1}{2}\pi < \text{Im}\,\{w\} < \frac{1}{2}\pi$.

6. Show that the transformation

$$w = \log \frac{z^2+1}{2z}$$

maps the half circle $|z| < 1$, $\text{Im}\,\{z\} > 0$ onto the infinite strip $0 < \text{Im}\,\{w\} < \pi$.

7. Show that the transformation

$$w = e^{\lambda z} - e^{(\lambda-1)z}, \qquad 0 < \lambda < 1,$$

maps the infinite strip $-\pi < \text{Im}\,\{z\} < \pi$ onto the full w-plane which is cut along the rays $\arg\,\{w\} = \pm\pi\lambda$, $\lambda^{-\lambda}(1-\lambda)^{\lambda-1} \leq |w| \leq \infty$.

8. Show that the transformation

$$w = z + e^z$$

maps the infinite strip $-\pi < \text{Im}\,\{z\} < \pi$ onto the full w-plane which is cut along the rays $\text{Im}\,\{w\} = \pm\pi i$, $-\infty \leq \text{Re}\,\{w\} \leq -1$.

9. Show that $w = \sin z$ maps the rectangle $-\frac{1}{2}\pi < \text{Re}\,\{z\} < \frac{1}{2}\pi$, $-c < \text{Im}\,\{z\} < c$ onto the interior of the ellipse

$$\frac{u^2}{\cosh^2 c} + \frac{v^2}{\sinh^2 c} = 1$$

which has been cut along the linear segments $-\cosh c \leq w \leq -1$, $1 \leq w \leq \cosh c$.

10. If D denotes the domain obtained by cutting the full w-plane along the rays $-\infty \leq w \leq -1$, $1 \leq w \leq \infty$, show that the transformation

$$w_1 = a\sqrt{z^2-1} + z\sqrt{1+a^2}, \qquad a \text{ real},$$

maps D onto itself. Use your result and the fact that $w = \sin z$ maps the infinite strip $-\frac{1}{2}\pi < \text{Re}\,\{z\} < \frac{1}{2}\pi$ onto D, in order to derive the addition theorem of the function $\sin z$. *Hint:* Employ a technique similar to that used in Exercise 3.

11. Show that the transformation

$$w = \log \sin z$$

maps the half-strip $-\frac{1}{2}\pi < \text{Re}\,\{z\} < \frac{1}{2}\pi$, $\text{Im}\,\{z\} > 0$ onto the strip $0 < \text{Im}\,\{w\} < \pi$.

12. Show that the transformation

$$w = \tan^2 \sqrt{z}$$

maps the interior of the parabola

$$64y^2 = \pi^2(\pi^2 - 16x)$$

onto the unit circle $|w| < 1$.

13. Show that the transformation

$$w = \sin \sqrt{z}$$

maps the interior of the parabola

$$4y^2 = \pi^2(\pi^2 - 4x),$$

which has been cut along the linear segment $0 < z < \frac{1}{4}\pi^2$, onto the upper half-plane $\text{Im}\,\{w\} > 0$.

14. Show that the transformation

$$w = \sin\left[\frac{\pi z}{1 + z^2}\right]$$

maps the unit circle $|z| < 1$ onto a domain D of the following description: D covers the entire w-plane on infinity of times, with the exception of the point $w = \infty$ and the linear segment $-1 \leq w \leq 1$; the points $w = \infty$ and $w = \pm 1$ are not covered at all and the segment $-1 < w < 1$ is covered by D exactly once.

3. Elliptic Functions. In the two preceding sections we were studying the conformal mapping properties of analytic functions which were known a priori. In this section we shall adopt a somewhat different procedure. Since it is assumed that the reader is not acquainted with the theory of elliptic functions, we shall define some of the fundamental elliptic functions by means of certain conformal mappings effected by them, and we shall then proceed to derive some of their other properties.

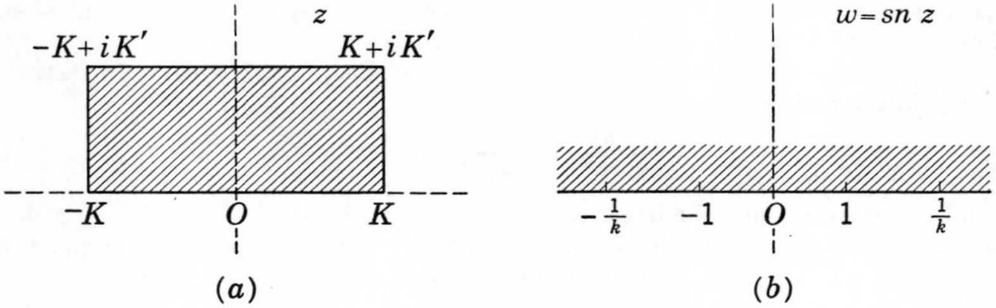

FIG. 34.

Our point of departure is the conformal mapping of the half-plane $\text{Im}\{w\} > 0$ onto a rectangle in the z-plane, where the points $w = \pm 1$, $w = \pm(1/k)$ $(0 < k < 1)$ are to correspond to the corners of the rectangle. By the Schwarz-Christoffel formula (50) of Sec. 6, Chap. V, this mapping is effected by the function

$$z = F(w) = \int_0^w \frac{dw}{\sqrt{(1 - w^2)(1 - k^2w^2)}}. \tag{13}$$

In order to determine the position of the rectangle in the z-plane, we observe that $F(w)$ is real for real w and that $F(-w) = -F(w)$. It follows that one of the sides of the rectangle coincides with part of the real axis and is situated symmetrically with respect to the origin. If we agree to take in (13) that branch of the square root for which $\sqrt{1} = 1$ and denote the height and width of the rectangle by K' and $2K$, respectively, the position of the rectangle will therefore be as indicated in Fig. 34*a*.

We now define the elliptic function $w = \text{sn}\, z$ as the analytic function for which $\text{sn}'(0) = 1$ and which maps the rectangle of Fig. 34*a* onto the upper half-plane indicated in Fig. 34*b* in such a way that the points

(14)

z	$-K + iK'$	$-K$	0	K	$K + iK'$	iK'
$\operatorname{sn} z$	$-k^{-1}$	-1	0	1	k^{-1}	∞

correspond to each other. We might also have defined $w = \operatorname{sn} z$ as the inverse of the function $z = F(w)$ introduced in (13). Obviously, the function $w = \operatorname{sn} z$ depends not only on z but also on the parameter k. It is, however, customary to regard one of the quantities

$$\tau = \frac{iK'}{K}, \tag{15}$$

$$q = e^{\pi i \tau} = e^{-\frac{K'}{K}}, \tag{16}$$

rather than k, as the parameter on which $\operatorname{sn} z$ depends. If it is desired to indicate the parameter, the symbol $\operatorname{sn} z$ is replaced by $\operatorname{sn}(z,\tau)$ or $\operatorname{sn}(z,q)$. We shall see later that k is uniquely determined if either τ or q are given.

The symbol $\operatorname{sn} z$ is used because of certain analogies between this function and the trigonometric function $\sin z$. It is also easy to see that the function $\sin z$ corresponds to a degenerate case of the function $\operatorname{sn} z$. This follows either by letting $k \to 0$ in (13) or by observing that, for $q \to 0$, the rectangle of Fig. 34*a* becomes an infinite half-strip. This analogy is carried further by the definition

$$\operatorname{cn} z = \sqrt{1 - \operatorname{sn}^2 z}, \qquad \operatorname{cn} 0 = 1, \tag{17}$$

of the elliptic function $\operatorname{cn} z$. Another elliptic function is introduced by

$$\operatorname{dn} z = \sqrt{1 - k^2 \operatorname{sn}^2 z}, \qquad \operatorname{dn} 0 = 1. \tag{18}$$

The functions $\operatorname{sn} z$, $\operatorname{cn} z$, $\operatorname{dn} z$ are referred to as the *Jacobian elliptic functions.* We add that the name "elliptic functions" is due to the fact that the integral (13) was first encountered in connection with the problem of finding the length of an arc of an ellipse. This led to the appellation "elliptic integrals" for a class of integrals related to (13) and, subsequently, to the name "elliptic functions" for the functions inverse to these integrals if the latter are regarded as functions of their upper limits.

With the help of (13) and (14), both K and K' can be expressed in terms of k. We have

$$K = \int_0^1 \frac{dt}{\sqrt{(1 - t^2)(1 - k^2t^2)}}, \tag{19}$$

and

$$iK' = \int_1^{k^{-1}} \frac{ds}{\sqrt{(1 - s^2)(1 - k^2s^2)}}.$$

The latter integral can be brought into a more elegant form. If we make the substitution

$$s = (1 - k'^2t^2)^{-\frac{1}{2}},$$

where

$$k' = \sqrt{1 - k^2}, \tag{20}$$

we obtain by an elementary computation

$$K' = \int_0^1 \frac{dt}{\sqrt{(1 - t^2)(1 - k'^2t^2)}}. \tag{21}$$

It is interesting to observe that the functional dependence of K with respect to k is the same as that of K' with respect to k', where k' is defined in (20), that is, $K'(k) = K(\sqrt{1 - k^2})$.

The fundamental property of the elliptic functions is their *double periodicity*, that is, the fact that such a function has two different periods which are not integral multiples of the same number. In the case of the function sn z, these periods are $4K$ and $2iK'$. In other words, the function sn z satisfies the identities

$$\begin{aligned} \text{sn}(z + 4K) &= \text{sn}\, z, \\ \text{sn}(z + 2iK') &= \text{sn}\, z. \end{aligned} \tag{22}$$

(22) can be proved by suitable application of the symmetry principle. Since $w = \text{sn}\, z$ maps the rectangle R of Fig. 35 onto the half-plane of Fig. 34*b*, sn z can be continued beyond the boundary of the rectangle by symmetry. Inverting the rectangle with respect to its upper side, we find that $w = \text{sn}\, z$ maps the rectangle R_1 of Fig. 35 onto the lower half-plane. Inverting, in turn, the rectangle R_1 with respect to its upper side, we see that the rectangle R_2 of Fig. 35 is again mapped by $w = \text{sn}\, z$ onto the upper half-plane. If z_1 and z_2 denote the points into which a point z of R is successively carried by these inversions, it is clear that $z_2 = z + 2iK'$. Since the image point of z is returned to its original position by the two inversions with respect to the real axis, it follows from the symmetry principle that $\text{sn}(z + 2iK') = \text{sn}\, z$. This proves the second identity (22). The proof of the first inequality (22) follows in the same way by considering the inversions R_3 and R_4 of R (see Fig. 35) and is left as an exercise to the reader.

Next, we show that *the function $w = \text{sn}\, z$ is single-valued at all finite points of the z-plane.* To this end, we divide the z-plane into a network of congruent rectangles by means of the lines Re $\{z\} = K(2n + 1)$, $n = 0, \pm 1, \pm 2, \ldots$ and Im $\{z\} = mK'$, $m = 0, \pm 1, \pm 2, \ldots$. All these rectangles can be obtained from the rectangle R of Fig. 35 by a suitable number of inversions. Although there are many different possibilities to get from R to a given rectangle R' in this fashion, a moment's reflection shows that the number of inversions connecting R and R' is always even if it is even for one particular chain of inversions, or else it is always odd if it is odd for one particular chain of inversions. On the other hand, an even number of inversions of a point w with respect to the real axis in the w-plane returns the point to its original position, while an odd number of such inversions carries the point w into the point $\bar{w}$. Hence, the analytic continuation of $w = \text{sn } z$ by means of the symmetry principle leads to a uniquely determined value of sn z at each point of the z-plane, regardless of the path along which sn z has been continued. This proves the above statement.

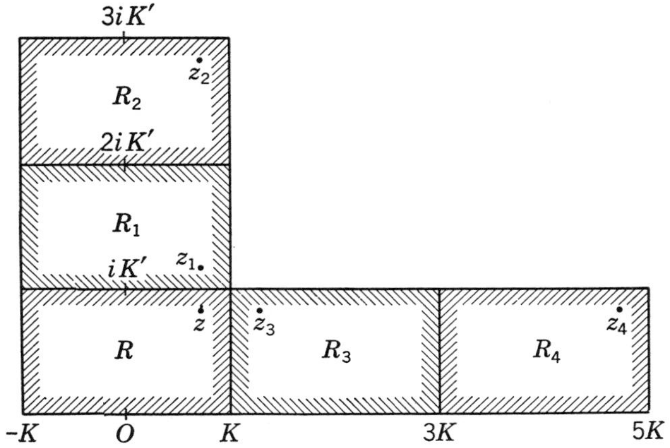

FIG. 35.

In view of the double periodicity (22), it is sufficient to know the values which the function $w = \text{sn } z$ takes in a rectangle of sides $4K$ and $2K'$, which are parallel to the x-axis and the y-axis, respectively. This is entirely analogous to the case of the function $w = \sin z$ in which it is sufficient to know the values of the function in a strip of width 2π which is parallel to the y-axis; the other values of $w = \sin z$ are then obtained from the relation $\sin (z + 2\pi n) = \sin z$, $n = \pm 1, \pm 2, \ldots$. We add that, in order to cover the entire z-plane by the homologues of the original *period rectangle*, it is necessary to add to the interior of the rectangle two of its sides; clearly, the latter have to be adjacent to each other. In the case of the function $w = \text{sn } z$, a convenient period rectangle will be the rectangle $-K \leq \text{Re } \{z\} < 3K$, $0 \leq \text{Im } \{z\} < 2K'$.

The period rectangle of $w = \text{sn } z$ consists of four homologues of the rectangle R in Fig. 35. As shown before, two of these rectangles are each mapped onto the half-plane Im $\{w\} > 0$ and the other two are mapped onto the half-plane Im $\{w\} < 0$. It follows that the function $w = \operatorname{sn} z$ maps the period rectangle onto a domain which covers the entire w-plane exactly twice. In other words, *if w_0 is an arbitrary complex number, then the equation* $\operatorname{sn} z = w_0$ *has exactly two solutions in each period rectangle.* This is also true of the equation $\operatorname{sn} z = \infty$. By (14), $\operatorname{sn} iK' = \infty$. If we invert the rectangle R in Fig. 35 with respect to its right side, the point $z = iK'$ is carried into the point $z = 2K + iK'$. Since the corresponding inversion with respect to the real axis in the w-plane leaves the point $w = \infty$ unchanged, it follows that $\operatorname{sn}(2K + iK') = \infty$. The singularities of $\operatorname{sn} z$ at $z = iK'$ and $z = 2K + iK'$ must be simple poles. This follows either by observing that the mapping at these points is conformal, or else by showing that otherwise there would exist values near $w = \infty$ which are taken in the period rectangle more than twice. Hence, the only singularities of $w = \operatorname{sn} z$ in our period rectangle are two simple poles at the points $z = iK'$ and $z = 2K + iK'$.

Since, in view of (22), the values of the function $w = \operatorname{sn} z$ in the entire z-plane are periodic repetitions of its values in a single period rectangle, it follows that *the only finite singularities of* $w = \operatorname{sn} z$ *are simple poles at the points*

$$(23) \quad z = 2nK + (2m + 1)iK', \qquad n = 0, \pm 1, \pm 2, \ldots, \quad m = 0, \pm 1, \pm 2, \ldots.$$

The zeros of $w = \operatorname{sn} z$ are found in a similar fashion. By (14), $\operatorname{sn} 0 = 0$. One inversion of the rectangle R in Fig. 35 with respect to its right side shows that also $\operatorname{sn} 2K = 0$. These are all the zeros in the period rectangle. It follows therefore from (22) that *the only zeros of* $w = \operatorname{sn} z$ *are simple zeros at the points*

$$(24) \quad z = 2nK + 2miK', \qquad n = 0, \pm 1, \pm 2, \ldots, \quad m = 0, \pm 1, \pm 2, \ldots.$$

We shall use (23) and (24) in order to set up an infinite product for the function $w = \operatorname{sn} z$. However, before we do so, we insert a few remarks concerning infinite products. An infinite product

$$P = \prod_{n=1}^{\infty} (1 + a_n), \qquad a_n \neq -1,$$

is defined as the limit of the products

$$P_n = (1 + a_1)(1 + a_2) \cdots (1 + a_n) = \prod_{\nu=1}^{n} (1 + a_\nu)$$

for $n \to \infty$. If this limit exists and is different from zero, we say that the infinite product converges. The product is said to converge absolutely, if the product

$$p = \prod_{n=1}^{\infty} (1 + |a_n|)$$

converges. An absolutely convergent product converges also in the usual sense. To show that this is true, write

$$p_n = \prod_{\nu=1}^{n} (1 + |a_\nu|)$$

and consider the expressions

$$P_n - P_{n-1} = (1 + a_1) \cdots (1 + a_{n-1})a_n$$

and

$$p_n - p_{n-1} = (1 + |a_1|) \cdots (1 + |a_{n-1}|)|a_n|.$$

We obviously have

$$|P_n - P_{n-1}| \leq p_n - p_{n-1}.$$

Since $\lim_{n\to\infty} p_n = p$, $\sum_n (p_n - p_{n-1})$ converges. Hence, the same is true of $\sum_n |P_n - P_{n-1}|$ and therefore also of $\sum_n (P_n - P_{n-1})$. But the convergence of the latter series is identical with the existence of $\lim_{n\to\infty} P_n$.

It remains to be shown that this limit cannot be zero. Before we do so, we first prove that the infinite product $\prod_n (1 + |a_n|)$ and the infinite series $\sum_n |a_n|$ converge and diverge together. Obviously,

$$e^{|a_n|} = 1 + |a_n| + \frac{|a_n|^2}{2!} + \cdots \geq 1 + |a_n|.$$

On the other hand, we have

$$|a_1| + \cdots + |a_n| \leq (1 + |a_1|) \cdots (1 + |a_n|),$$

as is seen by multiplying out the product. Hence,

$$\sum_{\nu=1}^{n} |a_\nu| \leq \prod_{\nu=1}^{n} (1 + |a_\nu|) \leq e^{\sum_{\nu=1}^{n} |a_\nu|},$$

which shows that for the convergence of $\prod_n (1 + |a_n|)$ it is necessary and sufficient that the series $\sum_n |a_n|$ converge. To show now that $\lim_{n\to\infty} P_n \neq 0$, we remark that since $\sum_n |a_n|$ converges and $1 + a_n \to 1$, the series

$$\sum_n \left| \frac{a_n}{1 + a_n} \right|$$

is also convergent. Hence, in view of the result just proved, the product

$$\prod_{\nu=1}^{n} \left(1 - \frac{a_\nu}{1 + a_\nu}\right) = \frac{1}{\prod\limits_{\nu=1}^{n} (1 + a_\nu)} = \frac{1}{P_n}$$

tends to a finite limit. Therefore, $\lim_{n\to\infty} P_n \neq 0$.

We now consider the infinite product

$$f(z) = \zeta \frac{\prod\limits_{m=0}^{\infty} (1 - q^{2m}\zeta^{-2}) \prod\limits_{m=1}^{\infty} (1 - q^{2m}\zeta^{2})}{\prod\limits_{m=0}^{\infty} (1 - q^{2m+1}\zeta^{-2}) \prod\limits_{m=0}^{\infty} (1 - q^{2m+1}\zeta^{2})}, \tag{25}$$

where

$$\zeta = e^{\frac{\pi i z}{2K}} \tag{26}$$

and q is defined by (16). The product (25) converges absolutely at all points of the z-plane at which none of the terms in the numerator and denominator of (25) vanishes. Indeed, it follows from (16) that $0 < q < 1$. Hence, the series $\sum_m q^m$ converges; in view of the criterion just developed, the four products in (25) will therefore converge absolutely. We add that, by taking logarithms, an infinite product of the type

$$Q = \prod_{m=1}^{\infty} (1 - q^m\zeta)$$

can be transformed into the series

$$\log Q = \sum_{m=1}^{\infty} \log (1 - q^m\zeta),$$

if the proper values of the logarithm are taken on both sides. The series will obviously converge if ultimately, that is, for large enough m, we take the value of log $(1 - q^m\zeta)$ which reduces to 0 if $(1 - q^m\zeta) \to 1$. Since in any closed domain in which $q^m\zeta \neq 1$, $m = 1, 2, \ldots$, the series is converging, it converges there uniformly. As shown in Sec. 3, Chap. III, it therefore represents there a regular analytic function of ζ. Taking exponentials, we find that the same is true of $Q = Q(\zeta)$. In view of (26), it therefore follows that (25) is an analytic function of z which is regular at all finite points of the z-plane at which none of the factors in the denominator vanish. Obviously, all finite singularities of the function (25) are poles which are caused by the zeros of the denominator.

We now determine the zeros and poles of the function (25). In view of (16) and (26), the zeros coincide with those points z at which either

$$e^{-\frac{2\pi mK'}{K} - \frac{\pi iz}{K}} = 1, \qquad m = 0, 1, 2, \ldots, \tag{27'}$$

or

$$e^{-\frac{2\pi mK'}{K} + \frac{\pi iz}{K}} = 1, \qquad m = 1, 2, \ldots. \tag{27''}$$

Taking logarithms and observing the indeterminacy of the logarithmic function, we find that the zeros of (25) coincide with the points z for which

$$-\frac{2\pi mK'}{K} - \frac{\pi iz}{K} = 2\pi in, \qquad m = 0, 1, 2, \ldots, n = 0, \pm 1, \pm 2, \ldots,$$

or

$$-\frac{2\pi mK'}{K} + \frac{\pi iz}{K} = 2\pi in, \qquad m = 1, 2, \ldots, n = 0, \pm 1, \pm 2, \ldots.$$

Obviously, the last two relations are equivalent to

$$z = 2nK + 2imK', \qquad n = 0, \pm 1, \pm 2, \ldots, m = 0, \pm 1, \pm 2, \ldots.$$

A comparison with (24) shows that these are precisely the zeros of sn z. We add that these zeros of the function (25) are all simple; this is an immediate consequence of the fact that the derivatives of the left-hand sides of (27) do not vanish for any finite value of z. The poles of (25) are situated at the points z for which

$$e^{-\frac{\pi(2m+1)K'}{K} - \frac{\pi iz}{K}} = 1, \qquad m = 0, 1, 2, \ldots,$$

or

$$e^{-\frac{\pi(2m+1)K'}{K} + \frac{\pi iz}{K}} = 1, \qquad m = 0, 1, 2, \ldots.$$

As before, we find that these are the points

$$z = 2nK + (2m + 1)iK', \qquad n = 0, \pm 1, \pm 2, \ldots, \quad m = 0, \pm 1, \pm 2, \ldots.$$

Since all these are simple zeros of the denominator, it follows that the only finite singularities of the function $f(z)$ in (25) are simple poles whose location is the same as that of the simple poles (23) of sn z.

We next show that the function (25) is a doubly periodic function whose periods are $4K$ and $2iK'$. In view of

$$e^{\frac{\pi i}{2K}(z+4K)} = e^{\frac{\pi i z}{2K}} \cdot e^{2\pi i} = e^{\frac{\pi i z}{2K}},$$

the quantity ζ defined in (26) does not change if z is replaced by $z + 4K$. Hence, the function (25) has the period $4K$. If z is replaced by $z + 2iK'$, we have

$$e^{\frac{\pi i}{2K}(z+2iK')} = e^{\frac{\pi i z}{2K}} e^{-\pi \frac{K'}{K}} = q e^{\frac{\pi i z}{2K}},$$

where q is defined by (16). As a result, the quantity ζ in (25) has now to be replaced by $q\zeta$. We obtain

$$f(z + 2iK') = q\zeta \frac{\prod_{m=0}^{\infty} (1 - q^{2(m-1)}\zeta^{-2}) \prod_{m=1}^{\infty} (1 - q^{2(m+1)}\zeta^{2})}{\prod_{m=0}^{\infty} (1 - q^{2m-1}\zeta^{-2}) \prod_{m=0}^{\infty} (1 - q^{2m+3}\zeta^{2})}$$

$$= qf(z) \frac{(1 - q^{-2}\zeta^{-2})(1 - q\zeta^{2})}{(1 - q^{2}\zeta^{2})(1 - q^{-1}\zeta^{-2})} = f(z).$$

Consider now the function

$$g(z) = \frac{\text{sn } z}{f(z)},$$

where $f(z)$ is the function defined in (25). Since the poles and zeros of sn z coincide with the poles and zeros, respectively, of $f(z)$, the function $g(z)$ is regular at all finite points of the z-plane. In view of the fact that both sn z and $f(z)$ have the periods $4K$ and $2iK'$, it further follows that $g(z)$ likewise has these periods. The values of $g(z)$ throughout the z-plane are therefore repetitions of the values taken in a single period rectangle. Since $g(z)$ is regular in the closure of such a rectangle, we have there $|g(z)| < M$, where M is a suitable constant. Hence, the inequality $|g(z)| < M$ must hold in the entire z-plane. In view of Liouville's theorem (Sec. 7, Chap. III), this means that $g(z)$ reduces to a constant, say C.

We have thus shown that

$$\text{sn } z = Cf(z), \tag{28}$$

where $f(z)$ is defined in (25). To determine the constant C, we use the fact that, by (14),

$$\operatorname{sn} K = 1, \qquad \operatorname{sn}(K + iK') = k^{-1}.$$

In these cases, the quantity ζ defined in (26) takes the values

$$e^{\frac{\pi i}{2K}K} = e^{\frac{\pi i}{2}} = i$$

and

$$e^{\frac{\pi i}{2K}(K+iK')} = e^{\frac{\pi i}{2}} e^{-\pi\frac{K'}{2K}} = i\sqrt{q},$$

respectively. Inserting these values in (25) and using (28) we obtain, after some manipulation,

$$(29) \qquad 1 = 2iC \prod_{n=1}^{\infty} \left(\frac{1 + q^{2n}}{1 + q^{2n-1}}\right)^2$$

and

$$\frac{1}{k} = \frac{iC}{2\sqrt{q}} \prod_{n=1}^{\infty} \left(\frac{1 + q^{2n-1}}{1 + q^{2n}}\right)^2.$$

Eliminating C from these two identities, we have

$$(30) \qquad k^2 = 16q \prod_{n=1}^{\infty} \left(\frac{1 + q^{2n}}{1 + q^{2n-1}}\right)^8,$$

whence $C^4k^2 = q$. Since $q > 0$, it follows from (29) that $iC > 0$. Hence, finally

$$(31) \qquad C = -i\frac{\sqrt[4]{q}}{\sqrt{k}},$$

where both radicals take their positive values. In view of (25), (28), and (31), we thus have proved the expansion

$$(32) \qquad \operatorname{sn}(z;q) = -i\frac{\sqrt[4]{q}}{\sqrt{k}}\zeta\frac{\prod_{n=0}^{\infty}(1 - q^{2n}\zeta^{-2})\prod_{n=1}^{\infty}(1 - q^{2n}\zeta^{2})}{\prod_{n=0}^{\infty}(1 - q^{2n+1}\zeta^{-2})\prod_{n=0}^{\infty}(1 - q^{2n+1}\zeta^{2})},$$

where ζ is defined by (26) and k, the *modulus* of the function sn z, is expressed in terms of q by means of (30).

Although so far our considerations have been confined to the case in which one of the periods of the function sn z is real and the other is pure imaginary, our results are of much wider application. The infinite product (32) converges for all values of q for which $|q| < 1$. Hence, if ω_1 and ω_2 are two complex numbers such that

$$\text{Re}\left\{\frac{\omega_1}{\omega_2}\right\} > 0 \tag{33}$$

and, in analogy to (16) and (26), q and ζ are defined by

$$q = e^{-2\pi\frac{\omega_1}{\omega_2}}, \qquad \zeta = e^{\frac{\pi i z}{\omega_2}}, \tag{34}$$

the product (32) will converge for this value of q. The reader will also confirm without difficulty that this function $\text{sn}\, z = \text{sn}(z;q)$ has the periods $2\omega_1$ and $2\omega_2$, that is, it satisfies the relations $\text{sn}(z + 2\omega_1) = \text{sn}\, z$, $\text{sn}(z + 2\omega_2) = \text{sn}\, z$. The period rectangle will now become a *period parallelogram* which, because of (33), cannot degenerate into a linear segment. For $|q| < 1$, the expression (32) is also a regular analytic function of the variable q. By the principle of permanence (Sec. 5, Chap. III) all analytic identities which were shown to hold in the case $0 < q < 1$ will therefore persist for all q such that $|q| < 1$. We thus have the following more general result.

The analytic function $\text{sn}(z;q)$ *of* (32), *where* ω_1, ω_2 *satisfy* (33) *and* q *and* ζ *are defined in* (34), *has the periods* $2\omega_1$ *and* $2\omega_2$; *the inverse of* $w = \text{sn}(z;q)$ *is the function* $z = F(w)$ *defined in* (13).

Since these general functions $\text{sn}\,(z;q)$ do not share the simple conformal mapping properties which are characteristic of the case $0 < q < 1$, we shall not pursue their study any further.

Infinite product expansions for the functions $\text{cn}\, z$ and $\text{dn}\, z$ defined in (17) and (18), respectively, can be obtained by suitable modification of the procedure employed in the case of $\text{sn}\, z$. In view of (17), the poles of $\text{cn}\, z$ coincide with those of $\text{sn}\, z$, while its zeros are situated at the points at which $\text{sn}\, z = \pm 1$. By (14), these are the points

$$z = \pm K + 4Kn + 2iK'm, \qquad n, m = 0, \pm 1, \pm 2, \ldots . \tag{35}$$

It should be noted that the equation $\text{sn}^2 z - 1 = 0$ has double roots at these points, or, what amounts to the same thing, that the derivative of $\text{sn}\, z$ vanishes there. This is an immediate consequence of the fact (see Fig. 34) that $w = \text{sn}\, z$ transforms right angles whose vertices are at these points into the angle π. Hence, $\sqrt{1 - \text{sn}^2 z}$ is regular at these points and $\text{cn}\, z$ is single-valued for all z. Using the fact that $\text{cn}\, z$ has simple zeros at the points (35) and simple poles at the points (23), and employing the same procedure as in the case of the function $\text{sn}\, z$, we finally arrive, after some manipulation, at the infinite product expansion

$$(36)\quad \operatorname{cn}(z;q) = \frac{\sqrt[4]{q(1-k^2)}}{\sqrt{k}}\,\zeta\,\frac{\prod_{n=0}^{\infty}(1+q^{2n}\zeta^{-2})\prod_{n=1}^{\infty}(1+q^{2n}\zeta^{2})}{\prod_{n=0}^{\infty}(1-q^{2n+1}\zeta^{-2})\prod_{n=0}^{\infty}(1-q^{2n+1}\zeta^{2})}.$$

The function dn z of (18) has again the same poles as sn z, while its zeros coincide with the points at which sn $z = \pm k^{-1}$. By (14), these are the points

$$z = (2n+1)K + (2m+1)iK'.$$

This leads to the product expansion

$$(37)\qquad \operatorname{dn}(z;q) = \sqrt[4]{1-k^2}\,\frac{\prod_{n=0}^{\infty}(1+q^{2n+1}\zeta^{-2})\prod_{n=0}^{\infty}(1+q^{2n+1}\zeta^{2})}{\prod_{n=0}^{\infty}(1-q^{2n+1}\zeta^{-2})\prod_{n=0}^{\infty}(1-q^{2n+1}\zeta^{2})}.$$

Details of the derivation of (36) and (37) are left as an exercise to the reader.

Our next objective is to derive a relation between the functions sn$(z;q)$ and sn$(z;q^2)$. Our point of departure is again the conformal mapping of Fig. 34, by means of which the function sn$(z;q)$ was originally defined.

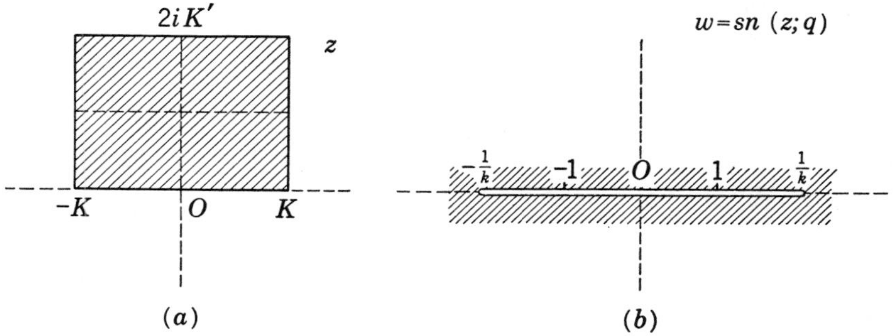

FIG. 36.

Applying the symmetry principle to an inversion of the rectangle in Fig. 34*a* with respect to its upper side, we find that the function $w = \operatorname{sn}(z;q)$ maps the rectangle of Fig. 36*a* onto the full w-plane which is furnished with a slit as indicated in Fig. 36*b*. Consider now the function sn αz (where α is such that sn $\alpha K = 1$) which maps the rectangle in Fig. 36*a* onto the upper half-plane. Since the sides of the rectangle are $2K$ and

$2K'$, respectively, the parameter q_1 belonging to this function is, by (16),

$$q_1 = e^{-\pi\frac{2K'}{K}} = q^2.$$

It follows that the function $w_1 = \mathrm{sn}(\alpha z; q^2)$ maps the rectangle of Fig. 36*a* onto the upper half-plane. Since, as the reader will easily confirm, the mapping

$$w = \frac{2}{k}\frac{\beta w_1}{\beta^2 + w_1^2}, \qquad \beta > 0,$$

transforms the upper half-plane $\mathrm{Im}\,\{w_1\} > 0$ into the slit domain of Fig. 36*b*, we find that the function

$$w = \frac{2}{k}\frac{\beta\,\mathrm{sn}(\alpha z; q^2)}{\beta^2 + \mathrm{sn}^2(\alpha z; q^2)} \tag{38}$$

maps the rectangle of Fig. 36*a* onto the slit domain of Fig. 36*b*. As shown above, the same conformal mapping is effected by the function $w = \mathrm{sn}(z;q)$. By the results of Sec. 4, Chap. V, two analytic functions which perform the same conformal mapping of a simply-connected domain are identical if their values agree in three boundary points. Now both the function (38) and $w = \mathrm{sn}(z;q)$ transform $z = 0$ into $w = 0$. If the constant β in (38) is so chosen that

$$\frac{2\beta}{1 + \beta^2} = k, \tag{39}$$

it follows further that the mapping (38) transforms the points $z = \pm K$ into the points

$$w = \frac{2}{k}\frac{\beta\,\mathrm{sn}(\pm\alpha K; q^2)}{\beta^2 + \mathrm{sn}^2(\pm\alpha K; q^2)} = \pm\frac{2}{k}\frac{\beta}{1 + \beta^2} = \pm 1.$$

Since also $\mathrm{sn}(\pm K;q) = \pm 1$, the two functions must therefore be identical. We have thus proved the identity

$$\mathrm{sn}(z;q) = \frac{2}{k}\frac{\beta\,\mathrm{sn}(\alpha z; q^2)}{\beta^2 + \mathrm{sn}^2(\alpha z; q^2)}, \tag{40}$$

where β is given by (39). The parameter k is the modulus of the elliptic function $\mathrm{sn}(z;q)$. To distinguish k from the modulus of the function $\mathrm{sn}(z;q^2)$, we shall denote the former by $k(q)$ and the latter by $k(q^2)$. In terms of this notation, the identity (30) reads

$$k^2(q) = 16q\prod_{n=1}^{\infty}\left(\frac{1 + q^{2n}}{1 + q^{2n-1}}\right)^8. \tag{41}$$

The constant β in (40) can be simply expressed in terms of the modulus

$k(q^2)$. Since $\operatorname{sn}[\alpha(K + 2iK'); q^2] = k^{-1}(q^2)$ and

$$\operatorname{sn}(K + 2iK'; q) = \operatorname{sn}(K;q) = 1,$$

it follows from (40) that

$$k(q) = \frac{2\beta k(q^2)}{1 + \beta^2 k^2(q^2)}. \tag{42}$$

Now the equation

$$\frac{x}{1 + x^2} = \frac{y}{1 + y^2}$$

is identical with

$$(y - x)(1 - xy) = 0$$

and its only two solutions are therefore $x = y$ and $xy = 1$. Comparing (39) and (42), we thus find that we have either $\beta k(q^2) = \beta$ or $\beta^2 k(q^2) = 1$. Since the first possibility is absurd, it follows that

$$\beta = \frac{1}{\sqrt{k(q^2)}}.$$

Hence, in view of (39), (40), and the fact that $\operatorname{sn}'(0) = 1$,

$$k(q) = \frac{2\sqrt{k(q^2)}}{1 + k(q^2)}, \tag{43}$$

$$\operatorname{sn}(z;q) = \frac{[1 + k(q^2)] \operatorname{sn}(\alpha z;q^2)}{1 + k(q^2) \operatorname{sn}^2(\alpha z;q^2)}, \qquad \alpha = [1 + k(q^2)]^{-1}. \tag{44}$$

By combining the mapping of Fig. 34 with some of the transformations discussed in the preceding sections, a number of important conformal

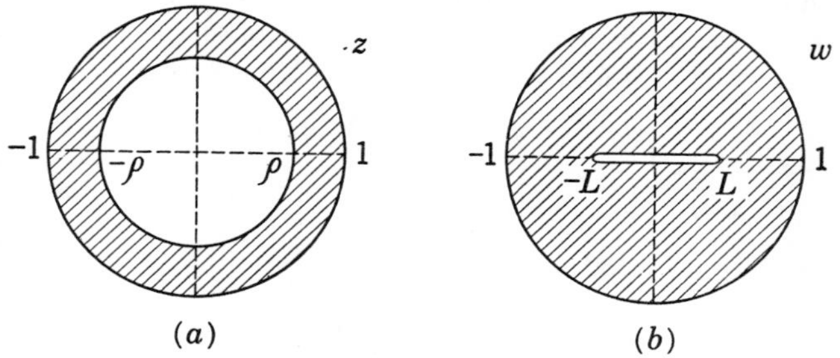

FIG. 37.

mappings can be obtained. As an example, we consider the conformal mapping of the circular ring of Fig. 37*a* onto the unit circle furnished with a symmetrical slit as indicated in Fig. 37*b*. It should be noted that the existence of such a mapping does not follow from the Riemann mapping theorem, since the domains are doubly-connected. It will be shown in

Chap. VII that any doubly-connected domain can be mapped onto a circular ring, where the ratio of the radii of the ring is completely determined by the original domain. In our case, this means that, given the length L in Fig. 37b, the radius ρ in Fig. 37a is completely determined, and vice versa.

Applying the symmetry principle to the circle $|z| = 1$, we find that the function $w = f(z)$ which effects the mapping of Fig. 37 will also map the circular ring $\rho < |z| < \rho^{-1}$ onto the entire w-plane which is furnished with the slits $-\infty \leq w \leq -L^{-1}$, $-L \leq w \leq L$, $L^{-1} \leq w \leq \infty$. Both these domains are symmetrical with respect to the real axis in their respective planes. If we can find a function which maps the upper half of the circular ring $\rho < |z| < \rho^{-1}$ onto the upper half-plane in such a way that the points

(45)

z	ρ	$-\rho$	ρ^{-1}	$-\rho^{-1}$
w	L	$-L$	L^{-1}	$-L^{-1}$

correspond to each other, it will therefore follow from the symmetry principle that this function is identical with $w = f(z)$. We now use the fact that the transformation $z = i\rho e^{-i\zeta}$ maps the rectangle $-\frac{1}{2}\pi < \text{Re}\,\{\zeta\} < \frac{1}{2}\pi$, $0 < \text{Im}\,\{\zeta\} < -2 \log \rho$ onto the upper half of the circular ring $\rho < |z| < \rho^{-1}$ in such a way that the points

(46)

ζ	$\frac{1}{2}\pi$	$-\frac{1}{2}\pi$	$\frac{1}{2}\pi - 2i \log \rho$	$-\frac{1}{2}\pi - 2i \log \rho$
z	ρ	$-\rho$	ρ^{-1}	$-\rho^{-1}$

correspond to each other (see Fig. 32, Sec. 2). Hence, the function $w = f(i\rho e^{-i\zeta})$ maps the rectangle in question onto the upper half-plane. Since, in view of (45) and (46), we have the correspondence

ζ	$\frac{1}{2}\pi$	$-\frac{1}{2}\pi$	$\frac{1}{2}\pi - 2i \log \rho$	$-\frac{1}{2}\pi - 2i \log \rho$
$L^{-1}f(i\rho e^{-i\zeta})$	1	-1	L^{-2}	$-L^{-2}$

it follows from (14) that

$$L^{-1}f(i\rho e^{-i\zeta}) = \text{sn}\,\frac{2K}{\pi}\,\zeta,$$

where the constants K and K' associated with the function sn ζ satisfy

$$\frac{K'}{K} = -\frac{4}{\pi} \log \rho, \tag{47}$$

and where the modulus of the function sn is

$$k = L^2. \tag{48}$$

Since, in view of (16) and (47), the parameter q associated with sn $2\pi^{-1}K\zeta$ is ρ^4, we thus find that the analytic function effecting the conformal mapping of Fig. 37 is of the form

$$w = f(z) = \sqrt{k(\rho^4)}\ \mathrm{sn}\left(\frac{2iK}{\pi}\log\frac{z}{\rho} + K;\ \rho^4\right). \tag{49}$$

We also note that, by (41) and (48), the functional dependence of the length L in Fig. 37b with respect to the radius is given by

$$L(\rho) = 2\rho\prod_{n=1}^{\infty}\left(\frac{1+\rho^{8n}}{1+\rho^{8n-4}}\right)^2.$$

As another example of a conformal mapping which can be carried out by means of elliptic functions, we consider the function $w = f(z)$ which maps the interior of an ellipse onto the unit circle. In order to avoid unnecessary parameters, we assume that the foci of the ellipse are situated at the points ± 1. If we further suppose—as we may, by the Riemann mapping theorem—that $f(0) = 0, f'(0) > 0$ then, for reasons of symmetry, $w = f(z)$ will map the upper half of the ellipse onto the upper half of the unit circle. We now recall from Sec. 2 that the function $z = \sin\zeta$ maps the rectangle $-\frac{1}{2}\pi < \mathrm{Re}\{\zeta\} < \frac{1}{2}\pi, 0 < \mathrm{Im}\{\zeta\} < c$ onto the upper half of an ellipse of semiaxes cosh c and sinh c. If the semiaxes of our ellipse are $a = \cosh c$, $b = \sinh c$, it follows therefore that $w = f(\sin\zeta)$ maps this rectangle onto the upper half of the unit circle $|w| < 1$. But this half circle is transformed by the mapping

$$w_1 = \frac{2w}{1+w^2}$$

onto the upper half-plane $\mathrm{Im}\{w_1\} > 0$. Hence, the function

$$w_1 = \frac{2f(\sin\zeta)}{1+f^2(\sin\zeta)}$$

will map the rectangle $-\frac{1}{2}\pi < \mathrm{Re}\{\zeta\} < \frac{1}{2}\pi$, $0 < \mathrm{Im}\{\zeta\} < c$ onto the upper half-plane. If $f(1) = \alpha$, the reader will confirm that the corners of the rectangle correspond to the points

ζ	$\frac{1}{2}\pi$	$-\frac{1}{2}\pi$	$\frac{1}{2}\pi + ic$	$-\frac{1}{2}\pi + ic$
w_1	$\dfrac{2\alpha}{1+\alpha^2}$	$-\dfrac{2\alpha}{1+\alpha^2}$	1	-1

In view of (14), it follows therefore that

$$w_1 = \frac{2f(\sin\zeta)}{1+f^2(\sin\zeta)} = \frac{2\alpha}{1+\alpha^2}\ \mathrm{sn}\ \frac{2K}{\pi}\zeta, \tag{50}$$

where the constants K and K' associated with the function sn satisfy $\pi K' = 2cK$. By (16), we have $q = e^{-2c}$, and the modulus $k = k(e^{-2c})$ is

$$k(e^{-2c}) = \frac{2\alpha}{1 + \alpha^2}.$$

In view of (43), it follows that

$$\alpha = \sqrt{k(e^{-4c})}.$$

If we compare (50) with (44) and use (43), we find that

$$f(\sin \zeta) = \sqrt{k(e^{-4c})}\, \mathrm{sn}\left(\frac{2K}{\pi}\zeta; e^{-4c}\right), \qquad K = K(e^{-4c}).$$

If we recall that the semiaxes of the ellipse were $a = \cosh c$ and $b = \sinh c$, we thus obtain the following result.

The interior of the ellipse

$$\frac{x^2}{a^2} + \frac{y^2}{b^2} = 1, \qquad a^2 - b^2 = 1, \qquad z = x + iy$$

is mapped onto the unit circle $|w| < 1$ *by the function*

$$(51) \qquad w = f(z) = \sqrt{k(\rho)}\, \mathrm{sn}\left(\frac{2K}{\pi}\sin^{-1} z; \rho\right), \qquad \rho = \left(\frac{a - b}{a + b}\right)^2.$$

In this mapping, the foci of the ellipse correspond to the points

$$w = \pm \sqrt{k(\rho)}.$$

EXERCISES

1. Show that the elliptic functions sn z, cn z, dn z have the differentiation formulas

$$\frac{d}{dz}\,\mathrm{sn}\, z = \mathrm{cn}\, z\, \mathrm{dn}\, z,$$
$$\frac{d}{dz}\,\mathrm{cn}\, z = -\,\mathrm{sn}\, z\, \mathrm{dn}\, z,$$
$$\frac{d}{dz}\,\mathrm{dn}\, z = -k^2\, \mathrm{sn}\, z\, \mathrm{cn}\, z.$$

2. Show that $w = \mathrm{sn}\, z$ satisfies the differential equation

$$w'^2 = (1 - w^2)(1 - k^2w^2),$$

and deduce similar differential equations for the functions $w = \mathrm{cn}\, z$ and $w = \mathrm{dn}\, z$.

3. With the help of (41), (37), and the fact that dn $0 = 1$, prove the identity

$$\prod_{n=1}^{\infty} (1 + q^{2n-1})^8 - \prod_{n=1}^{\infty} (1 - q^{2n-1})^8 = 16q \prod_{n=1}^{\infty} (1 + q^{2n})^8, \qquad |q| < 1.$$

4. Show that the function $w = \operatorname{cn} z$ maps the rectangle $-K < \operatorname{Re}\{z\} < K$, $0 < \operatorname{Im}\{z\} < K'$ onto the right half-plane $\operatorname{Re}\{w\} > 0$ which has been cut along the linear segment $0 < w \leq 1$.

5. Show that the function

$$w = \sqrt{\frac{1 - \operatorname{cn} z}{1 + \operatorname{cn} z}}$$

maps the rectangle $-K < \operatorname{Re}\{z\} < K$, $-K' < \operatorname{Im}\{z\} < K'$ onto the unit circle $|w| < 1$.

6. Show that the function

$$w = \operatorname{sn}\left(\frac{4K}{\pi}\tan^{-1} z\right)$$

maps the unit circle $|z| < 1$ onto a domain D with the following properties: D covers the entire w-plane (including $w = \infty$) an infinity of times, with the exception of the linear segments $-k^{-1} \leq w \leq -1$ and $1 \leq w \leq k^{-1}$ which are not covered at all; D has no branch points and it has no boundary points other than the points of the above linear segments.

7. If $w = f(z,\rho)$ denotes the function effecting the conformal mapping of Fig. 37, use the fact—following from the symmetry principle—that $w = f(z,\rho)$ maps the circular ring $\rho < |z| < \rho^{-1}$ onto the full w-plane with the slits $-\infty \leq w \leq -L^{-1}$, $-L \leq w \leq L$, $L^{-1} \leq w \leq \infty$, in order to deduce the relations

$$f(z\rho^{-1},\rho) = \frac{2}{L(\rho)}\frac{f(z,\rho^2)}{1 + f^2(z,\rho^2)}$$

and

$$L^2(\rho) = \frac{2L(\rho^2)}{1 + L^2(\rho^2)}.$$

Using the form (49) of the mapping function $f(z,\rho)$, show that these two relations are identical with the identities (44) and (43) respectively.

8. Show that the function $w = f(z,\rho)$ of the preceding exercise maps the circular ring $\rho^2 < |z| < 1$ onto a domain which consists of two replicas of the unit circle $|w| < 1$ and has two simple branch points at $w = \pm L(\rho)$.

9. If $f(z,\rho)$ is the function of Exercise 7, show that the transformation

$$w = \frac{1 + f\left(\dfrac{1+z}{1-z}, \rho\right)}{1 - f\left(\dfrac{1+z}{1-z}, \rho\right)}$$

maps the z-plane with two equally large circular holes indicated in Fig. 38*a* onto the

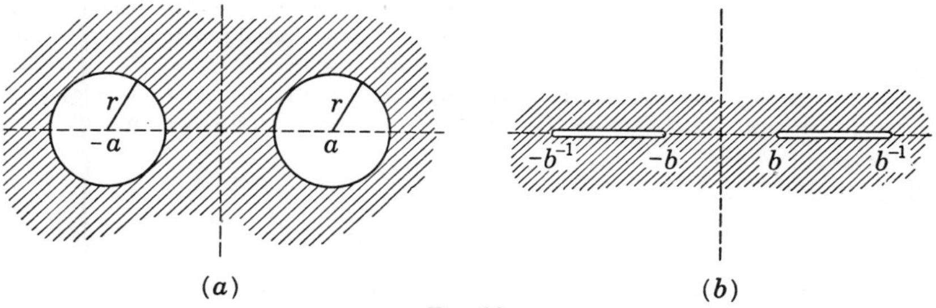

FIG. 38.

w-plane with two equal collinear slits indicated in Fig. 38b, where

$$a = \frac{1 + \rho^2}{1 - \rho^2}, \qquad r = \frac{2\rho}{1 - \rho^2}, \qquad b = \frac{1 - L(\rho)}{1 + L(\rho)}.$$

10. Using the fact that the function (13) yields the conformal mapping of Fig. 34, prove the identity

$$\int_0^\pi \int_0^\infty \frac{\rho \, d\rho \, d\theta}{\sqrt{1 - 2\rho^2 \cos 2\theta + \rho^4} \sqrt{1 - 2\rho^2 k^2 \cos 2\theta + \rho^4 k^4}} = 2KK'.$$

11. If $K = K'$, it follows from (19), (20), and (21) that $k = 2^{-\frac{1}{2}}$. Deduce that $k(e^{-\pi}) = 2^{-\frac{1}{2}}$ and show further, with the help of (43), that

$$k(e^{-\frac{1}{2}\pi}) = 2 \sqrt[4]{2} (\sqrt{2} - 1).$$

12. Show that the transformation

$$w = \operatorname{sn}^2 z$$

maps the rectangle $0 < \operatorname{Re}\{z\} < K$, $-K' < \operatorname{Im}\{z\} < K'$ onto the full w-plane with the slits $-\infty \leq w \leq 0$, $1 \leq w \leq \infty$.

13. Show that the series

$$\vartheta_1(z,q) = 2 \sum_{n=0}^{\infty} (-1)^n q^{(n+\frac{1}{2})^2} \sin (2n + 1)z$$

and

$$\vartheta_4(z,q) = 1 + 2 \sum_{n=1}^{\infty} (-1)^n q^{n^2} \cos 2nz, \qquad |q| < 1,$$

converge uniformly in any closed finite domain and thus are entire functions of z. Show further that

$$\vartheta_1(z + \pi) = -\vartheta_1(z), \qquad \vartheta_4(z + \pi) = \vartheta_4(z),$$
$$\vartheta_1(z + \pi\tau) = -q^{-1}e^{-2iz}\vartheta_1(z),$$
$$\vartheta_4(z + \pi\tau) = -q^{-1}e^{-2iz}\vartheta_4(z),$$

($q = e^{\pi i \tau}$) and that therefore the function

$$P(z) = \frac{\vartheta_1(z)}{\vartheta_4(z)}$$

satisfies the relations $P(z + \pi\tau) = P(z)$, $P(z + \pi) = -P(z)$ and thus is a doubly periodic function of z with the periods 2π and $\pi\tau$. *Hint:* Express the trigonometric functions in the definitions of ϑ_1 and ϑ_4 by means of exponential functions and replace the summations from 0 to ∞ by summations from $-\infty$ to ∞.

14. If C is the boundary of the parallelogram with the corners z_0, $z_0 + \pi$, $z_0 + \pi + \pi\tau$, $z_0 + \pi\tau$ and $\vartheta(z)$ denotes either of the functions $\vartheta_1(z)$, $\vartheta_4(z)$, show that

$$\frac{1}{2\pi i} \int_C \frac{\vartheta'(z)}{\vartheta(z)} dz = \frac{1}{2\pi i} \int_{z_0}^{z_0+\pi} \left[\frac{\vartheta'(z)}{\vartheta(z)} - \frac{\vartheta'(z + \pi\tau)}{(z + \pi\tau)} \right] dz$$
$$- \frac{1}{2\pi i} \int_{z_0}^{z_0+\pi\tau} \left[\frac{\vartheta'(z)}{\vartheta(z)} - \frac{\vartheta'(z + \pi)}{\vartheta(z + \pi)} \right] dz = \frac{1}{2\pi i} \int_{z_0}^{z_0+\pi} 2i \, dz = 1.$$

Hint: Use the "quasi periodicity" of the function $\vartheta(z)$ derived in the preceding exercise.

15. Show that

$$\vartheta_1(z,q) = -ie^{iz+\frac{\pi i\tau}{4}}\vartheta_4\left(z+\frac{\pi\tau}{2}, q\right),$$

and deduce from the result of the preceding exercise that the only zeros and poles of the doubly periodic function $P(z)$ of Exercise 13 in a period parallelogram are situated at the points congruent to $z = 0, \pi$ and $z = \frac{1}{2}\pi\tau, \frac{1}{2}\pi\tau + \pi$, respectively.

16. Show that the zeros and poles of the function $P(z)$ coincide with those of $\operatorname{sn}(2K/\pi)z$ and that both functions have the same periods, and conclude that

$$P(z) = A \operatorname{sn}\frac{2K}{\pi}z,$$

where A is a constant.

17. Using the fact that $\operatorname{sn} K = 1$, $\operatorname{sn}(K + iK') = k^{-1}$, show that the value of the constant A in the preceding exercise is

$$A = \frac{2\sum_{n=0}^{\infty} q^{(n+\frac{1}{2})^2}}{1 + 2\sum_{n=1}^{\infty} q^{n^2}}$$

and that

$$k^2 = 16q\left[\frac{\sum_{n=0}^{\infty} q^{n(n+1)}}{1 + 2\sum_{n=1}^{\infty} q^{n^2}}\right]^4$$

Hint: Use the identity of Exercise 15. *Remark:* Because of the extremely rapid convergence of the series involved, this expression is used for the practical computation of k^2.

4. Domains Bounded by Arcs of Confocal Conics. In this section we consider the conformal mapping of domains whose boundaries consist of a number of confocal elliptic or hyperbolic arcs, where the common foci of these arcs may be assumed to be situated at the points ± 1 without restricting the generality of our considerations. Examples of such domains are the interior or exterior of an ellipse, the interior or exterior of one branch of a hyperbola, the entire plane slit along an elliptic or hyperbolic arc, a domain bounded by an elliptic arc and a hyperbolic arc intersecting it at right angles, and so forth.

Let now $w = f(z)$ be the analytic function which maps the unit circle $|z| < 1$ onto the domain D which is bounded by arcs of confocal conics whose common foci are at $w = \pm 1$. Since, as shown in Sec. 2, the mapping $\zeta = \sin^{-1} w$ transforms all ellipses and hyperbolas with the foci $w = \pm 1$ into linear segments parallel to the real and imaginary axes in the ζ-plane, respectively, it follows that

$$\zeta = \sin^{-1} f(z)$$

transforms the circumference $|z| = 1$ into such linear segments. As a result, the differential

$$d\zeta = \frac{f'(z)\,dz}{\sqrt{1 - f^2(z)}}$$

will be either real or pure imaginary if $|z| = 1$. On the other hand, the differential

$$\frac{dz}{z} = \frac{ie^{i\theta}}{e^{i\theta}}\,d\theta = i\,d\theta, \qquad z = e^{i\theta},$$

is pure imaginary for $|z| = 1$. Combining this with the above expression for $d\zeta$, we find that

$$\frac{zf'(z)}{\sqrt{1 - f^2(z)}}$$

is either real or pure imaginary for $|z| = 1$. Hence, the square of this expression is real for $|z| = 1$. Thus,

$$\frac{z^2 f'^2(z)}{1 - f^2(z)} = \text{real}, \qquad |z| = 1. \tag{52}$$

This relation holds for all functions mapping the unit circle onto domains bounded by arcs of confocal conics. The further discussion depends on the particular shape of the domain D in question. To illustrate the method, we first treat the case, already discussed in the preceding section, in which D is the interior of an ellipse. In this case, the expression (52) will have two singularities, namely, two simple poles at the points $z = \pm\alpha$ which correspond to the foci $f(z) = \pm 1$ of the ellipse. Consider now the function

$$p(z) = \frac{z^2}{(\alpha^2 - z^2)(1 - \alpha^2 z^2)}, \qquad 0 < \alpha < 1.$$

In view of

$$p(e^{i\theta}) = \frac{1}{(\alpha^2 - e^{2i\theta})(e^{-2i\theta} - \alpha^2)} = -\frac{1}{|\alpha^2 - e^{2i\theta}|^2} < 0,$$

it follows from (52) that

$$\frac{f'^2(z)(\alpha^2 - z^2)(1 - \alpha^2 z^2)}{1 - f^2(z)} = \text{real}, \qquad |z| = 1. \tag{53}$$

The expression (53) is regular in $|z| \leq 1$. Indeed, the two zeros of the denominator at $z = \pm\alpha$ are compensated by the factor $(\alpha^2 - z^2)$ in the numerator; on the circumference $|z| = 1$, $f(z)$, and therefore also the expression (53), is regular since the ellipse is an analytic curve. We now recall from Sec. 10, Chap. III, that an analytic function which is regular in the closure of a domain and which takes real values on the boundary reduces to a constant. Hence, we may conclude from (53) that

$$\frac{f'^2(z)}{1 - f^2(z)} = \frac{\alpha^2\beta^2}{(\alpha^2 - z^2)(1 - \alpha^2 z^2)},$$

where, in view of $f(0) = 0$, $\beta = f'(0)$. With

$$g(z) = \frac{1}{\alpha\beta} \sin^{-1} [f(\alpha z)], \tag{54}$$

this may be brought into the form

$$g'^2(z) = \frac{1}{(1 - z^2)(1 - \alpha^4 z^2)}.$$

A comparison with (13) shows that

$$z = \operatorname{sn}[g(z)],$$

where the modulus of the elliptic function is $k = \alpha^2$. Hence, replacing αz by z, we conclude from (54) that

$$z = \sqrt{k} \operatorname{sn} \left(\frac{1}{\beta \sqrt{k}} \sin^{-1} w \right), \qquad w = f(z).$$

To determine the constant β, we observe that the point $z = \alpha = \sqrt{k}$ corresponds to the point $w = 1$. Hence,

$$\operatorname{sn} \left(\frac{1}{\beta \sqrt{k}} \frac{\pi}{2} \right) = 1,$$

whence, by (14),

$$\frac{1}{\beta \sqrt{k}} = \frac{2K}{\pi}.$$

Thus, finally,

$$z = \sqrt{k} \operatorname{sn} \left(\frac{2K}{\pi} \sin^{-1} w \right), \qquad k = \alpha^2,$$

in agreement with (51).

The mapping of $|z| < 1$ onto the exterior of an ellipse of foci $w = \pm 1$ is effected by the elementary function

$$w = \frac{1}{2}\left(\rho z + \frac{1}{\rho z}\right), \quad a = \frac{1}{2}\left(\frac{1}{\rho} + \rho\right), \quad b = \frac{1}{2}\left(\frac{1}{\rho} - \rho\right), \quad \rho < 1, \tag{55}$$

as shown in Sec. 1. It is interesting that this mapping function can also be obtained with the help of the procedure employed in the case of the interior of the ellipse. If $w = f(z)$ is the desired mapping function, then $f(z)$ must satisfy the condition (52). Since the values of $f(z)$ are now situated in the exterior of the ellipse, the only possible singularity of the expression (52) is the point at which $f(z) = \infty$. If we assume (as we may, in view of the Riemann mapping theorem) that $f(0) = \infty$, then $f(z)$ will have an expansion

$$f(z) = \frac{C}{z} + a_0 + a_1 z + \cdots, \qquad |z| < 1.$$

Inserting this in (52), we find that the latter expression has the value -1 for $z = 0$. Hence the expression (52) is regular for $|z| \leq 1$. Since it is real for $|z| = 1$, it therefore must reduce to the constant -1. Thus

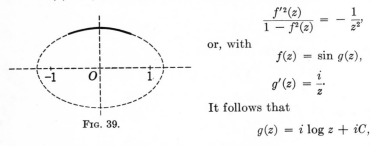

$$\frac{f'^2(z)}{1 - f^2(z)} = -\frac{1}{z^2},$$

or, with

$$f(z) = \sin g(z),$$

$$g'(z) = \frac{i}{z}.$$

It follows that

$$g(z) = i \log z + iC,$$

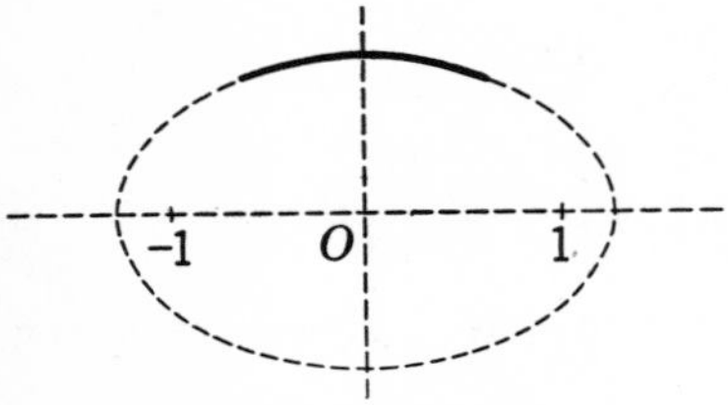

FIG. 39.

where C is an integration constant. Hence,

$$\begin{aligned} f(z) &= \sin (i \log z + iC) \\ &= \frac{1}{2i} (e^{-\log z - C} - e^{\log z + C}) \\ &= \frac{1}{2i} \left(\frac{e^{-C}}{z} - e^{C} z \right). \end{aligned}$$

For $C = \log (-i\rho)$, this takes the form (55).

As a more complicated example, we next treat the case in which the image domain D is the full w-plane which has been slit along an arc of an ellipse of foci $w = \pm 1$, which, for the sake of simplicity, will be assumed to be symmetric with respect to the imaginary axis (see Fig. 39). We shall require that the function $w = f(z)$ which maps $|z| < 1$ onto this domain D transforms the points $z = \pm i$ into the intersections of both edges of the slit with the imaginary axis and that, moreover, $f(0) = \infty$. That this is permissible is seen as follows. If $w = f_1(z)$ is the function which maps the left half of the unit circle onto the left half of the domain of Fig. 39, it follows from the Riemann mapping theorem that we can make the three boundary points $z = 0, \pm i$ of the half circle correspond to $w = \infty$ and the intersection of the two edges of the slit with the imaginary axis, respectively. Applying then the symmetry principle, we find that $w = f_1(z)$ also maps the full circle $|z| < 1$ onto the domain D; hence, $f_1(z)$ is identical with the above function $f(z)$.

From the symmetry of $w = f(z)$ it also follows that the two points for which $f(z) = \pm 1$ are symmetric with respect to the imaginary axis. If one of these points is denoted by α, the other must therefore be situated at $z = -\bar{\alpha}$. This symmetry also applies to the two points on $|z| = 1$ which correspond to the two ends of the slit. If z_1 and z_2 denote these two points, then $z_1 = e^{i\theta_0}$ and $z_2 = -e^{-i\theta_0}$ $(0 \leq \theta_0 < 2\pi)$. We now apply the condition (52), which must be satisfied by the mapping function $f(z)$. The singularities of the expression

$$F(z) = \frac{z^2 f'^2(z)}{1 - f^2(z)}$$

in $|z| < 1$ are, clearly, two simple poles at the points $z = \alpha$ and $z = -\bar{\alpha}$ at which $f^2(z) = 1$ and, possibly, the point $z = 0$ at which $f(z)$ has a simple pole. However, at $z = 0$, $F(z)$ is regular and $F(0) = -1$, as shown before in a similar case. Consider now the function

$$p(z) = \frac{z}{(\alpha - z)(1 - \bar{\alpha}z)}, \qquad |\alpha| < 1.$$

Since

$$p(e^{i\theta}) = \frac{1}{(\alpha - e^{i\theta})(e^{-i\theta} - \bar{\alpha})} = -\frac{1}{|\alpha - e^{i\theta}|^2} < 0,$$

it follows from (52) that

$$(56) \quad F_1(z) = \frac{f'^2(z)(\alpha - z)(1 - \bar{\alpha}z)(\bar{\alpha} + z)(1 + \alpha z)}{1 - f^2(z)} = \text{real}, \qquad |z| = 1.$$

Although we have now removed the simple poles at the points $z = \alpha$, $z = -\bar{\alpha}$, we have at the same time introduced a double pole at $z = 0$. In order to remove this singularity of the function $F_1(z)$, we observe that $F_1(z)$ has double zeros at the points $z = e^{i\theta_0}$ and $z = -e^{-i\theta_0}$. Indeed, these points correspond to the ends of the slit; hence the angle π is transformed into the angle 2π, which shows that we must have $f'(z) = 0$ at these points. Since $F_1(z)$ has the factor $f'^2(z)$, these zeros of $f'(z)$ give rise to double zeros of $F_1(z)$.

Since, as the reader will easily confirm,

$$\frac{z^2}{(z - e^{i\theta_0})^2(z + e^{-i\theta_0})^2} < 0, \qquad |z| = 1,\ z \neq e^{i\theta_0},\ -e^{-i\theta_0},$$

it follows from (56) that

$$F_2(z) = \frac{z^2 f'^2(z)(\alpha - z)(1 - \bar{\alpha}z)(\bar{\alpha} + z)(1 + \alpha z)}{[1 - f^2(z)](z - e^{i\theta_0})^2(z + e^{-i\theta_0})^2} = \text{real}, \qquad |z| = 1.$$

Since the factor z^2 cancels the double pole of $F_1(z)$ and since the newly introduced double poles at $z = e^{i\theta_0}$ and $z = -e^{-\theta_0}$ are canceled by the double zeros of $F_1(z)$ at these points, it follows that $F_2(z)$ is regular in $|z| \leq 1$. Being real for $|z| = 1$, $F_2(z)$ must therefore reduce to a constant. Hence, in view of $F_2(0) = -|\alpha|^2$,

$$\frac{f'^2(z)}{1 - f^2(z)} = -\frac{|\alpha|^2(z - e^{i\theta_0})^2(z + e^{-i\theta_0})^2}{z^2(\alpha - z)(1 - \bar{\alpha}z)(\bar{\alpha} + z)(1 + \alpha z)},$$

or, with

$$f(z) = \sin g(z),$$

$$g'(z) = \frac{i|\alpha|(z - e^{i\theta_0})(z + e^{-i\theta_0})}{z\sqrt{(\alpha - z)(1 - \bar{\alpha}z)(\bar{\alpha} + z)(1 + \alpha z)}}.$$

We thus obtain the final formula

$$f(z) = i \sinh\left[c + \int_i^z \frac{|\alpha|(z - e^{i\theta_0})(z + e^{-i\theta_0})\,dz}{z\sqrt{(\alpha - z)(1 - \bar{\alpha}z)(\bar{\alpha} + z)(1 + \alpha z)}}\right],$$

where $2 \sinh c$ $(c > 0)$ is the minor axis of the ellipse.

It is worth noting that the relation (52) is the analytic expression of the so-called "optical property" of the ellipse or the hyperbola, and that it is therefore possible to derive (52) in a purely geometric way, without using the mapping properties of the function $\sin z$. In the case of the ellipse, the optical property is the fact that, if the ellipse is made of reflecting material, all light rays emanating from one focus must pass through the other focus. Geometrically speaking, the lines connecting a point of the ellipse with the two foci make equal angles with the tangent to the ellipse at that point. As shown in Sec. 8, Chap. V, the angle between the positive axis and the tangent to the conformal image of $|z| = 1$ by means of $w = f(z)$ is $\arg\{izf'(z)\}$. Since the foci are at $w = \pm 1$, the angles between the positive axis and the lines connecting the point $w = f(z)$ with $w = \pm 1$ are $\arg\{f(z) - 1\}$ and $\arg\{f(z) + 1\}$, respectively. The optical property of the ellipse is therefore expressed by the relation

$$\arg\left\{\frac{izf'(z)}{f(z) - 1}\right\} = \arg\left\{-\frac{f(z) + 1}{izf'(z)}\right\}.$$

This is equivalent to

$$\arg\left\{-\frac{z^2f'^2(z)}{1 - f^2(z)}\right\} = 0,$$

that is,

$$\frac{z^2f'^2(z)}{1 - f^2(z)} \leq 0.$$

This proves (52) in the case of the ellipse. The analogue of the optical property in the case of the hyperbola is the fact that the tangent at a point of the hyperbola bisects the angle between the two lines connecting the point with the foci. As before, it follows that

$$\arg\left\{\frac{f(z)-1}{izf'(z)}\right\} = \arg\left\{\frac{izf'(z)}{f(z)+1}\right\},$$

whence

$$\frac{z^2f'^2(z)}{1-f^2(z)} \geq 0.$$

This procedure can also be generalized to the case of a parabolic arc. In this case the angle between the tangent at a point of the parabola and the parallel to the axis of the parabola is equal to the angle between the tangent and the line connecting the point with the focus. If the axis of the parabola coincides with the positive axis in the w-plane and the focus is situated at the origin, we have therefore

$$\arg\left\{\frac{f(z)}{izf'(z)}\right\} = \arg\,\{izf'(z)\},$$

and thus

$$\arg\left\{\frac{z^2f'^2(z)}{f(z)}\right\} = 0.$$

It follows that the condition

$$\frac{z^2f'^2(z)}{f(z)} \geq 0 \tag{57}$$

must be satisfied on any arc of $|z| = 1$ which is mapped by $w = f(z)$ onto a parabolic arc whose focus is at the origin. As an example, consider the mapping of $|z| < 1$ onto the interior of a parabola, *i.e.*, the region containing the focus. If the focus is situated at the origin, the condition (57) must be satisfied at all points of $|z| = 1$. By Exercise 3, Sec. 10, Chap. III, the number of poles of the expression (57) in $|z| \leq 1$ is equal to the number of its zeros in this region, where zeros or poles on the boundary are counted with half their multiplicities. If the parabola opens toward the left, we may assume that the mapping is symmetric with respect to the real axis and, hence, that $f(-1) = \infty$. If, as we may, we further assume that $f(0) = 0$, the expression (57) has clearly only one simple zero in $|z| \leq 1$, namely, at $z = 0$. It therefore has either a simple pole in $|z| < 1$ or a double pole on $|z| = 1$. Since $z = -1$ is the only possible singularity of (57) in $|z| \leq 1$, we therefore have a double pole at that point. Now the function

$$p(z) = \frac{z}{(1+z)^2}$$

has also a double pole at $z = -1$ and it has a simple zero at $z = 0$. Since, moreover, $p(z) > 0$ for $|z| = 1$, it follows from (57) that

$$\frac{z(1+z)^2 f'^2(z)}{f(z)}$$

is regular in $|z| \leq 1$ and positive for $|z| = 1$. It therefore reduces to a positive constant, to be denoted by C^2. Hence,

$$\frac{f'^2(z)}{f(z)} = \frac{C^2}{z(1+z)^2},$$

or, with

$$f(z) = g^2(z),$$
$$2g'(z) = \frac{C}{\sqrt{z}\,(1+z)}.$$

By integration,

$$g(z) = C \tan^{-1} \sqrt{z},$$

and thus

$$w = f(z) = C^2(\tan^{-1} \sqrt{z})^2.$$

The inverse function, mapping the inside of the parabola onto the unit circle, is therefore

$$z = \tan^2 C' \sqrt{w}, \qquad CC' = 1,$$

in agreement with Exercise 12, Sec. 2.

To map $|z| < 1$ onto the outside of a parabola, suppose that the parabola opens toward the right and that $f(-1) = \infty$. Since $f(z) \neq 0$ in $|z| \leq 1$, the expression (57) will now have a double zero at $z = 0$. Accordingly, it must have a pole of fourth order at $z = -1$. The function

$$\frac{(1+z)^4 f'^2(z)}{f(z)}$$

is therefore regular for $|z| \leq 1$ and positive for $|z| = 1$. It thus reduces to a constant, say $4C^2$. Hence,

$$\frac{f'^2(z)}{f(z)} = \frac{4C^2}{(1+z)^4},$$

whence

$$[\sqrt{f(z)}]' = \frac{C}{(1+z)^2},$$

and finally

$$f(z) = \frac{C^2}{(1+z)^2},$$

in agreement with Exercise 3, Sec. 1.

As a last application, we determine the function $w = f(z)$ which maps $|z| < 1$ onto the entire w-plane which has been slit along a finite parabolic arc; the focus of the parabolic arc is again at the origin. If $z = 0$ is to correspond to $w = \infty$, the expression (57) clearly has a simple pole at $z = 0$; the only other pole of (57) is the point $z = \alpha$ at which $f(z) = 0$. On the other hand, (57) has double zeros at the points $z = a$, $z = b$ ($|a| = |b| = 1$) which correspond to the ends of the slit. Since the functions

$$\frac{\sqrt{ab}\, z}{(z - a)(z - b)}, \qquad |a| = |b| = 1,$$

and

$$\frac{z}{(z - \alpha)(1 - \bar{\alpha}z)}, \qquad |\alpha| < 1,$$

are real for $|z| = 1$, it thus follows from (57) that

$$\frac{z^3 f'^2(z)(z - \alpha)(1 - \bar{\alpha}z)ab}{f(z)(z - a)^2(z - b)^2}$$

is regular for $|z| \leq 1$ and real for $|z| = 1$. It therefore must be equal to a constant, which we denote by $4abC^2$. Solving for $f(z)$, we find that the desired mapping function is of the form

$$f(z) = C^2 \left[\int_\alpha^z \frac{(z - a)(z - b)\, dz}{z \sqrt{z(z - \alpha)(1 - \bar{\alpha}z)}} \right]^2.$$

EXERCISES

1. If D denotes the finite domain enclosed by the two parabolas

$$v^2 = 4a(a - u), \qquad v^2 = 4a(a + u), \qquad a > 0,\ w = u + iv,$$

show that the conformal mapping of the unit circle onto D is effected by the function

$$w = f(z) = C^2 \left[\int_0^z \frac{dz}{\sqrt{z(1 + z^2)}} \right]^2, \qquad \sqrt{a} = C \int_0^1 \frac{dz}{\sqrt{z(1 + z^2)}}.$$

Hint: Note that the two parabolas intersect at right angles and that consequently $[f'(z)]^{-2}$ has simple poles at the points corresponding to the points of intersection.

2. Show that the mapping

$$w = \sin [C \tan^{-1} z]$$

transforms $|z| < 1$ into the region between the two branches of the same hyperbola, if the foci of the latter are at $w = \pm 1$.

3. Let D be one of the two domains into which an ellipse of foci ± 1 is divided by one branch of a hyperbola whose foci are also at ± 1. Show that

$$w = f(z) = \sin \left[C \int_0^z \frac{dz}{\sqrt{(1 + z^2)(a - z)(1 - az)}} \right]$$

maps $|z| < 1$ onto D, where a is real and $f(a)$ is either 1 or -1, according as D contains the point $w = 1$ or $w = -1$.

4. Let D be that part of an ellipse of foci ± 1 which is obtained by intersecting the ellipse with both branches of a hyperbola of foci ± 1, and which contains the origin. Show that the unit circle $|z| < 1$ is transformed into this domain D by the function

$$w = f(z) = \sin\left[C \int_0^z \frac{dz}{\sqrt{(z^2 - \alpha^2)(z^2 - \bar{\alpha}^2)}}\right],$$

where $|\alpha| = 1$ and α is neither real nor pure imaginary. Show also that this is equivalent to

$$z = \alpha \operatorname{sn}[C' \sin^{-1} w],$$

where $CC' = \bar{\alpha}$ and the modulus of the elliptic function sn is $k = \alpha^2$.

5. If, in the mappings of this section, the fundamental domain is taken to be the upper half-plane $\operatorname{Im}\{z\} > 0$ instead of the unit circle $|z| < 1$, show that (52) and (57) have to be replaced by the conditions

$$\frac{f'^2(z)}{1 - f^2(z)} = \text{real}, \qquad z = \text{real},$$

and

$$\frac{f'^2(z)}{f(z)} = \text{real}, \qquad z = \text{real},$$

respectively.

6. Let $w = f(z)$ be regular on a part L of the real axis, and let $f(z)$ satisfy there the condition

$$\frac{f'^4(z)}{[f(z) - a][f(z) - b][f(z) - c][f(z) - d]} \geq 0.$$

Show that $w = f(z)$ maps L onto an arc C with the following geometric property: If w is on C and β_1 and β_2 are the lines bisecting the angles between the lines connecting w with a, b and b, c, respectively, then the tangent to C at w makes equal angles with β_1 and β_2.

5. The Schwarzian s-functions. In Sec. 7, Chap. V, we discussed the analytic functions which map the upper half-plane onto a triangle whose sides are circular arcs. These functions are known by the name of the *Schwarzian s-functions* or the *Schwarzian triangle functions.* In the present section we shall be concerned with a further study of these functions, where the emphasis will be on the inverse to a given s-function rather than on the function itself. The reasons for doing so are similar to those for considering the elliptic function sn z rather than the elliptic integral (13) of which it is the inverse. While the integral (13) is an infinitely many-valued function, depending on the integration path connecting 0 and w, its inverse $w = \operatorname{sn} z$ is single-valued in its entire domain of existence, and this single-valuedness greatly facilitates all operations involving these functions. The s-functions are also infinitely many-valued. While it is not always true that the inverse of such a function is single-valued, there are important special cases in which the inverse has this property. It is to these cases that we shall devote our particular attention.

Taking account of the fact that we are now interested in the inverse of the s-function, we assume in this section that the curvilinear triangle with the angles $\pi\alpha$, $\pi\beta$, $\pi\gamma$ is situated in the z-plane. In this notation, the function $z = s(w)$ maps the upper half-plane Im $\{w\} > 0$ onto the curvilinear triangle. The inverse to $z = s(w)$ will be denoted by $w = S(z)$ or, if it is desired to mention the angles of the triangle, by $w = S(\alpha,\beta,\gamma;z)$. In this definition, the function $w = S(z)$ is determined only up to an arbitrary linear transformation of the z-plane onto itself. To make things definite, we shall, by a suitable linear substitution, transform the triangle into such a position that the vertex with the angle $\pi\alpha$ is at the origin, while the two circular arcs meeting there become linear segments. Obviously, this is always possible if the angle $\pi\alpha$ is different from zero. Indeed, let C_1 and C_2 be the two circles which meet at $z = A$ under the angle $\pi\alpha \neq 0$. Since $\alpha \neq 0$, C_1 and C_2 intersect also at another point, say B. The linear substitution

$$z^* = \frac{z - A}{z - B}$$

transforms all circles through B into straight lines; in particular, the circles C_1 and C_2 are transformed into two straight lines through the origin (both C_1 and C_2 pass through A). By an additional rotation, we may make one of these lines, say the line connecting $z = 0$ with the vertex at the angle $\pi\gamma$, coincide with the real axis.

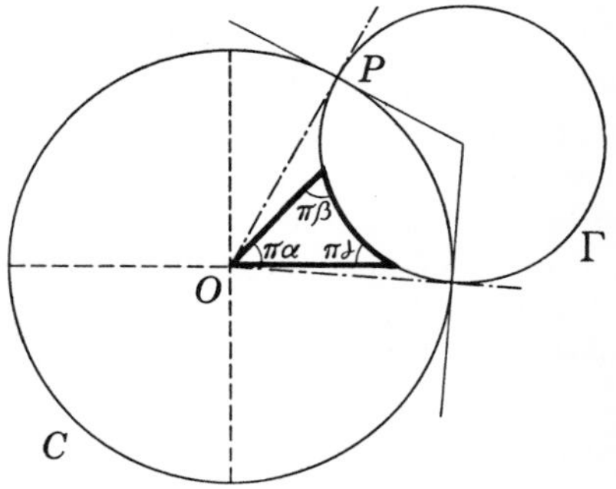

FIG. 40.

In the following, we shall confine ourselves to triangles for which $\alpha + \beta + \gamma < 1$, that is, to triangles the sum of whose angles is less than 2π. Any such triangle has an *orthogonal circle*, that is, a circle which intersects the three circles making up the triangle at right angles. To show this, we bring the triangle into the position just mentioned and we observe that any circle about the origin is obviously orthogonal to the two rectilinear sides of the triangle. If Γ denotes the circle which forms the third side of the triangle, it follows from elementary considerations that the origin is in the exterior of Γ if $\alpha + \beta + \gamma < 1$. Hence, it is possible to draw a tangent from the origin to Γ (see Fig. 40). If P denotes the point of contact of this tangent, then the circle C about the origin which passes through P will obviously be the desired orthogonal circle. It is also clear that the triangle is entirely contained in the interior of the orthogonal circle. If one of the angles $\pi\beta$, $\pi\gamma$ is zero, then the corresponding vertex must be situated on the orthogonal circle. Indeed, the circle about the origin which passes through the vertex is orthogonal to one of the sides of the triangle meeting there; since the angle between the sides is 0, this circle is also orthogonal to the other side. If all three angles are zero, a case to be discussed in detail in the following section, all three vertices of the triangle are situated on the orthogonal circle.

We are mainly interested in those cases in which the functions

$$w = S(\alpha,\beta,\gamma;z)$$

are single-valued functions of the variable z. To find these cases, we observe that $w = S(z)$ maps two linear segments which meet at the origin under the angle $\pi\alpha$ onto a part of the real axis. As shown in detail in Sec. 6, Chap. V, this means that $S(z)$ is of the form

$$S(z) = z^{\frac{1}{\alpha}}S_1(z), \qquad S_1(0) \neq 0, \tag{58}$$

where $S_1(z)$ is regular at $z = 0$. But a power of z cannot be single-valued in the neighborhood of the origin unless its exponent is an integer. In view of $\alpha > 0$, this integer must be positive. $S(z)$ will therefore be single-valued in the neighborhood of $z = 0$ if, and only if, α is the reciprocal of a positive integer. The same result is obtained for β and γ by observing that, by suitable linear transformations, each of the other two vertices can be brought into the center of the orthogonal circle. Hence, a necessary condition for $w = S(z)$ to be single-valued is that α, β, γ be reciprocals of positive integers.

To show that this condition is also sufficient, we observe that, in view of the symmetry principle, all possible analytic continuations of $w = S(z)$ to points outside the original triangle can be obtained by successive inversions of the triangle in the z-plane and by corresponding inversions of the half-plane Im $\{w\} > 0$ with respect to the real axis. Now an inversion with respect to a circular arc transforms circles into circles; moreover, it preserves angles, although it reverses their orientation. Hence, any number of inversions of a circular triangle with respect to various circular arcs will again lead to a circular triangle with the same angles; if the number of inversions is even, the orientation of the angles will be the same as in the original triangle, while the orientation will be reversed if the number of inversions is odd. Since the corresponding inversions in the w-plane are simple symmetries with respect to the real axis, it follows that the boundaries of all these triangles are mapped onto parts of the real axis. By the argument employed in the case of the original triangle, we find therefore that $S(z)$ will be single-valued near the vertices of the inverted triangles if α, β, γ are the reciprocals of positive integers. On the other hand, these vertices are clearly the only possible singularities of $w = S(z)$. If α, β, γ are reciprocals of positive integers, it follows from (58) that $S(z)$ is regular at the vertices at which $S(z) = 0, 1$, while $S(z)$ has a pole of order $1/\beta$ at the vertices at which $S(z) = \infty$.

Summing up our results, we thus find that for all possible analytic continuations of $S(z)$, that is, in the entire domain of existence of this function, $S(z)$ has no singularities except poles of order β^{-1}. If we can show that the domain of existence of $S(z)$ is simply-connected, it will therefore follow from the monodromy theorem (Sec. 5, Chap. III) that $S(z)$ is indeed single-valued. (The fact that the function $S(z)$ has poles does not invalidate the reasoning leading to the proof of the monodromy theorem.) To find the domain of existence of $S(z)$ in the case $\alpha + \beta + \gamma < 1$, we depart from the observation made above that the fundamental triangle is situated within the orthogonal circle C. Since an inversion preserves the magnitudes of angles, it is clear that the sides of an inverted triangle are orthogonal to the image of C yielded by the inversion. But this image coincides with C, since an inversion with respect to a circular arc η transforms any circle orthogonal to η into itself. If we continue in this fashion, it is therefore clear that all the triangles which are obtained from the original triangle by successive inversions have the same orthogonal circle.

It follows that all these triangles are situated in the interior of the orthogonal circle C. This shows that $S(z)$ cannot be continued to points outside C. Within C, however, every point can be reached by a sufficient number of inversions. To see this, suppose that all possible successive inversions have been carried out. At the boundary of the domain covered by these triangles there can be no circular arcs of positive radius, since otherwise it would be possible to enlarge the domain by another inversion. It follows that the boundary of the domain is occupied by limit points of circular arcs whose radii tend to zero. But all these arcs are orthogonal to C, and therefore the circles to which they belong must intersect C. Since their radii tend to zero, the points of these arcs necessarily converge to C. This shows that any point within C can be reached by a sufficient number of inversions. Hence, *in the case* $\alpha + \beta + \gamma < 1$, *the domain of definition of* $S(z)$ *coincides with the interior of the orthogonal circle.* Since $S(z)$ cannot be continued beyond the circumference C of the orthogonal circle, C is a *natural boundary* of the function $S(z)$ as defined in Sec. 5, Chap. III.

As mentioned before, the function $S(z)$ has no singularities but poles if α, β, γ are the reciprocals of positive integers. Since, for $\alpha + \beta + \gamma < 1$, the domain of existence of $S(z)$ is a circle, *i.e.*, a simply-connected domain, it follows from the monodromy theorem that $S(z)$ is single-valued throughout its domain of existence. We thus have the following result.

If

$$\alpha = \frac{1}{m}, \qquad \beta = \frac{1}{n}, \qquad \gamma = \frac{1}{p}, \tag{59}$$

where

$$\frac{1}{m} + \frac{1}{n} + \frac{1}{p} < 1, \tag{60}$$

and m, n, p are positive integers, then the function $w = S(\alpha,\beta,\gamma;z)$ *is single-valued in the interior of the orthogonal circle C of the fundamental triangle.* $S(z)$ *cannot be continued beyond C.*

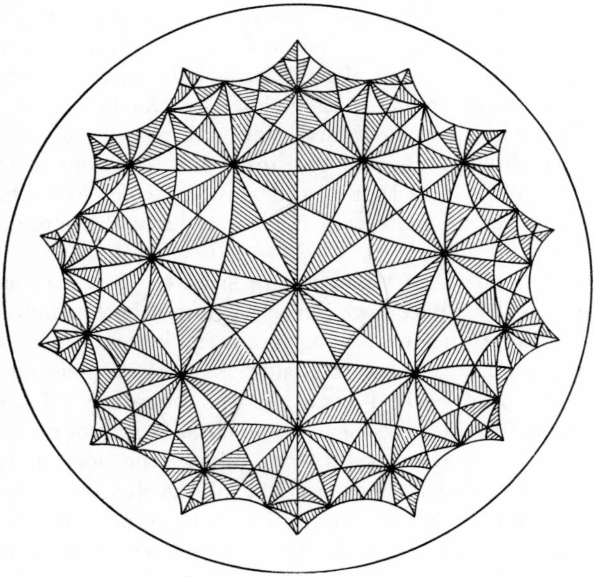

FIG. 41.

The geometric expression of the single-valuedness of the function $S(z)$ in these cases is the fact that the totality of the inversions of the fundamental triangle yields a simple covering of the interior of C. Figure 41 shows part of the triangular net obtained in the case $m = 2$, $n = 7$, $p = 3$, where the shaded and white triangles correspond to the upper and lower half-planes, respectively.

An inversion with respect to a circular arc preserves the magnitude of an angle but inverts its orientation. Hence, an even number of inversions preserves both the magnitude of an angle and its orientation, and it yields therefore a conformal mapping. Since, moreover, an inversion transforms circles into circles, it follows from the results of Sec. 2, Chap. V, that *an even number of inversions is equivalent to a linear transformation.* Since, in the w-plane, the corresponding inversions are simple symmetries with respect to the real axis, an even number of inversions in the w-plane will return us to our point of departure. If the linear transformation corresponding to an even number of inversions in the z-plane is of the form

$$z_1 = \frac{az + b}{cz + d},$$

it follows therefore that the function $w = S(z)$ satisfies the functional relation

$$S\left(\frac{az + b}{cz + d}\right) = S(z), \tag{61}$$

that is, the value of $w = S(z)$ is reproduced if the argument z is made subject to certain linear transformations. Functions with this property are known as *automorphic functions.* The various linear substitutions which reproduce the value of an automorphic function form a group. Indeed, if T and T' denote two such substitutions, it follows from $S(Tz) = S(z)$, $S(T'z) = S(z)$ that

$$S(TT'z) = S[T(T'z)] = S(T'z) = S(z).$$

Moreover, if T^{-1} is the substitution inverse to T, we have

$$S(z) = S(TT^{-1}z) = S(T^{-1}z).$$

The group belonging to a given function $S(z)$ can be constructed from three fundamental substitutions. Since all analytic continuations of $S(z)$ beyond the boundary of the fundamental triangle are obtained by inversions with respect to one of the sides of the triangle or by their combinations, it follows that any even number of inversions can be obtained by a suitable combination of the three linear transformations equivalent to the consecutive performance of two of the three fundamental inversions. The group can therefore be *generated* by three suitable substitutions (and, of course, the inverses of the latter). Every inversion with respect to one of the sides of a triangle leaves the orthogonal circle invariant, and the same is therefore true of the linear substitutions obtained by an even number of inversions. If we normalize the functions $S(z)$ by the requirement that the radius of the orthogonal circle be 1, and if we observe that the most general linear transformation of the unit circle onto itself is

$$z_1 = \delta\left(\frac{a - z}{1 - \bar{a}z}\right), \qquad |\delta| = 1,\ |a| < 1, \tag{62}$$

we arrive at the following result:

If the conditions (59) *and* (60) *are satisfied, then* $S(z)$ *is a single-valued automorphic function in* $|z| < 1$ *which is invariant under a group of linear substitutions of the form* (61), *i.e., we have*

$$f\left[\delta\left(\frac{z-a}{1-\bar{a}z}\right)\right] = f(z) \tag{63}$$

for the substitutions of this group. The group in question can be generated by three particular substitutions and their inverses.

We add a few remarks concerning the analytic expression of the function $w = S(z)$. It was shown in Sec. 7, Chap. V, that the inverse $z = s(w)$ of $w = S(z)$ is, up to an arbitrary linear transformation, of the form (75), Sec. 7, Chap. V. This expression does not, however, yield the inverse of the function $w = S(z)$ as normalized in the present section. To obtain the latter function—corresponding to the position of the fundamental triangle as indicated in Fig. 40—we have to go back to the fact, proved in Sec. 7, Chap. V, that the function mapping the upper half-plane onto a circular triangle with the angles $\pi\alpha$, $\pi\beta$, $\pi\gamma$ can be represented as the quotient of two linearly independent solutions of the hypergeometric differential equation

$$w(1-w)\zeta'' + [c - (1+a+b)w]\zeta' - ab\zeta = 0, \qquad \zeta' = \frac{d\zeta}{dw}, \tag{64}$$

where

$$\alpha = 1 - c, \qquad \beta = b - a, \qquad \gamma = c - a - b. \tag{65}$$

As also shown in Sec. 7, Chap. V, (64) is solved by the hypergeometric series

$$F(a,b,c;w) = 1 + \frac{ab}{1!c}w + \frac{a(a+1)b(b+1)}{2!c(c+1)}w^2 + \cdots, \tag{66}$$

which converges for $|w| < 1$. A suitable second solution is obtained by the remark that by substituting

$$\zeta = w^{1-c}\zeta_1$$

in (64) we arrive at a differential equation for ζ_1 which differs from that for ζ only by the fact that a, b, c are now replaced by $a - c + 1$, $b - c + 1$, $2 - c$, respectively. Since (64) is solved by the function (66), it follows that another solution of (64) is given by

$$w^{1-c}F(a-c+1,\, b-c+1,\, 2-c;\, w), \qquad 0 < c < 1. \tag{67}$$

Since, for $w = 0$, this solution reduces to 0, while the solution (66) reduces to 1, these two solutions are linearly independent.

Consider now the quotient of the two solutions (66) and (67), *i.e.*, the function

$$z = s(w) = \frac{w^{1-c}F(a - c + 1, b - c + 1, 2 - c; w)}{F(a,b,c;w)}, \tag{68}$$

where a, b, c are connected with α, β, γ by (65), or, if we solve (65) for a, b, c, by

$$a = \tfrac{1}{2}(1 - \alpha - \beta - \gamma), \qquad b = \tfrac{1}{2}(1 - \alpha + \beta - \gamma), \qquad c = 1 - \alpha. \tag{69}$$

Since $c < 1$, it follows from (68) that $s(0) = 0$, which shows that one vertex of the circular triangle is situated at $z = 0$. The hypergeometric function (66) is obviously real if all its arguments are real. If we choose that determination of w^{1-c} which is real for positive w, it follows therefore from (68) that $s(w)$ is real if w varies along the real axis from $w = 0$ to $w = 1$. Hence, one side of the circular triangle is a part of the positive axis terminating at the origin. A second side of the circular triangle is obtained as the conformal image of the negative axis $-\infty < w < 0$. While the two hypergeometric functions in (68) are also real for these values of w, the factor w^{1-c} is now equal to

$$\begin{aligned} w^{1-c} &= (-|w|)^{1-c} = |w|^{1-c}(e^{\pi i})^{1-c} = |w|^{1-c}e^{\pi i(1-c)} \\ &= |w|^{1-c}e^{\pi i\alpha}. \end{aligned}$$

This shows that $-\infty < w < 0$ is mapped by $z = s(w)$ onto a linear segment which makes the angle $\pi\alpha$ with the real axis at the origin. We have thus proved that the circular triangle yielded by the function (68) has indeed the position indicated in Fig. 40.

We finally compute the coordinates of the vertices of the circular triangle upon which $\operatorname{Im}\{w\} > 0$ is mapped by (68), that is, the values of $s(0)$, $s(1)$, $s(\infty)$. To find $w(1)$, we observe that, by (73), Sec. 7, Chap V,

$$\begin{aligned} F(a,b,c;1) &= \frac{\Gamma(c)}{\Gamma(b)\Gamma(c-b)} \int_0^1 t^{b-1}(1-t)^{c-a-b-1}\,dt \\ &= \frac{\Gamma(c)\Gamma(c-a-b)}{\Gamma(c-a)\Gamma(c-b)}. \end{aligned}$$

The condition $c - a - b > 0$, which is required for the existence of the integral, is satisfied since, by (65), $\gamma = c - a - b$. Using (68) we thus find

$$s(1) = \frac{\Gamma(2-c)\Gamma(c-a)\Gamma(c-b)}{\Gamma(c)\Gamma(1-a)\Gamma(1-b)}. \tag{70}$$

This is the vertex with the angle $\pi\gamma$. Since $s(0) = 0$, it thus remains to find the vertex $s(\infty)$, corresponding to the angle $\pi\beta$. For this purpose we use the fact, for which the reader is referred to Whittaker-Watson's "Modern Analysis," that the analytic continuation of the function $F(a,b,c;w)$ to points near $w = \infty$ of the cut plane $|\arg\{-w\}| < \pi$ is given by

$$(71)\quad F(a,b,c;w) = \frac{\Gamma(c)\Gamma(a-b)}{\Gamma(b)\Gamma(c-a)}(-w)^{-a}F(a, 1-c+a, 1-b+a; w^{-1}) + \frac{\Gamma(c)\Gamma(b-a)}{\Gamma(a)\Gamma(c-b)}(-w)^{-b}F(b, 1-c+b, 1-a+b; w^{-1}).$$

Using the fact that, by (65), $b - a = \beta > 0$, we find from (71) and (68) by a formal computation that

$$(72)\qquad s(\infty) = e^{\pi i(1-c)}\frac{\Gamma(b)\Gamma(c-a)\Gamma(2-c)}{\Gamma(c)\Gamma(b-c+1)\Gamma(1-a)}.$$

If $\alpha = 0$, that is, $c = 1$, the representation (68) breaks down since the two solutions (66) and (67) of the hypergeometric equation (64) cease to be independent. This corresponds to the fact that in this case the triangle degenerates into a linear segment. The other two angles may, of course, take the value 0. Hence, the only case in which the representation (68) seems to be unable to yield the mapping function of a circular triangle with given angles is that in which all the angles are zero. This difficulty can, however, be circumvented by means of a simple artifice, indicated in Exercise 2. A more detailed discussion of the case $\alpha = \beta = \gamma = 0$ will be found in the following section.

EXERCISES

1. If (68) maps $\text{Im}\{w\} > 0$ onto a circular triangle whose angles $\pi\beta$ and $\pi\gamma$ are zero, show that the radius of the orthogonal circle of the triangle is

$$R = \frac{\Gamma(1+\alpha)\Gamma^2\left(\frac{1}{2}-\frac{\alpha}{2}\right)}{\Gamma(1-\alpha)\Gamma^2\left(\frac{1}{2}+\frac{\alpha}{2}\right)},$$

which, in view of the identities

$$x\Gamma(x) = \Gamma(x+1), \qquad \Gamma(x)\Gamma(1-x) = \frac{\pi}{\sin \pi x},$$

can also be brought into the form

$$R = \frac{\alpha}{\pi^3}\sin \pi\alpha \sin^2\frac{\pi\alpha}{2}\,\Gamma^2(\alpha)\Gamma^4\left(\frac{1}{2}-\frac{\alpha}{2}\right).$$

Hint: Use the fact that a vertex with the angle zero must lie on the orthogonal circle.

2. Show that the inverse of the function

$$z = s(\tfrac{1}{2},0,0;w) = \sqrt{w}\,\frac{F(\frac{3}{4},\frac{3}{4},\frac{3}{2};w)}{F(\frac{1}{4},\frac{1}{4},\frac{1}{2};w)}$$

maps the zero-angle triangle ABA' of Fig. 42 onto the full z-plane which has been cut along the rays $-\infty \leq z \leq 0$, $1 \leq z \leq \infty$. Show also that the radius of the orthogonal circle is

$$\frac{4}{\pi^2}\Gamma^4\left(\frac{1}{4}\right),$$

and that the mapping of the upper half-plane onto the triangle ABA' is effected by the function $z = s[0,0,\frac{1}{2};(1 - w^2)^{-1}]$. *Hint:* Consider the triangle OBA, and apply the symmetry principle with respect to the linear segment OB.

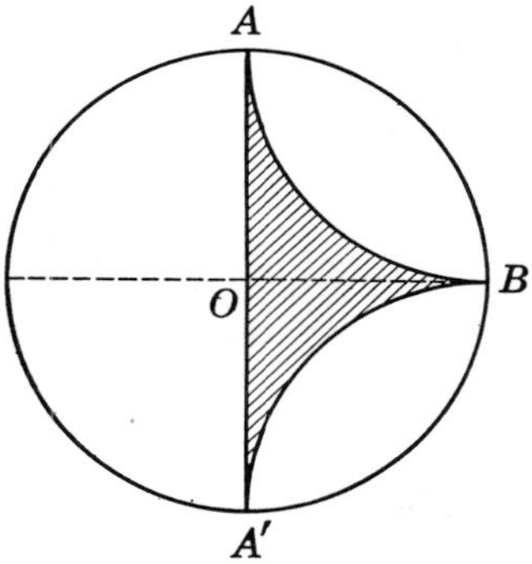

FIG. 42.

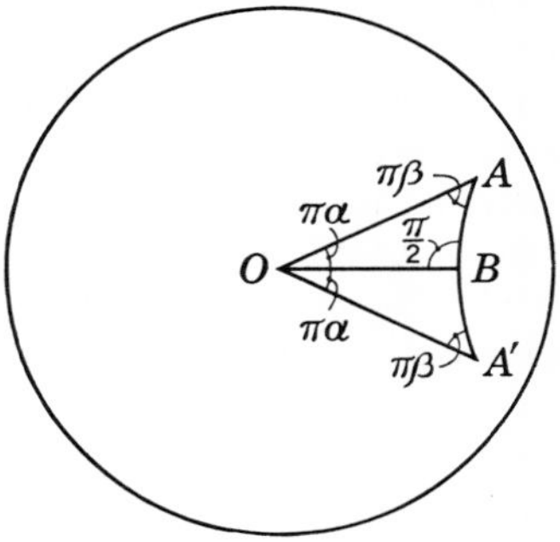

FIG. 43.

3. Using the fact that the triangle OAA' (see Fig. 43) with the angles $2\pi\alpha$, $\pi\beta$, $\pi\beta$ can be obtained from the triangle OAB with the angles $\pi\alpha$, $\pi\beta$, $\frac{1}{2}\pi$ by means of an inversion with respect to OB, prove the identity

$$4S[\alpha,\beta,\tfrac{1}{2};(-\tfrac{1}{4})^{\alpha}z] = \frac{S^2(2\alpha,\beta,\beta;z)}{S(2\alpha,\beta,\beta;z) - 1}.$$

Hint: Observe that the mapping

$$w_1 = \frac{w^2}{4(w - 1)}$$

transforms the half-plane $\text{Im}\,\{w\} > 0$ into the w_1-plane which is slit along $-\infty \leq w_1 \leq 0$, $1 \leq w_1 \leq \infty$, and that both $w = 1$ and $w = \infty$ correspond to $w_1 = \infty$ while the origin is mapped onto itself. The constant factor of z in the left-hand side of the above identity is most conveniently determined by comparing the expressions (68) for the inverses of the two S-functions at $z = w = 0$.

4. If $\alpha + \beta + \gamma > 1$, show that the function $S(\alpha,\beta,\gamma;z)$ can be single-valued only in the following cases:

α	β	γ	
$\frac{1}{2}$	$\frac{1}{2}$	n^{-1}	(n arbitrary)
$\frac{1}{2}$	$\frac{1}{3}$	$\frac{1}{3}$	
$\frac{1}{2}$	$\frac{1}{3}$	$\frac{1}{4}$	
$\frac{1}{2}$	$\frac{1}{3}$	$\frac{1}{5}$	

5. If $\alpha + \beta + \gamma = 1$, show that the only single-valued functions $S(\alpha,\beta,\gamma;z)$ are those corresponding to the values

α	β	γ
$\frac{1}{2}$	$\frac{1}{4}$	$\frac{1}{4}$
$\frac{1}{2}$	$\frac{1}{3}$	$\frac{1}{6}$
$\frac{1}{3}$	$\frac{1}{3}$	$\frac{1}{3}$

Show also that in these cases the function $S(\alpha,\beta,\gamma;z)$ is single-valued at all finite points of the z-plane.

6. The Elliptic Modular Function. We devote this section to a more detailed study of the Schwarzian s-function for which $\alpha = \beta = \gamma = 0$, that is, the function which maps the upper half-plane onto a circular triangle with zero angles. In order to obtain an explicit expression for this function, we might use the result of Exercise 2 of the preceding section. It is, however, more convenient to utilize the result of Exercise 3, Sec. 7, Chap. V. As shown there, the function $f(w)$ mapping Im $\{w\} > 0$ onto a zero-angle triangle is of the form

$$z = f(w) = \frac{K(1-w)}{K(w)}, \tag{73}$$

where

$$K(w) = \int_0^1 \frac{dt}{\sqrt{(1-t^2)(1-wt^2)}}. \tag{74}$$

In order to determine the exact location of the triangle in the z-plane, we observe from (74) that $K(0) = \frac{1}{2}\pi$ and that $K(w)$ increases from this value to $+\infty$ if w varies from 0 to 1. It therefore follows from (73) that $f(0) = \infty$ and $f(1) = 0$, while $f(w)$ is positive for $0 < w < 1$. Hence, the side of the triangle which corresponds to the segment $0 < w < 1$ coincides with the positive axis. Since the angles of the triangle are zero, the other two sides must be a half circle touching the real axis at the origin and a parallel to the real axis. We thus obtain the zero-angle triangle of Fig. 44. It is clear that the triangle must lie in the lower half-plane since its interior must be surrounded in the positive sense if z traverses the positive axis from $+\infty$ to 0. Finally, the fact that the vertex corresponding to $w = \infty$ is at $z = -i$ follows from the relation

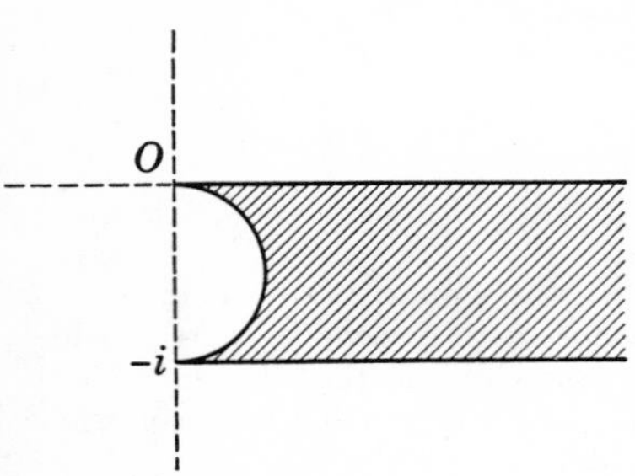

FIG. 44.

$$f(w)f(1-w) = 1$$

which is an immediate consequence of (73). Setting, in particular, $w = \frac{1}{2} + it$, we obtain $f(\frac{1}{2} + it)f(\frac{1}{2} - it) = 1$. Since $f(w)$ is real for $0 < w < 1$, it can be continued across $0 < w < 1$ by the relation $f(\bar{w}) = \overline{f(w)}$. Hence, $f(\frac{1}{2} - it) = \overline{f(\frac{1}{2} + it)}$ and therefore $|f(\frac{1}{2} + it)|^2 = 1$. For $t \to \infty$, it follows that $|f(\infty)| = 1$. Since $f(\infty)$ must also lie on the lower section of the imaginary axis, we arrive at $f(\infty) = -i$.

There exists an interesting connection between this mapping function and the elliptic function sn u discussed in Sec. 3. A comparison of (73) and (74) with (19), (20), and (21) shows that if w is identified with the square of the modulus k of the elliptic function sn u, then $f(w)$ is equal to K'/K. We thus have the parametric representation

$$w = k^2, \qquad z = f(w) = \frac{K'(k)}{K(k)}.$$

It is because of this connection that the inverse of the function $z = f(w)$ is known as the *elliptic modular function*. Because of the resulting simplification of many formulas it is customary to give this name to the inverse of $z = if(w)$ rather than to that of $z = f(w)$. This function is therefore defined by

$$w = k^2, \qquad z = i\frac{K'}{K} = \tau,$$

where τ is the quantity defined in (15). If we use the symbol $w = J(z)$ for the elliptic modular function, $J(z)$ is thus defined by

$$z = i\frac{K'}{K} = \tau, \qquad J(z) = k^2. \tag{75}$$

In view of (16) and (30), $J(z)$ is represented by the infinite product

$$J(z) = 16e^{\pi iz} \prod_{n=1}^{\infty} \left(\frac{1 + e^{2\pi inz}}{1 + e^{(2n-1)\pi iz}}\right)^8, \qquad \text{Im }\{z\} > 0. \tag{76}$$

In view of what was said before, $w = J(z)$ maps the zero-angle triangle of Fig. 45 onto the upper half-plane Im $\{w\} > 0$. Obviously, the orthogonal circle of this triangle is the real axis Im $\{z\} = 0$ and its interior is the upper half-plane Im $\{z\} > 0$. In view of the results of the preceding section, $J(z)$ may therefore be continued to all points of the half-plane Im $\{z\} > 0$ and $J(z)$ is regular and single-valued there. The real axis is a natural boundary of the function $J(z)$ and $J(z)$ cannot be continued to the lower half-plane Im $\{z\} < 0$. As likewise shown in the preceding section, the function $J(z)$, being a particular case of the triangle functions considered there, is an automorphic function, *i.e.*, we have

$$J\left(\frac{\alpha z + \beta}{\gamma z + \delta}\right) = J(z)$$

for a certain group of linear substitutions which is generated by three fundamental substitutions. As the latter we may take the substitutions equivalent to the successive performance of two different inversions with respect to the sides of the triangle of Fig. 45. If z_1, z_2, z_3 are the points inverse to a point z with respect to Re $\{z\} = 0$, Re $\{z\} = 1$, $|z - \frac{1}{2}| = \frac{1}{2}$, respectively, the reader will verify that the relations

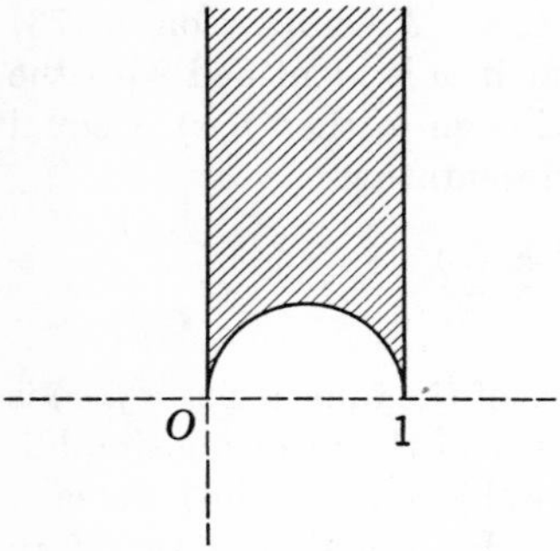

FIG. 45.

$$z_1 = -\bar{z}, \qquad z_2 = 2 - \bar{z}, \qquad z_3 = \frac{\bar{z}}{2\bar{z} - 1}$$

hold. Performing two of these inversions in succession, we obtain the six points

$$z \pm 2, \qquad \frac{z}{1 \pm 2z}, \qquad \frac{2 - z}{3 - 2z}, \qquad \frac{2 - 3z}{1 - 2z}.$$

It follows that

$$(77) \qquad \begin{aligned} J(z \pm 2) &= J(z), & J\left(\frac{z}{1 \pm 2z}\right) &= J(z), \\ J\left(\frac{2 - z}{3 - 2z}\right) &= J(z), & J\left(\frac{2 - 3z}{1 - 2z}\right) &= J(z). \end{aligned}$$

By suitable combination of these substitutions it can be shown that

$$(78) \qquad J\left(\frac{\alpha z + \beta}{\gamma z + \delta}\right) = J(z),$$

where β, γ are even integers and α, δ are odd integers which are subject to the condition $\alpha\delta - \beta\gamma = 1$ and are otherwise arbitrary. The proof of this result is left for Exercises 1 to 5 at the end of this section.

If the half-plane Im $\{z\} > 0$ is filled by an infinity of successive inversions of the original triangle, the w-plane is covered by an infinity of upper and lower half-planes which are the conformal images of the inverted triangles. Each half-plane has three "neighbors" which are connected with it along the stretch $0 < w < 1$ and the rays $-\infty < w < 0$ and $1 < w < \infty$, respectively. The totality of half-planes which are connected with each other in the manner indicated is known as the *modular surface.* It is clear that the modular surface M, which, in view of the foregoing, is the conformal image of $\text{Im}\{z\} > 0$ as given by $w = J(z)$, has the following properties:

(*a*) M has no branch points.

(*b*) The only boundary points of M are the points $w = 0, 1, \infty$. These points do not belong to M. Indeed, each of the half-planes making up M is connected with its neighbors along all points of the real axis with the exception of the points $w = 0, 1, \infty$. The latter points do not belong to any of the half-planes, which shows that they do not belong to M.

(*c*) Every point of the w-plane, with the exception of $w = 0, 1, \infty$, is covered by M an infinity of times.

The fact that the function $w = J(z)$ maps $\text{Im}\{z\} > 0$ onto the modular surface described above can be utilized for a short proof of a famous result on entire functions, known as *Picard's theorem*. We recall that an entire function $f(z)$ is an analytic function which is regular for all finite values of z. As shown in Sec. 7, Chap. III, such a function must have a singularity at $z = \infty$ unless it reduces to a constant. Leaving aside the uninteresting case in which this singularity is a pole and, consequently, $f(z)$ is a polynomial, we consider the case in which $z = \infty$ is an essential singularity. We saw in Sec. 6, Chap. III, that in the neighborhood of an essential singularity an analytic function will approach any given value arbitrarily closely. Hence, an entire function will come arbitrarily near any given value [if $f(z)$ is a polynomial, it follows from the fundamental theorem of algebra that every value is actually taken by $f(z)$]. We shall now show that even more is true and that the following theorem holds.

An entire function takes every finite value, with one possible exception.

That such an exceptional value can occur is shown by the function e^z which never takes the value 0. To prove Picard's theorem, suppose that $f(z)$ is not a constant and that it leaves out the values a and b. Obviously, the function

$$g(z) = \frac{f(z) - a}{b - a} \tag{79}$$

is also an entire function and $g(z)$ leaves out the values 0 and 1. Consider now the function

$$u(z) = J^{-1}[g(z)], \tag{80}$$

where $J^{-1}(w)$ is the inverse of the elliptic modular function $J(z)$. $J^{-1}(w)$ maps the modular surface M onto the upper half-plane and it is therefore defined and regular at all points of the w-plane except at the points $w = 0, 1, \infty$. $J^{-1}(w)$ is, of course, infinitely many-valued. However, once a definite branch has been chosen at a given point w, the analytic continuation of $J^{-1}(w)$ along any curve which avoids the points $w = 0, 1, \infty$ is uniquely determined. (Note that M has no branch points!) Since the function $g(z)$ does not take the values $w = 0, 1$ and, because of its regularity, is also different from ∞, it follows therefore that the function $u(z)$ defined in (80) is regular at all finite points of the plane. By the monodromy theorem it is thus also single-valued. Moreover, since the values of $u(z)$ are a subset of the values of $J^{-1}(w)$, we have $\text{Im }\{u(z)\} > 0$ for all finite z. The transformation

$$u \to \frac{u - i}{u + i}$$

carries the upper half-plane into the interior of the unit circle. It follows that the function

$$v(z) = \frac{u(z) - i}{u(z) + i}$$

is regular and satisfies $|v(z)| \leq 1$ for all finite z. By Liouville's theorem (Sec. 7, Chap. III), $v(z)$ therefore reduces to a constant. Hence, the same is true for $v(z)$ and, in view of (79) and (80), also for $f(z)$. Our assumption that $f(z)$ is not a constant and leaves out the values a and b has thus led to a contradiction. This proves Picard's theorem.

We now introduce the analytic function $Q(z)$ which is related to $J(z)$ by the identity

$$J(z) = Q(e^{\pi i z}), \tag{81}$$

which can also be written

$$Q(z) = J\left(\frac{1}{\pi i} \log z\right). \tag{82}$$

Since the transformation $z \to e^{\pi i z}$ carries the half-plane $\text{Im }\{z\} > 0$ into the (infinitely often covered) interior of the unit circle, the domain of definition of $Q(z)$ will be the interior of the unit circle. In view of (76), $Q(z)$ has the product expansion

$$Q(z) = 16z \prod_{n=1}^{\infty} \left(\frac{1 + z^{2n}}{1 + z^{2n-1}}\right)^8, \qquad |z| < 1. \tag{83}$$

This formula shows that $Q(z)$ is regular in the unit circle and vanishes at the origin.

The surface M^* onto which the unit circle is mapped by $w = Q(z)$ is closely related to the modular surface M. In order to find the structure of M^*, we observe that the function $z = e^{\pi i w}$ maps $\text{Im }\{w\} > 0$ onto a domain D which covers the unit circle $|z| < 1$ an infinity of times, with the exception of the point $z = 0$ which is not covered at all. Furthermore, D has no branch points and all boundary points of D satisfy either $|z| = 1$ or $z = 0$. The inverse function, that is, $w = (1/\pi i) \log z$, maps D onto the upper half-plane $\text{Im}\,\{w\} > 0$. Hence, if z approaches a point of $|z| = 1$, $w = (1/\pi i) \log z$ will approach a point of the real axis. In view of (82) and the fact that $J(w)$ maps $\text{Im}\,\{w\} > 0$ onto the modular surface, $Q(z)$ will then approach a boundary point of the modular surface, *i.e.*, one of the points $0, 1, \infty$. It is further clear that $Q(z)$ does not take the values $0, 1, \infty$ if z is a point such that $0 < |z| < 1$, since these values must coincide with points of the modular surface.

The point $z = 0$ requires a special discussion, since $z = 0$ is a boundary point of D. If we write $z = re^{i\theta}$, we have

$$\frac{1}{\pi i} \log z = \frac{1}{\pi i} \log r + \frac{\theta}{\pi} = -\frac{i}{\pi} \log r + \frac{\theta}{\pi}.$$

Hence, by (82),

$$\lim_{z \to 0} Q(z) = \lim_{r \to 0} J\left(-\frac{i}{\pi} \log r + \frac{\theta}{\pi}\right).$$

We now keep θ fixed and let $r \to 0$. Since $\log r \to -\infty$, it follows that the required limit is equal to the limiting value of $J(w)$ if $\text{Re}\,\{w\} = \text{const.}$ and $\text{Im}\,\{w\} \to +\infty$. But this value is 0. Indeed, as shown by Fig. 45, such a path leads to the third vertex of the fundamental triangle or of one of the triangles obtained from it by inversion, and this vertex is mapped onto the point 0. Hence, $\lim_{z \to 0} Q(z) = 0$. in accordance with (83).

The surface M^* onto which $|z| < 1$ is mapped by the function $w = Q(z)$ may now be described as follows.

(*a*) M^* has no branch points.

(*b*) Every point of the w-plane, with the exception of $w = 0, 1, \infty$, is covered by M^* an infinity of times.

(*c*) The only boundary points of M^* are the points $w = 0, 1, \infty$. The points $w = 1, \infty$ do not belong to M^*, while the point $w = 0$ is covered by one sheet of M^* only.

With the help of the function $Q(z)$ we can prove the following striking result on analytic functions regular in the unit circle.

Let $f(z) = z + a_2z^2 + \cdots$ be regular in the unit circle and let $f(z) \neq 0$ at all points of the unit circle other than its center. If α is a value such that

$$|\alpha| < \tfrac{1}{16},$$

then the equation $f(z) = \alpha$ has a solution in $|z| < 1$. The constant $\frac{1}{16}$ is sharp, i.e., it cannot be replaced by a larger value.

The conclusion of the theorem may also be expressed by saying that the conformal image of the unit circle as given by $w = f(z)$ completely covers the interior of a circle about the origin whose radius is $\frac{1}{16}$. To prove the theorem, suppose β is a value not taken by $f(z)$ in $|z| < 1$ and consider the function

$$q(z) = Q^{-1}\left[\frac{f(z)}{\beta}\right]. \tag{84}$$

Since $Q(w) = 0$ for $w = 0$ and $Q(w) \neq 0$ for $0 < |w| < 1$, the inverse of $Q(w)$ is uniquely defined in a small neighborhood of $w = 0$, and we have $q(0) = 0$. Starting from $z = 0$, $q(z)$ can be analytically continued along any curve contained in the unit circle. Indeed, $\beta^{-1}f(z)$ does not take the values $0, 1, \infty$ for $0 < |z| < 1$, and these are the only points at which the analytic continuation of the inverse of $Q(w)$ would become impossible. Hence, the function $q(z)$ defined by (84) is regular in $|z| < 1$. Moreover, since the values of $q(z)$ are a subset of the values of $Q^{-1}(W)$, it follows that $|q(z)| < 1$. In view of $q(0) = 0$, this entails $|q'(0)| \leq 1$, where equality can hold only if $q(z) \equiv \gamma z, |\gamma| = 1$. Expressing $q(z)$ by means of (84), we obtain

$$\frac{|f'(0)|}{|\beta||Q'(0)|} \leq 1.$$

Since $f'(0) = 1$ and, by (83), $Q'(0) = 16$, this yields

$$|\beta| \geq \tfrac{1}{16}, \tag{85}$$

where equality is possible only if $f(z) \equiv \frac{1}{16}\bar{\gamma}Q(\gamma z), |\gamma| = 1$. A value β which is not taken by $f(z)$ in $|z| < 1$ must therefore satisfy (85). This proves our theorem. That the constant $\frac{1}{16}$ cannot be improved upon is shown by the function

$$w = f(z) = \tfrac{1}{16}Q(z) = z + \cdots$$

which satisfies the hypotheses of the theorem and leaves out the value $w = \frac{1}{16}$.

The value $\frac{1}{16}$ of the constant in the preceding theorem was taken from the expression (83) for $Q(z)$, which, in turn, was derived from the theory of elliptic functions. It thus would appear that previous knowledge of elliptic functions and of their connection with the elliptic modular function is required for an adequate theoretical and numerical treatment of the latter function. This, however, is not the case. We shall show that the main properties of the function $Q(z)$ can be easily derived from the geometric properties of the surface M^*.

Let then $w = Q(z)$ be the function which maps $|z| < 1$ onto M^* in such a way that $Q(0) = 0$ and $Q'(0) > 0$, and consider the function

$Q_1(z) = \sqrt{Q(z^2)}$. $Q_1(z)$ is also regular in $|z| < 1$. Indeed, its only possible singularities are the zeros of $Q(z^2)$, and $Q(z)$ does not vanish in $|z| < 1$ except at $z = 0$ (only one sheet of M^* covers the origin). At $z = 0$, $Q_1(z)$ is obviously regular. $Q_1(z)$ is, moreover, an odd function, that is, $Q_1(-z) = -Q_1(z)$, as the reader will verify without difficulty. The surface M_1 onto which $w_1 = Q_1(z)$ maps the unit circle clearly has the following properties.

(*a*) All boundary points of M_1 are situated at $w_1 = -1, 0, 1, \infty$. Indeed, because of $Q_1{}^2(z) = Q(z^2)$, any boundary point of M_1 other than those mentioned would give rise to a boundary point of M^* other than those at $w = 0, 1, \infty$.

(*b*) The points $w_1 = \pm 1, \infty$ are not covered by M_1, while the point $w_1 = 0$ is covered by one sheet of M_1 only. This follows in the same way as property (*a*).

(*c*) M_1 has no branch points, that is, $Q_1'(z) \neq 0$ for $|z| < 1$. This is a consequence of $Q_1(z)Q_1'(z) = zQ'(z^2)$ and the fact that M^* has no branch points [that is, $Q'(z) \neq 0$].

Consider now the function

$$w_2 = Q_2(z) = \frac{4Q_1(z)}{[1 + Q_1(z)]^2}.$$

The transformation

$$w_2 = \frac{4w_1}{(1 + w_1)^2}$$

carries the points $w_1 = 0, \infty, 1, -1$ into the points $w_2 = 0, 0, 1, \infty$, respectively. Moreover, no point w_1 other than $0, \infty, 1, -1$ is transformed into one of the points $w_2 = 0, 1, \infty$. It follows that the boundary points of the surface M_2 onto which $|z| < 1$ is mapped by $w_2 = Q_2(z)$ can be situated only at the points $w_2 = 0, 1, \infty$. The points $w_2 = 1, \infty$ are clearly not covered by M_2, while $w_2 = 0$ is covered by one sheet of M_2 only (for $z = 0$). M_2 has no branch points. Indeed,

$$Q_2'(z) = 4\,\frac{[1 - Q_1(z)]}{[1 + Q_1(z)]^3}\,Q_1'(z),$$

and since $Q_1(z) \neq 1$, $Q_1'(z) \neq 0$, we also have $Q_2'(z) \neq 0$.

If we compare this with the description of the surface M^* given above, we find that M_2 has precisely those properties which characterized M^*. From this it is not difficult to conclude that M^* and M_2 must be identical, or, what amounts to the same thing, that $Q_2(z) = Q(z)$. If we examine the argument leading to the fact that the function $q(z)$ of (84) satisfies $|q(z)| < 1$, provided $f(z) \neq 0$ for $0 < |z| < 1$ and $f(z) \neq \beta$ for $|z| < 1$, we find that the only properties of $Q(z)$ used were the properties of the surface M^* listed above. Since these are also properties of M_2, the above argument will therefore remain valid if $Q(z)$ is replaced by $Q_2(z)$. Furthermore, both $Q(z)$ and $Q_2(z)$ may take the role of the function $\beta^{-1}f(z)$ in (84). It follows that

$$|q_1(z)| = |Q^{-1}[Q_2(z)]| \leq 1, \qquad |z| < 1,$$

and

$$|q_2(z)| = |Q_2^{-1}[Q(z)]| \leq 1, \qquad |z| < 1.$$

Since, moreover, $q_1(0) = q_2(0) = 0$, we have $|q_1'(0)| \leq 1$, $|q_2'(0)| \leq 1$, where equality is possible only if $q_1(z) \equiv \gamma_1 z$ and $q_2(z) \equiv \gamma_2 z$, respectively, where $|\gamma_1| = |\gamma_2| = 1$. In view of

$$q_1'(0) = \frac{Q_2'(0)}{Q'(0)}, \qquad q_2'(0) = \frac{Q'(0)}{Q_2'(0)} = \frac{1}{q_1'(0)},$$

we necessarily have $|q_1'(0)| = |q_2'(0)| = 1$, and thus $q_1(z) = \gamma_1 z$. Because of

$$Q_2(z) = \frac{4\sqrt{Q(z^2)}}{[1 + \sqrt{Q(z^2)}]^2}, \tag{86}$$

we shall have $Q_2'(0) > 0$ if the positive value of the radical is taken and if $Q'(0) > 0$. Hence, $\gamma_1 = 1$ and therefore

$$Q^{-1}[Q_2(z)] = q_1(z) = z.$$

It follows that indeed $Q_2(z) = Q(z)$. Applying this result to (86), we find that the function $Q(z)$ satisfies the remarkable functional equation

$$Q(z) = \frac{4\sqrt{Q(z^2)}}{[1 + \sqrt{Q(z^2)}]^2}. \tag{87}$$

If we differentiate (87) and set $z = 0$ in the result, we obtain

$$Q'(0) = 4\sqrt{Q'(0)},$$

whence $Q'(0) = 16$. This yields the constant in the above "$\frac{1}{16}$-theorem," which has thus been proved without loans from the theory of elliptic functions. More generally, we can use the functional equation (87) in order to obtain all the coefficients of the power series expansion of $Q(z)$. If we write

$$Q(z) = 16[z + a_2 z^2 + a_3 z^3 + \cdots], \tag{88}$$

we have

$$\sqrt{Q(z^2)} = 4z\sqrt{1 + a_2 z^2 + a_3 z^4 + \cdots}$$
$$= 4z + 2a_2 z^3 + \left(2a_3 - \frac{a_2^2}{2}\right) z^5 + \cdots.$$

Hence,

$$(89)\qquad \frac{4\sqrt{Q(z^2)}}{[1+\sqrt{Q(z^2)}]^2} = 4\sum_{n=1}^{\infty} (-1)^{n-1} n[\sqrt{Q(z^2)}]^n$$
$$= 4\left[4z - 32z^2 + (2a_2 + 192)z^3 - (32a_2 + 1{,}024)z^4 + \left(2a_3 - \frac{a_2{}^2}{2} + 288a_2 + 5{,}120\right)z^5 + \cdots\right].$$

Equating this to $Q(z)$, we obtain

$$a_2 = -8, \qquad a_3 = 44, \qquad a_4 = -192, \qquad a_5 = 718, \; . \; . \; . \; .$$

The first terms of the power series expansion of $Q(z)$ are therefore

$$(90)\qquad Q(z) = 16[z - 8z^2 + 44z^3 - 192z^4 + 718z^5 - \cdots].$$

From this procedure we may draw yet another conclusion. Clearly, the coefficient of z^{2n-1} in the expansion of the odd function $\sqrt{Q(z^2)}$ is a polynomial in the coefficients $a_2, \ldots, a_n$. It follows that both the coefficient of z^{2n-1} and the coefficient of z^{2n} in the expansion (89) are polynomials in $a_2, \ldots, a_n$. Since (89) is equal to (88) we thus find that both a_{2n-1} and a_{2n} can be expressed as polynomials of the coefficients $a_2, \ldots, a_n$. This shows that all the coefficients of $Q(z)$ are uniquely determined by our procedure. Hence, the function $Q(z)$ is the only solution of the functional equation (87) which has a finite, nonvanishing derivative at the origin. The latter condition is, of course, essential, since we determined $Q'(0)$ from the equation $Q'(0) = 4\sqrt{Q'(0)}$. If this condition is dropped, the equation (87) will obviously also admit the solutions $Q(z^\nu)$, where ν is an arbitrary complex number.

The fact that $Q(z)$ is the only solution of (87) makes it possible to identify $Q(z)$ with the square of the modulus $k = k(q)$ of the Jacobian elliptic functions. Formula 43 (Sec. 3) shows that $k^2(q)$, taken as a function of q, satisfies the functional equation (87). Since, in view of (41), the derivative of this function has a finite, nonzero value for $q = 0$, it follows therefore that $k^2(z) = Q(z)$ and that $Q(z)$ can be computed by the formula (83) which is identical with the formula (41) for k^2. For the practical computation of $Q(z)$ it is, however, better to use the result of Exercise 17, Sec. 3, *i.e.*, the formula

$$(91)\qquad Q(z) = 16z\left[\frac{\sum_{n=0}^{\infty} z^{n(n+1)}}{1 + 2\sum_{n=1}^{\infty} z^{n^2}}\right]^4, \qquad |z| < 1,$$

since the two series involved converge with extraordinary rapidity, especially if $|z|$ is not too close to 1. If $|z|$ is close to 1, the convergence can be further improved by first computing $Q(z^2)$ and then using the identity (87).

There exists a simple relation between $Q(z)$ and $Q(-z)$, namely,

$$Q(-z) = \frac{Q(z)}{Q(z) - 1}. \tag{92}$$

To prove (92), we observe that $\sqrt{Q(z^2)}$ is an odd function of z and that therefore, by (87),

$$Q(-z) = -\frac{4\sqrt{Q(z^2)}}{[1 - \sqrt{Q(z^2)}]^2}.$$

Combining this and (87), we obtain

$$\begin{aligned}\frac{Q(z)}{Q(-z)} &= -\left[\frac{1 - \sqrt{Q(z^2)}}{1 + \sqrt{Q(z^2)}}\right]^2 = \frac{4\sqrt{Q(z^2)}}{[1 + \sqrt{Q(z^2)}]^2} - 1 \\ &= Q(z) - 1,\end{aligned}$$

which is identical with (92).

We end this section with the proof of a theorem on analytic functions which utilizes the properties of the function $J(z)$.

If the odd analytic function

$$f(z) = z + a_3z^3 + \cdots + a_{2n+1}z^{2n+1} + \cdots, \qquad |z| < 1 \tag{93}$$

is regular in $|z| < 1$, *then the values taken by* $w = f(z)$ *in* $|z| < 1$ *fully cover the circle*

$$|w| < \frac{4\pi^2}{\Gamma^4(\frac{1}{4})} = 0.228 \cdots.$$

Again this is a "best possible" theorem, *i.e.*, the theorem would not be true if the constant involved were replaced by a larger value. To prove the theorem, we recall that $w = J(z)$ maps Im $\{z\} > 0$ onto the modular surface M described above. Since the transformation

$$z \to i\frac{1 + z}{1 - z}$$

carries the unit circle into the upper half-plane, it follows therefore that the function

$$w = F(z) = 2J\left[i\left(\frac{1 + z}{1 - z}\right)\right] - 1 \tag{94}$$

maps $|z| < 1$ onto a surface M_1 which is similar to M, the only difference being that the two finite "holes" of M_1 are situated at $w = \pm 1$ and not, as those of M, at $w = 0, 1$. By (94), we have $F(0) = 2J(i) - 1$. It can easily be shown that this vanishes. Indeed, since the inverse of $w = J(z)$ is $z = -if(w)$, where $f(w)$ is defined by (73) and (74), it follows from (73) that

$$J^{-1}\left(\frac{1}{2}\right) = i\frac{K(\frac{1}{2})}{K(\frac{1}{2})} = i,$$

and thus $J(i) = \frac{1}{2}$, whence $F(0) = 2J(i) - 1 = 0$.

Suppose now that the odd function $f(z)$ defined in (93) omits the value α for $|z| < 1$. Since $f(z)$ is an odd function, that is, $f(-z) = -f(z)$, $f(z)$ necessarily also omits the value $-\alpha$. Hence the function $\alpha^{-1}f(z)$ does not take the values ± 1 for $|z| < 1$. In view of the properties of the surface M_1, it follows therefore that

$$\frac{f(z)}{\alpha} = F[\omega(z)], \tag{95}$$

where $|\omega(z)| \leq |z|$. The detailed reasoning leading to (95) is analogous to that used above in the proof of the $\frac{1}{16}$-theorem, and it is therefore left as an exercise to the reader. It follows from (93) and (95) that

$$\frac{1}{\alpha} = \frac{f'(0)}{\alpha} = F'(0)\omega'(0).$$

Since $|\omega'(0)| \leq 1$ and, by (94), $F'(0) = 4iJ'(i)$, we have

$$\frac{1}{|\alpha|} \leq 4|J'(i)|,$$

whence

$$|\alpha| \geq \frac{1}{4|J'(i)|}. \tag{96}$$

A value α which is not taken by $w = f(z)$ in $|z| < 1$ thus must have a distance from the origin which is not less than $[4|J'(i)|]^{-1}$. This proves our theorem, provided we can show that

$$|J'(i)| = \frac{\Gamma^4(\frac{1}{4})}{16\pi^2}. \tag{97}$$

Now the inverse of $w = J(z)$ is $z = -if(w)$, where $f(w)$ is defined by (73) and (74). Hence

$$\begin{aligned}\frac{1}{J'(z)} &= -i\frac{d}{dw}\left[\frac{K(1-w)}{K(w)}\right] \\ &= \frac{i}{K^2(w)}[K(w)K'(1-w) + K'(w)K(1-w)].\end{aligned}$$

In view of $J(i) = \frac{1}{2}$, this yields

$$\frac{1}{J'(i)} = \frac{2iK'(\frac{1}{2})}{K(\frac{1}{2})}.$$

By (74),

$$K(\tfrac{1}{2}) = \int_0^1 \frac{dt}{\sqrt{(1 - t^2)(1 - \frac{1}{2}t^2)}}$$

and

$$K'(\tfrac{1}{2}) = \frac{1}{2}\int_0^1 \frac{t^2\,dt}{\sqrt{(1 - t^2)(1 - \frac{1}{2}t^2)(1 - \frac{1}{2}t^2)}}.$$

With the substitution

$$t = \left(\frac{2}{s}\right)^{\frac{1}{2}} (1 - s)^{\frac{1}{4}}(1 - \sqrt{1 - s})^{\frac{1}{2}},$$

this becomes

$$K(\tfrac{1}{2}) = \frac{1}{2\sqrt{2}} \int_0^1 s^{-\frac{1}{2}}(1 - s)^{-\frac{3}{4}}\,ds,$$

$$K'(\tfrac{1}{2}) = \frac{1}{2\sqrt{2}} \int_0^1 s^{-\frac{1}{2}}(1 - s)^{-\frac{1}{4}}\,ds.$$

Since

$$\int_0^1 s^{\alpha-1}(1 - s)^{\beta-1}\,ds = \frac{\Gamma(\alpha)\Gamma(\beta)}{\Gamma(\alpha + \beta)},$$

and $\alpha\Gamma(\alpha) = \Gamma(\alpha + 1)$, we thus obtain

$$\frac{1}{J'(i)} = 8i\left(\frac{\Gamma(\frac{3}{4})}{\Gamma(\frac{1}{4})}\right)^2,$$

or, in view of

$$\Gamma(\alpha)\Gamma(1 - \alpha) = \frac{\pi}{\sin \pi\alpha},$$

$$\frac{1}{J'(i)} = \frac{16\pi^2 i}{\Gamma^4(\frac{1}{4})},$$

in agreement with (97).

Since the surface M_1 is symmetric with respect to the origin and since $F(0) = 0$, $w = F(z)$ is an odd function of z. The function

$$f(z) = [F'(0)]^{-1}F(z)$$

thus satisfies the hypotheses of the theorem and leaves out the values $w = \pm[F'(0)]^{-1} = \pm[4J'(i)]^{-1}$. In view of (97), the theorem is therefore sharp.

EXERCISES

1. A linear substitution

$$z' = \frac{az + b}{cz + d}$$

is said to be unimodular if $\begin{vmatrix} a & b \\ c & d \end{vmatrix} = ad - bc = 1$. If S and S' are unimodular substitutions, show that the same is true for the combined substitution SS'. Show also that the inverse of a unimodular substitution is again unimodular. *Hint:* If S is represented by the symbol $\begin{pmatrix} a & b \\ c & d \end{pmatrix}$, show that

$$\begin{pmatrix} a & b \\ c & d \end{pmatrix}\begin{pmatrix} a' & b' \\ c' & d' \end{pmatrix} = \begin{pmatrix} a'a + b'c & a'b + b'd \\ c'a + d'c & c'b + d'd \end{pmatrix}$$

and use the multiplication law for determinants.

2. Show that all substitutions $\begin{pmatrix} a & b \\ c & d \end{pmatrix}$ of the group generated by the substitutions $S_1 = \begin{pmatrix} 1 & 2 \\ 0 & 1 \end{pmatrix}$ and $S_2 = \begin{pmatrix} 1 & 0 \\ 2 & 1 \end{pmatrix}$ are unimodular and are, moreover, such that b, c are even integers and a, d are odd integers.

3. If S_1 and S_2 are the substitutions of the preceding exercise, show that

$$S_1{}^n = \begin{pmatrix} 1 & 2n \\ 0 & 1 \end{pmatrix}, \qquad S_2{}^m = \begin{pmatrix} 1 & 0 \\ 2m & 1 \end{pmatrix}.$$

4. If $S = \begin{pmatrix} a & b \\ c & d \end{pmatrix}$ is unimodular, a, d odd, b, c even, $b \neq 0$ and $|a| > |b|$, show that $S = S'S_2{}^m$, where $S' = \begin{pmatrix} a' & b' \\ c' & d' \end{pmatrix}$ is such that $|a'| < |b'|$ and $b' = b$, and m is a suitable integer. Similarly, if $|a| < |b|$, show that $S = S'S_1{}^n$, where $a' = a$, $|b'| < |a'|$, and n is again a suitable integer.

5. Using the results of the preceding exercise, show that by a suitable combination of substitutions S_1 and S_2 the initial substitution S can finally be reduced to a substitution $S' = \begin{pmatrix} a' & b' \\ c' & d' \end{pmatrix}$ for which either $a' = d' = 1$ or $a' = d' = -1$, and where either b' or c' is zero. Conclude that the group of unimodular substitutions

$$S = \begin{pmatrix} a & b \\ c & d \end{pmatrix}$$

(a, d odd, b, c even) is identical with the group of substitutions which leave the elliptic modular function invariant.

6. Using the fact that the function $F(z)$ of (94) is an odd function of z, show that the function $Q(z)$ satisfies the identity

$$Q(e^{-\pi z}) + Q(e^{-\frac{\pi}{z}}) = 1.$$

7. Combining (92) and the result of the preceding exercise, show that

$$Q(-e^{-\pi z})Q(-e^{-\frac{\pi}{z}}) = 1.$$

8. Show that all coefficients of the power series expansion $-Q(-z) = \sum_{n=1}^{\infty} b_n z^n$ are nonnegative and deduce that

$$\max_{|z|=\rho} |Q(z)| \geq 16\rho.$$

Hint: Use the product expansion (83).

9. Use the results of Exercises 7 and 8 to show that

$$|Q(z)| \leq \tfrac{1}{16} e^{-\frac{\pi^2}{\log |z|}}.$$

10. If $Q(z) = \sum_{n=1}^{\infty} c_n z^n$ is the power series expansion of $Q(z)$, show that

$$|c_n| \leq \tfrac{1}{16} e^{2\pi\sqrt{n}}.$$

Hint: Use Cauchy's inequality (Sec. 4, Chap. III) and the result of the preceding exercise.

11. Using (92), prove the identities

$$16z \prod_{n=1}^{\infty} (1 + z^{2n})^8 = \prod_{n=1}^{\infty} (1 + z^{2n-1})^8 - \prod_{n=1}^{\infty} (1 - z^{2n-1})^8$$

and

$$16z \left(\sum_{n=0}^{\infty} z^{n(n+1)} \right)^4 = \left(1 + 2 \sum_{n=1}^{\infty} z^{n^2}\right)^4 - \left[1 + 2 \sum_{n=1}^{\infty} (-1)^n z^{n^2}\right].$$

12. Show that the second identity of the preceding exercise is equivalent to the following theorem in the theory of numbers: Let N_1 and N_2 denote the number of times the number $8N + 4$ (N being a positive integer) can be written in different ways as the sum of 4 odd squares and 4 even squares, respectively (where the squares of 0, positive, and negative numbers are admitted, and permutations of the same four squares are considered different from each other). Then $N_1 = 2N_2$. *Hint:* Replace z by z^4 in the identity in question and compare the coefficients of equal powers.

13. Show that the function

$$P(z) = \frac{2}{\pi} \text{arc sin} \sqrt{Q(z^2)}$$

maps $|z| < 1$ onto a surface R_3 of the following description: (*a*) R_3 has no branch points; (*b*) all boundary points of R_3 are situated at the points $w = 0, \pm n, n = 1, 2, 3, \ldots$; (*c*) the point $w = 0$ is covered by one sheet of R_3 (for $z = 0$), while the points $\pm 1, \pm 2, \ldots$ are not covered at all. Use this function in order to prove the following theorem: If $w = f(z) = a_1 z + a_2 z^2 + \cdots$ is regular in $|z| < 1$ and does not take the values $w = 0, \pm 1, \pm 2, \ldots$ in $0 < |z| < 1$, then $|a_1| \leq \frac{8}{\pi}$. Show also that this inequality is sharp.

14. Let $f(z)$ be an analytic function which is regular and $\neq 0$, 1 at all finite points of the z-plane with the exception of the points $z = \ldots, -2, -1, 0, 1, 2, \ldots$. If $f(0) = 0$, show that $|f'(0)| \leq 2\pi$. *Hint:* Consider $f[P(z)]$, where $P(z)$ is the function defined in the preceding exercise.

15. Show that the function $w = T(z) = 4Q(z)[1 - Q(z)]$ maps $|z| < 1$ onto a surface R_4 with the following properties: (*a*) all boundary points of R_4 are situated either at $w = 0$ or at $w = \infty$; (*b*) $w = 0$ is covered by one sheet of R_4 (for $z = 0$), while $w = \infty$ does not belong to R_4; (*c*) all sheets of R_4 have branch points of the first order at $w = 1$.

CHAPTER VII

CONFORMAL MAPPING OF MULTIPLY-CONNECTED DOMAINS

1. Canonical Domains. By the Riemann mapping theorem, all simply-connected domains with more than one boundary point are conformally equivalent to each other. Both for theoretical and practical purposes we may therefore confine ourselves to the study of the conformal mappings of arbitrary simply-connected domains onto one and the same standard domain. The ideal standard domain of this type is the interior of a circle. As we had ample opportunity to see in the two preceding chapters, the geometric simplicity of the circle is reflected in a corresponding simplicity of results and in a comparative ease of analytic manipulation.

If we try to introduce the concept of the standard domain, or, as we shall also say, the *canonical domain*, into the theory of conformal mapping of multiply-connected domains, we meet two initial obstacles. The first, and less serious, one is due to the fact that a conformal mapping is continuous and thus preserves the order of connectivity of a domain. For example, a conformal map of a doubly-connected domain is again a doubly-connected domain. It therefore becomes necessary to introduce distinct canonical domains for each order of connectivity. The second difficulty is caused by the fact that no exact equivalent of the Riemann mapping theorem holds in the multiply-connected case. It is not true that any two domains of the same order of connectivity can be conformally mapped onto each other. To put it differently, not all domains of the same order of connectivity are of the same *conformal type.* That this is the case can be shown by a very simple example. Suppose there exists a function $w = f(z)$ which maps the circular ring $1 < |z| < r$ onto the circular ring $1 < |w| < R$, where $r \neq R$, and consider the harmonic function $h(z) = \log r \log |f(z)| - \log R \log |z|$. Clearly, $h(z)$ is harmonic in $1 < |z| < r$ and has vanishing boundary values. $h(z)$ is therefore identically zero, and the analytic function $g(z) = \log r \log f(z) - \log R \log z$, of which $h(z)$ is the real part, reduces to an imaginary constant, *i.e.*,

$$g(z) = \log r \log f(z) - \log R \log z = i\alpha. \tag{1}$$

If z describes the circle $|z| = r$ in the positive sense, $w = f(z)$ describes $|w| = R$ in the positive sense, and both arg $\{z\}$ and arg $\{f(z)\}$ grow by 2π. Since arg $\{z\} = \text{Im}\{\log z\}$, it follows from (1) that the imaginary part of

$g(z)$ grows by the amount $2\pi(\log r - \log R)$ if z describes $|z| = r$. But $\text{Im}\,\{g(z)\} = \alpha = \text{const.}$, and we must therefore have $\log r = \log R$, that is, $r = R$, contrary to our assumption that $r \neq R$.

This shows that two concentric circular rings are of different conformal type unless the ratio of the two radii is the same for both rings. This ratio, that is, the quantity r_2/r_1 if the ring in question is $r_1 < |z| < r_2$, is known as the *modulus* of the ring. We shall see later that any doubly-connected domain D can be conformally mapped onto a circular ring. The conformal type of a doubly-connected domain is thus completely determined by the number r_2/r_1, which is therefore also called *the modulus of the doubly-connected domain D*. Using this terminology, we may say that two doubly-connected domains are of the same conformal type if, and only if, their moduli are the same. This situation may also be described by saying that the modulus of a doubly-connected domain is a conformal invariant, *i.e.*, the modulus does not change if the domain is made subject to a conformal transformation.

A similar situation obtains in the case of domains of higher connectivity. We shall see (Exercise 13, Sec. 2) that the conformal type of a domain of connectivity n ($n > 2$) is determined by $3n - 6$ real numbers which are again called the moduli, or *Riemann moduli*, of the domain. Two domains of connectivity n may therefore be conformally mapped onto each other if, and only if, they agree in all $3n - 6$ moduli. Hence, we cannot expect to find a canonical domain onto which all domains of a given order of connectivity can be mapped. This difficulty is overcome by giving a more flexible definition of a canonical domain which determines the general geometric character of the domain without fixing its moduli. We give in the following a list of the more important of these domains.

a. The Parallel Slit Domain (*Fig.* 46*a*). This domain consists of the entire plane, including the point at infinity, to which a number of parallel rectilinear slits have been applied. We shall see in the following section that any multiply-connected domain D can be conformally mapped onto a domain of this type. Moreover, if $w = \varphi(z)$ denotes the function effecting this mapping, we may arbitrarily assign a point ζ of D such that ζ is carried into the point at infinity. Furthermore, we may require that the parallel slits form a given angle θ with the real axis and that the Laurent expansion of $\varphi(z) = \varphi_\theta(z,\zeta)$ near $z = \zeta$ be of the form

$$(2)\qquad \varphi_\theta(z,\zeta) = \frac{1}{z - \zeta} + a_\theta(z - \zeta) + b_\theta(z - \zeta)^2 + \cdots .$$

We shall see that these requirements determine the function $\varphi_\theta(z,\zeta)$ uniquely.

b. The Circular Slit Domain (*Fig.* 46*b*). This domain consists of the entire plane which has been slit along a number of concentric circular arcs whose common center is the origin. We shall see that any multiply-connected domain D can be mapped onto a domain of this type and that, moreover, it is possible to choose two arbitrary points u and v of D which are to be carried into the origin and the point at infinity, respectively. If $P(z) = P(z;u,v)$ denotes the mapping function in question, $P(z)$ may be normalized by the requirement that its residue at $z = v$ be equal to 1, that is,

$$P(z) = P(z;u,v) = \frac{1}{z - v} + b_0 + b_1(z - v) + \cdots . \tag{3}$$

This normalization determines the function $P(z;u,v)$ uniquely.

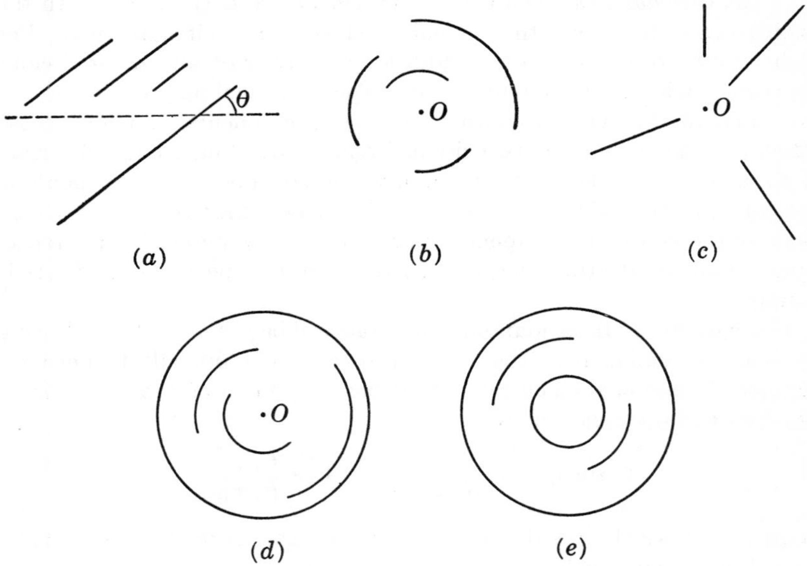

FIG. 46.

c. The Radial Slit Domain (*Fig.* 46*c*). This is obtained by furnishing the entire plane with rectilinear slits pointing at the origin. As in the preceding case, two arbitrarily chosen points of D may be made to correspond to the origin and the point at infinity. If we denote the function effecting this mapping by $Q(z) = Q(z;u,v)$, we have $Q(u) = 0$ and, with the same normalization as in (3),

$$Q(z) = Q(z;u,v) = \frac{1}{z - v} + c_0 + c_1(z - v) + \cdots . \tag{4}$$

This function is again uniquely determined.

d. The Circle with Concentric Circular Slits (Fig. 46d). This is the interior of a circle, which may be taken to be the unit circle, which has been slit along a number of circular arcs concentric with the outer circle. The conformal mapping transforming D into this domain distinguishes one of the boundary components of D, namely, the boundary component which is carried into the outer circumference. Accordingly, if the boundary components of D are denoted by $C_1, C_2, \ldots, C_n$, there will be n essentially different mapping functions of this type, depending upon the boundary continuum C_ν which is distinguished. These functions will be denoted by $R_\nu(z) = R_\nu(z;\zeta)$, $\nu = 1, \ldots, n$, where the parameter ζ indicates the arbitrarily chosen point ζ of D for which $R_\nu(\zeta;\zeta) = 0$. This, and the condition $R_\nu'(\zeta;\zeta) > 0$, determines $R_\nu(z;\zeta)$ uniquely.

e. The Circular Ring with Concentric Circular Slits (Fig. 46e). In the doubly-connected case, this domain reduces to a circular ring. For higher connectivity, we have in addition a number of slits along circular arcs concentric with the circular ring. The conformal mapping of D onto this domain distinguishes two boundary components, namely, those which are carried into the two circumferences bounding the circular ring. If C_ν and C_μ are these two boundary components, the corresponding mapping function will be denoted by $S_{\nu\mu}(z)$, where the first subscript may indicate the boundary component mapped onto the outer circumference. Apart from an arbitrary multiplicative constant, the function $S_{\nu\mu}(z)$ is unique.

The existence of these mapping functions will be proved in the following section. Assuming here their existence, it is not difficult to derive a number of relations which must hold between some of these functions. The first of these relations is

$$(5) \qquad P(z;u,v) = A\,\frac{R_\nu(z;u)}{R_\nu(z;v)}, \qquad A = \frac{R_\nu'(v;v)}{R_\nu(v;u)}.$$

To prove (5), we observe that the functions $R_\nu(z;u)$ defined in Sec. 1*d* have a constant absolute value on each boundary component C_μ, $\mu = 1, \ldots, n$. The same is therefore true of the function $P(z;u,v)$ defined in (5). This latter function further vanishes at $z = u$ and shows the behavior indicated in (3) at $z = v$. We next observe that the function $w = R_\nu(z;\zeta)$ maps C_ν onto the circumference $|w| = 1$. If C_ν is described in the positive direction with respect to D, $|w| = 1$ will therefore be described in the positive direction and $\arg\{w\} = \arg\{R_\nu(z;\zeta)\}$ will increase by 2π. Since this holds for $\zeta = u$ and $\zeta = v$, it follows that $\Delta_\nu \arg\{P(z;u,v)\}$, the corresponding variation of the argument of the function defined in (5), is zero. On C_μ $(\mu \neq \nu)$, we have $\Delta_\mu \arg\{R_\nu(z;\zeta)\} = 0$. Indeed if z describes C_μ, $w = R_\nu(z,\zeta)$ describes "both edges" of a circular slit and returns to its initial value. Doing so, the point w obviously cannot surround the origin. It follows that arg $\{R_\nu(z,\zeta)\}$ returns to its initial value, that is, Δ_μ arg $\{R_\nu(z,\zeta)\} = 0$. By (5), we therefore also have

$$\Delta_\mu \arg \{P(z;u,v)\} = 0.$$

As pointed out before, the function defined by (5) has a constant absolute value on each C_μ, $\mu = 1, \ldots, n$, that is,

$$(6) \qquad |P(z;u,v)| = |P(z)| = \gamma_\mu, \qquad z \in C_\mu,\ \mu = 1, \ldots, n.$$

Let now α be an otherwise arbitrary finite complex number which is subject to the condition that $|\alpha| \neq \gamma_\mu$ ($\mu = 1, \ldots, n$), and consider the variation of arg $\{P(z) - \alpha\}$ if z traverses the entire boundary C of D. Since C is composed of the components C_μ, we have

$$(7) \qquad \Delta_C \arg \{P(z) - \alpha\} = \sum_{\mu=1}^{n} \Delta_\mu \arg \{P(z) - \alpha\}.$$

Since $|\alpha| \neq \gamma_\mu$, $\mu = 1, \ldots, n$, we have on each individual C_μ either $|P(z)| = \gamma_\mu < |\alpha|$ or $|P(z)| = \gamma_\mu > |\alpha|$. In the first case we write $P(z) - \alpha = -\alpha[1 - \alpha^{-1}P(z)]$, whence

$$\Delta_\mu \arg \{P(z) - \alpha\} = \Delta_\mu \arg \left\{1 - \frac{P(z)}{\alpha}\right\},$$

$-\alpha$ being a constant. Since $|\alpha^{-1}P(z)| < 1$, the point $1 - \alpha^{-1}P(z)$ clearly cannot surround the origin if z describes C_μ. Hence Δ_μ arg $\{P(z) - \alpha\} = 0$. If $|\alpha| < |P(z)|$ on C_μ, we write

$$\Delta_\mu \arg \{P(z) - \alpha\} = \Delta_\mu \arg \{P(z)\} + \Delta_\mu \arg \left\{1 - \frac{\alpha}{P(z)}\right\}.$$

Since $|\alpha[P(z)]^{-1}| < 1$ on C_μ, the last term is again equal to 0. As shown above, Δ_μ arg $\{P(z)\} = 0$ for all C_μ, and it follows therefore that Δ_μ arg $\{P(z) - \alpha\} = 0$. Inserting this in (7), we see that the total variation of arg $\{P(z) - \alpha\}$ along the boundary C of D is zero. By the argument principle (Sec. 10, Chap. III), this variation is equal to 2π times the difference between the number of zeros and the number of poles of $P(z) - \alpha$. Since $P(z)$, and therefore also $P(z) - \alpha$, has precisely one pole in D (at $z = v$), it follows therefore that $P(z) - \alpha$ has exactly one zero in D. In other words, $P(z)$ takes the value α once, and once only. Although this has so far been proved only for values α such that $|\alpha| \neq \gamma_\mu$, $\mu = 1, \ldots, n$, this result can be extended immediately to these excluded values. Suppose $|\alpha| = \gamma_\mu$ and $P(z_1) = P(z_2) = \alpha$, $z_1, z_2 \in D$. In this case, a small circle about $w = \alpha$ will be the conformal image of two distinct small subdomains of D containing the points z_1 and z_2, respectively. Hence, there exist two points z_1^* and z_2^* near z_1 and z_2, respectively, such that $P(z_1^*) = P(z_2^*) = \alpha^*$, where α^* is arbitrarily close to $|\alpha| = \gamma_\mu$. But this contradicts our previous result.

Summing up, we have shown that the function $w = P(z)$ takes, in D, no value more than once and that this function actually takes any value α with the possible exception of values situated on the circles $|w| = \gamma_\mu$. But this means that $w = P(z)$ maps D onto the circular slit domain defined in Sec. 1*b*. The identification of our function $P(z)$ with the mapping function $P(z;u,v)$ listed in Sec. 1*b* is therefore complete. We add that a similar reasoning also establishes the uniqueness of the mapping function $P(z;u,v)$. If there existed another function, say $P_1(z;u,v)$, then the function

$$F(z) = \frac{P(z;u,v)}{P_1(z;u,v)}$$

would be regular in D (the poles and the zeros cancel) and $|F(z)|$ would be constant on each C_ν. Since, moreover, $\Delta_C \arg \{F(z) - \alpha\} = 0$, we conclude as before that $F(z) - \alpha$ has as many zeros in D as it has poles. But $F(z)$ is regular in D, which shows that $F(z) \neq \alpha$ in D where α is any complex number in D which is not situated on one of n concentric circumferences. But this means $F(z)$ reduces to a constant which, in view of the fact that both P and P_1 have the normalization (3), must be equal to 1. Hence, $P(z) = P_1(z)$.

By a similar procedure it is possible to show that the mapping functions listed in Sec. 1*d* and *e* are connected by the relation

$$S_{\nu\mu}(z) = \frac{R_\nu(z;\zeta)}{R_\mu(z;\zeta)}. \tag{8}$$

The method of proof differs from that employed in the preceding case only in some minor details, and the proof is therefore left as an exercise for the reader.

The functions (2) effecting the parallel slit mappings described in Sec. 1*a* can be expressed in terms of any two of them. The formulas become particularly simple if we choose as the two representatives the functions

$$\varphi(z,\zeta) = \varphi_0(z,\zeta), \qquad \psi(z,\zeta) = \varphi_{\frac{\pi}{2}}(z,\zeta), \tag{9}$$

that is, the functions mapping D onto domains whose slits are parallel to the real axis and the imaginary axis, respectively. With this notation, we have

$$\varphi_\theta(z,\zeta) = e^{i\theta}[\cos\theta\varphi(z,\zeta) - i\sin\theta\psi(z,\zeta)]. \tag{10}$$

To prove (10), we observe that $\varphi(z_1,\zeta) - \varphi(z_2,\zeta)$ is real and $\psi(z_1,\zeta) - \psi(z_2,\zeta)$ is pure imaginary if z_1 and z_2 are any two points on the same boundary component C_ν. Hence, it is clear from (10) that

$$\arg\{[\varphi_\theta(z_1,\zeta) - \varphi_\theta(z_2,\zeta)]^2\} = 2\theta,$$

that is, all points $\varphi_\theta(z,\zeta)$ $(z \in C_\nu)$ lie on a straight line which forms the angle θ with the positive axis. Furthermore, the function $\varphi_\theta(z,\zeta)$ defined by (10) obviously has the normalization (2). These two facts are sufficient to identify $\varphi_\theta(z,\zeta)$ with the function yielding the parallel slit mapping described in Sec. 1*a*. Indeed, all that is required for this purpose is to show that the function defined by (10) is univalent in D. To derive this property, let α be a complex number which does not lie on one of the n parallel lines on which the conformal images of $C_1, \ldots, C_n$ are situated, and consider $\Delta_C \arg\{\varphi_\theta(z,\zeta) - \alpha\} = \sum_{\nu=1}^{n} \Delta_\nu \arg\{\varphi_\theta(z,\zeta) - \alpha\}$. If z describes C_ν, the point $\varphi_\theta(z,\zeta)$ is confined to a straight line and it therefore cannot surround a point α which is not on this line. Hence,

$$\Delta_\nu \arg\{\varphi_\theta(z,\zeta) - \alpha\} = 0,$$

and the total variation of $\arg\{\varphi_\theta(z,\zeta) - \alpha\}$ along C vanishes. By the argument principle, $\varphi_\theta(z,\zeta) - \alpha$ therefore has as many zeros as it has poles. Since $\varphi_\theta(z,\zeta)$ has but one pole (at $z = \zeta$), it follows that $\varphi_\theta(z,\zeta)$ takes the value α exactly once. This proves the univalence of $\varphi_\theta(z,\zeta)$, and thus establishes the identity (10). A similar reasoning also proves the uniqueness of the mapping function $\varphi_\theta(z,\zeta)$ with the normalization (2); details of this proof are left to the reader.

We close this section with a general remark regarding analytic functions defined in multiply-connected domains. The monodromy theorem (Sec. 5, Chap. III), which guarantees the single-valuedness of functions which are regular in a simply-connected domain, does not hold in multiply-connected domains. In the latter, a function may be regular without being single-valued, as illustrated by the function $w = \log z$ in the ring $1 < |z| < 2$. This remark applies in particular to functions $F(z) = \int_{z_0}^{z} f(z)\,dz$ which are indefinite integrals of functions $f(z)$ which are regular and single-valued in a multiply-connected domain D. $F(z)$ is certainly regular in D, but it will not be single-valued there unless the integral $\int_C f(z)\,dz$ vanishes for every closed contour C within D. As shown in detail in Sec. 9, Chap. I, all these contours can be built up from

$n - 1$ particular closed contours. These may be taken to be such as to surround $n - 1$ of the n "holes" of D (if D is finite, one of these holes contains the point at infinity). $F(z)$ may therefore have $n - 1$ independent periods which are equal to the values of $\int f(z)\,dz$, taken over these $n - 1$ contours.

EXERCISES

1. If D is the unit circle, show that

$$P(z;u,v) = \beta \left(\frac{z - u}{1 - \bar{u}z}\right)\left(\frac{1 - \bar{v}z}{z - v}\right), \qquad \beta = \frac{1 - \bar{u}v}{(v - u)(1 - |v|^2)}$$

and

$$Q(z;u,v) = \gamma \left(\frac{z - u}{z - v}\right)\left(\frac{1 - \bar{u}z}{1 - \bar{v}z}\right), \qquad \gamma = \frac{1 - |v|^2}{(v - u)(1 - \bar{u}v)}.$$

2. If D is the unit circle, show that

$$\varphi_\theta(z,\zeta) = \frac{1}{1 - |\zeta|^2}\left[\frac{1 - \bar{\zeta}z}{z - \zeta} + e^{2i\theta}\left(\frac{z - \zeta}{1 - \bar{\zeta}z}\right)\right].$$

3. If $\varphi(z,\zeta) = \varphi_0(z,\zeta)$ and $\psi(z,\zeta) = \varphi_{\frac{1}{2}\pi}(z,\zeta)$, show that the function

$$\sigma(z) = \frac{\varphi'(z,\zeta)}{\psi'(z,\zeta)}$$

does not take pure imaginary values in D, and conclude that $\sigma(z)$ is regular and $\operatorname{Re}\{\sigma(z)\} > 0$ in D, while $\operatorname{Re}\{\sigma(z)\} = 0$ on the boundary C of D. *Hint:* Use the fact that the derivative of the univalent function $\varphi_\theta(z,\zeta)$ cannot vanish in D and employ the identity (10).

4. If the functions $M(z,\zeta)$ and $N(z,\zeta)$ are defined by

$$M(z,\zeta) = \tfrac{1}{2}[\varphi(z,\zeta) - \psi(z,\zeta)],$$
$$N(z,\zeta) = \tfrac{1}{2}[\varphi(z,\zeta) + \psi(z,\zeta)],$$

where $\varphi(z,\zeta)$ and $\psi(z,\zeta)$ are the functions defined in the preceding exercise, show that $M(z,\zeta)$ is regular in D while $N(z,\zeta)$ has a pole with the residue 1 at $z = \zeta$. Show further that the function

$$F(z) = \frac{M'(z,\zeta)}{N'(z,\zeta)}$$

is regular in D and maps D onto the (necessarily multiply covered) interior of the unit circle. *Hint:* Express $F(z)$ in terms of the function $\sigma(z)$ of the preceding exercise.

5. Show that $M'(z,\zeta)$ and $N'(z,\zeta)$ cannot vanish simultaneously and conclude that $N'(z,\zeta)$ does not vanish in D. *Hint:* Use (10) and the properties of the function $F(z)$ of the preceding exercise.

6. If s is the length parameter on the boundary C of D and if $z' = z'(s) = dz/ds$, show that

$$M'(z,\zeta)N'(z,\zeta)z'^2 > 0, \qquad z \in C,$$

and conclude, by means of the argument principle and the result of the preceding exercise, that $M'(z,\zeta)$ has precisely $2n - 2$ zeros in D, where n is the connectivity of D. *Hint:* Use the fact that $\varphi'(z,\zeta)z'$ and $i\psi'(z,\zeta)z'$ are real on C and employ the method used in a similar case in Sec. 10, Chap. III.

7. Show that the function $F(z)$ of Exercise 4 maps D onto a domain which covers the interior of the unit circle $2n$ times.

8. Show that

$$M'(z,\zeta)z' = \overline{N'(z,\zeta)z'}, \qquad z \in C,$$

and conclude that

$$\arg\left\{\frac{\partial N(z,\zeta)}{\partial s}\right\} = \arg\{N'(z,\zeta)z'\} = -\frac{1}{2}\arg\{F(z)\},$$

where $F(z)$ is defined in Exercise 4.

9. Using the results of Exercises 4 and 8, prove that the function $w = N(z,\zeta)$ maps the boundary components $C_1, \ldots, C_n$ of D onto simple convex curves. Combining this with the result of Exercise 5, conclude that the function $N(z,\zeta)$ is univalent in D.

2. Characterization of Canonical Mappings by Extremal Problems. In proving the existence of the various canonical conformal mappings described in the preceding section we may confine ourselves to the case in which the original domain D is bounded by n closed analytic curves. This is a consequence of the fact, to be demonstrated presently, that an arbitrary domain D of connectivity n can always be mapped onto a domain D' which is bounded by n closed analytic curves. If $z' = h(z)$ is the analytic function effecting this mapping and if $w = F(z')$ is the function mapping D' onto the canonical domain D^*, then the function $w = F[h(z)]$ yields the desired canonical mapping of D onto D^*.

To show that any domain D of connectivity n can be mapped onto a domain bounded by n closed analytic curves, we assume, in order to exclude trivial cases, that each of the n boundary components $C_1, \ldots, C_n$ consists of more than one boundary point. We denote by A the domain bounded by C_1 which is obtained from D by "filling in" the $n - 1$ "holes" bounded by $C_2, C_3, \ldots, C_n$ (if D is finite and C_1 is not the outer boundary of D, one of these "holes" will contain the point at infinity). By the Riemann mapping theorem, A may be mapped conformally onto the interior of the unit circle. By this mapping, D is transformed into a domain D_1 which is bounded by the unit circle and by $n - 1$ continua, say $C_2', C_3', \ldots, C_n'$, which are the conformal images of $C_2, \ldots, C_n$. Next, denote by B the simply-connected domain bounded by C_2' which is obtained by filling in all the other holes of D_1 (including the exterior of the unit circle). This domain may again be mapped onto the interior of the unit circle, and this mapping will transform D_1 into a domain D_2 which is bounded by the unit circle and by $n - 1$ other continua. One of these continua is the conformal image of the unit circle as yielded by an analytic function which is regular at all points of the unit circle (this circle is in the interior of B). Hence, this continuum is a closed analytic curve. Two of the boundary components of D_2, namely, this curve and the unit circle, are therefore closed analytic curves. Continuing in this fashion, we arrive after another $n - 2$ auxiliary mappings at a conformal map of D which is bounded by n closed analytic curves.

All the canonical conformal mappings we shall consider may be characterized by the fact that they solve certain extremal problems within the class of univalent functions in D. These extremal properties may be used in order to prove the existence of these mappings. In fact, this was precisely the procedure adopted in the proof of the Riemann mapping theorem (Sec. 4, Chap. V). There, we considered an extremal problem the existence of whose solution was guaranteed by the compactness of the family of functions involved. We then showed that this solution must be the mapping onto the unit circle, as any other assumption involved a contradiction.

Before we apply this idea to the existence proofs for our canonical conformal mappings, we derive a formula which will be useful for our purposes. Let $p = p(z)$ and $q = q(z)$ be two analytic functions which have the following properties: p and q are single-valued in a multiply-connected domain D which is bounded by n analytic curves C; p and q are regular in $D + C$ with the possible exception of a finite number of poles in D, while $p - q$ is regular in $D + C$. We then have the identity

$$\iint_D |p' - q'|^2 \, dx \, dy = \operatorname{Re} \left\{ \frac{1}{i} \int_C (\bar{p} - \bar{q}) p' \, dz \right\} + \frac{1}{2i} \int_C \bar{q} q' \, dz - \overline{\frac{1}{2i} \int_C \bar{p} p' \, dz}. \tag{11}$$

(11) is true regardless of whether the domain is finite or not. In the latter case, the function $p - q$, which is regular in $D + C$, has an expansion of the type $p - q = a_0 + a_1 z^{-1} + a_2 z^{-2} + \cdots$ near $z = \infty$. Hence, $p' - q' = -a_1 z^{-2} - 2a_2 z^{-3} - \cdots$, which makes it obvious that the area integral in (11) has a finite value.

(11) is an immediate consequence of Green's formula (106) (Sec. 10, Chap. V). Since $p - q$ is regular and single-valued in D, it follows from the latter formula that

$$\begin{aligned}
\iint_D |p' - q'|^2 \, dx \, dy &= \iint_D (p' - q')(\bar{p}' - \bar{q}') \, dx \, dy \\
&= \frac{1}{2i} \int_C (\bar{p} - \bar{q})(p' - q') \, dz \\
&= \frac{1}{2i} \int_C (\bar{p} - \bar{q}) p' \, dz + \frac{1}{2i} \int_C \bar{q} q' \, dz \\
&\quad - \frac{1}{2i} \int_C \bar{p} q' \, dz.
\end{aligned} \tag{12}$$

The last line integral can be transformed by integration by parts. We have

$$-\frac{1}{2i}\int_C \bar{p}q'\,dz = -\frac{1}{2i}\int_C \bar{p}\,dq = -\frac{1}{2i}[\bar{p}q]_C + \frac{1}{2i}\int_C q\,\overline{dp}$$
$$= -\frac{1}{2i}[\bar{p}q]_C - \overline{\frac{1}{2i}\int_C \bar{q}p'\,dz}.$$

The integrated part vanishes. Indeed, the integration is carried out, in turn, over the closed curves $C_1, \ldots, C_n$. Since p and q are single-valued in D, the expression $\bar{p}q$ returns to its initial value if z describes any of these closed curves. Hence,

$$-\frac{1}{2i}\int_C \bar{p}q'\,dz = -\overline{\frac{1}{2i}\int_C \bar{q}p'\,dz}$$
$$= \overline{\frac{1}{2i}\int_C (\bar{p} - \bar{q})p'\,dz} - \overline{\frac{1}{2i}\int_C \bar{p}p'\,dz}.$$

Inserting this in (12), we obtain the identity (11).

Let now S_ζ denote the class of analytic functions $f(z)$ which are univalent in D and have a simple pole of residue 1 at a point $z = \zeta$ of D, that is, $f(z)$ has the Laurent expansion

$$f(z) = \frac{1}{z - \zeta} + a_0 + a_1(z - \zeta) + \cdots \tag{13}$$

near $z = \zeta$. For S_∞ we shall also write S. The expansion of a function of S near $z = \infty$ will be of the form

$$f(z) = z + a_0 + \frac{a_1}{z} + \frac{a_2}{z^2} + \cdots \tag{14}$$

Within the class S_ζ, we consider the extremal problem

$$\text{Re}\,\{e^{-2i\theta}a_1\} = \text{max.}, \qquad f(z) \in S_\zeta, \tag{15}$$

where a_1 is defined by (13). We shall show that *the problem* (15) *is solved by the function* $\varphi_\theta(z,\zeta)$ *of* (2) *which maps* D *onto the parallel slit domain indicated in Fig.* 46*a*, that is,

$$\text{Re}\,\{e^{-2i\theta}a_1\} \leq \text{Re}\,\{e^{-2i\theta}a_\theta\}, \tag{16}$$

where a_θ is defined in (2).

We first show that, provided the mapping function $\varphi_\theta(z,\zeta)$ exists, the inequality (16) holds for all functions $f(z)$ of S_ζ which are regular on C. Since $\varphi = \varphi_\theta(z,\zeta)$ maps the analytic curves C onto rectilinear slits, it follows from the symmetry principle that φ is regular on C (see Sec. 5, Chap. V). By (2) and (13), the function $\varphi - f\,[f = f(z)]$ is thus regular in $D + C$, and we may therefore identify φ and f with the functions p and q in (11). Hence,

$$(17)\qquad \iint_D |\varphi' - f'|^2\,dx\,dy = \operatorname{Re}\left\{\frac{1}{i}\int_C (\bar{\varphi} - \bar{f})\varphi'\,dz\right\} + \frac{1}{2i}\int_C \bar{f}f'\,dz - \overline{\frac{1}{2i}\int_C \bar{\varphi}\varphi'\,dz}.$$

$w = \varphi_\theta(z,\zeta) = \varphi$ maps C onto a linear segment which forms the angle θ with the positive axis. Hence, the differential $e^{-i\theta}\,d\varphi = e^{-i\theta}\varphi'\,dz$ is real on C and we may write

$$e^{-i\theta}\varphi'\,dz = \overline{e^{-i\theta}\varphi'\,dz} = e^{i\theta}\overline{\varphi'\,dz},$$

whence

$$(18)\qquad \varphi'\,dz = e^{2i\theta}\overline{\varphi'\,dz}.$$

This identity can be used for the evaluation of the first line integral in (17). We have

$$I = \operatorname{Re}\left\{\frac{1}{2i}\int_C (\bar{\varphi} - \bar{f})\varphi'\,dz\right\}$$
$$= \operatorname{Re}\left\{\frac{e^{2i\theta}}{2i}\overline{\int_C (\varphi - f)\varphi'\,dz}\right\} = -\operatorname{Re}\left\{\frac{e^{-2i\theta}}{2i}\int_C (\varphi - f)\varphi'\,dz\right\}.$$

This is a contour integral which can be evaluated by the residue theorem. The only singularity of the integrand occurs at $z = \zeta$, and the residue at that point is, in view of (2) and (13), $a_1 - a_\theta$. Hence,

$$(19)\qquad I = \operatorname{Re}\,\{\pi e^{-2i\theta}(a_\theta - a_1)\}.$$

By (18), we further have

$$e^{2i\theta}\overline{\int_C \bar{\varphi}\varphi'\,dz} = \int_C \varphi\varphi'\,dz = \tfrac{1}{2}\int_C d(\varphi^2) = \tfrac{1}{2}[\varphi^2]_C,$$

and this is zero since φ^2 is single-valued on C. This shows that the last line integral in(17) has the value zero.

To evaluate the second line integral in (17), we write $w = f(z)$ and introduce polar coordinates $w = Re^{i\phi}$ in the w-plane. We obtain

$$\frac{1}{2i}\int_C \bar{f}f'\,dz = \frac{1}{2i}\int_C \bar{w}\,dw$$
$$= \frac{1}{2i}\int_C R\,dR + \frac{1}{2}\int_C R^2\,d\phi.$$

The first integral is zero since $\int_C R\,dR = \frac{1}{2}\int_C d(R^2)$, and this again vanishes as R is a single-valued function on C. Hence, in view of

$$C = C_1 + \cdots + C_n,$$

$$\frac{1}{2i}\int_C \bar{f}f'\,dz = \sum_{\nu=1}^{n}\frac{1}{2}\int_{C_\nu} R^2\,d\phi = -\sum_{\nu=1}^{n} A_\nu = -A,$$

where A_ν is the area enclosed by the image curve of C_ν in the w-plane. The negative sign is due to the fact that this area is surrounded in the negative sense if C_ν is described in the positive sense with respect to D.

Collecting our results, we thus find that (17) is equivalent to

$$A + \iint_D |\varphi' - f'|^2\,dx\,dy = \pi\,\mathrm{Re}\,\{e^{-2i\theta}(a_\theta - a_1)\}. \tag{20}$$

Now the area integral is obviously positive unless φ' and f' are identical. A, being an area, is likewise nonnegative. It follows therefore from (20) that

$$\mathrm{Re}\,\{e^{-2i\theta}(a_\theta - a_1)\} \geq 0,$$

which is identical with (16).

If $f(z)$ belongs to the class S, that is, $f(z)$ has the normalization (14), and a_θ is defined by

$$\varphi_\theta(z, \infty) = \varphi_\theta(z) = z + \frac{a_\theta}{z} + \cdots, \tag{21}$$

the validity of the inequality (16) follows in the same way as before. The only change occurs in the evaluation of the contour integral in the expression

$$I = -\mathrm{Re}\left\{\frac{e^{-2i\theta}}{2i}\int_C (\varphi - f)\varphi'\,dz\right\}.$$

By (14) and (21), the integrand is near $z = \infty$ of the form

$$\left[-a_0 + \frac{(a_\theta - a_1)}{z} + \cdots\right]\left(1 - \frac{c_\theta}{z^2} + \cdots\right).$$

Since only the terms with z^{-1} can give a contribution to the integral, we obtain

$$I = -\mathrm{Re}\left\{e^{-2i\theta}(a_\theta - a_1)\frac{1}{2i}\int_C \frac{dz}{z}\right\}.$$

If D' is a subdomain of D which is bounded by C and a sufficiently large circle α, then $\int z^{-1}\,dz = 0$ if the integration is extended over the boundary of D'. Hence,

$$\int_C \frac{dz}{z} = -\int_\alpha \frac{dz}{z} = -2\pi i,$$

and thus

$$I = \text{Re}\,\{\pi e^{-2i\theta}(a_\theta - a_1)\},$$

which is identical with (19).

(16) has been proved under the assumption that there exists a function $\varphi_\theta(z,\zeta)$ which maps D onto the parallel slit domain indicated in Fig. 46a. This assumption is certainly justified if D is simply-connected. By the Riemann mapping theorem, D may then be mapped onto the interior of the unit circle, and the mapping of the latter domain onto the exterior of a rectilinear slit is elementary (see Exercise 2 of the preceding section). (16) is therefore proved for simply-connected domains D and for functions $f(z)$ which are regular on the boundary C of D. The latter restriction on the functions $f(z)$ is easily lifted. Indeed, suppose $f(z)$ is of the form (14) (the case in which $\zeta \neq \infty$ is entirely analogous and is left as an exercise to the reader) and denote by $z = h(u)$ the function mapping $|u| > 1$ onto D in such a way that $h(\infty) = \infty$ and $h'(\infty) > 0$, that is,

$$z = h(u) = \alpha u + b_0 + \frac{b_1}{u} + \cdots, \qquad \alpha > 0.$$

The function

$$f_1(u) = \frac{1}{\alpha} f[h(u)] = u + \frac{a_0 + b_0}{\alpha} + \frac{a_1 + \alpha b_1}{\alpha^2 u} + \cdots$$

will then belong to the class S associated with the domain $|u| > 1$. Since α and b_1 are independent of the function $f(z)$ and $\alpha > 0$, the mapping which maximizes Re $\{e^{-2i\theta}a_1\}$ will also maximize Re $\{e^{-2i\theta}\alpha^{-2}(a_1 + \alpha b_1)\}$, and it is therefore sufficient to consider the case in which D is the domain $|z| > 1$. Let then $f(z)$ be univalent in $|z| > 1$, and consider the function $f_R(z) = R^{-1}f(Rz)$, $(R > 1)$. Obviously, $f_R(z)$ is also univalent in $|z| > 1$ and it has the expansion

$$f_R(z) = z + \frac{a_0}{R} + \frac{a_1}{R^2 z} + \cdots.$$

Since, moreover, $f_R(z)$ is regular for $|z| = 1$, we may apply the inequality (16). This yields

$$\frac{1}{R^2}\,\text{Re}\,\{e^{-2i\theta}a_1\} \leq \text{Re}\,\{e^{-2i\theta}a_\theta\}.$$

Letting $R \to 1$, we find that (16) holds for *all* functions of S in the simply-connected case.

Applying (16) in particular to the function $f(z) = z$, which certainly belongs to S, we obtain

$$\text{Re}\,\{e^{-2i\theta}a_\theta\} \geq 0, \tag{22}$$

an inequality satisfied by the coefficient a_θ of the function mapping any simply-connected domain D onto the exterior of a slit of inclination θ. In view of what was shown above, equality in (22) is possible only if D coincides with this slit domain.

Let now D be a multiply-connected domain, and consider the extremal problem (15) within the family S of univalent functions $f(z)$ with the normalization (14). As shown in Sec. 8, Chap. V, the family of functions which are univalent in a domain is compact, provided constant limits are excluded. In view of the normalization (14), the latter is the case and, by the results of Sec. 3, Chap. IV, our extremal problem has therefore a solution within the family S. In other words, there exists a univalent function

$$F(z) = z + A_0 + \frac{A_1}{z} + \cdots$$

such that

$$\text{Re}\,\{e^{-2i\theta}a_1\} \leq \text{Re}\,\{e^{-2i\theta}A_1\}, \tag{23}$$

if a_1 is the coefficient of z^{-1} of any other function of S. We shall show that $w = F(z)$ necessarily maps D onto a parallel slit domain of the type indicated in Fig. 46*a*.

Indeed, suppose this is not the case. The domain D^* onto which D is mapped by $w = F(z)$ will then have at least one boundary component, say β, which is not a rectilinear slit with the inclination θ. The exterior of β is a simply-connected domain which may be mapped conformally onto the exterior of a rectilinear slit with the inclination θ. If

$$w' = p(w) = w + \frac{a_\theta}{w} + \cdots$$

is the function effecting this mapping, then the combined mapping function

$$w' = F^*(z) = p[F(z)] = z + A_0 + \frac{A_1 + a_\theta}{z} + \cdots$$

clearly transforms D into another schlicht domain and thus belongs to S. Hence, by (23),

$$\text{Re}\,\{e^{-2i\theta}(A_1 + a_\theta)\} \leq \text{Re}\,\{e^{-2i\theta}A_1\},$$

which shows that Re $\{e^{-2i\theta}a_\theta\} \leq 0$. On the other hand we have, by (22), Re $\{e^{-2i\theta}a_\theta\} > 0$ unless β reduces to a rectilinear slit of inclination θ. The assumption that not all boundary components of D^* are slits of this type has thus led to a contradiction. This finally proves both the possibility of the parallel slit mapping and the extremal property of the function $\varphi_\theta(z)$ effecting this mapping.

The methods of proof which yield both the existence and the extremal properties of the other canonical mapping functions discussed in the preceding section are very similar to that used in the case of the parallel slit mapping, and we may therefore treat these cases more briefly. First, consider the functions $P = P(z;u,v)$ and $Q = Q(z;u,v)$ which yield the circular slit mapping and the radial slit mapping, respectively. P and Q are normalized by (3) and (4) and both P and Q are zero if $z = u$. We denote by $f = f(z)$ a function which is univalent in D, has a pole with the normalization

$$f(z) = \frac{1}{z - v} + b_0 + b_1(z - v) + \cdots \tag{24}$$

at $z = v$, and vanishes for $z = u$. We, moreover, assume that $f(z)$ is regular on the boundary C of D. The functions

$$p = \log P, \qquad p_1 = \log Q, \qquad q = \log f \tag{25}$$

are not regular in D, since they have logarithmic poles at $z = u$ and $z = v$. Although not single-valued in D, these functions have no periods about the boundary components C_ν, that is, $\int_{C_\nu} dp = \int_{C_\nu} dp_1 = \int_{C_\nu} dq = 0$. Indeed,

$$\int_{C_\nu} dp = \int_{C_\nu} d[\log P] = \int_{C_\nu} d[\log |P|] + i \int_{C_\nu} d[\arg \{P\}].$$

Since $\log |P|$ is single-valued in D, the first integral on the right-hand side is zero. The second integral vanishes since the conformal image of C_ν does not surround the origin, and therefore $\arg \{P\}$ returns to its initial value. The results for p_1 and q follow in the same way.

The functions

$$p - q = \log\left(\frac{P}{f}\right), \qquad p_1 - q = \log\left(\frac{Q}{f}\right)$$

are regular and single-valued in D. Indeed, the singularities of p and q and of p_1 and q cancel each other, and $p - q$ and $p_1 - q$ are free of periods about the boundary components since the same is true of p, p_1, q. Hence, the pairs of functions p, q and p_1, q of (25) may be identified with the pair p, q in the formula (11). We thus obtain

$$\iint_D \left| \frac{P'}{P} - \frac{f'}{f} \right|^2 dx\, dy = \operatorname{Re} \left\{ \frac{1}{i} \int_C \overline{\log \left(\frac{P}{f} \right)} \frac{P'}{P} dz \right\} \tag{26}$$

$$+ \frac{1}{2i} \int_C \overline{\log f} \frac{f'}{f} dz - \frac{1}{2i} \int_C \overline{\log P} \frac{P'}{P} dz,$$

and a similar identity in which P is replaced by Q. Since P and Q map the boundary components C_ν onto circular slits and radial slits, respectively, we have $\log |P| = \operatorname{Re} \{\log P\} = \text{const.}$ and

$$\arg \{Q\} = \operatorname{Im} \{\log Q\} = \text{const.}$$

if z is on C_ν. Hence $d(\log P)$ is pure imaginary and $d(\log Q)$ is real on C, that is,

$$\frac{1}{i} \frac{P'}{P} dz = \text{real}, \qquad \frac{Q'}{Q} dz = \text{real}, \qquad z \in C.$$

It follows that

$$\frac{1}{i} \frac{P'}{P} dz = \overline{\frac{1}{i} \frac{P'}{P} dz}, \qquad \frac{1}{i} \frac{Q'}{Q} dz = - \overline{\frac{1}{i} \frac{Q'}{Q} dz}. \tag{27}$$

In view of (27), the first and third line integrals on the right-hand side of (26) take the forms

$$\pm \operatorname{Re} \left\{ \frac{1}{i} \int_C \log \left(\frac{P}{f} \right) \frac{P'}{P} dz \right\} \tag{28}$$

and

$$\mp \frac{1}{2i} \int_C \log P \frac{P'}{P} dz = \mp \frac{1}{4i} \int_C d[(\log P)^2], \tag{29}$$

respectively. For brevity, (28) and (29) have been written so as to apply to both P and Q, the upper sign referring to P and the lower sign to Q.

The integral (29) vanishes since $\log P$ has no periods about the C_ν. The value of (28) is, by the residue theorem,

$$\pm 2\pi \operatorname{Re} \left\{ \log \frac{P'(u)}{f'(u)} \right\} = \pm 2\pi \log \left| \frac{P'(u)}{f'(u)} \right|,$$

where the fact that $P(u) = f(u) = 0$ and that $\dfrac{P}{f} \to 1$ for $z \to v$ has been taken into account. The integral

$$\frac{1}{2i} \int_C \overline{\log f} \frac{f'}{f} dz \tag{30}$$

can be expressed as a negative area. The reader will confirm that the mapping $z \to \log f(z)$ transforms the C_ν into n simple closed curves C_ν', and that the integral (30) is equal to the negative value of the combined area A enclosed by the C_ν'. Collecting our results, we thus find that

$$A + \iint_D \left| \frac{P'}{P} - \frac{f'}{f} \right|^2 dx\,dy = \pm 2\pi \log \left| \frac{P'(u)}{f'(u)} \right|.$$

The left-hand side is obviously positive unless f is identical with P. Remembering that the lower sign stands for the function Q, we thus obtain

$$|Q'(u)| \leq |f'(u)| \leq |P'(u)|, \tag{31}$$

where equality is possible only for $f = Q$ or $f = P$.

In the simply-connected case, the existence of the circular and radial slit mappings is an immediate consequence of the Riemann mapping theorem, and (31) is therefore proved for functions $f(z)$ which, in addition to satisfying the other required conditions, are also regular on the boundary of the simply-connected domain D. We may, however, free ourselves from this latter condition by repeating the procedure employed in the case of the function $\varphi_\theta(z)$. (31) is therefore true for any function $f(z)$ which is univalent in a simply-connected domain D, has the normalization (24)—or, if $v = \infty$, the normalization (14)—and satisfies $f(u) = 0$.

If $v = \infty$, we may take $f(z) = z - u$. In this case, we therefore obtain from (31)

$$|Q'(u;u,\infty)| \leq 1 \leq |P'(u;u,\infty)|. \tag{32}$$

With the help of (32) we now prove both the existence of the slit mappings indicated in Fig. 46*b* and *c* and the following extremal properties characteristic of these mappings:

Let $S(u,v)$ denote the family of functions $f(z)$ which are univalent in a multiply-connected domain D, are normalized by (24), *and satisfy $f(u) = 0$. Then the problem*

$$|f'(u)| = \max., \qquad f(z) \in S(u;v), \tag{33}$$

is solved by the function $P(z;u,v)$ which maps D onto the circular slit domain indicated in Fig. 46*b; the problem*

$$|f'(u)| = \min., \qquad f(z) \in S(u,v), \tag{34}$$

is solved by the function $Q(z;u,v)$ which yields the radial slit domain indicated in Fig. 46*c.*

The existence of a function $F(z)$ solving (33) is guaranteed by the fact that the family $S(u,v)$ is compact. If $w = F(z)$ does not yield the circular slit mapping, then there exists at least one boundary component, say C_ν, whose image C_ν' is not a circular slit. Let $P(z) = P(z;0,\infty)$ be the function mapping the exterior of C_ν' onto the exterior of a circular slit. Then, by (32), $|P'(0)| > 1$. Clearly, the function

$$F_1(z) = P[F(z)]$$

also belongs to $S(u,v)$. But, in view of $|P'(0)| > 1$,

$$|F_1'(u)| = |P'(0)||F'(u)| > |F'(u)|,$$

contradicting the fact that $F(z)$ solves the problem (33). The assumption that $F(z)$ does not yield the circular slit mapping has thus led to an absurdity. This proves the existence of the circular slit mapping. The existence proof for the radial slit mapping follows in the same way from the extremal problem (34), and it is left as an exercise to the reader.

The existence of the function $R_\nu(z)$ which maps D onto a circle with circular slits as indicated in Fig. 46*d* can be easily inferred from that of the function $P(z;u,v)$. If D is the given multiply-connected domain, we first map the outside of the hole corresponding to the boundary component C_ν onto the interior of the unit circle. By the Riemann mapping theorem, this can be done in such a way that a given point ζ of D is mapped onto the origin. This new domain, say D', is contained in $|z| < 1$, and its outer boundary is $|z| = 1$. Let now D'' be the domain which is inverse to D' with respect to $|z| = 1$, and denote by D^* the domain containing D', D'' and the points of $|z| = 1$. Clearly, D^* contains both the origin and the point $z = \infty$. As shown before, there exists a function $P(z;0,\infty)$ which maps D^* onto a circular slit domain. Consider now the function

$$P_1(z) = \frac{1}{\overline{P[(\bar{z}^{-1}),0,\infty]}}. \tag{35}$$

In view of its definition, the domain D^* is transformed into itself by the inversion $z \to \bar{z}^{-1}$. The values of $P(\bar{z}^{-1})$ in D will therefore also fill the above circular slit domain. Since the transformations $w \to \bar{w}$ and $w \to w^{-1}$ obviously carry circular arcs about the origin into circular arcs of the same kind, it thus follows that the analytic function P_1 likewise maps D^* onto a circular slit domain. By (35), we have $P_1(0) = 0$, $P_1(\infty) = \infty$, $P_1'(\infty) = 1$. As pointed out in Sec. 1, such a circular slit mapping is unique. We must therefore have $P_1(z) = P(z)$ and thus, by (35),

$$P(z)\overline{P\left(\frac{1}{\bar{z}}\right)} = 1.$$

Taking, in particular, values of z such that $|z|^2 = z\bar{z} = 1$, we find that

$$|P(z)| = 1, \qquad |z| = 1.$$

Since $P(z)$ is univalent, it maps $|z| = 1$ onto a simple closed curve. Hence, $P(z)$ maps the domain D' onto the interior of the unit circle which is furnished with a number of circular slits about the origin. Going back to the original domain D, we see that we have shown the existence of a function $R_\nu(z)$ which maps D onto a domain as indicated in Fig. 46d in such a way that $R_\nu(\zeta) = 0$ and that the boundary component C_ν corresponds to the circumference of the unit circle.

The function $R_\nu(z)$ may be characterized by the following extremal problem.

Let $E(\zeta)$ *denote the family of functions* $f(z)$ *which are bounded—that is,* $|f(z)| \leq 1$*—and univalent in* D*, satisfy* $f(\zeta) = 0$ *and, moreover, are such that* C_ν *corresponds to the outer boundary* γ *of the domain* Δ *onto which* D *is mapped by* $w = f(z)$*. Then the problem*

$$|f'(\zeta)| = \text{max.}, \qquad f(z) \in E(\zeta), \tag{36}$$

is solved by the function $R_\nu(z)$ *which maps* D *onto the unit circle with circular slits.*

If $w' = p(w)$ is the function which maps the interior of γ onto $|w'| < 1$ so that $p(0) = 0$, then, by the Schwarz lemma, $|p'(0)| \geq 1$. Hence, the function $f_1(z) = p[f(z)]$ of $E(\zeta)$ satisfies $|f_1'(\zeta)| = |f'(\zeta)||p'(0)| \geq |f'(\zeta)|$, which shows that the function solving the problem (36) must map C_ν onto the circumference of the unit circle. For functions of this type, the extremal property (36) can easily be reduced to the extremal property (33) of the circular slit mapping. The procedure is similar to that used above in connection with the existence proof for the function $R_\nu(z)$, and it is left as an exercise to the reader.

EXERCISES

1. Show that there exists a function $T_\nu(z)$ which maps a multiply-connected domain D onto the interior of the unit circle which is furnished with a number of rectilinear slits pointing at the origin, such that $T_\nu(\zeta) = 0$ ($\zeta \in D$), and that the boundary component C_ν is transformed into the circumference of the unit circle. *Hint:* Generalize the procedure employed in the existence proof for the function $R_\nu(z)$.

2. Show that the function $T_\nu(z)$ of the preceding exercise solves the extremal problem

$$|f'(\zeta)| = \text{min.},$$

where $|f(z)| < 1$ and $f(z)$ is univalent in D and $f(z)$, moreover, satisfies the conditions $f(\zeta) = 0$, $|f(z)| = 1$ for $z \in C_\nu$. *Hint:* Generalize the method of proof indicated in the case of the extremal problem (36).

3. If $\zeta = \xi + i\eta$ and if $\partial/\partial\zeta$ and $\partial/\partial\bar{\zeta}$ denote the differential operators defined in (103), Sec. 10, Chap. V, show that the canonical mapping functions $\varphi_\theta(z,\zeta)$ and $P(z;\zeta,v)$ are related by the identity

$$\varphi_\theta(z,\zeta) = -\frac{\partial \log P(z;\zeta,v)}{\partial \zeta} + e^{2i\theta}\frac{\partial \log P(z;\zeta,v)}{\partial \bar{\zeta}}.$$

Show also that this identity remains true if $P(z;\zeta,v)$ is replaced by one of the mapping functions $R_\nu(z,\zeta)$. *Hint:* Show that the right-hand expression has both the singularity and the boundary behavior characteristic of $\varphi_\theta(z,\zeta)$.

4. If $Q(z;u,v)$ is the radial slit mapping function and $N(z,\zeta)$ is the Neumann function of D, discussed in Sec. 7, Chap. I, show that

$$\log |Q(z;u,v)| = N(z,v) - N(z,u).$$

Hint: Complete the harmonic function $N(z,v) - N(z,u)$ to an analytic function and show that the resulting function is free of periods about the boundary components of D and that it has both the singularities and the boundary behavior characteristic of $\log Q(z;u,v)$.

5. Show that the circular slit function $P = P(z;\zeta,v)$ and the radial slit function $Q = Q(z;\zeta,v)$ are related by the identities

$$\frac{\partial \log P}{\partial \zeta} = \frac{\partial \log Q}{\partial \zeta}, \qquad \frac{\partial \log P}{\partial \bar{\zeta}} = -\frac{\partial \log Q}{\partial \bar{\zeta}}.$$

6. Using the extremal property of the mapping function $\varphi_\theta(z,\zeta)$, show that any function

$$f(z) = \frac{1}{z} + b_0 + b_1 z + \cdots,$$

which is univalent in $|z| < 1$, satisfies

$$|b_1| \leq 1.$$

7. Using the extremal properties of the circular and radial slit mappings, show that the function $f(z)$ of the preceding exercise is subject to the sharp inequalities

$$\frac{1 - |z|^2}{|z|^2} \leq |f'(z)| \leq \frac{1}{|z|^2(1 - |z|^2)}.$$

Hint: Use the results of Exercise 1, Sec. 1.

8. Show that the function $P_\theta(z;u,v)$ defined by

$$\log P_\theta(z;u,v) = e^{i\theta}[\cos\theta \log Q(z;u,v) - i \sin\theta \log P(z;u,v)]$$

is univalent in D and that $w = P_\theta(z;u,v)$ maps D onto a slit domain bounded by arcs of logarithmic spirals which intersect the rays $\arg\{w\} = \text{const.}$ with the angle θ. *Hint:* Study the boundary behavior of $P_\theta(z;u,v)$ and apply the argument principle.

9. Generalizing the procedure used in the proof of the extremal properties of the functions $P(z;u,v)$ and $Q(z;u,v)$, derive the following result: If $f(z)$ is univalent in D, $f(u) = 0$, and $f(z)$ is normalized by (24), then

$$\text{Re}\,\{e^{-2i\theta}\log f'(u)\} \geq \text{Re}\,\{e^{-2i\theta}\log P_\theta'(u;u,v)\},$$

where $P_\theta(z;u,v)$ is the spiral slit function defined in the preceding exercise.

10. Using the results of the preceding exercise (for $\theta = \frac{1}{4}\pi, \frac{3}{4}\pi$) and of Exercise 1, Sec. 1, prove the following sharp result: If $f(z)$ is the function of Exercise 6, then

$$\text{Re}\,\{z^2 f'(z)\} \leq 0 \qquad \text{if } |z|^2 \leq 1 - e^{-\frac{1}{2}\pi}.$$

11. Let $f(z) = z + a_2z^2 + \cdots$ be regular and univalent in $|z| < 1$, and let $g(z)$ denote the odd univalent function $g(z) = \sqrt{f(z^2)}$. Applying the result of Exercise 9 to the function

$$f_1(z) = \frac{g'(v)}{g(z) - g(v)} + \frac{g'(v)}{2g(v)}$$

and setting $u = -v$, $\theta = \pm\frac{1}{4}\pi$, prove the sharp result

$$\left| \arg \left\{ \frac{zf'(z)}{f(z)} \right\} \right| \leq \log \left(\frac{1 + \rho}{1 - \rho} \right), \qquad |z| = \rho < 1.$$

Deduce that

$$\text{Re} \left\{ \frac{zf'(z)}{f(z)} \right\} > 0$$

for $\rho < \tanh \frac{1}{4}\pi$ and that, therefore, the function $w = f(z)$ maps the circle $|z| < \tanh \frac{1}{4}\pi$ onto a star-shaped domain. Show also that $\tanh \frac{1}{4}\pi$ is the exact value of the "star radius" for the family of functions in question. *Hint:* Obtain the explicit form of $P_\theta(z;u,v)$ for the unit circle from the result of Exercise 1, Sec. 1.

12. Using (11), show that

$$\iint_D \left| \varphi_\theta'(z,\zeta) + \frac{1}{(z - \zeta)^2} \right|^2 dx\, dy = \text{Re}\, \{2\pi e^{-2i\theta} a_\theta\} - A,$$

where

$$A = \frac{1}{2i} \int_C \frac{dz}{(\bar{z} - \bar{\zeta})(z - \zeta)^2}$$

is the area of the complement of the domain into which D is transformed by the substitution $z \to \dfrac{1}{z - \zeta}$.

13. From the fact that a domain of connectivity n can be mapped conformally onto a parallel slit domain of the type indicated in Fig. 46a, deduce that a domain of connectivity n is conformally characterized by $3n - 6$ real numbers (the Riemann moduli). *Hint:* Count the number of real parameters determining the slit domain and deduct the arbitrary real parameters entering the definition of the parallel slit mapping function.

3. The Green's Function and the Dirichlet Problem. Some of the developments of Chap. I—especially in the later sections—were carried out under the assumption that the Dirichlet problem of Sec. 5, Chap. I, that is, the problem of constructing a harmonic function with given boundary values, always has a solution. In the case of a simply-connected domain, the truth of this assumption was shown in Sec. 4, Chap. V. We saw there that if the Dirichlet problem can be solved for a domain D, it can also be solved for all domains conformally equivalent to D. By the Riemann mapping theorem, all simply-connected domains with more than one boundary point are conformally equivalent to the unit circle. Since, for the latter, the Dirichlet problem is solved by the Poisson integral (Sec. 6, Chap. I), this disposes of the simply-connected case.

With the help of some of the canonical mappings whose existence was demonstrated in the preceding section, we are now in a position to prove the existence of a solution of the Dirichlet problem in the case of a general domain of finite connectivity. It is sufficient to do so for domains bounded by simple closed analytic curves. As shown at the beginning of Sec. 2, any domain of finite connectivity can be conformally mapped onto a domain of this type and, as pointed out above, it is sufficient to solve the Dirichlet problem for one particular representative of a given conformal class.

Our point of departure is the analytic function $w = S_{\nu\mu}(z)$ which maps D onto a ring with circular slits as indicated in Fig. 46*e*. The two subscripts indicate that the outer and inner circles of the ring are the conformal images of the boundary components C_ν and C_μ, respectively. On the boundary component C_k ($k = 1, 2, \ldots, n$) we have $|S_{\nu\mu}(z)| = \alpha_{k\nu} =$ const.; if we normalize $S_{\nu\mu}(z)$ by making the radius of the outer circle equal to 1, we have $\alpha_{\nu\nu} = 1$. It follows that

$$\text{Re}\,\{\log S_{\nu\mu}(z)\} = \log |S_{\nu\mu}(z)| = \log \alpha_{k\nu}, \qquad z \in C_k. \tag{37}$$

Now the function $\log |S_{\nu\mu}(z)|$, being the real part of an analytic function, is a harmonic function which, clearly, is single-valued in D. Taking, in particular, $\mu = n$ and employing the notation $\sigma_\nu(z) = \log |S_{\nu n}(z)|$, we have thus shown the existence of single-valued harmonic functions $\sigma_\nu(z)$ which take constant boundary values on the boundary components $C_1, \ldots,$ C_n. In view of our normalization, we have $\sigma_\nu(z) = 0$ if z is on C_n.

We now show that we may choose constant coefficients

$$a_{k\nu} \ (k,\nu = 1, \ldots, n-1)$$

such that the harmonic functions

$$\omega_\nu(z) = \sum_{k=1}^{n-1} a_{k\nu}\sigma_k(z), \qquad \nu = 1, \ldots, n-1, \tag{38}$$

will have the boundary values $\delta_{\nu\mu}$ on the boundary component C_μ, that is, $\omega_\nu(z) = 1$ for $z \in C_\nu$ and $\omega_\nu(z) = 0$ for $z \in C_\mu$, $\mu \neq \nu$. Setting $\log \alpha_{k\nu} = \beta_{k\nu}$ in (37), we find that the function (38) has the boundary values

$$\sum_{k=1}^{n-1} a_{k\nu}\beta_{\mu k}$$

for $z \in C_\mu$. The existence of the functions (38) with the required boundary values will therefore be proved if we can show that, for every ν, the system of $n - 1$ linear equations

$$\sum_{k=1}^{n-1} a_{k\nu}\beta_{\mu k} = \delta_{\nu\mu}, \qquad \mu = 1, 2, \ldots, n-1,$$

has a solution. By the theory of linear equations, this will be the case if, and only if, the corresponding homogeneous system

$$\sum_{k=1}^{n-1} a_{k\nu}\beta_{\mu k} = 0, \qquad \mu = 1, 2, \ldots, n-1,$$

does not possess a nontrivial solution. To prove the latter, suppose that this is not true and that such a solution does exist. In view of what was said above, this entails the existence of coefficients $a_{1\nu}, a_{2\nu}, \ldots, a_{n-1,\nu}$, not all zero, such that the harmonic function $\sigma(z) = \sum_{k=1}^{n-1} a_{k\nu}\sigma_k(z)$ has the boundary values zero if z approaches any point of $C_1, \ldots, C_{n-1}$. On C_n, $\sigma(z)$ is likewise zero, since this is true of all functions $\sigma_k(z)$. Hence, the harmonic function $\sigma(z)$ is zero at all points of the boundary of D, and it is therefore identically zero.

In other words, there will exist constant coefficients $A_1, \ldots, A_{n-1}$, not all which are zero, such that the harmonic function

$$\sigma(z) = \sum_{\nu=1}^{n-1} A_\nu \log |S_{\nu n}(z)|$$

vanishes identically. Its harmonic conjugate $\sigma^*(z)$ must therefore reduce to a constant, that is,

$$\sigma^*(z) = \sum_{\nu=1}^{n-1} A_\nu \arg \{S_{\nu n}(z)\} = \text{const.}$$

Now a glance at Fig. 46*e* shows that $\arg \{S_{\nu n}(z)\}$ grows by the amounts 2π and -2π if z traverses C_ν and C_n, respectively, while this expression returns to its initial value if any of the other boundary components are described. It follows therefore that $\sigma^*(z)$ grows by the amount $2\pi A_\nu$ if z describes the boundary component C_ν ($\nu = 1, \ldots, n-1$). But since $\sigma^*(z)$ is constant, this is absurd unless all the A_ν are zero.

We have thus proved the existence of the harmonic functions $\omega_\nu(z)$ ($\nu = 1, \ldots, n-1$) which have the boundary values 1 on C_ν and take the values 0 at all other points of the boundary C of D. A corresponding function $\omega_n(z)$ is defined by

$$\omega_n(z) = 1 - \omega_1(z) - \cdots - \omega_{n-1}(z).$$

Clearly, $\omega_n(z) = 1$ for $z \in C_n$ and $\omega_n(z) = 0$ for $z \in C_\nu$, $\nu \neq n$. A comparison with Sec. 10, Chap. I, shows that these functions are identical with the *harmonic measures* of $C_1, \ldots, C_n$ considered there.

We next consider the harmonic function

$$G(z,\zeta) = \text{Re}\,\{\log R_n(z)\} = \log |R_n(z)|,$$

where $R_n(z)$ maps D onto the unit circle with circular slits indicated in Fig. 46*d*. Since $|R_n(z)| =$ const. on each boundary component, we have

$$\text{(39)} \qquad G(z,\zeta) = \gamma_\nu, \qquad z \in C_\nu,\ \nu = 1, \ldots, n-1,$$

and $G(z,\zeta) = 0$ for $z \in C_n$. $R_n(z)$ is regular in D and its only zero is located at $z = \zeta$. Hence, the function $G(z,\zeta)$ is harmonic at all points of D, with the exception of the point $z = \zeta$ where $G(z,\zeta) - \log |z - \zeta|$ is harmonic. Consider now the function

$$g(z,\zeta) = -G(z,\zeta) + \sum_{\nu=1}^{n-1} \gamma_\nu \omega_\nu(z),$$

where $\omega_\nu(z)$ is the harmonic measure of C_ν. In view of (39) and the boundary behavior of the $\omega_\nu(z)$, it follows that the function $g(z,\zeta)$ has the boundary values 0 at all points of C. $g(z,\zeta)$ is harmonic throughout D except at $z = \zeta$, where

$$g(z,\zeta) = -\log |z - \zeta| + g_1(z,\zeta),$$

and $g_1(z,\zeta)$ is harmonic. A comparison with Sec. 5, Chap. I, shows that $g(z,\zeta)$ coincides with the *Green's function* of D introduced there. The existence of the Green's function has thus been proved in the case of a general multiply-connected domain D.

Once the existence of the Green's function is known, it is not difficult to show that the general boundary value problem always has a solution, provided the given boundary value function satisfies certain smoothness requirements. If s is a length parameter on C which grows monotonically if the point $z(s)$ describes the entire boundary C of D in the positive sense, and if $U(z) = U[z(s)]$ is a bounded and integrable function of s, we shall show that *there exists a harmonic function* $u(\zeta) = u(\xi,\eta)$ $(\zeta = \xi + i\eta)$ *which has the prescribed boundary values* $U(z)$ *at all points of* C *at which* $U(z)$ *is continuous.* The latter restriction is unavoidable since at the discontinuities of $U(z)$ the boundary values cease to be uniquely defined. *If the Green's function* $g(z,\zeta)$ *of* D *is known, then the harmonic function* $u(\zeta)$ *can be represented in the form*

$$(40) \qquad u(\zeta) = -\frac{1}{2\pi}\int_C U(z)\,\frac{\partial g(z,\zeta)}{\partial n}\,ds,$$

where $\partial/\partial n$ denotes differentiation with respect to the outwards pointing normal. We add that the normal derivative of $g(z,\zeta)$ exists and is continuous on all boundary components of C since the latter are closed analytic curves (see Sec. 5, Chap. V).

As shown in Sec. 5, Chap. I, we have the identity

$$(41) \qquad v(\zeta) = -\frac{1}{2\pi}\int_C v(z)\,\frac{\partial g(z,\zeta)}{\partial n}\,ds$$

for any function $v(z)$ which is harmonic in $D + C$. Let now be z_0 be an arbitrary point of C, and let $v(z)$ be a harmonic function such that $v(z) - v(z_0) > 0$ if z is any point of C which does not coincide with z_0. The reader will easily confirm that a suitable function $v(z)$ will be $\log|z - a|$ or $-\log|z - a|$, where a is an appropriately chosen point outside D. We denote by α a small arc of C which contains z_0 and is taken small enough so that

$$(42) \qquad |v(z) - v(z_0)| < \epsilon, \qquad z \in \alpha,$$

where ϵ is an arbitrarily small positive parameter. Setting, in (41), $v(z) \equiv 1$, we obtain

$$(43) \qquad 1 = -\frac{1}{2\pi}\int_C \frac{\partial g(z,\zeta)}{\partial n}\,ds.$$

Using this, we derive from (41) the identity

$$(44) \qquad \begin{aligned} v(\zeta) - v(z_0) &= -\frac{1}{2\pi}\int_C [v(z) - v(z_0)]\,\frac{\partial g(z,\zeta)}{\partial n}\,ds \\ &= -\frac{1}{2\pi}\int_\alpha [v(z) - v(z_0)]\,\frac{\partial g(z,\zeta)}{\partial n}\,ds \\ &\qquad -\frac{1}{2\pi}\int_{C-\alpha} [v(z) - v(z_0)]\,\frac{\partial g(z,\zeta)}{\partial n}\,ds. \end{aligned}$$

By Exercise 3, Sec. 5, Chap. I, $\dfrac{\partial g}{\partial n} \leq 0$ on C. It thus follows from (42) and (43) that

$$(45) \qquad \begin{aligned} \left| -\frac{1}{2\pi}\int_\alpha [v(z) - v(z_0)]\,\frac{\partial g(z,\zeta)}{\partial n}\,ds \right| &\leq \frac{\epsilon}{2\pi}\int_\alpha \left|\frac{\partial g(z,\zeta)}{\partial n}\right| ds \\ &\leq \frac{\epsilon}{2\pi}\int_C \left|\frac{\partial g(z,\zeta)}{\partial n}\right| ds = -\frac{\epsilon}{2\pi}\int_C \frac{\partial g(z,\zeta)}{\partial n}\,ds = \epsilon. \end{aligned}$$

Let now β denote a fixed small arc of C which contains z_0, and take the arc α in (44) small enough so that $\alpha \subset \beta$. Since there exists a positive constant m such that $v(z) - v(z_0) > m > 0$ on $C - \beta$, we have

$$\left| -\frac{1}{2\pi} \int_{C-\alpha} [v(z) - v(z_0)] \frac{\partial g(z,\zeta)}{\partial n} ds \right| = \frac{1}{2\pi} \int_{C-\alpha} |v(z) - v(z_0)| \left| \frac{\partial g(z,\zeta)}{\partial n} \right| ds$$

$$\geq \frac{1}{2\pi} \int_{C-\beta} |v(z) - v(z_0)| \left| \frac{\partial g(z,\zeta)}{\partial n} \right| ds \geq \frac{m}{2\pi} \int_{C-\beta} \left| \frac{\partial g(z,\zeta)}{\partial n} \right| ds$$

Applying (45) and this inequality to (44), we obtain

$$\frac{1}{2\pi} \int_{C-\beta} \left| \frac{\partial g(z,\zeta)}{\partial n} \right| ds \leq \frac{\epsilon}{m} + \frac{1}{m} |v(\zeta) - v(z_0)|.$$

Now ϵ is arbitrarily small, while m is fixed. Since $v(\zeta) \to v(z_0)$ if $\zeta \to z_0$, it follows therefore that

$$\lim_{\zeta \to z_0} \int_{C-\beta} \left| \frac{\partial g(z,\zeta)}{\partial n} \right| ds = 0, \tag{46}$$

where β is any arc of C which contains the boundary point z_0.

Consider now the function $u(\zeta)$ defined by (40). Since the Green's function is harmonic in both arguments (see Sec. 5, Chap. I), $u(\zeta)$ is a harmonic function of ζ. If z_0 is a point at which the boundary function $U(z)$ is continuous, we may choose a small subarc β of C such that

$$|U(z) - U(z_0)| < \epsilon, \qquad z \in \beta,$$

for given positive ϵ. Writing (40) in the form

$$u(\zeta) - U(z_0) = -\frac{1}{2\pi} \int_{\beta} [U(z) - U(z_0)] \frac{\partial g(z,\zeta)}{\partial n} ds$$

$$- \frac{1}{2\pi} \int_{C-\beta} [U(z) - U(z_0)] \frac{\partial g(z,\zeta)}{\partial n} ds$$

and observing that, analogously to (45),

$$\left| -\frac{1}{2\pi} \int_{\beta} [U(z) - U(z_0)] \frac{\partial g(z,\zeta)}{\partial n} ds \right| \leq \epsilon,$$

we find that

$$|u(\zeta) - U(z_0)| \leq \epsilon + \frac{1}{2\pi} \int_{C-\beta} |U(z) - U(z_0)| \left| \frac{\partial g(z,\zeta)}{\partial n} \right| ds.$$

The function $U(z)$ is bounded on C, that is, there is a constant M such that $|U(z)| < M$. Hence, $|U(z) - U(z_0)| < 2M$ and therefore

$$|u(\zeta) - U(z_0)| \le \epsilon + \frac{M}{\pi} \int_{C-\beta} \left| \frac{\partial g(z,\zeta)}{\partial n} \right| ds.$$

By (46), it follows that

$$\limsup_{\zeta \to z_0} |u(\zeta) - U(z_0)| \le \epsilon$$

and thus, since ϵ is arbitrarily small,

$$\lim_{\zeta \to z_0} u(\zeta) = U(z_0).$$

This shows that at all points of continuity of $U(z)$ the harmonic function $u(\zeta)$ defined by (40) has indeed the prescribed boundary values $U(z)$.

EXERCISES

1. If $p(z,\zeta) = g(z,\zeta) + ih(z,\zeta)$ and $w_\nu(z) = \omega_\nu(z) + i\omega_\nu^*(z)$ are the analytic functions whose real parts are the Green's function and the harmonic measures, respectively, show that the differentials $iw_\nu'\,dz$ and $ip'\,dz$ are real on C and that the latter is, moreover, positive.

2. If the function $f(z)$ is regular and single-valued in D, show that

$$\iint_D \overline{f'(z)} w_\nu'(z)\, dx\, dy = 0, \qquad 1, \ldots, n,$$

where $w_\nu(z)$ is defined in the preceding exercise. *Hint:* Use Green's formula (106) (Sec. 10, Chap. V) and the result of the preceding exercise.

3. It was pointed out at the end of Sec. 1 that not every regular and single-valued function $f(z)$ is the derivative of a single-valued function. Show that a function $f(z)$ which is regular and single-valued in D can be written in the form

$$f(z) = g'(z) + \sum_{\nu=1}^{n-1} \alpha_\nu w_\nu'(z),$$

where $g(z)$ is regular and single-valued in D and $\alpha_1, \ldots, \alpha_{n-1}$ are appropriate constants. *Hint:* Use the properties of the harmonic measures discussed in Sec. 10, Chap. I, in particular, formula (61).

4. Let $u(z)$ be the harmonic function defined by

$$u(z) = \operatorname{Re}\left\{\frac{1}{z-\zeta}\right\} + \frac{1}{2\pi} \int_C \operatorname{Re}\left\{\frac{1}{\eta - \zeta}\right\} \frac{\partial g(z,\eta)}{\partial n_\eta}\, ds_\eta,$$

where ζ is a point of D. Show that $u(z)$ vanishes at all points of C. Show further that the constants $a_1, \ldots, a_{n-1}$ may be so chosen that the harmonic conjugate of the function

$$u_1(z) = u(z) + \sum_{\nu=1}^{n-1} a_\nu \omega_\nu(z)$$

is single-valued in D, and conclude that $u_1(z) = \operatorname{Re}\{\psi(z,\zeta)\}$, where $\psi(z,\zeta)$ is the vertical slit mapping function (9).

5. Show that the function $u(z)$ of the preceding exercise is related to the Green's function $g(z,\zeta)$ by means of the formula

$$u(z) = \frac{g(z,\zeta)}{\partial \xi}, \qquad \xi = \text{Re}\,\{\zeta\}.$$

6. Let $f(z)$ be regular and single-valued in $D + C$ and let $\partial/\partial z$ and $\partial/\partial \bar{z}$ denote the differential operators (103) (Sec. 10, Chap. V). Applying Green's formula (104) (Sec. 10, Chap. V) to the domain obtained from D by deleting from it a small circle of radius ϵ and center ζ, show that

$$\iint_D \frac{\partial^2 g(z,\zeta)}{\partial \bar{z}\, \partial \zeta} f(z)\, dx\, dy = -\frac{\pi}{2} f(\zeta)$$

and that, therefore, the function

$$k(z,\zeta) = -\frac{2}{\pi} \frac{\partial^2 g(z,\zeta)}{\partial z\, \partial \bar{\zeta}}$$

has the "reproducing property"

$$\iint_D \overline{k(z,\zeta)} f(z)\, dx\, dy = f(\zeta)$$

[compare formula (128), Sec. 10, Chap. V]. Show also that $k(z,\zeta)$ is regular at all points of D. *Hint:* Use the fact that $\dfrac{\partial g(z,\zeta)}{\partial \zeta} = 0$ for $z \in C$, and let $\epsilon \to 0$.

7. Show that the reproducing property and the regularity of the function $k(z,\zeta)$ characterize it uniquely.

4. Area Problems. In this section, we shall solve two extremal problems involving areas related to the conformal maps of a given multiply-connected domain D. The first problem refers to schlicht mappings of D which are effected by univalent functions $f(z)$ with the normalization

$$f(z) = \frac{1}{z - \zeta} + a_0 + a_1(z - \zeta) + \cdots, \tag{47}$$

where ζ is a given point of D. If $\zeta = \infty$, (47) is replaced by

$$f(z) = z + a_0 + \frac{a_1}{z} + \cdots. \tag{47'}$$

Since the domain D' onto which D is mapped by $w = f(z)$ contains the point at infinity, the complement D'' of D' has a finite area, where the area of a point set is defined as the greatest lower bound of the areas of all polygons which contain the set. The first problem now consists in finding the particular function $f(z)$ for which this "outer area" is the largest possible among all univalent functions normalized by (47) or (47').

The second problem refers to functions $g(z)$ which are regular and single-valued in D and are normalized by the condition $g'(\zeta) = 1$. The (finite or infinite) area of the domain D^* onto which D is mapped by $w = g(z)$ is

$$A = \iint_D |g'(z)|^2 \,dx\,dy; \tag{48}$$

if D^* is not schlicht, those parts of the w-plane which are covered by D^* more than once are counted in (48) with their multiplicities. The problem is to find the particular function $g(z)$ for which this "inner area" is the smallest possible among all functions $g(z)$ satisfying $g'(\zeta) = 1$.

We shall see that the two area problems are closely related to each other and to the parallel slit mappings discussed earlier. If $\varphi(z,\zeta)$ and $\psi(z,\zeta)$ are the functions (9) which yield, respectively, the horizontal slit mapping and the vertical slit mapping of D, we introduce the functions $M(z,\zeta)$ and $N(z,\zeta)$ by the definitions

$$\begin{aligned} M(z,\zeta) &= \tfrac{1}{2}[\varphi(z,\zeta) - \psi(z,\zeta)], \\ N(z,\zeta) &= \tfrac{1}{2}[\varphi(z,\zeta) + \psi(z,\zeta)]. \end{aligned} \tag{49}$$

In view of (2) and (9), the function $M(z,\zeta)$ is regular in D while $N(z,\zeta)$ has a simple pole with the normalization (47), or (47′), as the case may be, at $z = \zeta$. By the results of Exercises 4 to 9, Sec. 1, the functions $M(z,\zeta)$ and $N(z,\zeta)$ are connected by the relation

$$\overline{M'(z,\zeta)\,dz} = N'(z,\zeta)\,dz, \qquad z \in C, \tag{50}$$

and $N(z,\zeta)$ is univalent in D. We shall prove that *the function* $N(z,\zeta)$ *yields the largest outer area among all univalent functions in D which are normalized by* (47), and that *the function* $[M'(\zeta,\zeta)]^{-1}M(z,\zeta)$ *yields the smallest inner area among all functions* $g(z)$ *which are regular in D and for which* $g'(\zeta) = 1$.

Suppose first that both $f(z)$ and $g(z)$ are regular on the boundary C of D. In this case, it follows from the identity (11) that

$$\iint_D |N' - f'|^2 \,dx\,dy = \operatorname{Re}\left\{\frac{1}{i}\int_C (\bar N - \bar f)N'\,dz\right\} + \frac{1}{2i}\int_C \bar f f'\,dz - \overline{\frac{1}{2i}\int_C \bar N N'\,dz} \tag{51}$$

and

$$\iint_D |M^{*\prime} - g'|^2 \,dx\,dy = \operatorname{Re}\left\{\frac{1}{i}\int_C (\overline{M^*} - \bar g)M^{*\prime}\,dz\right\} + \frac{1}{2i}\int_C \bar g g'\,dz - \overline{\frac{1}{2i}\int_C \overline{M^*}M^{*\prime}\,dz}, \tag{52}$$

where

$$M^*(z,\zeta) = \frac{M(z,\zeta)}{M'(\zeta,\zeta)}. \tag{53}$$

In view of (50), the first integral on the right-hand side of (51) is equivalent to a contour integral whose integrand is regular in D. By Cauchy's theorem, this integral is therefore zero and (51) takes the form

$$\iint_D |N' - f'|^2 \, dx \, dy = \frac{1}{2i}\int_C \bar{f}f' \, dz - \overline{\frac{1}{2i}\int \bar{N}N' \, dz}.$$

As shown in Sec. 2, we have

$$\frac{1}{2i}\int_C \bar{f}f' \, dz = -A_e(f),$$

where $A_e(f)$ is the outer area associated with $f(z)$. Hence

$$\iint_D |N' - f'|^2 \, dx \, dy = A_e(N) - A_e(f),$$

and therefore

$$A_e(f) < A_e(N), \tag{54}$$

unless $f(z)$ and $N(z,\zeta)$ coincide. This shows that among all univalent functions in D which are normalized by (47) and are regular on C, $N(z,\zeta)$ indeed yields the largest outer area.

Before we generalize this result to functions $f(z)$ which are not regular on C, we simplify the identity (52). In view of (50), the first term on the right-hand side of (52) can be brought into the form

$$-\text{Re}\left\{\frac{1}{i}\int_C (M^* - g)N' \, dz\right\}.$$

By (53), $M^{*\prime}(\zeta,\zeta) = 1$. Since also $g'(\zeta) = 1$, the derivative of $M^* - g$ vanishes for $z = \zeta$. The only singularity of $N(z,\zeta)$ in D is a double pole with the principal part $-(z - \zeta)^2$. By the residue theorem, the value of the contour integral is therefore zero and (52) reduces to

$$\iint_D |M^{*\prime} - g'|^2 \, dx \, dy = \frac{1}{2i}\int_C \bar{g}g' \, dz - \overline{\frac{1}{2i}\int_C \overline{M^*}M^{*\prime} \, dz}.$$

Both $g(z)$ and $M^*(z,\zeta)$ are regular in D and we may therefore transform the two line integrals by Green's formula (106) (Sec. 10, Chap. V). This yields

$$\iint_D |M^{*\prime} - g'|^2\,dx\,dy = \iint_D |g'|^2\,dx\,dy - \iint_D |M^{*\prime}|^2\,dx\,dy.$$

By (48), the two integrals on the right-hand side are the inner areas $A_i(g)$ and $A_i(M^*)$ associated with the functions $g(z)$ and $M^*(z,\zeta)$, respectively. Hence,

$$\iint_D |M^{*\prime} - g'|^2\,dx\,dy = A_i(g) - A_i(M^*),$$

and therefore

$$A_i(g) > A_i(M^*), \tag{55}$$

unless $g(z)$ and $M^*(z,\zeta)$ coincide. We have thus proved that $M^*(z,\zeta)$ yields the smallest inner area among all functions $g(z)$ which are regular and single-valued in $D + C$ and are normalized by $g'(\zeta) = 1$.

Both $A_e(N)$ and $A_i(M^*)$ can be expressed in terms of $M'(\zeta,\zeta)$. In view of (50), we have

$$A_e(N) = -\frac{1}{2i}\int_C \bar{N}N'\,dz = \overline{\frac{1}{2i}\int_C NM'\,dz},$$

whence, by the residue theorem, $A_e(N) = \pi\overline{M'(\zeta,\zeta)}$. Since $A_e(N)$ is necessarily positive, this shows that $M'(\zeta,\zeta) > 0$ and therefore

$$A_e(N) = \pi M'(\zeta,\zeta). \tag{56}$$

Using (50) and (53), we similarly obtain

$$A_i(M^*) = \frac{1}{2i}\int_C \overline{M^*}M^{*\prime}\,dz = -\frac{1}{M'(\zeta,\zeta)}\overline{\frac{1}{2i}\int_C M^*N'\,dz},$$

whence, by the residue theorem,

$$A_i(M^*) = \frac{\pi}{M'(\zeta,\zeta)}. \tag{57}$$

A comparison of (56) and (57) reveals the remarkable relation

$$A_e(N)\cdot A_i(M^*) = \pi^2. \tag{58}$$

It remains to be shown that the extremal properties of $N(z,\zeta)$ and $M^*(z,\zeta)$ are preserved if we lift the restriction that the functions $f(z)$ and $g(z)$ be regular on C. Since the univalent functions $f(z)$ form a normal and compact family in D, it is clear that there exists a function $f_0(z)$, which is univalent in D and has the normalization (47), such that

$$A_e(f) \le A_e(f_0), \tag{59}$$

where $f(z)$ is any other function of the same family. Let now $D_1, D_2, \ldots$ be a sequence of analytically bounded domains such that $D_k \subset D_{k+1}$ and $D_k \to D$ for $k \to \infty$ and denote by $N_k = N_k(z,\zeta)$ the function $N(z,\zeta)$ associated with the domain D_k. Since $f_0(z)$ is regular in the closure of D_k, it follows from (54) that

$$A_e^{(k)}(f_0) < A_e(N_k),$$

whence

$$A_e(f) \leq \lim_{k \to \infty} A_e(N_k). \tag{60}$$

The existence of the limit follows from the fact that, in view of (54), $A_e(N)$ decreases monotonically if the domain associated with $N(z,\zeta)$ increases. Now for any given closed subdomain Δ of D there exists a k_0 such that $N_k(z,\zeta)$ is univalent in Δ if $k > k_0$. Hence we may extract from the sequence $N_k(z,\zeta)$ a converging subsequence whose limit $N_0(z,\zeta)$ is univalent in D. By (60), we thus have $A_e(f_0) \leq A_e(N_0)$. On the other hand, it follows from (59) that $A_e(f_0) \geq A_e(N_0)$, and therefore

$$A_e(f_0) = A_e(N_0).$$

This shows that the function $N_0(z,\zeta)$ yields the largest possible external area.

Our proof will therefore be completed if we can show that

$$A_e(N_0) = A_e(N),$$

that is,

$$\lim_{k \to \infty} A_e(N_k) = A_e(N). \tag{61}$$

In view of (56), (49), (9), and (2), we have

$$A_e(N) = \pi M'(\zeta,\zeta) = \tfrac{1}{2}\pi[a_0 - a_{\frac{1}{2}\pi}],$$

or, since $M'(\zeta,\zeta)$ is positive,

$$A_e(N) = \tfrac{1}{2}\pi \operatorname{Re} \{a_0 - a_{\frac{1}{2}\pi}\}. \tag{62}$$

As will be recalled from Sec. 2, the parallel slit function

$$\varphi_\theta(z,\zeta) = \frac{1}{z - \zeta} + a_\theta(z - \zeta) + \cdots$$

has the extremal property

$$\operatorname{Re} \{e^{-2i\theta}b_1\} \leq \operatorname{Re} \{e^{-2i\theta}a_\theta\}, \tag{63}$$

where b_1 is the coefficient of $(z - \zeta)$ in the expansion

$$f(z) = \frac{1}{z - \zeta} + b_0 + b_1(z - \zeta) + \cdots$$

of a univalent function $f(z)$. If $a_\theta^{(k)}$ denotes the coefficient of $(z - \zeta)$ in the expansion of the function $\varphi_\theta(z,\zeta)$ associated with D_k, it follows from (63) that

$$\text{Re}\,\{e^{-2i\theta}a_\theta\} \leq \text{Re}\,\{e^{-2i\theta}a_\theta^{(k+1)}\} \leq \text{Re}\,\{e^{-2i\theta}a_\theta^{(k)}\}.$$

Hence, lim Re $\{e^{-2i\theta}a_\theta^{(k)}\}$ for $k \to \infty$ exists and

$$\lim_{k\to\infty} \text{Re}\,\{e^{-2i\theta}a_\theta^{(k)}\} \geq \text{Re}\,\{e^{-2i\theta}a_\theta\}. \tag{64}$$

On the other hand, there exists a subsequence of the functions $\varphi_\theta(z,\zeta)$ associated with the domains D_k which converges to a univalent function in D, the proof being the same as further above in the case of the functions $N_k(z,\zeta)$. It therefore follows from (63) that

$$\lim_{k\to\infty} \text{Re}\,\{e^{-2i\theta}a_\theta^{(k)}\} \leq \text{Re}\,\{e^{-2i\theta}a_\theta\},$$

and thus, in view of (64),

$$\lim_{k\to\infty} \text{Re}\,\{e^{-2i\theta}a_\theta^{(k)}\} = \text{Re}\,\{e^{-2i\theta}a_\theta\}.$$

Setting $\theta = 0$ and $\theta = \frac{1}{2}\pi$ in this identity and adding the results obtained, we find that

$$\lim_{k\to\infty} \text{Re}\,\{a_0^{(k)} - a_{\frac{1}{2}\pi}^{(k)}\} = \text{Re}\,\{a_0 - a_{\frac{1}{2}\pi}\}.$$

In view of (62), this is equivalent to (61). This completes the proof of the extremal property of the function $N(z,\zeta)$.

The corresponding completion of the proof in the case of the inner area problem is now an easy matter. Let $g'(\zeta) = 1$, and let $g(z)$ be regular in D. Since $g(z)$ is regular in the closure of D_k, it follows from (55) that

$$A_i^{(k)}(g) > A_i(M_k^*).$$

Hence,

$$A_i(g) \geq \limsup_{k\to\infty} A_i(M_k^*).$$

A glance at (56), (57), and (61) shows that lim $A_i(M_k^*)$ exists and is equal to $A_i(M^*)$. It follows therefore that

$$A_i(g) \geq A_i(M^*),$$

which shows that the function $M^*(z,\zeta)$ indeed solves the inner area problem.

EXERCISES

1. Construct the functions $M(z,\zeta)$ and $N(z,\zeta)$ in the case in which D is the unit circle. Show that the two functions map D onto the interior and the exterior of a circle, respectively, and verify the relation (58). *Hint:* Use the result of Exercise 2, Sec. 1.

2. If the function

$$f(z) = \frac{1}{z-\zeta} + b_0 + b_1(z-\zeta) + \cdots$$

is univalent in D, show that the point b_1 must lie within or on a circle of radius $M'(\zeta,\zeta)$ and center B, where B is defined by the expansion

$$N(z,\zeta) = \frac{1}{z-\zeta} + B(z-\zeta) + \cdots .$$

Hint: Use (16) and (10).

3. Generalizing the methods indicated in Exercises 4 to 9, Sec. 1, show that the function

$$N_\lambda(z,\zeta) = N(z,\zeta) + \lambda M(z,\zeta) = \frac{1}{z-\zeta} + B_\lambda(z-\zeta) + \cdots$$

is univalent in D if $0 \leq \lambda \leq 1$.

4. If the function

$$f(z) = \frac{1}{z-\zeta} + B_\lambda(z-\zeta) + \cdots$$

is univalent in D, show that

$$A_e(f) \leq A_e(N_\lambda),$$

where A_e denotes the outer area and $N_\lambda = N_\lambda(z,\zeta)$ is the function defined in the preceding exercise.

5. The Kernel Function and Orthonormal Sets.

Let $f(z)$ be a function which is regular and single-valued in a domain D and on its boundary C. If $M(z,\zeta)$ is the function defined by (49), it follows from (50) and Green's formula that

$$\iint_D f'\overline{M'}\,dx\,dy = \overline{\iint_D \bar{f}'M'\,dx\,dy} = \overline{\frac{1}{2i}\int_C \bar{f}M'\,dz}$$
$$= \overline{\frac{1}{2i}\int_C \bar{f}\overline{N'}\,dz} = -\frac{1}{2i}\int_C fN'\,dz.$$

The contour integral can be evaluated with the help of the residue theorem. Since the only singularity of $N(z,\zeta)$ in D is a double pole with the principal part $-(z-\zeta)^{-2}$, we obtain

$$\iint_D f'\overline{M'}\,dx\,dy = \pi f'(\zeta).$$

The function

$$K(z,\zeta) = \pi^{-1}M'(z,\zeta) \tag{65}$$

has therefore the "reproducing property"

$$\iint_D \overline{K(z,\zeta)} f'(z)\, dx\, dy = f'(\zeta) \tag{66}$$

with respect to all functions which are derivatives of regular and single-valued functions in $D + C$. $K(z,\zeta)$ is known as the *kernel function* or *Bergman kernel function* of D.

The assumption that $f(z)$ be regular at the points of C is not essential for the validity of the identity (66), and it can be replaced by the much weaker condition

$$\iint_D |f'(z)|^2\, dx\, dy < \infty. \tag{67}$$

This condition is, of course, inevitable since otherwise the existence of the integral in (66) would be in doubt. The class of functions $f(z)$ which are regular and single-valued in D and for which (67) holds will be denoted by $L^2(D)$. In order to show that (66) is valid for this wider class of functions, we recall that the function $M^*(z,\zeta)$ defined in (53) solves the minimum problem for the inner area, *i.e.*, we have

$$\iint_D |M^{*\prime}|^2\, dx\, dy \leq \iint_D |g'|^2\, dx\, dy,$$

where $g(z)$ is any function of $L^2(D)$ for which $g'(\zeta) = 1$. Setting

$$g(z) = M^*(z,\zeta) + \epsilon[f(z) - zf'(\zeta)],$$

where $f(z)$ is an arbitrary function of $L^2(D)$ and ϵ a fixed complex number, we obtain

$$\begin{aligned} \iint_D |M^{*\prime}|^2\, dx\, dy &\leq \iint_D |M^{*\prime} + \epsilon[f' - f_\zeta']|^2\, dx\, dy \\ &= \iint_D |M^{*\prime}|^2\, dx\, dy + 2\,\mathrm{Re}\left\{\epsilon \iint_D \overline{M^{*\prime}}(f' - f_\zeta')\, dx\, dy\right\} \\ &\qquad + |\epsilon|^2 \iint_D |f' - f_\zeta'|^2\, dx\, dy, \end{aligned}$$

where $f_\zeta' = f'(\zeta)$. It follows that

$$0 \leq 2\,\mathrm{Re}\left\{\epsilon \iint_D \overline{M^{*\prime}}(f' - f_\zeta')\, dx\, dy\right\} + |\epsilon|^2 \iint_D |f' - f_\zeta'|^2\, dx\, dy.$$

Since we may take $|\epsilon|$ as small as we please and since $\arg\{\epsilon\}$ is arbitrary, this entails

$$\iint_D \overline{M^{*\prime}}(f' - f_\zeta') \, dx \, dy = 0,$$

whence

$$\iint_D \overline{M^{*\prime}} f' \, dx \, dy = f'(\zeta) \iint_D \overline{M^{*\prime}} \, dx \, dy. \tag{68}$$

In view of (53) and (65), we have

$$M^{*\prime}(z,\zeta) = \frac{K(z,\zeta)}{K(\zeta,\zeta)},$$

which shows that (68) is equivalent to

$$\iint_D \overline{K(z,\zeta)} f'(z) \, dx \, dy = f'(\zeta) \iint_D \overline{K(z,\zeta)} \, dx \, dy. \tag{69}$$

Now the identity (66) has been proved for functions $f(z)$ which are regular in $D + C$, and we may therefore apply it to the function $f(z) = z$. This yields

$$\iint_D \overline{K(z,\zeta)} \, dx \, dy = 1.$$

Inserting this in (69), we obtain the identity (66), which has thus been shown to be valid for any function $f(z)$ of $L^2(D)$.

The reproducing property (66) characterizes the function $K(z,\zeta)$ uniquely within the class of derivatives of functions of $L^2(D)$. If there were another function, say $K_1(z,\zeta)$, for which (66) holds, it would follow that

$$\iint_D \overline{(K - K_1)} f' \, dx \, dy = 0$$

for every $f(z)$ of $L^2(D)$. Setting, in particular, $f'(z) = K(z,\zeta) - K_1(z,\zeta)$, we obtain

$$\iint_D |K - K_1|^2 \, dx \, dy = 0,$$

which shows that $K_1(z,\zeta)$ must be identical with $K(z,\zeta)$.

Another immediate consequence of (66) is the fact that $K(z,\zeta)$ is "Hermitian," *i.e.*,

$$K(\zeta,z) = \overline{K(z,\zeta)}. \tag{70}$$

Setting $f'(z) = K(z,\eta)$ $(\eta \in D)$ in (66), we obtain

$$\iint_D \overline{K(z,\zeta)} K(z,\eta) \, dx \, dy = K(\zeta,\eta).$$

Since, on the other hand,

$$\iint_D \overline{K(z,\zeta)}K(z,\eta)\,dx\,dy = \overline{\iint_D \overline{K(z,\eta)}K(z,\zeta)\,dx\,dy},$$

this is also equal to $\overline{K(\eta,\zeta)}$, and this proves (70).

The reader will observe that these developments are very similar to those of Sec. 10, Chap. V, and indeed the kernel function introduced there is identical with the kernel function of the present section if the domain D is simply-connected. Formulas (128) of Sec. 10, Chap. V, and (66) of this section are seen to coincide if it is remembered that, because of the monodromy theorem, any function which is regular and single-valued in a simply-connected domain is a derivative of a function of the same character. The reader will verify that the formal connection between the kernel function and a complete set of functions $\{u_\nu(z)\}$ of $L^2(D)$, orthonormalized by the conditions

$$\iint_D u_\nu'(z)\overline{u_\mu'(z)}\,dx\,dy = \delta_{\nu\mu}, \tag{71}$$

is the same as that discussed in detail in Sec. 10, Chap. V. In analogy to (132) (Sec. 10, Chap. V), we shall therefore have the expansion

$$K(z,\zeta) = \sum_{\nu=1}^{\infty} u_\nu'(z)\overline{u_\nu'(\zeta)}, \tag{72}$$

provided the set of functions $\{u_\nu(z)\}$, which is orthonormalized by (71), is complete with respect to the class $L^2(D)$. The existence of complete sets of functions of $L_2(D)$ is easy to show. In view of the general properties of sets of orthogonal functions discussed in Sec. 10, Chap. V, a set of functions $\{v_\nu(z)\}$ of $L^2(D)$ (which is not necessarily orthonormalized) is complete if there exists no nontrivial function of $L^2(D)$ which is orthogonal to all the functions of the set. Consider now the set $\{v_\nu(z)\}$ defined by

$$v_\nu'(z) = K(z,\zeta_\nu), \qquad \nu = 1, 2, \ldots,$$

where the ζ_ν are points of D such that $\lim_{\nu\to\infty} \zeta_\nu = \zeta_0$, $\zeta_0 \in D$. If the set is not complete, there exists a function $f(z)$ of $L^2(D)$ with $f'(z) \not\equiv 0$ such that

$$\iint_D \overline{K(z,\zeta_\nu)}f'(z)\,dw = 0, \qquad \nu = 1, 2, \ldots.$$

In view of (66), this is equivalent to $f'(\zeta_\nu) = 0$, $\nu = 1, 2, \ldots$. But since the points ζ_ν have the limit point ζ_0 at which $f'(z)$ is regular, it follows from the results of Sec. 5, Chap. III, that $f'(z)$ vanishes identically. This shows that the set $K(z,\zeta_\nu)$ is complete.

For the practical computation of the kernel function by means of the bilinear expansion (72) it is, of course, essential to have a complete set of functions of $L^2(D)$ which is known a priori. Such a set is obtained by a suitable generalization of the fact proved in Sec. 10, Chap. V, that the functions $1, z, z^2, \ldots$ form a complete set with respect to a finite simply-connected domain whose complement is a closed domain. In the case in which D is a simply-connected domain of this type which contains the point at infinity, we may choose a point α such that the points of the interior of a small circle of center α do not belong to D. The functions

$$1, \frac{1}{z-\alpha}, \frac{1}{(z-\alpha)^2}, \frac{1}{(z-\alpha)^3}, \cdots \tag{73}$$

will then form a complete set in D. Indeed, if $f(z)$ is regular and single-valued in D, then the function $g(z') = g[(z-\alpha)^{-1}] = f(z)$ is regular and single-valued in a finite domain D' which corresponds to D by means of the mapping $z' = (z-\alpha)^{-1}$. The completeness of the set (73) with respect to D is therefore an immediate consequence of the completeness of the powers of z' with respect to D'.

Let now D be a domain of connectivity n which is bounded by the simple closed analytic curves $C_1, \ldots, C_n$, and let $f(z)$ be regular and single-valued in D. If $f(z)$ is also regular on the boundary C of D, we have, by the Cauchy integral formula,

$$f(z) = \frac{1}{2\pi i}\int_C \frac{f(\zeta)}{\zeta - z}\,d\zeta.$$

This may also be written in the form

$$f(z) = f_1(z) + f_2(z) + \cdots + f_n(z), \tag{74}$$

where

$$f_\nu(z) = \frac{1}{2\pi i}\int_{C_\nu} \frac{f(\zeta)}{\zeta - z}\,d\zeta. \tag{75}$$

The simple closed analytic curve C_ν divides the z-plane into two simply-connected domains. One of these two domains has points in common with D and the other has not. If we denote the former of these domains by D_ν', it is clear that the function $f_\nu(z)$ defined in (75) is a regular and single-valued analytic function at all points of D', irrespective of whether these points do or do not belong to D.

(74) shows therefore that $f(z)$ may be decomposed into n functions each of which is regular and single-valued in a certain simply-connected domain.

This result can also be extended to the case in which no assumption is made regarding the regularity of $f(z)$ at the points of C. In this case, we replace C_ν by a closed contour C_ν^* which is inside D and arbitrarily close to C_ν. We again have the decomposition (74), where now

$$f_\nu(z) = \int_{C_\nu^*} \frac{f(\zeta)}{\zeta - z}\, d\zeta. \tag{76}$$

This expression obviously does not depend on the curve C_ν^* if C_ν^* is varied continuously in such a manner as not to cross the point z. Since C_ν^* may be taken arbitrarily close to C_ν, the function $f_\nu(z)$ defined by (76) is therefore regular in the entire domain D_ν'.

Let now α_ν be a point in the interior of the complement of D_ν', or, intuitively speaking, a point inside the "hole" surrounded by C_ν. Since C_ν is a simple closed analytic curve, the complement of C_ν' is a closed domain and we may apply the above result according to which the functions

$$1, \frac{1}{z - \alpha_\nu}, \frac{1}{(z - \alpha_\nu)^2}, \ldots$$

form a complete set with respect to C_ν'. In view of (74), we thus arrive at the following result:

Let D be a multiply-connected domain bounded by the simple closed analytic curves $C_1, \ldots, C_n$, and let α_ν be a point inside the "hole" of D which is surrounded by C_ν. Then the functions

$$1, \qquad \frac{1}{(z - \alpha_\nu)^m}, \qquad \nu = 1, \ldots, n, \qquad m = 1, 2, \ldots, \tag{77}$$

form a complete set with respect to the class of functions $L^2(D)$.

If D is finite, one of the points α_ν will be in the exterior of the domain bounded by the outer boundary of D. In this case it will be convenient to place the point in question at $z = \infty$; this will result in the functions $(z - \alpha_\nu)^{-m}$ being replaced by z^m. For example, if D is a finite doubly-connected domain whose inner boundary surrounds the origin, a convenient complete set of functions is provided by the set

$$z^\nu, \qquad \nu = \ldots -1, 0, 1, \ldots. \tag{78}$$

The practical computation of the kernel function is thus reduced to the orthonormalizing of the rational functions (77) by means of (71) and the Schmidt process described in Sec. 10, Chap. V. Since the computation of line integrals is easier than that of area integrals, it is advisable to transform the orthonormalization conditions (71), by means of Green's formula, into

(79) $$\frac{1}{2i}\int_C u_\nu'(z)\overline{u_\mu(z)}\,dz = \delta_{\nu\mu}.$$

As an example, consider the case in which D is the circular ring $0 < \rho < |z| < 1$. For this domain, the set (78) is complete. It is, moreover, also orthogonal. Since C consists of the circles $|z| = 1$, $|z| = \rho$, the former described in the positive direction and the latter in the negative direction, we have

$$a_{\nu\mu} = \frac{\nu}{2i}\int_C z^{\nu-1}\bar{z}^\mu\,dz = \frac{\nu}{2i}\int_{|z|=1} z^{\nu-1}\bar{z}^\mu\,dz - \frac{\nu}{2i}\int_{|z|=\rho} z^{\nu-1}\bar{z}^\mu\,dz.$$

On $|z| = 1$, we may replace $\bar{z}$ by z^{-1} and, on $|z| = \rho$, we have $\bar{z} = \rho^2 z^{-1}$. Hence,

$$a_{\nu\mu} = \frac{\nu}{2i}\int_{|z|=1} z^{\nu-\mu-1}\,dz - \frac{\nu\rho^{2\mu}}{2i}\int_{|z|=\rho} z^{\nu-\mu-1}\,dz.$$

By the residue theorem, both integrals vanish unless $\nu = \mu$. This proves the orthogonality. To find the normalization factor, we need the value of $a_{\nu\nu}$. We obtain

$$a_{\nu\nu} = \pi\nu(1 - \rho^{2\nu}),$$

which shows that the orthonormal set is

(80) $$u_\nu(z) = \frac{z^\nu}{\sqrt{\pi\nu(1-\rho^{2\nu})}}, \qquad \nu = \ldots, -1, 1, \ldots.$$

In view of (72), the kernel function of the circular ring $\rho < |z| < 1$ is thus of the form

$$K(z,\zeta) = \frac{1}{\pi}\sum_{\nu=-\infty}^{\infty}\frac{\nu(z\bar{\zeta})^{\nu-1}}{1-\rho^{2\nu}}.$$

If $\{u_\nu(z)\}$ is a complete orthonormal set with respect to $L^2(D)$, then any function $f(z)$ of $L^2(d)$ has an expansion

(81) $$f(z) = \sum_{\nu=1}^{\infty} a_\nu u_\nu(z)$$

which converges uniformly and absolutely in any closed subdomain of D. The proof of this fact is identical with that carried out in Sec. 10, Chap. V, and is therefore not repeated here. The coefficients a_ν are the Fourier coefficients of the function $f'(z)$ corresponding to the orthonormalization conditions (71), and they are therefore of the form

$$a_\nu = \iint_D f'(z)\overline{u_\nu'(z)}\,dx\,dy.$$

If $f(z)$ is regular on C, this may be replaced by

$$a_\nu = \frac{1}{2i}\int_C f'(z)\overline{u_\nu(z)}\,dz. \tag{82}$$

Formulas (81) and (82) may be used for the practical computation of the various mapping functions discussed in Secs. 1 and 2. As an example, consider the function $\varphi_\theta(z,\zeta)$ which maps D onto the parallel slit domain indicated in Fig. 46a. The only singularity of $\varphi_\theta(z,\zeta)$ is a simple pole of residue 1 at $z = \zeta$. Hence, the function

$$f(z) = \varphi_\theta(z,\zeta) - \frac{1}{z-\zeta}$$

is regular in D. Since C consists of analytic curves, $f(z)$ is, moreover, regular on C, and it thus certainly is a function of $L^2(D)$. It therefore follows from (81) and (82) that

$$\varphi_\theta(z,\zeta) = \frac{1}{z-\zeta} + \sum_{\nu=1}^{\infty} a_\nu u_\nu(z), \tag{83}$$

where

$$a_\nu = \frac{1}{2i}\int_C \varphi_\theta'(z,\zeta)\overline{u_\nu(z)}\,dz + \frac{1}{2i}\int_C \frac{\overline{u_\nu(z)}}{(z-\zeta)^2}\,dz.$$

The first integral can be evaluated by means of the identity (18). We obtain

$$\frac{1}{2i}\int_C \varphi_\theta'(z,\zeta)\overline{u_\nu(z)}\,dz = -e^{2i\theta}\,\overline{\frac{1}{2i}\int_C \varphi_\theta'(z,\zeta)u_\nu(z)\,dz},$$

which, in view of the residue theorem, is equal to $\pi e^{2i\theta}\overline{u_\nu'(\zeta)}$. Hence,

$$a_\nu = \pi e^{2i\theta}\overline{u_\nu'(\zeta)} + \frac{1}{2i}\int_C \frac{\overline{u_\nu(z)}}{(z-\zeta)^2}\,dz.$$

The last integral may be simplified by partial integration. We have

$$\int_C \frac{\overline{u_\nu(z)}}{(z-\zeta)^2}\,dz = -\left[\frac{\overline{u_\nu(z)}}{z-\zeta}\right]_C + \int_C \frac{\overline{u_\nu'(z)\,dz}}{z-\zeta}.$$

The integrated part vanishes since $u_\nu(z)$ is single-valued. Therefore

$$a_\nu = \pi e^{2i\theta}\overline{u_\nu'(\zeta)} + \frac{1}{2i}\int_C \frac{\overline{u_\nu'(z)\,dz}}{z-\zeta}. \tag{84}$$

(83) and (84) show that the mapping function $\varphi_\theta(z,\zeta)$ can be immediately computed if a complete orthonormal set of the class $L^2(D)$ is known.

As an illustration, consider the circular ring $\rho < |z| < 1$. As shown above, the functions (80) form a complete orthonormal set with respect to this domain. We therefore conclude that the function $\varphi_\theta(z,\zeta)$ which maps the ring onto the parallel slit domain of inclination θ is of the form

$$\varphi_\theta(z,\zeta) = \frac{1}{z-\zeta} + \sum_{\nu=-\infty}^{\infty} a_\nu z^\nu,$$

where

$$a_\nu = \frac{1}{\sqrt{\pi\nu(1-\rho^{2\nu})}}\left[\frac{\sqrt{\pi\nu}}{\sqrt{1-\rho^{2\nu}}}\, e^{2i\theta}\bar{\zeta}^{\nu-1} + \frac{\sqrt{\pi\nu}}{\sqrt{1-\rho^{2\nu}}}\,\frac{1}{2\pi i}\int_C \frac{\bar{z}^{\nu-1}\,\overline{dz}}{z-\zeta}\right],$$

that is,

$$a_\nu = \frac{1}{1-\rho^{2\nu}}\left[e^{2i\theta}\bar{\zeta}^{\nu-1} + \frac{1}{2\pi i}\int_C \frac{\bar{z}^{\nu-1}\,\overline{dz}}{z-\zeta}\right].$$

To evaluate the line integral, we remember that $z\bar{z} = 1$ on the outer circle while $z\bar{z} = \rho^2$ on the inner circle. Hence

$$b_\nu = \frac{1}{2\pi i}\int_C \frac{\bar{z}^{\nu-1}\,\overline{dz}}{z-\zeta} = -\overline{\left[\frac{1}{2\pi i}\int_{|z|=1}\frac{z^{\nu-1}\,dz}{(1/z)-\bar{\zeta}}\right]} + \overline{\left[\frac{1}{2\pi i}\int_{|z|=\rho}\frac{z^{\nu-1}\,dz}{(\rho^2/z)-\bar{\zeta}}\right]},$$

which may also be written in the form

$$\overline{\zeta b_\nu} = \frac{1}{2\pi i}\int_{|z|=1}\frac{z^\nu\,dz}{z-\bar{\zeta}^{-1}} - \frac{1}{2\pi i}\int_{|z|=\rho}\frac{z^\nu\,dz}{z-\rho^2\bar{\zeta}^{-1}}.$$

The possible residues of these integrals are at $z = 0$ (if ν is negative) and at $z = \rho^2\bar{\zeta}^{-1}$; the pole at $z = \bar{\zeta}^{-1}$ does not make a contribution since $|\zeta^{-1}| > 1$. For $z = \rho^2\bar{\zeta}^{-1}$, we have a residue in the second integral since $|\rho^2\bar{\zeta}^{-1}| < \rho$. If $\nu > 0$, we thus obtain

$$\overline{\zeta b_\nu} = -\rho^{2\nu}\bar{\zeta}^{-\nu}, \qquad \nu > 0. \tag{85}$$

If $\nu = -\mu$ $(\mu > 0)$, both integrals have residues at $z = 0$ which give the contributions $-\bar{\zeta}^\mu$ and $\rho^{-2\mu}\bar{\zeta}^\mu$ to the values of the integrals. Adding this to (85), we thus obtain

$$\overline{\zeta b_{-\mu}} = -\rho^{-2\mu}\bar{\zeta}^\mu - \bar{\zeta}^\mu + \rho^{-2\mu}\bar{\zeta}^\mu = -\bar{\zeta}^\mu, \qquad \mu > 0.$$

Hence, $b_\nu = -\rho^{2\nu}\zeta^{-\nu-1}$ if $\nu > 0$ and $b_\nu = -\zeta^{-\nu-1}$ if $\nu < 0$. Using these values, we arrive—after some manipulation—at the expression

$$\varphi_\theta(z,\zeta) = \frac{1}{z-\zeta} + \frac{e^{2i\theta}}{\bar{\zeta}}\sum_{\nu=1}^{\infty}\frac{(z\bar{\zeta})^\nu - \rho^{2\nu}(z\bar{\zeta})^{-\nu}}{1-\rho^{2\nu}} + \frac{1}{\zeta}\sum_{\nu=1}^{\infty}\frac{\rho^{2\nu}[(z/\zeta)^{-\nu} - (z/\zeta)^\nu]}{1-\rho^{2\nu}}$$

for the function mapping the circular ring $\rho < |z| < 1$ onto a parallel slit domain.

EXERCISES

1. Transforming the contour integral

$$\int_C M(z,\eta)M'(z,\zeta)\,dz, \qquad \zeta \in D,\ \eta \in D$$

by means of (50) and the relation

$$\overline{M(z,\zeta)} = N(z,\zeta) + \text{const.}, \qquad z \in C_\nu,$$

obtained by integrating (50), prove the symmetry property

$$N'(z,\zeta) = N'(\zeta,z).$$

2. Show that the function $K(z,\zeta)$ and the function $k(z,\zeta)$ of Exercise 6, Sec. 3, are connected by the relation

$$K(z,\zeta) = k(z,\zeta) + \sum_{\nu=1}^{n-1} \alpha_\nu w_\nu'(z),$$

where the α_ν are suitable constants and the functions $w_\nu(z)$ are defined in Exercise 1, Sec. 3. *Hint:* Use the results of Exercises 2, 3, and 6, Sec. 3, and identify $K(z,\zeta)$ by the reproducing property (66) and the fact that $K(z,\zeta)$ is the derivative of a single-valued function.

3. Let $q_1(z), \ldots, q_{n-1}(z)$ be the functions obtained by orthonormalizing the functions $w_1'(z), \ldots, w_{n-1}'(z)$ by means of the Schmidt process (Sec. 10, Chap. V) and denote by $k_0(z,\zeta)$ the function

$$k_0(z,\zeta) = \sum_{\nu=1}^{n-1} q_\nu(z)\overline{q_\nu(\zeta)}.$$

Show that

$$k(z,\zeta) = k_0(z,\zeta) + K(z,\zeta),$$

where $k(z,\zeta)$ is defined in Exercise 6, Sec. 3. *Hint:* Same as in the preceding exercise.

4. By applying the formula (66) to the function

$$f(z) = \varphi_\theta(z,\eta) - \frac{1}{z-\eta},$$

show that the parallel slit mapping function $\varphi_\theta(z,\zeta)$ has the representation

$$\varphi_\theta'(z,\zeta) = -\frac{1}{(z-\zeta)^2} + \pi e^{2i\theta}K(z,\zeta) + \frac{1}{2i}\int_C \frac{K(z,t)\,\overline{dt}}{t-\zeta}.$$

5. If $P(z) = P(z;u,v)$ and $Q(z) = Q(z;u,v)$ are the functions mapping D onto a circular slit domain and a radial slit domain (see Fig. 46*b* and *c*), respectively, and if $\{u_\nu(z)\}$ is a complete set of functions normalized by (79), show that

$$\log P(z) = \log\left(\frac{z-u}{z-v}\right) + \sum_{\nu=1}^{\infty} a_\nu u_\nu(z)$$

and

$$\log Q(z) = \log\left(\frac{z-u}{z-v}\right) + \sum_{\nu=1}^{\infty} b_\nu u_\nu(z),$$

where

$$a_\nu = \pi[\overline{u_\nu(u)} - \overline{u_\nu(v)}] + \frac{1}{2i}\int_C \overline{u_\nu(z)}\left[\frac{1}{z-v} - \frac{1}{z-u}\right] dz$$

and

$$b_\nu = -\pi[\overline{u_\nu(u)} - \overline{u_\nu(v)}] + \frac{1}{2i}\int_C \overline{u_\nu(z)}\left[\frac{1}{z-v} - \frac{1}{z-u}\right] dz.$$

6. If D is the circular ring $\rho < |z| < 1$, show that the circular and radial slit mappings of D are yielded by the functions $P(z)$ and $Q(z)$ defined by

$$\log P(z;u,v) = \log\left(\frac{z-u}{z-v}\right) + \sum_{\nu=1}^{\infty} \frac{A_\nu - B_\nu\rho^{2\nu}}{\nu(1-\rho^{2\nu})}$$

and

$$\log Q(z;u,v) = \log\left(\frac{z-u}{z-v}\right) - \sum_{\nu=1}^{\infty} \frac{A_\nu + B_\nu\rho^{2\nu}}{\nu(1-\rho^{2\nu})},$$

where

$$A_\nu = (\bar{u}z)^\nu - (\bar{v}z)^\nu,$$

$$B_\nu = \left(\frac{z}{u}\right)^\nu - \left(\frac{z}{v}\right)^\nu - \left(\frac{u}{z}\right)^\nu + \left(\frac{v}{z}\right)^\nu.$$

7. Let D be a finite domain and let $R(z,\zeta)$ denote the function mapping D onto the circle with circular slits (see Fig. 46*d*) in such a way as to make the outer boundary of D correspond to the full circumference. If $\{u_\nu(z)\}$ is a complete set of functions normalized by (79), show that

$$\log R(z,\zeta) = \log(z-\zeta) + \sum_{\nu=1}^{\infty} a_\nu u_\nu(z),$$

where

$$a_\nu = \pi\overline{u_\nu(\zeta)} - \frac{1}{2i}\int_C \overline{u_\nu(z)}\frac{dz}{z-\zeta}.$$

8. Let C_ν and C_μ be two inner boundary components of a domain D, and let $S_{\nu\mu}(z)$ denote the function mapping D onto a ring with circular slits (see Fig. 46*e*) so that C_ν and C_μ correspond, respectively, to the inner and outer circumference. If α and β are points surrounded by the closed curves C_ν and C_μ, respectively, show that

$$\log S_{\nu\mu}(z) = \log\left(\frac{z-\alpha}{z-\beta}\right) + \sum_{k=1}^{\infty} a_k u_k(z),$$

where

$$a_k = \frac{1}{2i}\int_C \overline{u_k(z)}\left[\frac{1}{z-\beta} - \frac{1}{z-\alpha}\right] dz$$

and $\{u_k(z)\}$ is a complete set of functions normalized by (79). How has this formula to be modified if either C_ν or C_μ is the outer boundary of D?

9. Show that the coefficient (84) can be brought into the form

$$a_\nu = -e^{i\theta} \int_C \operatorname{Im} \left\{ \frac{e^{-i\theta}}{z - \zeta} \right\} \overline{u_\nu'(z)\, dz}.$$

10. Using the results of the preceding exercise and of Exercise 12, Sec. 2, prove the following theorem:

If $f(z)$ is regular and single-valued in $D + C$ and $f(z)$ is normalized by

$$\iint_D |f'(z)|^2 \, dx\, dy = 1,$$

then

$$\left| \int_C \operatorname{Im} \left\{ \frac{e^{-i\theta}}{z - \zeta} \right\} f'(z)\, dz \right|^2 \leq \operatorname{Re} \{2\pi e^{-2i\theta} a_\theta\} - A,$$

where

$$A = \frac{1}{2i} \int_C \frac{dz}{(\bar{z} - \bar{\zeta})(z - \zeta)^2}$$

and a_θ is the coefficient of the parallel slit mapping function $\varphi_\theta(z,\zeta)$ indicated in (2). Equality will hold only in the case in which $f(z)$ is a constant multiple of the function $\varphi_\theta(z,\zeta) - (z - \zeta)^{-1}$.

Hint: Apply Bessel's inequality (Sec. 10, Chap. V) to the coefficient (84).

6. Bounded Functions. In Sec. 3, Chap. V, we showed by elementary means that an analytic function which is regular and bounded in the unit circle is subject to the inequality

$$|f'(\zeta)| \leq \frac{1}{1 - |\zeta|^2}, \qquad |\zeta| < 1,$$

where the sign of equality can hold only in the case in which $f(z)$ is of the form

$$f(z) = \gamma \left(\frac{z - \zeta}{1 - \bar{\zeta} z} \right), \qquad |\gamma| = 1.$$

This result may also be expressed by saying that, within the family of functions which are regular and bounded in the unit circle, the problem $|f'(\zeta)| = \text{max.}$ is solved by the function which maps the unit circle onto itself and carries the point $z = \zeta$ into the origin.

Our aim is to generalize this problem and its solution to the case in which the unit circle is replaced by a general multiply-connected domain D. We are thus concerned with the family B of analytic functions which are regular and single-valued in D and are bounded there, that is, $|f(z)| \leq 1$, $z \in D$. Within this family, we consider the extremal problem

$$(86) \qquad |f'(\zeta)| = \text{max.}, \qquad f(z) \in B,\ \zeta \in D.$$

We shall show that *the function* $w = F(z)$ *solving the problem* (86) *maps* **D** *onto a domain which covers the entire unit circle* $|w| < 1$ *precisely* n *times,* and we shall also develop an algorithm for the actual computation of the largest value of $|f'(\zeta)|$.

Since, clearly, the extremal map is the same for all domains which are conformally equivalent to D, we may restrict ourselves to the case in which D is bounded by simple closed analytic curves. A further simplification is introduced by the fact that we may confine ourselves to functions $f(z)$ of B for which $f'(\zeta) > 0$ and $f(\zeta) = 0$. If $f(z)$ is in B, the same is true of $\gamma f(z)$ where $|\gamma| = 1$, and this justifies the assumption $f'(\zeta) > 0$. If $f(\zeta) \neq 0$, then the function

$$f_1(z) = \frac{f(z) - f(\zeta)}{1 - \overline{f(\zeta)}f(z)}$$

is also in B. In view of

$$|f_1'(\zeta)| = \frac{|f'(\zeta)|}{1 - |f(\zeta)|^2} > |f'(\zeta)|,$$

it is clear that the function solving the problem (86) must vanish at the point $z = \zeta$.

If $f(z)$ is a function of B, then the function

$$\varphi(z) = \frac{4}{\pi} \tan^{-1} f(z) \tag{87}$$

satisfies the condition

$$-1 \leq \text{Re } \{\varphi(z)\} \leq 1, \qquad z \in D; \tag{88}$$

this is an immediate consequence of the fact that the mapping

$$w' = \tan^{-1} w$$

transforms the circle $|w| < 1$ into the infinite strip $-\frac{1}{4}\pi < \text{Re } \{w'\} < \frac{1}{4}\pi$. In view of $f(\zeta) = 0$, it follows from (87) that

$$\varphi'(\zeta) = \frac{4}{\pi} f'(\zeta).$$

Since, moreover, $\varphi'(\zeta) > 0$, the problem (86) is therefore equivalent to the extremal problem Re $\{\varphi'(\zeta)\}$ = max. within the family of functions satisfying (88). If we write

$$\varphi(z) = u(z) + iv(z),$$

we have, by the Cauchy-Riemann equations,

$$\text{Re } \{\varphi'(z)\} = \frac{\partial u(z)}{\partial x}, \qquad z = x + iy.$$

(86) is therefore equivalent to the problem

(89) $$\frac{\partial u(\zeta)}{\partial \xi} = \max., \qquad \zeta = \xi + i\eta,$$

where $u(z)$ is a harmonic function which, by (88), satisfies

(90) $$|u(z)| \leq 1$$

and, furthermore, possesses a harmonic conjugate which is single-valued in D.

If $g(z,\zeta)$ denotes the Green's function of D, then we have, by (41),

(91) $$u(\zeta) = -\frac{1}{2\pi}\int_C u(z)\frac{\partial g(z,\zeta)}{\partial n}\,ds.$$

In order to avoid difficulties which may be caused by the irregular boundary behavior of $u(z)$, we temporarily assume that $u(z)$ is also harmonic at the points of C. Differentiating under the integral sign in (91), we obtain

(92) $$u_\xi(\zeta) = \frac{\partial u(\zeta)}{\partial \xi} = -\frac{1}{2\pi}\int_C u(z)\frac{\partial^2 g(z,\zeta)}{\partial n\,\partial \xi}\,ds.$$

The function $u(z)$ has a single-valued harmonic conjugate $v(z)$. If $P_\nu(v)$ denotes the period of $v(z)$ with respect to a circuit about the boundary component C_ν, we thus have $P_\nu(v) = 0$, $\nu = 1, \ldots, n$. With the help of the Cauchy-Riemann equations, the harmonic measures $\omega_\nu(z)$ (which have the boundary values $\delta_{\nu\mu}$ on C_μ), and Green's formula, this can be transformed into a condition involving $u(z)$. We have

$$P_\nu(v) = \int_{C_\nu} dv = \int_{C_\nu}\frac{\partial v}{\partial s}\,ds = \int_{C_\nu}\frac{\partial u}{\partial n}\,ds$$
$$= \int_C \omega_\nu \frac{\partial u}{\partial n}\,ds = \int_C u\frac{\partial \omega_\nu}{\partial n}\,ds,$$

and therefore

(93) $$P_\nu(v) = \int_C u(z)\frac{\partial \omega_\nu(z)}{\partial n}\,ds = 0, \qquad \nu = 1, \ldots, n-1.$$

The condition $P_n(v) = 0$ has been omitted, since, in view of $\omega_1(z) + \cdots + \omega_n(z) \equiv 1$, it follows from $P_1(v) = \cdots = P_{n-1}(v) = 0$. Taking into account the conditions (93), we conclude from (92) that

(94) $$u_\xi(\zeta) = -\frac{1}{2\pi}\int_C u(z)\left[\frac{\partial^2 g(z,\zeta)}{\partial n\,\partial \xi} + \sum_{\nu=1}^{n-1}\lambda_\nu\frac{\partial \omega_\nu(z)}{\partial n}\right]ds,$$

where $\lambda_1, \ldots, \lambda_{n-1}$ are arbitrary real parameters.

From (90) and (94), we obtain the inequality

$$(95) \qquad |u_\xi(\zeta)| \leq \frac{1}{2\pi} \int_C \left| \frac{\partial^2 g}{\partial n\, \partial \xi} + \sum_{\nu=1}^{n-1} \lambda_\nu \frac{\partial \omega_\nu}{\partial n} \right| ds = \frac{1}{2\pi} \int_C |R|\, ds,$$

where the abbreviation

$$(96) \qquad R = \frac{\partial^2 g(z,\zeta)}{\partial n\, \partial \xi} + \sum_{\nu=1}^{n-1} \lambda_\nu \frac{\partial \omega_\nu(z)}{\partial n}$$

has been used. (95) is true for arbitrary values of the real parameters λ_ν. In order to obtain the best inequality of this type, we choose the values of the parameters λ_ν such as to give to the right-hand side of (95) its smallest possible value. Such a minimizing set $\lambda_1, \ldots, \lambda_{n-1}$ exists. Indeed, since by (95) the expression in question is bounded from below, we have only to show that

$$A = \int_C |R|\, ds$$

cannot approach its greatest lower bound if one or more of the parameters λ_ν tend to $\pm\infty$. Suppose, then, this is the case and choose a sequence of sets $\{\lambda_\nu^{(m)}\}$, $m = 1, 2, \ldots$, for which A tends to its greatest lower bound. Taking, if necessary, a subsequence, we must have a value of ν, say $\nu = 1$, such that $|\lambda_\nu^{(m)}| \leq |\lambda_1^{(m)}|$ for $\nu \neq 1$ and m sufficiently large (and, of course $|\lambda_1^{(m)}| \to \infty$ for $m \to \infty$). Dividing A by $|\lambda_1^{(m)}|$ and letting $m \to \infty$, we find that

$$\lim_{m \to \infty} \int_C \left| \frac{\partial \omega_1}{\partial n} + \sum_{\nu=2}^{n-1} \frac{\lambda_\nu^{(m)}}{\lambda_1^{(m)}} \frac{\partial \omega_\nu}{\partial n} \right| ds = 0.$$

Since $|\lambda_\nu^{(m)}/\lambda_1^{(m)}| \leq 1$, we may choose a sequence of integers $m_1, m_2, \ldots$ such that the limits

$$\lim_{k \to \infty} \frac{\lambda_\nu^{(m_k)}}{\lambda_1^{(m_k)}} = \lambda_\nu'$$

exist. Hence,

$$\int_C \left| \frac{\partial \omega_1}{\partial n} + \sum_{\nu=2}^{n-1} \lambda_\nu' \frac{\partial \omega_\nu}{\partial n} \right| ds = 0,$$

whence

$$\frac{\partial \omega_1}{\partial n} + \sum_{\nu=2}^{n-1} \lambda_\nu' \frac{\partial \omega_\nu}{\partial n} \equiv 0, \qquad z \in C.$$

Since a harmonic function can have an identically vanishing normal derivative only if it reduces to a constant (see Exercise 3, Sec. 4, Chap. I), it follows that

$$\omega(z) = \omega_1(z) + \sum_{\nu=2}^{n-1} \lambda_\nu \omega_\nu(z) = \text{const.}$$

This, however is absurd. In view of the properties of the functions $\omega_\nu(z)$, the function $\omega(z)$ is zero on C_n and has the value 1 on C_1, which shows that it cannot be a constant. This completes the proof for the existence of a set $\lambda_1, \ldots, \lambda_{n-1}$ which gives to the right-hand side of (95) its smallest possible value.

If this minimizing set is again denoted by $\lambda_1, \ldots, \lambda_{n-1}$, it follows from the minimum property that

$$\int_C \left| R \pm \epsilon \frac{\partial \omega_\nu}{\partial n} \right| ds \geq \int_C |R|\, ds,$$

where R is defined by (96) and ϵ is a small positive parameter. In view of

$$\begin{aligned} \int_C \left| R \pm \epsilon \frac{\partial \omega_\nu}{\partial n} \right| ds &= \int_C |R| \left| 1 \pm \frac{\epsilon}{R} \frac{\partial \omega_\nu}{\partial n} \right| ds \\ &= \int_C |R| \left[1 \pm \frac{\epsilon}{R} \frac{\partial \omega_\nu}{\partial n} + o(\epsilon) \right] ds \\ &= \int_C |R|\, ds \pm \epsilon \int_C \frac{|R|}{R} \frac{\partial \omega_\nu}{\partial n}\, ds + o(\epsilon), \end{aligned}$$

this is equivalent to

$$\pm \epsilon \int_C \frac{|R|}{R} \frac{\partial \omega_\nu}{\partial n}\, ds + o(\epsilon) \geq 0,$$

where $\epsilon^{-1} o(\epsilon) \to 0$ for $\epsilon \to 0$. Hence,

$$\pm \int_C \frac{|R|}{R} \frac{\partial \omega_\nu}{\partial n}\, ds \geq 0,$$

and therefore

$$\int_C \frac{|R|}{R} \frac{\partial \omega_\nu}{\partial n}\, ds = 0, \qquad \nu = 1, \ldots, n-1. \tag{97}$$

The expression $\dfrac{|R|}{R}$ is equal to 1 if R is positive and to -1 if R is negative; at the points at which $R = 0$ it ceases to be defined (we shall see later that there are $2n$ such points).

We now define a harmonic function $U(z)$ by the boundary value problem

$$U(z) = -\frac{|R|}{R}, \qquad z \in C. \tag{98}$$

As shown in Sec. 3, such a function exists and may be represented in the form

$$U(\eta) = \frac{1}{2\pi} \int_C \frac{|R|}{R} \frac{\partial g(z,\eta)}{\partial n} \, ds.$$

In view of (97), we have

$$\int_C U(z) \frac{\partial \omega_\nu}{\partial n} \, ds = 0, \qquad \nu = 1, \ldots, n-1.$$

As a comparison with (93) shows, this means that the harmonic conjugate $V(z)$ of $U(z)$ is free of periods about the various boundary components of D. In other words, the analytic function

$$\phi(z) = U(z) + iV(z) \tag{99}$$

is regular and single-valued in D. By (98) and the maximum principle for harmonic functions, we further have $|U(z)| \leq 1$ for $z \in D$.

(98) may be written in the form

$$|R| = -U(z)R. \tag{100}$$

If we combine this with (95) and (96), we obtain

$$|u_\xi(\zeta)| \leq -\frac{1}{2\pi} \int_C U(z) \left[\frac{\partial^2 g(z,\zeta)}{\partial n} + \sum_{\nu=1}^{n-1} \lambda_\nu \frac{\partial \omega_\nu(z)}{\partial n} \right] ds,$$

whence, in view of (94) and the fact that $U(z)$ has a single-valued conjugate,

$$|u_\xi(\zeta)| \leq U_\xi(\zeta). \tag{101}$$

(101) has been proved for functions $\varphi(z)$ which are regular on C. In order to free ourselves from this restriction, we note that the harmonic function

$$h(z) = \frac{\partial g(z,\zeta)}{\partial \xi} + \sum_{\nu=1}^{n-1} \lambda_\nu \omega_\nu(z)$$

has the boundary values λ_ν for $z \in C_\nu$ $(\nu = 1, \ldots, n-1)$ and 0 for C_n. Given a sufficiently small positive ϵ, we can therefore find closed curves $C_1^\epsilon, \ldots, C_n^\epsilon$ bounding a domain $D_\epsilon \subset D$ such that $h(z) = \lambda_\nu + \epsilon$ or $h(z) = \lambda_\nu - \epsilon$ $(\lambda_n = 0)$ for $z \in C_\nu^\epsilon$ and $D_\epsilon \to D$ for $\epsilon \to 0$. If g_ϵ and $\omega_{\nu,\epsilon}$ denote, respectively, the Green's function and the harmonic measures of D_ϵ, we clearly have

$$(102)\quad \frac{\partial g(z,\zeta)}{\partial \xi} + \sum_{\nu=1}^{n-1} \lambda_\nu \omega_\nu(z)$$

$$= \frac{\partial g_\epsilon(z,\zeta)}{\partial \xi} + \sum_{\nu=1}^{n-1} (\lambda_\nu + \epsilon_\nu)\omega_{\nu,\epsilon}(z) + \epsilon_n \omega_{n,\epsilon}(z),$$

where $\epsilon_\nu = \epsilon$ or $\epsilon_\nu = -\epsilon$, as the case may be. Now if the harmonic function $u(z)$ is regular in D, it is regular in the closure $D_\epsilon + C_\epsilon$ of D_ϵ, and we may apply (95) to this domain (where g and ω_ν have, of course, to be replaced by g_ϵ and $\omega_{\nu,\epsilon}$, respectively). In view of (102) and the fact that the parameters λ_ν in (95) are arbitrary, we obtain

$$|u_\xi(\zeta)| \leq \frac{1}{2\pi} \int_{C_\epsilon} \left| \frac{\partial g(z,\zeta)}{\partial n} + \sum_{\nu=1}^{n-1} \lambda_\nu \frac{\partial \omega_\nu(z)}{\partial n} \right| ds.$$

Since $C_\epsilon \to C$ for $\epsilon \to 0$, it follows that (95), and therefore also (101), remains true if no restrictive assumption regarding the boundary behavior of $u(z)$ is made. This finally proves that the function (99) indeed solves our extremal problem.

We now turn to a closer study of the properties of the extremal function $\phi(z)$. Since $|U(z)| = |\text{Re}\,\{\phi(z)\}| \leq 1$ for $z \in D$ and $|U(z)| = 1$—except at the zeros of R—for $z \in C$, it follows that $w = \phi(z)$ maps D onto a domain D' which consists of a number of replicas of the infinite strip $-1 < \text{Re}\,\{w\} < 1$. On the boundary of each sheet of D', the value of $U(z)$ evidently has two "jumps," namely, from 1 to -1 and from -1 to 1. In view of (100), these jumps coincide with the zeros of the expressions (96) on C. It follows therefore that the number of sheets of D' is equal to one-half the number of these zeros. We shall show that the number of the zeros of (96) on C is precisely $2n$ and that, therefore, D' has n sheets.

If $w_\nu(z) = \omega_\nu(z) + i\omega_\nu{}^*(z)$ is the analytic function whose real part is $\omega_\nu(z)$, we have

$$\frac{1}{i} w_\nu{}' \, dz = \frac{1}{i} \, dw_\nu(z) = \frac{1}{i} (d\omega + i \, d\omega^*)$$

$$= \frac{1}{i} \left(\frac{\partial \omega_\nu}{\partial s} + i \frac{\partial \omega_\nu{}^*}{\partial s} \right) ds = \frac{\partial \omega_\nu{}^*}{\partial s} \, ds = \frac{\partial \omega_\nu}{\partial n} \, ds,$$

where the Cauchy-Riemann equations and the fact that $\omega_\nu = \text{const.}$ on each boundary component have been used. Similarly, if

$$p(z) = g(z,\zeta) + ih(z,\zeta),$$

it follows from $g(z,\zeta) = 0$, $z \in C$, that

$$\frac{1}{i} p'(z)\, dz = \frac{\partial g(z,\zeta)}{\partial n}\, ds,$$

where $p'(z)$ has a simple pole with the residue -1 at $z = \zeta$. Hence,

$$\frac{1}{i} \frac{\partial p'(z)}{\partial \xi}\, dz = \frac{\partial^2 g(z,\zeta)}{\partial n\, \partial \xi}\, ds,$$

where

$$\frac{\partial p'(z)}{\partial \xi} = -\frac{1}{(z-\zeta)^2} + p_1(z)$$

and $p_1(z)$ is regular in D. Using these identities, we find that

$$\left[\frac{\partial^2 g(z,\zeta)}{\partial n} + \sum_{\nu=1}^{n-1} \lambda_\nu \frac{\partial \omega_\nu}{\partial n}\right] ds = \frac{1}{i}\left[-\frac{1}{(z-\zeta)^2} + p_1(z) + \sum_{\nu=1}^{n-1} \lambda_\nu w_\nu'(z)\right] dz,$$

or, in view of (96),

(103) $$R\, ds = -\frac{1}{i} q(z)\, dz, \qquad z \in C,$$

where

(104) $$q(z) = \frac{1}{(z-\zeta)^2} + q_1(z)$$

and $q_1(z)$ is regular in D.

(103) shows that the zeros of R coincide with the zeros of the analytic function $q(z)$ on C. To count these zeros, we observe that, by (103),

$$[q(z)\, dz]^2 \leq 0, \qquad z \in C.$$

Hence, the argument of this expression is constant on C, and we have

$$2\Delta_C \arg \{q(z)\} + 2\Delta_C \arg \{dz\} = 0.$$

If D is finite, then $\Delta_C \arg \{dz\} = -2\pi(n-2)$ (see Sec. 10, Chap. III). Hence, $\Delta_C \arg \{q(z)\} = 2\pi(n-2)$. Since, by (104), $q(z)$ has a double pole at $z = \zeta$, it follows therefore from the argument principle that $q(z)$ has n zeros in $D + C$, where the zeros of $q(z)$ on C have to be counted with half their multiplicities. This shows that the number of zeros of $q(z)$ on C cannot exceed $2n$. On the other hand, $q(z)$ cannot have less than two zeros on each boundary component C_ν, $\nu = 1, \ldots, n$. As pointed out before, each zero corresponds to a jump of $U(z)$ from 1 to -1, or vice versa, and there obviously must be at least one jump of each kind on C_ν. We thus conclude that all the zeros of $q(z)$ in $D + C$ lie on C and that their number is $2n$. This completes the proof of the fact that the function (99) maps D onto the n-times covered infinite strip $-1 < \operatorname{Re}\{w\} < 1$.

If, by means of the transformation (87), we return to the family B of bounded functions, we find that the function $F(z)$ solving the extremal problem (86) is of the form

$$F(z) = \tan \frac{\pi}{4} \phi(z) \tag{105}$$

and that $w = F(z)$ maps D onto the n-times covered unit circle $|w| < 1$.

The relation (100) gives rise to an interesting inequality. By (100) and (103), we have

$$\frac{1}{i} U(z)q(z)\,dz \geq 0, \qquad z \in C. \tag{106}$$

Now we have either $U(z) = 1$ or $U(z) = -1$ for $z \in C$ [except at the $2n$ zeros of $q(z)$]. By (99) and (105), it follows therefore that

$$F(z) = \tan\left[\pm \frac{\pi}{4} + it\right], \qquad z \in C,$$

where t is real. Hence

$$\frac{2F(z)}{1 + F^2(z)} = \sin\left[\pm \frac{\pi}{2} + 2it\right] = \pm \cosh 2t.$$

Since $\cosh 2t$ is positive, this expression is thus positive or negative according as $U(z) = 1$ or $U(z) = -1$. Combining this with (106), we obtain

$$\frac{1}{i} F(z) \frac{q(z)}{1 + F^2(z)} \geq 0, \qquad z \in C,$$

or, with the notation

$$g(z) = \frac{q(z)}{1 + F^2(z)}, \tag{107}$$

$$\frac{1}{i} F(z)g(z)\,dz > 0. \tag{108}$$

The equality sign in (108) has been omitted since, as the reader will deduce without difficulty from (105) and (107), the zeros of $q(z)$ are canceled by the zeros of $1 + F^2(z)$ and $g(z)$ is therefore regular and different from 0 on C. We also note that

$$g(z) = \frac{1}{(z - \zeta)^2} + g_1(z), \tag{109}$$

where $g_1(z)$ is regular in $D + C$; this follows from (104), (107), and the fact that $F(\zeta) = 0$.

The fact that the function $F(z)$ solves the extremal problem (86) can be read off from the inequality (108). If $f(z)$ is a function of B which is continuous in $D + C$, we have

$$|f'(\zeta)| = \left| \frac{1}{2\pi i} \int_C f(z)g(z)\,dz \right| \leq \frac{1}{2\pi} \int_C |g(z)\,dz|$$
$$= \frac{1}{2\pi} \int_C \left| \frac{1}{i} F(z)g(z)\,dz \right| = \frac{1}{2\pi i} \int_C F(z)g(z)\,dz = F'(\zeta),$$

where (108), the residue theorem, and the fact that $|F(z)| \equiv 1$ on C, have been used.

The inequality (108) also enables us to solve the extremal problem

$$\int_C |h(z)|\,ds = \text{min.}, \tag{110}$$

where $h(z)$ is of the form

$$h(z) = \frac{1}{(z - \zeta)^2} + h_1(z) \tag{111}$$

and $h_1(z)$ is regular in D. The boundary behavior of $h(z)$ has, of course, to be such that the integral in (110) exists. Now if $F(z)$ is the function just considered, it follows from (111) and the residue theorem that

$$2\pi F'(\zeta) = \left| \frac{1}{i} \int_C F(z)h(z)\,dz \right| \leq \int_C |h(z)|\,ds.$$

Since, by (108), (109), and the fact that $|F(z)| \equiv 1$ for $z \in C$,

$$2\pi F'(\zeta) = \left| \frac{1}{i} \int_C F(z)g(z)\,dz \right| = \int_C |g(z)|\,ds,$$

this shows that

$$\int_C |h(z)|\,ds \geq \int_C |g(z)|\,ds. \tag{112}$$

The function $g(z)$ is thus the solution of the extremal problem (110). We leave it as an exercise to the reader to show that the sign of equality in (112) is possible only if $h(z)$ is identical with $g(z)$.

We now introduce the function

$$\hat{L}(z,\zeta) = \frac{1}{2\pi(z - \zeta)} + \cdots \tag{113}$$

defined by

$$4\pi^2 \hat{L}^2(z,\zeta) = g(z). \tag{114}$$

The function $\hat{L}(z,\zeta)$ is single-valued and, apart from its simple pole at $z = \zeta$, regular in D. To show this, we recall that $g(z)$ has a double pole at $z = \zeta$ and is free of zeros in $D + C$. $\hat{L}(z,\zeta)$ is therefore free of singularities in $D + C$ except for the pole indicated in (113). It remains to show that $\hat{L}(z,\zeta)$ is single-valued in D, or—what amounts to the same thing—that the variation of arg $\{\hat{L}(z,\zeta)\}$ on each boundary component C_ν is an integral multiple of 2π. By (114), this will be the case if the corresponding variation of arg $\{g(z)\}$ is an integral multiple of 4π. But this is an immediate consequence of (108). Indeed, we have

$$\arg\{g(z)\} = -\arg\{dz\} - \arg\{F(z)\},$$

and since the increment of arg $\{dz\}$ on C_ν is either 2π or -2π and arg $\{F(z)\}$ grows by 2π [$w = F(z)$ describes $|w| = 1$ in the positive direction], we find that the increment of arg $\{g(z)\}$ on C_ν is either 0 or 4π.

We further introduce the function

$$\hat{K}(z,\zeta) = F(z)\hat{L}(z,\zeta). \tag{115}$$

Since the pole of $\hat{L}(z,\zeta)$ is compensated by the zero of $F(z)$ at $z = \zeta$, $\hat{K}(z,\zeta)$ is regular and single-valued in D. In view of (108) and (114), we have the relation

$$\frac{1}{i}\hat{K}(z,\zeta)\hat{L}(z,\zeta)\,dz > 0, \qquad z \in C. \tag{116}$$

This shows that

$$\arg\left\{\frac{1}{i}\hat{L}(z,\zeta)\,dz\right\} = -\arg\{\hat{K}(z,\zeta)\} = \arg\{\overline{\hat{K}(z,\zeta)}\}.$$

Since, by (115), $|\hat{K}(z,\zeta)| = |\hat{L}(z,\zeta)|$ for $z \in C$ and $\left|\frac{dz}{ds}\right| = 1$, we may therefore conclude that

$$\overline{\hat{K}(z,\zeta)}\,ds = \frac{1}{i}\hat{L}(z,\zeta)\,dz, \qquad z \in C. \tag{117}$$

The identity (117) has a number of important consequences. If $f(z)$ is an arbitrary regular and single-valued function in D for which

$$\int_C |f(z)|^2\,ds$$

exists, it follows from (117) that

$$\int_C \overline{\hat{K}(z,\zeta)}f(z)\,ds = \frac{1}{i}\int \hat{L}(z,\zeta)f(z)\,dz.$$

Using (113) and the residue theorem, we therefore find that $\hat{K}(z,\zeta)$ has the reproducing property

$$\int_C \overline{\hat{K}(z,\zeta)} f(z)\, ds = f(\zeta), \tag{118}$$

which is reminiscent of the reproducing property (66) of the Bergman kernel function $K(z,\zeta)$. $\hat{K}(z,\zeta)$ is known as the *Szegö kernel function* of D. Like the Bergman kernel function, $\hat{K}(z,\zeta)$ can be computed with the help of orthogonal functions. If $u_1(z)$, $u_2(z)$, . . . is a set of functions for which

$$\int_C |u_\nu(z)|^2\, ds < \infty$$

and which are orthonormalized by the conditions

$$\int_C \overline{u_\nu(z)} u_\mu(z)\, ds = \delta_{\nu\mu}, \tag{119}$$

it follows from the Bessel inequality that

$$\sum_\nu \left| \int_C \overline{\hat{K}(z,\zeta)} u_\nu(z)\, ds \right|^2 \leq \int_C |\hat{K}(z,\zeta)|^2\, ds.$$

In view of (118), this is equivalent to

$$\sum_\nu |u_\nu(\zeta)|^2 \leq \hat{K}(\zeta,\zeta). \tag{120}$$

If the set $\{u_\nu(z)\}$ is complete, that is, if any function $f(z)$ such that $\int_C |f(z)|^2\, ds < \infty$ can be approximated by linear combinations of the $u_\nu(z)$ so as to make the integral

$$\int_C \left| f(z) - \sum_\nu \alpha_\nu u_\nu(z) \right|^2 ds$$

arbitrarily small, then the sign of equality holds in (120). We thus have

$$\hat{K}(\zeta,\zeta) = \sum_{\nu=1}^{\infty} |u_\nu(\zeta)|^2, \tag{121}$$

provided the set of functions $\{u_\nu(z)\}$ is complete in the sense indicated.

From (121), we can derive the bilinear expansion

$$\hat{K}(z,\zeta) = \sum_{\nu=1}^{\infty} u_\nu(z)\overline{u_\nu(\zeta)}, \tag{122}$$

which converges absolutely and uniformly in any closed subdomain of D. The derivation of (122) from (121) is almost word for word the same as the derivation of the corresponding expansion of the function $K(z,\zeta)$ in Sec. 10, Chap. V, and it is therefore omitted here. By carefully examining the proof of the fact that the functions (77) form a complete set with respect to the class $L^2(D)$, the reader will further be able to show that these functions also form a complete set in the sense indicated above.

For the practical computation of the function $\hat{K}(z,\zeta)$ in the case in which D is bounded by simple closed analytic curves, we thus have to orthonormalize the rational functions (77) by means of the conditions (119). If this is done, $\hat{K}(z,\zeta)$ is obtained from (122). Since, by (115) and (113), $2\pi\hat{K}(\zeta,\zeta) = F'(\zeta)$, it follows that

$$F'(\zeta) = 2\pi \sum_{\nu=1}^{\infty} |u_\nu(\zeta)|^2.$$

This yields the exact value of the maximum in the extremal problem (86). If it is desired to compute the extremal function $F(z)$ itself, it is sufficient to find the function $\hat{L}(z,\zeta)$ since, by (115),

$$F(z) = \frac{\hat{K}(z,\zeta)}{\hat{L}(z,\zeta)}.$$

Now we have, by (117),

$$\int_C \overline{\hat{K}(z,\zeta)} \frac{ds}{z-\eta} = \frac{1}{i} \int_C \hat{L}(z,\zeta) \frac{dz}{z-\eta}, \qquad \eta \in D.$$

Hence, by (113) and the residue theorem,

$$\int_C \overline{\hat{K}(z,\zeta)} \frac{ds}{z-\eta} = \frac{1}{\zeta-\eta} + 2\pi\hat{L}(\eta,\zeta).$$

Changing the integration variable and replacing η by z, we finally obtain

$$\hat{L}(z,\zeta) = \frac{1}{2\pi(z-\zeta)} + \frac{1}{2\pi} \int_C \frac{\overline{\hat{K}(\zeta,t)}}{t-z}\, ds, \qquad ds = |dt|.$$

EXERCISES

1. Using (117), show that

$$\overline{\frac{1}{i} \hat{K}(z,\zeta)\hat{K}(z,\eta)\, dz} = \frac{1}{i} \hat{L}(z,\zeta)L(z,\eta)\, dz, \qquad z \in C,$$

where ζ and η are two points of D, and deduce the symmetry property

$$\hat{L}(\eta,\zeta) = -\hat{L}(\zeta,\eta).$$

2. Show that

$$\overline{\frac{1}{i}\hat{K}(z,\zeta)\hat{L}(z,\eta)\,dz} = \frac{1}{i}\hat{K}(z,\eta)\hat{L}(z,\zeta)\,dz, \qquad z \in C,$$

and conclude that

$$\overline{\hat{K}(\eta,\zeta)} = K(\zeta,\eta).$$

3. Show that the function $\hat{K}(z,\zeta)$ and the Bergman kernel function $K(z,\zeta)$ are related by the identity

$$4\pi\hat{K}^2(z,\zeta) = K(z,\zeta) + \sum_{\nu=1}^{n-1} \alpha_\nu w_\nu'(z),$$

where the functions $w_\nu(z)$ are defined in Exercise 1, Sec. 3, and the α_ν are appropriate constants; if D is simply-connected, show that this relation reduces to $4\pi\hat{K}^2(z,\zeta) = K(z,\zeta)$. *Hint:* Use (117) and the hint for Exercise 2, Sec. 5.

4. If D is the unit circle, show that the functions

$$\frac{z^n}{\sqrt{2\pi}}, \qquad n = 0, 1, \ldots,$$

form a complete set which is orthonormalized by the conditions (119). Deduce that, in this case,

$$\hat{K}(z,\zeta) = \frac{1}{2\pi(1 - \bar{\zeta}z)}, \qquad \hat{L}(z,\zeta) = \frac{1}{2\pi(z - \zeta)}$$

and, therefore,

$$F(z) = \frac{z - \zeta}{1 - \bar{\zeta}z},$$

in accordance with previous results.

5. If D is the circular ring $\rho < |z| < 1$, show that the functions

$$\frac{z^n}{\sqrt{2\pi(1 - \rho^{2n+1})}}, \qquad n = \ldots, -1, 0, 1, \ldots,$$

form a complete set orthonormalized by (119) and deduce that in this case

$$\hat{K}(z,\zeta) = \frac{1}{2\pi}\sum_{n=-\infty}^{\infty} \frac{(z\bar{\zeta})^n}{1 - \rho^{2n+1}},$$

$$\hat{L}(z,\zeta) = \frac{1}{2\pi(z - \zeta)} - \frac{1}{2\pi}\sum_{n=0}^{\infty} \frac{\rho^{2n+1}(z^{2n+1} - \zeta^{2n+1})}{(z\zeta)^{n+1}(1 - \rho^{2n+1})},$$

$$F'(\zeta) = \sum_{n=-\infty}^{\infty} \frac{|\zeta|^{2n}}{1 - \rho^{2n+1}}.$$

6. Using (116), show that

$$\frac{2\pi\hat{K}(z,\zeta)\hat{L}(z,\zeta)}{\hat{K}(\zeta,\zeta)} = -p'(z,\zeta) + \sum_{\nu=1}^{n-1} \beta_\nu w_\nu'(z),$$

where the functions $p(z,\zeta)$ and $w_\nu(z)$ are defined in Exercise 1, Sec. 3, and the β_ν are appropriate constants.

7. Using (118) show that the function

$$f_0(z) = \frac{\hat{K}(z,\zeta)}{K(\zeta,\zeta)}$$

solves the extremal problem

$$\int_C |f(z)|^2\,ds = \text{min.}, \qquad f(\zeta) = 1,$$

where $f(z)$ is regular and single-valued in D and is such that the integral $\int_C |f(z)|^2\,ds$ exists.

8. If $\{u_\nu(z)\}$ is a complete set of functions orthonormalized by (119), show that

$$\hat{L}(z,\zeta) = \frac{1}{2\pi(z-\zeta)} + \sum_{\nu=1}^{\infty} a_\nu u_\nu(z),$$

where

$$a_\nu = -\frac{1}{2\pi}\int_C \frac{\overline{u_\nu(z)}\,ds}{z-\zeta}.$$

9. Using (118) and the fact that $|\hat{K}(z,\zeta)| = |\hat{L}(z,\zeta)|$ for $z \in C$, show that

$$\hat{K}(\zeta,\zeta) = \frac{1}{4\pi^2}\int_C \frac{ds}{|z-\zeta|^2} - \sum_{\nu=1}^{\infty} |a_\nu|^2,$$

where a_ν is defined in the preceding exercise.

10. If the functions $u_1(z), \ldots, u_m(z)$ are orthonormalized by means of (119), show that

$$\sum_{\nu=1}^{m} |u_\nu(\zeta)|^2 \le \hat{K}(\zeta,\zeta) \le \frac{1}{4\pi^2}\int_C \frac{ds}{|z-\zeta|^2} - \sum_{\nu=1}^{m} |a_\nu|^2,$$

where a_ν is defined in Exercise 8 and both inequalities turn into equalities if $m \to \infty$ and the set $\{u_\nu(z)\}$ is complete.

11. If the functions $u_1(z), \ldots, u_m(z)$ are orthonormalized by (119), show that

$$\hat{K}(z,\zeta) = \sum_{\nu=1}^{m} u_\nu(z)\overline{u_\nu(\zeta)} + R_m(z,\zeta),$$

where

$$|R_m(z,\zeta)|^2 \le \rho_m(z)\rho_m(\zeta)$$

and

$$\rho_m(\eta) = \frac{1}{4\pi^2}\int_C \frac{ds}{|z-\eta|^2} - \sum_{\nu=1}^{m} [|u_\nu(\eta)|^2 + |a_\nu(\eta)|^2],$$

$$a_\nu(\eta) = \frac{1}{2\pi}\int_C \frac{\overline{u_\nu(z)}\,ds}{z-\eta}.$$

Show that $\rho_m(\eta) \to 0$ if $m \to \infty$ and the set $\{u_\nu(z)\}$ is complete. *Hint:* Apply the Schwarz inequality to the remainder of the series (122) and use the result of Exercise 9.

INDEX

S

T

U

V

W

Z

A CATALOGUE OF SELECTED DOVER BOOKS IN ALL FIELDS OF INTEREST

A CATALOGUE OF SELECTED DOVER BOOKS IN ALL FIELDS OF INTEREST

AMERICA'S OLD MASTERS, James T. Flexner. Four men emerged unexpectedly from provincial 18th century America to leadership in European art: Benjamin West, J. S. Copley, C. R. Peale, Gilbert Stuart. Brilliant coverage of lives and contributions. Revised, 1967 edition. 69 plates. 365pp. of text.
21806-6 Paperbound $3.00

FIRST FLOWERS OF OUR WILDERNESS: AMERICAN PAINTING, THE COLONIAL PERIOD, James T. Flexner. Painters, and regional painting traditions from earliest Colonial times up to the emergence of Copley, West and Peale Sr., Foster, Gustavus Hesselius, Feke, John Smibert and many anonymous painters in the primitive manner. Engaging presentation, with 162 illustrations. xxii + 368pp.
22180-6 Paperbound $3.50

THE LIGHT OF DISTANT SKIES: AMERICAN PAINTING, 1760-1835, James T. Flexner. The great generation of early American painters goes to Europe to learn and to teach: West, Copley, Gilbert Stuart and others. Allston, Trumbull, Morse; also contemporary American painters—primitives, derivatives, academics—who remained in America. 102 illustrations. xiii + 306pp. 22179-2 Paperbound $3.50

A HISTORY OF THE RISE AND PROGRESS OF THE ARTS OF DESIGN IN THE UNITED STATES, William Dunlap. Much the richest mine of information on early American painters, sculptors, architects, engravers, miniaturists, etc. The only source of information for scores of artists, the major primary source for many others. Unabridged reprint of rare original 1834 edition, with new introduction by James T. Flexner, and 394 new illustrations. Edited by Rita Weiss. 6⅝ x 9⅝.
21695-0, 21696-9, 21697-7 Three volumes, Paperbound $15.00

EPOCHS OF CHINESE AND JAPANESE ART, Ernest F. Fenollosa. From primitive Chinese art to the 20th century, thorough history, explanation of every important art period and form, including Japanese woodcuts; main stress on China and Japan, but Tibet, Korea also included. Still unexcelled for its detailed, rich coverage of cultural background, aesthetic elements, diffusion studies, particularly of the historical period. 2nd, 1913 edition. 242 illustrations. lii + 439pp. of text.
20364-6, 20365-4 Two volumes, Paperbound $6.00

THE GENTLE ART OF MAKING ENEMIES, James A. M. Whistler. Greatest wit of his day deflates Oscar Wilde, Ruskin, Swinburne; strikes back at inane critics, exhibitions, art journalism; aesthetics of impressionist revolution in most striking form. Highly readable classic by great painter. Reproduction of edition designed by Whistler. Introduction by Alfred Werner. xxxvi + 334pp.
21875-9 Paperbound $3.00

ALPHABETS AND ORNAMENTS, Ernst Lehner. Well-known pictorial source for decorative alphabets, script examples, cartouches, frames, decorative title pages, calligraphic initials, borders, similar material. 14th to 19th century, mostly European. Useful in almost any graphic arts designing, varied styles. 750 illustrations. 256pp. 7 x 10. 21905-4 Paperbound $4.00

PAINTING: A CREATIVE APPROACH, Norman Colquhoun. For the beginner simple guide provides an instructive approach to painting: major stumbling blocks for beginner; overcoming them, technical points; paints and pigments; oil painting; watercolor and other media and color. New section on "plastic" paints. Glossary. Formerly *Paint Your Own Pictures*. 221pp. 22000-1 Paperbound $1.75

THE ENJOYMENT AND USE OF COLOR, Walter Sargent. Explanation of the relations between colors themselves and between colors in nature and art, including hundreds of little-known facts about color values, intensities, effects of high and low illumination, complementary colors. Many practical hints for painters, references to great masters. 7 color plates, 29 illustrations. x + 274pp.
20944-X Paperbound $3.00

THE NOTEBOOKS OF LEONARDO DA VINCI, compiled and edited by Jean Paul Richter. 1566 extracts from original manuscripts reveal the full range of Leonardo's versatile genius: all his writings on painting, sculpture, architecture, anatomy, astronomy, geography, topography, physiology, mining, music, etc., in both Italian and English, with 186 plates of manuscript pages and more than 500 additional drawings. Includes studies for the Last Supper, the lost Sforza monument, and other works. Total of xlvii + 866pp. 7⅞ x 10¾.
22572-0, 22573-9 Two volumes, Paperbound $12.00

MONTGOMERY WARD CATALOGUE OF 1895. Tea gowns, yards of flannel and pillow-case lace, stereoscopes, books of gospel hymns, the New Improved Singer Sewing Machine, side saddles, milk skimmers, straight-edged razors, high-button shoes, spittoons, and on and on . . . listing some 25,000 items, practically all illustrated. Essential to the shoppers of the 1890's, it is our truest record of the spirit of the period. Unaltered reprint of Issue No. 57, Spring and Summer 1895. Introduction by Boris Emmet. Innumerable illustrations. xiii + 624pp. 8½ x 11⅝.
22377-9 Paperbound $8.50

THE CRYSTAL PALACE EXHIBITION ILLUSTRATED CATALOGUE (LONDON, 1851). One of the wonders of the modern world—the Crystal Palace Exhibition in which all the nations of the civilized world exhibited their achievements in the arts and sciences—presented in an equally important illustrated catalogue. More than 1700 items pictured with accompanying text—ceramics, textiles, cast-iron work, carpets, pianos, sleds, razors, wall-papers, billiard tables, beehives, silverware and hundreds of other artifacts—represent the focal point of Victorian culture in the Western World. Probably the largest collection of Victorian decorative art ever assembled—indispensable for antiquarians and designers. Unabridged republication of the Art-Journal Catalogue of the Great Exhibition of 1851, with all terminal essays. New introduction by John Gloag, F.S.A. xxxiv + 426pp. 9 x 12.
22503-8 Paperbound $5.00

THE ARCHITECTURE OF COUNTRY HOUSES, Andrew J. Downing. Together with Vaux's *Villas and Cottages* this is the basic book for Hudson River Gothic architecture of the middle Victorian period. Full, sound discussions of general aspects of housing, architecture, style, decoration, furnishing, together with scores of detailed house plans, illustrations of specific buildings, accompanied by full text. Perhaps the most influential single American architectural book. 1850 edition. Introduction by J. Stewart Johnson. 321 figures, 34 architectural designs. xvi + 560pp.
22003-6 Paperbound $5.00

LOST EXAMPLES OF COLONIAL ARCHITECTURE, John Mead Howells. Full-page photographs of buildings that have disappeared or been so altered as to be denatured, including many designed by major early American architects. 245 plates. xvii + 248pp. 7⅞ x 10¾. 21143-6 Paperbound $3.50

DOMESTIC ARCHITECTURE OF THE AMERICAN COLONIES AND OF THE EARLY REPUBLIC, Fiske Kimball. Foremost architect and restorer of Williamsburg and Monticello covers nearly 200 homes between 1620-1825. Architectural details, construction, style features, special fixtures, floor plans, etc. Generally considered finest work in its area. 219 illustrations of houses, doorways, windows, capital mantels. xx + 314pp. 7⅞ x 10¾. 21743-4 Paperbound $4.00

EARLY AMERICAN ROOMS: 1650-1858, edited by Russell Hawes Kettell. Tour of 12 rooms, each representative of a different era in American history and each furnished, decorated, designed and occupied in the style of the era. 72 plans and elevations, 8-page color section, etc., show fabrics, wall papers, arrangements, etc. Full descriptive text. xvii + 200pp. of text. 8⅜ x 11¼.
21633-0 Paperbound $5.00

THE FITZWILLIAM VIRGINAL BOOK, edited by J. Fuller Maitland and W. B. Squire. Full modern printing of famous early 17th-century ms. volume of 300 works by Morley, Byrd, Bull, Gibbons, etc. For piano or other modern keyboard instrument; easy to read format. xxxvi + 938pp. 8⅜ x 11.
21068-5, 21069-3 Two volumes, Paperbound $12.00

KEYBOARD MUSIC, Johann Sebastian Bach. Bach Gesellschaft edition. A rich selection of Bach's masterpieces for the harpsichord: the six English Suites, six French Suites, the six Partitas (Clavierübung part I), the Goldberg Variations (Clavierübung part IV), the fifteen Two-Part Inventions and the fifteen Three-Part Sinfonias. Clearly reproduced on large sheets with ample margins; eminently playable. vi + 312pp. 8⅛ x 11. 22360-4 Paperbound $5.00

THE MUSIC OF BACH: AN INTRODUCTION, Charles Sanford Terry. A fine, non-technical introduction to Bach's music, both instrumental and vocal. Covers organ music, chamber music, passion music, other types. Analyzes themes, developments, innovations. x + 114pp. 21075-8 Paperbound $1.95

BEETHOVEN AND HIS NINE SYMPHONIES, Sir George Grove. Noted British musicologist provides best history, analysis, commentary on symphonies. Very thorough, rigorously accurate; necessary to both advanced student and amateur music lover. 436 musical passages. vii + 407 pp. 20334-4 Paperbound $4.00

CATALOGUE OF DOVER BOOKS

MATHEMATICAL PUZZLES FOR BEGINNERS AND ENTHUSIASTS, Geoffrey Mott-Smith. 189 puzzles from easy to difficult—involving arithmetic, logic, algebra, properties of digits, probability, etc.—for enjoyment and mental stimulus. Explanation of mathematical principles behind the puzzles. 135 illustrations. viii + 248pp.
20198-8 Paperbound $2.00

PAPER FOLDING FOR BEGINNERS, William D. Murray and Francis J. Rigney. Easiest book on the market, clearest instructions on making interesting, beautiful origami. Sailboats, cups, roosters, frogs that move legs, bonbon boxes, standing birds, etc. 40 projects; more than 275 diagrams and photographs. 94pp.
20713-7 Paperbound $1.00

TRICKS AND GAMES ON THE POOL TABLE, Fred Herrmann. 79 tricks and games—some solitaires, some for two or more players, some competitive games—to entertain you between formal games. Mystifying shots and throws, unusual caroms, tricks involving such props as cork, coins, a hat, etc. Formerly Fun on the Pool Table. 77 figures. 95pp.
21814-7 Paperbound $1.25

HAND SHADOWS TO BE THROWN UPON THE WALL: A SERIES OF NOVEL AND AMUSING FIGURES FORMED BY THE HAND, Henry Bursill. Delightful picturebook from great-grandfather's day shows how to make 18 different hand shadows: a bird that flies, duck that quacks, dog that wags his tail, camel, goose, deer, boy, turtle, etc. Only book of its sort. vi + 33pp. 6½ x 9¼.
21779-5 Paperbound $1.00

WHITTLING AND WOODCARVING, E. J. Tangerman. 18th printing of best book on market. "If you can cut a potato you can carve" toys and puzzles, chains, chessmen, caricatures, masks, frames, woodcut blocks, surface patterns, much more. Information on tools, woods, techniques. Also goes into serious wood sculpture from Middle Ages to present, East and West. 464 photos, figures. x + 293pp.
20965-2 Paperbound $2.00

HISTORY OF PHILOSOPHY, Julián Marías. Possibly the clearest, most easily followed, best planned, most useful one-volume history of philosophy on the market; neither skimps on essential ideas nor buries reader in detail. [illegible] Strong coverage of non-Anglo-American philosophers and contemporary movements. 1957 edition. Translated by Stanley Appelbaum and Clarence Strowbridge. xviii + 505pp.
21739-6 Paperbound $3.00

YOGA: A SCIENTIFIC EVALUATION, Kovoor T. Behanan. Scientific but non-technical study of physiological results of yoga exercises; done under auspices of Yale U. Relations to Indian thought, to psychoanalysis, etc. 16 photos. xxiii + 270pp.
20505-3 Paperbound $2.50

Prices subject to change without notice.

Available at your book dealer or write for free catalogue to Dept. GI, Dover Publications, Inc., 180 Varick St., N. Y., N. Y. 10014. Dover publishes more than 150 books each year on science, elementary and advanced mathematics, biology, music, art, literary history, social sciences and other areas.